HUMAN ANATOMY AND PHYSIOLOGY

JAMES E. CROUCH *San Diego State College*

J. ROBERT McCLINTIC *Fresno State College*

Illustrations by URSULA WOLF-ROTTKAY

JOHN WILEY & SONS, INC. New York · London · Sydney · Toronto

Preface

It has been thirty-five years since Alexis Carrel published his best seller, *Man—The Unknown.* Today, though knowledge of the human body fills volumes, the average person—even the college student—knows very little about himself. Basic sciences such as those dealing with structure and function (anatomy and physiology) have too often escaped his attention. Even greater is our lack of understanding of our own behavior, a behavior that is endangering the survival, not only of our species, but of all life.

In presenting this work on human anatomy and physiology, we seek to fill a part of this knowledge gap and to aid students in a better understanding of themselves. Although the emphasis in the book is on gross anatomy and its functions, an effort is made to give a broader dimension by considering, at appropriate points, the structure of cells and tissues. A knowledge of cell components is necessary for an understanding of physiology; indeed, the cell provides a basis for unifying all concepts covered in the book.

The inclusion of information on embryology and comparative anatomy also broadens the view of human anatomy and the nature of man as an animal organism possessing his own unique characteristics.

We assume that each student has some familiarity with biology and chemistry—at least a grasp of some terminology and chemical symbolism. In a few instances, more detailed explanation of anatomy or of physiological process has been earmarked by black squares to satisfy the curiosity of the more interested student, but not to discourage those who are satisfied by the basic material of the text. The book should fulfill the needs of students going into nursing, the allied health fields, physical education, or art; or students who wish for a better understanding of their bodies and how they work. Dental and medical students will find it a valuable review or preview of anatomy and physiology.

The systemic rather than the regional approach is used in the presentation of anatomy because it seems to provide a more logical basis for understanding physiology. The information is developed in an ascending and broadening spiral. Basic concepts are presented in the early chapters of the book. Each succeeding chapter leads into another system of the body and, finally, in the discussion of the orienting and integrating functions of the nervous and endocrine systems, all is viewed as a total organism continuous with its external environment. In the introduction of each system, overall functions are recorded, followed by a description of basic anatomy which leads logically to the discussion of process, that is, of physiology.

Careful drawings not only provide clarification of the anatomy but also may extend one's knowledge and comprehension of structure beyond the descriptive words of the text. These drawings plus diagrams and photographs elucidate the anatomy and also the physiology. The summarizing charts and tables placed appropriately within the chapters are a valuable aid to gaining an organized view of the body and to accelerating the learning process.

Brief consideration is given to some of the deviations from the normal structure and function of the body. These demonstrate or emphasize to advantage the importance of normal structure and function, show the influence of our animal ancestry, and may help to understand and appreciate some of the current medical and health problems of our time.

Questions at the end of each chapter are provided to stimulate thought, to aid the student in synthesizing and reviewing the material, and to check his knowledge. Additional readings, at the end of the book for each chapter, enable the interested student to gain more from his experience in the course and also to gain a better understanding of the text.

The anatomical terminology is largely that from *Nomina Anatomica 1961,* second edition. In some instances, reference is made to older terminologies, since these terms are still used in much of the current literature. Important words or phrases are italicized in the text to give them emphasis and to aid the student in learning the subject. They are particularly useful in reviewing the material.

We gratefully acknowledge our indebtedness to all of those who have contributed to man's knowledge of human anatomy and physiology. In writing this book we have drawn freely from this body of common knowledge. We also thank the colleagues who have provided help and encouragement and, particularly, the students who have studied with us over the years and who have been a challenge and an inspiration.

Miss Ursula Wolf-Rottkay, through her excellent illustrations, has added immeasurably to the educational and esthetic values of the book. We thank her for the illustrations and for her continuous help and cooperation during the entire project.

We especially thank Leon L. Gardner and Roger K. Larson who read the manuscript and gave us their suggestions and criticism.

We express our appreciation to Mrs. Harriet Bell and Mrs. Helen Morris for a well-typed manuscript.

We also appreciate the help and guidance provided by William Bryden of Wiley.

Of course, we assume full responsibility for any errors or omissions in the book.

JAMES E. CROUCH
J. ROBERT McCLINTIC

Contents

CHAPTER **4** Body fluids and their regulation   89

CHAPTER **5** The body as a whole   103

CHAPTER **6** Integumentary system   123

CHAPTER 10 Blood   279

NOTE TO THE STUDENT

In a few instances, more detailed explanations of anatomy or of physi-ological process have been earmarked by black squares to satisfy the curiosity of the more interested student, but not to discourage students who are satisfied by the basic material of the text.

Important words or phrases are italicized in the text to give them emphasis and to aid the student in learning the subject.

J.E.C.
J.R. MCC.

HUMAN ANATOMY AND PHYSIOLOGY

CHAPTER 1

The Study of Man

The big question remains: What is man? It remains not because it is hopelessly insoluble, but because every generation must solve it in relation to the situation it faces. Biology is here relevant; a solution based only on biology may well be wrong, but, surely, no solution ignoring either the organismic or the molecular biology can be right and reasonable.
THEODOSIUS DOBZHANSKY.

THE STUDY OF MAN

QUESTIONS

Man's place in nature

What is man? To most biologists man is an animal organism, the result of a long evolution and related, therefore, to all other animals and indeed to all life. Much has been learned or better understood about man's anatomy, physiology, and behavior through comparative studies. It is a fortunate circumstance that man has been able to use other animals for experimentation and study and that the knowledge gained proves applicable to man in most instances. Man himself is used for experimentation in the disciplines where there is likely to be no damage to the individual. Knowledge is also gained by the study of sick and abnormal individuals or through surgery. The growing volume of research in the science of *ethology* (behavior) has contributed abundantly to our understanding of the behavior of animals, including, of course, man.

Studies made on cells and tissues reveal also the common characteristics of all organisms in structure and function. Much emphasis in modern biology is on the molecular and atomic structure of the cell, particularly on the nucleus and its chromosomes and genes. The DNA and RNA complex, and its significance for the understanding of genetics and heredity and of all life, open new possibilities for the further understanding of biology.

The question "What is man?" is not, however, fully answered in the above statements. Man, like all species of organisms, has his own

3

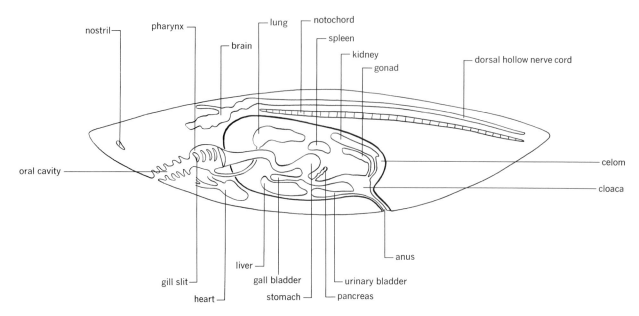

FIGURE 1–1. Schematic sketch of chordate anatomy.

unique characteristics not the least of which is that he is concerned with such matters as the nature of life, his relationship to it, and his relationship to the physical environment. As G. W. Corner has said, "After all, if he is an ape, he is the only ape that is debating what kind of ape he is."

Man's place in nature is objectively described by placing him, along with other living organisms, in the accepted scheme of *classification* as formulated by biologists. On this basis he is a member of the animal *kingdom* which, like the plant kingdom, is divided into a number of major categories, the *phyla.* The phyla are each critically defined and organisms are assigned to them on the basis of certain characteristics. The study of the phyla is called *phylogeny.* The animal phyla range from the relatively simple Protozoa, mostly single-celled forms, to the Chordata which encompasses back-boned animals, including man. This scheme of classification is based on the concept of the evolution of life and the placement of organisms indicates the closeness or the distance of their relationship.

The *phylum Chordata* includes all of the animals that have a notochord, a dorsal hollow nerve cord, and pharyngeal pouches (Fig. 1–1). These structures are apparent at some point in the life cycle *(ontogeny)* of these animals although they may be lost at other stages. Thus, it is important in defining an animal to consider not just the adult stage, but the embryo as well. The notochord, for example, an obvious structure in the embryo of man, is not apparent, as such, in the adult.

The *notochord* is a flexible rod along the dorsal side of the animal ventral to the dorsal hollow nerve cord. It is composed of vacuolated cells enclosed in a fibrous sheath and serves as a central axis of support for the animal. It persists throughout the life cycle of some of the lower chordates such as the lamprey eel, but is replaced structurally and functionally in most vertebrates by the vertebral column. In the adults of man and other mammals the notochord is represented only by a small mass of tissue in the center of the intervertebral discs known as the *nucleus pulposus* (Fig. 1–2).

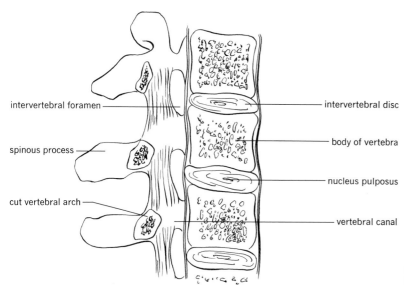

FIGURE 1-2. Median section of vertebrae and intervertebral disc.

intervertebral foramen

spinous process

cut vertebral arch

intervertebral disc

body of vertebra

nucleus pulposus

vertebral canal

The *dorsal hollow nerve cord* lies along the back or posterior side of the animal and comes about as an infolding of the mid-dorsal part of the ectoderm germ layer of the embryo. It becomes the spinal cord and brain, which remain throughout life as hollow organs, there being a central canal in the spinal cord and ventricles in the brain.

The segmentally arranged *pharyngeal pouches* are out-pocketings or evaginations of the lateral walls of the pharynx of chordates. Over each evagination of the pharyngeal wall the body wall invaginates and the two make contact and then break through to form a *gill slit* or *pharyn-*

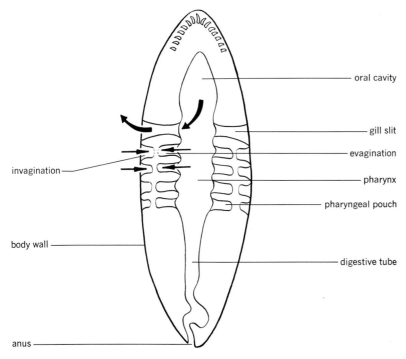

FIGURE 1-3. Schematic sketch showing origin of gill slits.

oral cavity

gill slit

evagination

pharynx

pharyngeal pouch

invagination

body wall

digestive tube

anus

geal cleft (Fig. 1–3). In lower vertebrates, such as fishes and some amphibians, gill slits are formed as described and gills for external respiration are produced in this area. In the lung-breathing vertebrates (reptiles, birds, and mammals), gills are not formed and gill slits are transitory structures in embryological development. In man and other higher vertebrates the pharyngeal pouches give rise to other structures such as the *auditory (Eustachian) tube* and *middle ear cavity.* In this case the corresponding invagination of the body wall becomes the *external auditory meatus* or canal; the membrane separating the two, the *eardrum.* Breaking an eardrum is analogous to forming a gill slit. Infants are sometimes born with openings or fistulae in the sides of the neck which are in a sense gill slits. Such reminders of one's vertebrate ancestry can be closed surgically.

Although the above characteristics of Chordates are said to be diagnostic for the phylum, there are others of importance which may be found also in members of other phyla. Body cavities are among these.

The body cavities are closed spaces within the body consisting of dorsal and ventral cavities (Fig. 1–4). The *dorsal cavity* is divided into *cranial* and *vertebral portions,* the former housing the brain, the latter the spinal cord.

The *ventral cavity,* also called the *celom,* is divided into a thoracic portion superior to the diaphragm and an abdominopelvic portion inferior to the diaphragm. The *thoracic cavity* is divided into two *pleural cavities,* one around each lung, and the *pericardial cavity* around the heart. Between the pleural cavities there is also a potential cavity, the *mediastinum.* It is not considered a celomic cavity. The *abdominopelvic cavity* is divided arbitrarily into *abdominal* and *pelvic portions* although no wall lies between them. Stomach, spleen, liver, pancreas, small intestine, and most of the large intestine lie within the abdominal cavity; the urinary bladder, sigmoid colon, rectum, and the ovaries, uterine tubes, and uterus of the female, and the prostate, seminal vesicles, and part of the ductus deferens of the male lie in the pelvic cavity.

The *phylum Chordata* is divided into a number of subphyla; the most important for our purposes is the vertebrata, characterized by the presence of a backbone or vertebral column. To this subphylum belong the *classes* of fish, amphibians, reptiles, birds, and mammals, the latter

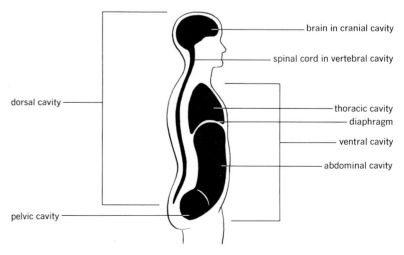

FIGURE 1–4. Body cavities, median section.

including man. The *class Mammalia* is characterized by the presence of hair and mammary glands. The class is in turn divided into orders, man being placed in the *order of primates*. Primates are characterized by the development of a large brain relative to the size of the body and particularly the growth of the cerebral cortex. In cerebral cortex growth and development man surpasses all other primates. Man's brain weighs about 1350 grams while that of the highest apes weighs about 400 grams.

Finally, man belongs to the *suborder Anthropoidea*, the *family Hominidae*, the *genus Homo* and the *species sapiens*. For further characterization of man, see R. J. Harrison's *Man the Peculiar Animal*, listed under references at the end of this chapter.

> "Man hath all that Nature hath, but more,
> And in that more lie all his hopes of good."
> *Arnold*

Man's relationship to other animals is summarized in the following classification:

Kingdom — Animalia
  Phylum — Chordata — notochord; dorsal hollow nerve cord; pharyngeal pouches.
   Subphylum — Vertebrata — backbone.
    Class — Mammalia — hair, mammary glands.
     Order — Primates — generalized anatomy; highly developed cerebrum; in some, five digits with nails; opposable thumb in some.
      Suborder — Anthropoidea — apes, monkeys, and men.
       Family — Hominidae — terrestrial biped, highly developed cerebrum and eyesight.
        Genus — Homo — steep facial angle, forehead, nose, and chin prominent.
         Species — sapiens — largest cerebrum; articulate speech.
          Scientific name — Homo sapiens.

Man, a unique animal

Man is related most closely to the vertebrates. It should be emphasized, however, that he has his own unique qualities and his phylogenetic position is the highest of all animals. His high position, the result of a long evolution, is due not only to the presence of a large cerebral cortex which surpasses that of all other animals, but to his bipedal locomotion and his upright posture which free his upper limbs and hands to better serve his cerebrum. He is the most adaptable of all the animals, capable of living in a wide range of environments. He is capable, too, more than any other animal, of changing the environment to satisfy his needs. His reproductive efficiency is the highest among animals, particularly in the area of postnatal care. He is reproductively active throughout the year. The female after menopause, at about the age of 50, is no longer capable of producing children, but may remain sexually active. The male shows no definite point of cessation of sexual activity, and is capable of producing viable sperm throughout most of his life.

George Gaylord Simpson in the *Meaning of Evolution* says, "In the basic diagnosis of *Homo sapiens* the most important features are probably interrelated factors of intelligence, flexibility, individualization, and socialization. . . . In man all four are carried to a degree incomparably greater than in any other sort of animal.'' Of these four features of man, intelligence is the most significant and the one on which the other three hinge. Man can learn; he is educable and thus is capable of the most varied responses to external and internal stimuli. These responses may be based on the experiences of others, even those who lived in the remote past and who left signs or manuscripts of their findings. The responses may be based on speculations into the future. Simpson refers to this as a ''new sort of evolution'' which has its basis in ''the inheritance of learning.'' Unlike organic evolution which rejects the inheritance of acquired characteristics, the new evolution operates by the inheritance of acquired characteristics—the knowledge and activities of those living in the past. Although in organic evolution new factors arise as mutations and occur purely by chance, in the new evolution the new factors arise consciously; they are influenced by the needs of the individual, and the individual in turn is influenced by his relationship to a social group. The new evolution is cultural.

While the ''old'' and the ''new'' evolutions are so different in kind—the one occurring without purpose or design and the other by conscious effort on the part of individuals and society—it does help to keep the comparison in mind as we attempt to understand man's uniqueness in the living world.

Man is said to be an ethical animal, indeed the only ethical animal. As such he is concerned with values; with right and wrong; with the ''good life,'' with individual worth and responsibility; with the relationship of men to each other and to the environment. Man has sought for an adequate ethic over the centuries and is still searching. Most of his attempts have resulted in systems of ethics built on revelation or intuitive thought. They have adhered to an ideal of absolutism. They have been considered as having eternal validity, for all forms of life. They have not satisfied the needs of mankind. Perhaps this is true because they fail to realize that change is the essence of life and that man, while a product of the old evolution, is subject also to the ''new'' evolution and it is only at this point that the need for an ethic appears. Perhaps Albert Schweitzer's definition of ethics is a good one. He said, ''Let me give you a definition of ethics: it is good to maintain life and further life; it is bad to damage and destroy life. And this ethic, profound, universal, has the significance of religion. It *is* religion.''

Man's problems center in three areas: overpopulation, destruction of the environment, and war. He can control population so that human life can continue on this earth and with some chance for a good life for all of the people. By using the knowledge and skills of our ecologists along with those of the economists, engineers, and others, he can use more wisely the natural resources of our world and avoid the destruction of the environment which goes on so rapidly today in the United States and around the world. If man cannot do these things, then what does he gain in his struggle to control human disease and to prolong human life? It is essential now, since we have neglected things for so long, to develop massive federal programs that use enormous amounts of money and the skills of our sociologists, psychologists, scientists, technologists, and intellectuals to attack these earthly problems.

We have only outlined man's nature, his ethics, and the problems he faces. It is not our purpose in this book to do more than attempt to place man in proper perspective.

Historical development of human anatomy and physiology

"More than two millennia ago, Greek sages discovered that to 'know thyself' is the foundation of all wisdom" (T. Dobzhansky, 1965).

To better understand modern man and his search for knowledge of himself, it is helpful to review the early history of his efforts. Since our concern in this book is with the anatomy and physiology of man, a review of the history of these sciences is most pertinent.

The study of anatomy originated in Greece and one of the most famous of the early anatomists and physicians was *Hippocrates (460– 370 B.C.).* He did little human dissection and his name is memorialized in the Hippocratic oath which is still recited by members of our medical profession. Hippocrates was born on the island of Cos. He studied and practiced medicine there and on other Greek islands as well as on the mainland of Greece. According to the Hippocratic works, all living bodies were made up of four humours: black bile (melancholia); yellow bile (cholera); phlegm (pituita); and blood (sanguis). The health of an individual, it was thought, depended on a proper mixing, "tempering," or "complexioning" of the four humours. If one of these was in excess in a patient, his condition or disease was classified according to that particular humour as melancholic, choleric, phlegmatic, or sanguine. These terms are still used to describe man's condition, although the concept of humours has long since been discarded. We still speak also of man's temperament and a lady's complexion. Each of the humours was associated with a particular organ: sanguine with the liver; melancholic, or black bile, with the spleen; phlegm with the lungs, and cholera, or yellow bile, with the gall bladder. The theory of the humours influenced medical thought for the next two millennia.

Perhaps Hippocrates' greatest contribution to the development of medical thought and the understanding of the structure and functions of the body was that he cast aside much of the superstition which had possessed the minds of his predecessors and contemporaries. They looked upon disease as a result of the displeasure of the gods, as the work of evil spirits, or as punishment for the sins of the patient. Hippocrates looked upon disease as having natural causes and as following a certain course which could be watched and described and even, in some cases, altered. This was the beginning of the rise of observational medicine.

Two men stand out above all those influenced by Hippocrates and they dominated human thought for over fifteen hundred years. They were Aristotle and Galen. *Aristotle (384–322 B.C.)* was the son of a physician and was a student of Plato (427–367 B.C.), whose work, the *Timaeus,* set forth a fanciful scheme of the world and the human body. The outer world and all matter were considered as living, the macrocosm; the human body was seen as a microcosm. This doctrine of a living macrocosm unfavorably influenced the development of anatomical thought for many years to come. Aristotle became a great systematizer of science and his influence continues even into the present time. He is known best for his three great zoological works, the *History of Animals,* the *Parts of Animals,* and the *Generation of Animals.* It is apparent from these works that Aristotle was thinking in terms of organic evolution or, as he termed it, the "scala naturae." He is often referred to as the founder of comparative anatomy, although he evidently never included the human body in his dissections. His anatomical illustrations are among the first on record and some of them, such as that of the mammalian urogenital system, are very accurate. Among his most outstanding anatomical accomplishments was his description of the "placental" development of the dogfish (Mustelus laevis).

Aristotle, as must be expected, had some erroneous views of anatomy. He placed great emphasis upon the heart, even considering it as the seat of intelligence. This position is particularly difficult to understand since his teacher *Plato* in his *Timaeus* had described the brain as the seat of feeling and thought. Even *Aristophanes* in his play *The Clouds* took the popular view of the day and wrote of concussion of the brain. Theophrastus of Eresus (370–287 B.C.), Aristotle's student and, after his death, his successor as head of his school, the Lyceum, departed from his teacher's position that the heart was the seat of the intellect. Aristotle considered the brain as an agent for cooling the heart. He also made no correct distinction between arteries and veins. The arteries, he thought, contained air as well as blood. A better morphologist than physiologist, he did good descriptions of the branches of the vena cava, and his writing on the anatomy of many animals—the cephalopod molluscs, for example—was excellent. While his descriptions of development of animals such as the chick were good, he had some peculiar ideas about procreation. He thought that the female provided the material substance of the embryo and that the contribution of the male was nonmaterial—it was the essential generative agent, the principle of life, the soul, or psyche. Aristotle was thus able to accept the idea of fertilization without material contact, in effect, parthenogenesis. He was also of the "vitalist" school, as is made clear in his book, *On soul (De anima).* These concepts influenced anatomical and physiological thought for many centuries.

About 325-300 B.C. the center of the scientific world shifted to Alexandria, Egypt, where anatomy first came into its own as a separate discipline. Herophilus and Erasistratus were the two greatest teachers of anatomy. *Herophilus,* sometimes called the "Father of Anatomy," worked on both human bodies and those of other animals. He wrote two books, *On Anatomy,* and *Of the Eyes,* as well as a book for midwives. Since these works of Herophilus and those of Erasistratus were lost, we depend largely on Galen's descriptions for our information about them.

Herophilus made the first clear distinction between arteries and veins but regarded the pulsations as an action of the arteries themselves. He first described the lacteals, lymph capillaries of the small intestine, and his observations were expanded by Erasistratus. Subsequently no further work was done on the lacteals for two thousand years, when Gasparo Aselli (1581-1626 A.D.) in 1622 observed them in a dog and described and illustrated them by large plates. Herophilus recognized the brain as the central organ of the nervous system and regarded it as the seat of intelligence. He was the first to divide nerves into sensory and motor, and he distinguished parts of the brain such as cerebrum, cerebellum, calamus scriptorius, and fourth ventricle. He described the meninges, and also made studies on the uterus and the liver.

Erasistratus (about 290 B.C.) was more physiologist than anatomist and is sometimes referred to as the "Father of Physiology," having founded it as a formal discipline. Erasistratus was more a "rationalist" than a "mystic." He thought, however, that vital spirit (pneuma) filled the left ventricle of the heart. This may have prevented him from discovering the circulation of blood, for his other ideas about blood and circulation were very modern. He knew, for example, that the right ventricle filled with blood which was discharged during heart contraction, but he believed that the left ventricle at the same time expelled vital spirit. He knew that semilunar valves prevented the return flow of blood or vital spirit into the ventricles. He is credited with naming the bicuspid valve of the heart, and claimed that it prevented the movement of vital spirit from the heart by any other route than the aorta. He misunderstood the auricles, regarding them as part of the pulmonary vessels. Erasistratus believed that blood was sucked into the arteries from the veins through very fine interconnecting vessels. He was thus aware of the existence of capillaries, although his understanding was limited by his confusion over the pneuma. He knew that cut arteries contained blood, but thought that otherwise they contained only air and pneuma; that it was the escape of pneuma from the cut vessel which caused a vacuum and that this vacuum sucked the blood from the veins through the interconnecting vessels.

Erasistratus was quite knowledgeable about the brain, recognizing not only cerebrum and cerebellum, but relating the extent of the cerebral convolutions to the relative intelligence of man and other animals. He described the cerebral ventricles and meninges but thought the ventricles were filled with animal spirit. He also studied cranial nerves in some detail but considered them as hollow tubes carrying animal spirits. He regarded the contraction of muscles as due to their distention by animal spirits. Even men like Descartes (1596–1650) and Borelli (1608–1679) as late as the 17th century were persuaded by such theories of muscle action.

Both Herophilus and Erasistratus were accused of dissection of living men. The evidence for such vivisection comes from Celsus (about 30 B.C.) and from Tertullian (about 155–222 A.D.). Tertullian called Herophilus a butcher of men who had dissected 600 living people. Although Herophilus was the first to dissect the living human body in public, he was probably not the first to have carried out such dissections.

While the Alexandrian School declined as a research center in anatomy about 250-50 B.C., it continued to be strong in teaching. Research activity remained strong in astronomy, mechanics, geography, and mathematics. For further consideration of the development of anatomy, we must turn to the Roman Empire.

The period of the Roman Empire saw a change in the character of science. Interest turned to "useful" knowledge and away from the theoretical. Science died as a creative force and intellectualism took a back seat to materialism. Instead of seeking causes of things as the Greeks had done, attention was directed toward the compilation of information from approved authorities. This decline in Rome had a depressing influence on schools in the eastern Mediterranean and Alexandria.

Although many Latin and Greek names are associated with anatomy, during this period few if any of them made any direct studies of human anatomy. Little more can be said for the work of Galen, even though he is generally considered the greatest physician since Hippocrates.

Galen (129–199 A.D.) was born in Pergamum. He became a prolific writer and was brilliant in his profession of medicine, where he developed the Galenic system. His philosophical view, however, was one of determinism of perfection in which all was fixed by a wise God. The body of man was considered perfect, as were all of God's creatures. This teleological position of Galen developed acceptance of dogma rather than the need for exploration in the minds of his students and, when he died, anatomy and physiology died also and were not revived for a thousand years.

The substance of Galen's work in anatomy was great and perhaps is best exemplified in his *On Anatomical Procedure*. In this work the sections on skeleton and muscles were most outstanding. His work on the circulatory system was less satisfactory, in part because he did not understand the movement of blood, and because he was influenced by the Hippocratic Collection of material. Since he dissected other animals rather than men, he ascribes to human anatomy parts which belong to lower animals. A good example is the bifid uterus which he dissected in other animals, and his illustrations of the human uterus show it as a bicornuate type. This view of the human uterus carried over to the 16th century and is but one of the many examples of errors which were perpetuated in the works of Galen. Yet his work was considered authoritative and was not questioned. So it remained throughout the Dark Ages as we see the rise and influence of religions, particularly Christianity, and of astrology. These influences superseded anatomy and physiology. Indeed, the human body was considered contemptible and unworthy of study.

After the downfall of the Roman Empire, intellectual leadership passed to the East, to people of Arabic language (Syria and Arabia). The old Greek medical texts were translated into Arabic and later, in the monasteries of Italy, into Latin. By the 12th and 13th centuries there were signs of revival of intellectual activity. In the 13th century, Albertus Magnus (1206-1280 A.D.) appeared as a real naturalist, and he added to the knowledge gained from his reading of Aristotle the knowledge that resulted from his own observations. The trend of awakening which he represented came to fruition in the next century with the development of the universities and the return to scholasticism. Yet Galen's work was still prominent. Even though human dissection was revived at the Medical School in Bologne (founded 1156 A.D.) it was done under serious disadvantages of poor preservatives, warm weather, and opposition by the church and others. Until the 16th century, Bologne remained the center of anatomy.

In the 15th century, artists began taking an interest in the human form and in using the scalpel as well. Names like Luca Signorelli (1444–1524), Andrea Verrocchio (1435–1488), Leonardo da Vinci (1452–1519), Albrecht Dürer (1471–1528), Michelangelo (1475–1564), and Raphael (1483–1521) are among the most famous. Their sound anatomical knowledge is seen in the many drawings of their dissections. *Leonardo da Vinci* was perhaps the most outstanding. He became so preoccupied with his anatomical work that it took precedence over his artistic interests. His anatomical notebooks attest to his eminence as an anatomist and biologist. He excelled in his drawings of muscles of the human body. Some of his errors can be attributed more to his imagination and to artist's license than to any deficiency in his observations. Particularly remarkable are his diagrams showing the action of muscles. While some of Leonardo's figures of circulatory structures show a Galenic influence, others demonstrate his keenness of observation and insight. He shows the distribution of blood vessels in the arms and legs with great accuracy and his drawings of the heart are superb. He introduced the technique of injection of anatomical structures with solidifying media and produced excellent casts of the ventricles of the brain. Leonardo's sketches probably had no direct influence on the work of Vesalius, but the naturalistic movement in art of which Leonardo was a part most certainly did.

The development of printing with movable type about 1450 brought, by the end of the 15th century, the production of illustrated anatomy texts. One of the first was that of *Jacob Berengar of Corpi,* who was professor of surgery at Bologne. He was first to describe many anatomical structures such as the vermiform appendix, the arytenoids as separate cartilages of the larynx, and the thymus. Many other anatomy texts appeared in this pre-Vesalian period.

Between 1450 and 1550 there was a great change from the practical anatomists to the medical scholars. These scholars were more concerned with writing than with dissection, and during this period many of the medical and anatomical classics were translated from Greek to Latin. Prior to this, during the Middle Ages, most of the anatomical works were translated from Arabic and did not do justice to anatomical facts to the same extent as the works that were translated from Greek. This intellectual movement was called *humanist.* Although many important anatomists belong to this period, the name of *Sylvius* is particularly prominent. He took his medical degree at Montpellier and came to the University of Paris in 1531, where he began to teach large audiences. Although he made important discoveries in anatomy, such as the first description of the sphenoid bone, perhaps his greatest contribution was in the systemization of anatomy and in improvements in nomenclature. Also, he was the teacher of Vesalius, with whom he frequently quarreled, but whom he also provided with great opportunity. Since Sylvius' experience was in the Paris school, he did not come under the influence of Renaissance Art in Italy and was out of sympathy with the use of illustrations in anatomy. On this matter Vesalius parted company with his teacher when he moved to Padua and art and illustration became a part of his equipment for the presentation of his anatomical studies.

Vesalius (1514–1564), born in Brussels, is often referred to as the *Reformer of Anatomy,* also as the *Father of Modern Topographical Anatomy.* He was a fortunate man in appearing on the scene when he did. Charles Singer in his book *A Short History of Anatomy and Physiology from the Greeks to Harvey* says of Vesalius, "His intellectual father was the Galenic Science that had gone before him. His mother was that fair creature, the New Art, then in the very bloom of youth. Until these two had come together there could be no Vesalius. When these two had come together there had to be a Vesalius."

Vesalius as a boy was constantly dissecting animals. It was at a time when such activity was not popular and there was little help because there was no great interest in nature and no books as we know them now. When ready for training, he studied at Louvain and later at Paris under Sylvius and Gunther. He thus came under the influence of Galenic Anatomy, which remained an influence in his life and work and is best expressed in his earliest works, a revision of Gunther's *Anatomical Institutions According to Galen* and the *Galenic Anatomical Institu-*

tions. He then went to Padua and by 1537 was appointed professor at the University. He developed his own techniques of teaching and a following of students who, by their interest and applause, stimulated him to put great energy into his work. He lectured to large audiences; his demonstrations were on the human body, and he used drawings and models and the bodies of other animals as additional aids. In five years (1543) he had completed his great work *On the Fabric of the Human Body* when he was only 28. Almost at the same time he completed a companion *Epitome,* and with these two books his life work was completed. He had in these few years broken with the authority and tradition of Galen, and anatomy remains until this day Vesalian. Vesalius himself was attacked severely for his work, even by those who had been his teachers. He was so incensed by this attack that he destroyed much of his unpublished work and gave up his dissections. He made no further contributions to anatomy and, in 1563, went to Jerusalem. The next year, on his return trip, he was taken ill and died on the Greek island of Zante on October 2, 1564.

As Andreas Vesalius brings us to modern anatomy, so *William Harvey (1578–1657)* brings us to the modern era in physiology. He introduced experimental procedures to the study of human anatomy and brought fame to Caius College at Cambridge, England, where he was graduated. He went to Padua where he was influenced by the work of Fabricius, a pupil of Vesalius. In 1615–1616, Harvey delivered anatomical lectures at the Royal College of Physicians in England in which he outlined his ideas about the circulation of blood by the heart. In 1628 he published in Frankfurt his famous though brief work called *On the Movement of the Heart and Blood in Animals.* Harvey not only advanced knowledge of circulation, but his methods are still recounted as an elucidation of method in science. Like Vesalius in anatomy, he broke with the authority of Galen and his followers. Like Vesalius also, he was severely criticized and his reputation suffered. Yet today, he is known as the *Father of Physiology.*

In closing this brief outline of the history of anatomy and physiology from the Greek to modern times, a quotation from William Harvey seems pertinent. "The blood is supposed to ooze through tiny pores in the septum of the heart from right to left ventricle, while the air is drawn from the lungs by the large pulmonary vein. According to this many little openings exist in the septum of the heart suited to the passage of blood. But, damn it, no such pores exist nor can they be demonstrated."

In reading and studying the chapters of this book, the student might now and then recall the amount of work and controversy, the attacks on reputations and lives, the persistence and dedication of many people, which have given us the knowledge and understanding recorded here for his education.

Man—a creature of dynamic balance—homeostasis

It is a well-known physiological principle that animals at all evolutionary levels maintain a certain constancy of the internal environment and that this is essential to the life of the cells and hence the life of the total organism. The internal environment is that portion of the body lying inside a living membrane, as distinguished from those parts of the body which have open passage to the external environment. The blood, the lymph, and tissue fluid are all a part of the internal environment, while the lumina of the bladder, of the alimentary tract, the uterus, and other organs all maintain connections to the body surface and materials leaving them need cross no living membranes. They are a part of the external environment.

Knowing that organisms are constantly taking in food, oxygen, water, and other materials, and further realizing that the cells are always producing carbon dioxide and other wastes, the question arises how the internal fluids of the body—the internal environment—can maintain nearly constant levels of sugars, salts, and water; and an osmotic balance and acid-base (pH) relationship which show only small variations. Add to this the stress from the environment to which all organisms are subject and which must be met if the organism is to survive, and you begin to realize the complexity of the adaptive and control processes which must be involved. This constancy or steady state of the internal environment of organisms was designated by Walter B. Cannon of Harvard University as *homeostasis.*

Absolute values cannot be given for this "constancy" of the internal environment. It is subject to variation within a species and even within one individual from one time to another. The values are more constant among organisms higher in the evolutionary scale. In general, higher animals can adjust more readily to a wider range of external environments than can the lower forms. At the same time, higher forms are more dependent on homeostasis for survival. Stress, such as disease which disrupts control systems, can be very dangerous. Man, one of these higher animals, is universal in distribution in the world. He can adapt to almost any external environment but he is always vulnerable to even the slightest deviations in the various physiological constants.

Physiological regulation

Homeostasis involves the action of regulatory systems and a large segment of physiology is devoted to the discovery and study of these systems. The study of control systems is called *cybernetics,* a term coined by Norbert Weiner in 1948. The basic concepts of this technology were developed and applied by Weiner to both artificial and biological systems, though mostly to the former. Modern automation was the outcome of this early work. A start has been made toward the application of this technology to the integrative systems of the living body but much remains to be accomplished.

An organism, to achieve homeostasis, must have the capacity to sense change, that is, to receive stimuli, and to take appropriate action as a result. Appropriate action, in this case, is to perpetuate the physiological constant whether it is the hydrogen ion concentration of the blood or the glucose level. This might be diagrammed as follows to show the direction of flow of information.

Change (stimulus) → | organism | → response
(input)   (system)   (output)

The principle is not different from the use of a thermostat to control the temperature in a house. The thermostat is set at a given temperature. When the temperature in the house falls below this level, the thermostat reports it to the furnace which responds by turning on. Conversely, if the temperature goes above the level for which the thermostat is set, it will cause the furnace to be shut off and may start an air conditioner if one is built into the system.

Feedback

In such systems a *feedback concept* is used. A feedback system is one in which the output of the system is fed back to the system. In the above example, the heat caused by the furnace being turned on (output) is fed back to the thermostat (system). It then becomes part of the input for a new stimulus-response cycle; that is, when the room gets too hot, the furnace will be turned off. Where, as in this particular example, the feedback change is in the reverse direction of the original condition, it is called *negative feedback*. Negative feedback makes for stability in the system or organism (in this case the maintenance of a constant temperature). This kind of feedback is found in most physiological regulatory systems. Positive feedback systems are those in which the feedback signal reinforces the input signal and the change in output is increased. Positive feedback results in instability rather than regulation and occurs infrequently in physiological systems. It is involved, for example, in the control of the pupil of the eye (pupillary reflex).

In homeostasis we are therefore concerned primarily with negative feedback systems. It is not our purpose at this point to look in depth at regulation but only to introduce the principles. Regulation will be treated in greater detail in the chapters on the nervous and endocrine systems, the regulating-integrating systems of the body. Abundant examples are also found in the chapters on circulation, respiration, and metabolism. One example, the regulation of blood sugar (glucose), may help to clarify our discussion of regulation up to this point. Glucose is the chief energy source of our cells and its concentration in the blood at any given time is critical. Excesses of glucose are converted into glycogen and stored in the liver, only to be reconverted to glucose and returned to the circulation when needed. Two hormones from the islets of Langerhans of the pancreas are the primary agents controlling these metabolic processes. A fall in blood-sugar level initiates secretion of the hormone *glucagon* from the alpha cells of the islets. This hormone influences the enzymes which stimulate conversion of glycogen to glucose in the liver and the concentration of blood sugar is raised. *Insulin,* a hormone from the beta cells of the islets, is secreted in response to an elevation in blood sugar (for example, after a meal) and stimulates the uptake of glucose by the cells as well as the conversion of glucose to glycogen for storage in the liver. These two hormones working together normally maintain the blood sugar at levels compatible with the health of the organism. Other factors, to be sure, are involved in this regulation. It is a well-known fact also that the metabolism of proteins and fats is related to and dependent upon the proper utilization and control of sugar metabolism. A more elaborate description of these relationships is given in chapter 22.

Levels of integration and control

The above section showed the importance of maintaining optimal operating conditions within the internal environment of the organism and suggested certain means of control. In addition, it should be pointed out that all changes, both external and internal, require control mechanisms for adaptation and that most parts of the organisms are involved in the process. An organism which cannot respond effectively to the ever-changing external environment has little chance of survival and a steady state internally will be of no help if the animal fails in this other, though not separate, area of activity. Life, in a very real sense, is the capacity to respond to change or stress in a controlled and effective way.

Control in organisms goes back basically to the function of the *gene.* Life began with the appearance of the first nucleic acids, of which genes are the modern descendants. Not only are nucleic acids the ultimate basis for control in all organisms, but they provide the blueprint for the perpetuation of life from generation to generation and for the creation of new life. Genes are variously defined: in the physical sense as simply a unit or portion of a chromosome; chemically as the *DNA* fraction of a *nucleoprotein;* or operationally as a unit of a chromosome which controls one reaction in metabolism. Genes may be considered as units which (by mutation) alter just one characteristic of an organism, or which in the reproductive process may cross from one chromosome to another. Genes account for the conservatism which results in the continuity of species from one generation to another, for the variation which causes individuality within species, and for the changes which, given time, are the basis for the evolution of life (Fig. 1–5).

Genes are not the only controlling agents of the cell. There are numerous such agents and they play many roles in the integrative processes of the cell. They act sometimes as *receptors;* other times as *effectors;* and sometimes as selective agents making decisions— *modulators* as they are often called in the cybernetic systems. The cell membrane is a good example. It is a receptor when stimulated by

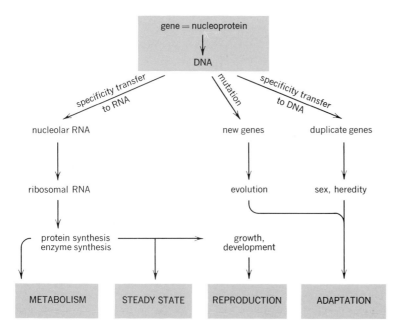

FIGURE 1–5. The role of the gene in controlling the organism's function.

a molecule; a modulator when it "decides" what to do about it; an effector when it responds to the stimulation. If the molecule is of glucose, the membrane will allow it to pass through into the cell. Many cell molecules free in the cytoplasm or complex cellular entities such as ribosomes, chloroplasts, and kinetosomes, play important control functions within the cell. The cell also extends its influence outward to affect its external environment or other cells with which it is associated in a tissue or in a more complex organism. The control and integration of many relatively simple organisms is managed on this cell-to-cell basis without the benefit of special integrative systems.

As multicellular organisms evolved, communication among cells was at first by cell-to-cell contact and their stimulus-modulator-response mechanism was essentially that within the cells. Some animals later evolved transportation systems (such as the circulatory system) which enabled cells to extend their influence more readily to distant parts of the organisms. But these were not controlling and integrating systems in the usual sense. These appeared later in the nervous and endocrine systems, which serve these functions primarily, though not exclusively. Locomotion places additional stresses upon most animals to which sedentary organisms are not subjected. These animals not only must adjust to a changing external environment as do other organisms, but they move quickly from one part of the external environment to another, placing extra stress upon integrative systems which must maintain the essential constancy of the internal environment.

The structural unit of the nervous system is the neuron, a specialized cell; the functional unit is the reflex arc. Reflex arcs have their receptors on the body surface or internally; sensory or afferent neurons transmit impulses (messages) to the brain or spinal cord which act as integrative centers or modulators; afferent neurons carry impulses to effectors or responding organs such as muscle tissue or glands.

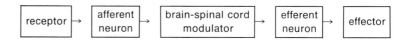

This system is presented in detail in Chapter 20 and illustrated in Fig. 20-7. The endocrine system, which produces hormones and releases them in the blood to reach their target organs, is elaborated in Chapter 22.

Even in the higher animals such as man, although special integrative systems are present, almost all organs of the body and their products exert some integrative or controlling influence. CO_2, looked upon primarily as a waste product eliminated from the lungs, is an important regulator of respiration. The kidneys are not just excretory organs but are necessary in regulating acid-base balance and osmotic pressure, and in other ways help to maintain the constancy of the internal environment. The liver is far more than an organ contributing to the digestion and absorption of food. It is involved in sugar, protein, and fat metabolism. One should remember, too, that there is a complex cooperative action among organs and tissues. What happens in one influences others. The control of body temperature in birds and mammals, which we take so much for granted, is a result of the co-ordinated actions of many organs of the body. One should never forget, either, that in all of this the basic controlling influence of cells is always going on.

One final level of integration and control must be emphasized—the brain, particularly man's brain. This brings us full circle to the questions discussed in the introductory part of this chapter: What is man? What is the nature of man? It brings us to the reason for the chapters that are to follow, which must help us to understand better the human organism, its anatomy and physiology in the broadest sense.

QUESTIONS

1. How would you define an animal of the phylum Chordata?
2. On what biological principle do we base our scheme of classification of animals and plants?
3. Where is man placed in this scheme of classification?
4. Why must one know the life history of an animal in order to classify it properly?
5. What is the nucleus pulposus and why is it significant in man?
6. What special characteristics of man make him unique and the species most capable of altering its environment?
7. Man is the only animal concerned with ethics. What does this mean and how is it pertinent to our worldwide problems?
8. What is homeostasis? What is cybernetics?
9. Define feedback as applied to biological or physical systems.
10. Distinguish between negative and positive feedback and state which is most widely used in the human body. Give examples of each as seen in living organisms.
11. What is the chief controlling agent in the cell and how is it defined?
12. Give some examples of the kinds of control exercised by the genes.
13. How is communication managed in multicellular animals?
14. Diagram a functional unit (reflex arc) of the controlling and integrative systems of vertebrates.
15. Some waste products of metabolism are also important regulating agents in the body. Name one.
16. The brain is sometimes referred to as the crowning glory of organic evolution. Why?
17. What contribution did each of the following make to the development of the science of anatomy: (a) Hippocrates, (b) Aristotle, (c) Herophilus?
18. Compare the work and influence of Galen, Vesalius, and William Harvey upon modern anatomy and physiology.

CHAPTER 2

The Basis of Structure and Function.
The Organization of the Body.

What a piece of work is man! How noble in reason!
How infinite in faculty, in form and moving, how express
and admirable! WM. SHAKESPEARE, IN *HAMLET*

THE CELL

Introduction

In 1663, Robert Hooke coined the term "cell" to describe the small rectangular units he observed under a crude microscope in a thin slice of cork (Fig. 2–1). The concept that cells formed the basis of structure in living organisms was stated by Schleiden and Schwann in 1838–1839. This concept is known as the *cell theory.* In the years following the introduction of the theory, the cell has been subjected to an increasingly sophisticated series of observations, tests, and analyses. Although a complete knowledge of a cell's life still eludes us, much has been learned and some useful generalizations may be made.

The cell is the *basis of life* itself. In single-celled organisms, one unit carries on all of the processes essential to the continued existence of the unit. The single cell *responds* to stimuli, *ingests materials, metabolizes* them to secure energy, *synthesizes* new materials, *rids itself of wastes,* and *reproduces* its kind. It is independent of other similar units. As the number of cells in an organism becomes larger, interdependence increases and diversity of structure and function appear. Although all cells must retain the ability to carry on those processes associated with continuance of life, some functions have become the special properties of those cells which can carry them out most efficiently. For example, muscle cells metabolize materials, but

23

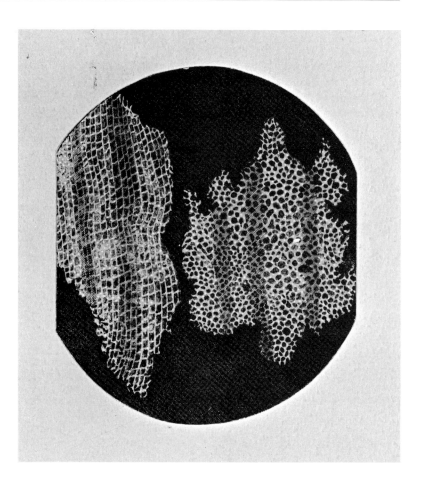

FIGURE 2–1. Drawing of cells in cork was published by Robert Hooke in 1665. Hooke called them cells, but the fact that all organisms are made of cells was not recognized until 19th century.

are specialized to contract or shorten. As certain functions are assumed by specialized cells, they become more dependent on the other cells of the body for such things as the supply of nutrients. A *division of labor* is established.

To a large degree form and function of a cell are related, particularly at the ultramicroscopic level. Muscle cells are typically elongated because in order to shorten effectively, they must possess considerable initial length. Figure 2–2 illustrates some of the diverse forms that cells assume in the human body.

The body is a vast agglomeration of individual cells. Whatever problems a cell faces must then become problems for the entire organism. While there is no such thing as a "typical cell," understanding of the structure and functions of a hypothetical unit should serve as a useful basis for understanding the functioning of the body as a whole.

The structure of a cell and the functions of its components

The three basic parts of a cell (membrane, cytoplasm, and nucleus) have been recognized for many years. Determination of the fine structure, and correlation of this structure with function, awaited the development of tools and techniques to magnify, isolate, and analyze individual cellular components. The modern view of the cell (Fig. 2–3) indicates that the historical subdivision into only three parts is far too simple to account for the many activities that the cell exhibits (Figs. 2–4 to 2–14).

The plasma membrane. The *plasma membrane (cell membrane)* is a folded three-layered structure (Fig. 2–4) serving to delimit a cell from its neighbors and from the environment. It also acts as the chief agent determining the passage of materials into and out of the cell. The term *permeability* is generally used to describe the ease with which a substance passes through the membrane. Plasma membranes are

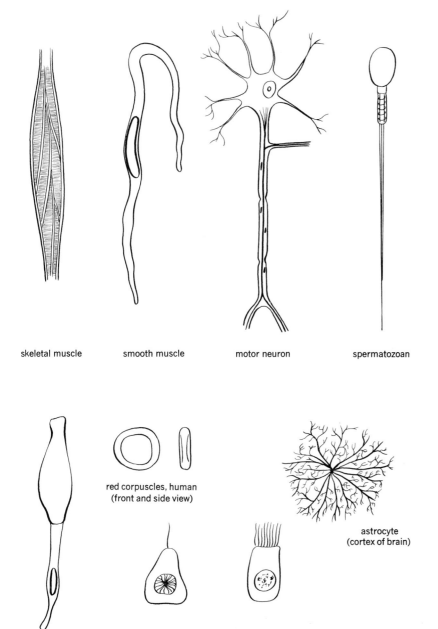

skeletal muscle   smooth muscle   motor neuron   spermatozoan

red corpuscles, human
(front and side view)

astrocyte
(cortex of brain)

secreting cell

flagellated
epithelial cell

ciliated
epithelial cell

FIGURE 2–2. Diverse forms of cells in the body (not to the same scale).

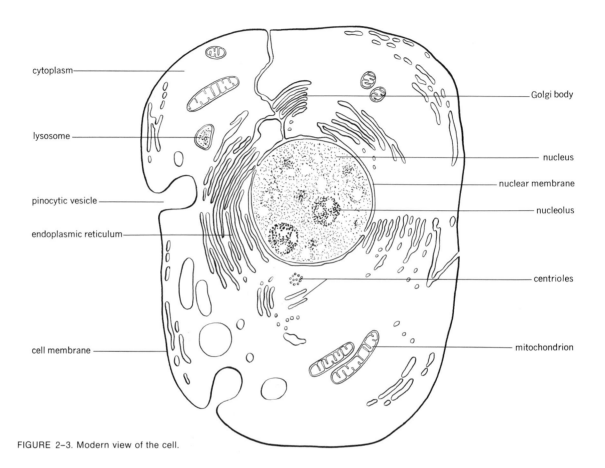

cytoplasm

lysosome

pinocytic vesicle

endoplasmic reticulum

cell membrane

Golgi body

nucleus

nuclear membrane

nucleolus

centrioles

mitochondrion

FIGURE 2–3. Modern view of the cell.

semipermeable, permitting some materials to enter easily, while restricting the passage of others. Such membranes are usually freely permeable to water, but act as a barrier to the unrestricted movement of almost all other materials. The membrane thus allows a great degree of control to be exerted upon the passage of materials through it. According to the most commonly accepted theory of ultrastructure, the membrane is composed of a double layer of lipid (fat) molecules, which is covered on the inner and outer surfaces with protein (Fig. 2–5). At intervals along the surface the lipid component is lacking and the protein forms a lining around the edges of the fat. A "pore" is thus created. Pores represent areas of greater permeability in the membrane. Logic would say that such areas cannot act as channels through which free flow of materials could occur, because the cell would then lose control over passage of materials. It has therefore been assumed that the pore is "covered" with a lid of protein or other material, or is lined with a protein which restricts passage of materials. In addition, the small diameter (7 to 10 Å) of the pore would act as a barrier to the free movement of larger substances.

The plasma membrane exhibits folds which may interdigitate with those of its neighbors and afford a means of holding cells together. Certain long cleft-like infoldings of the membrane of the cell are sites of *pinocytosis,* a process by which substances in solution are engulfed.

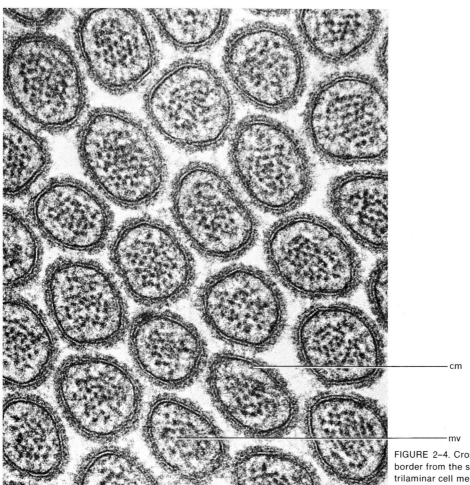

cm

mv

FIGURE 2–4. Cross section through microvillous brush border from the small intestine of lab mouse, showing trilaminar cell membrane. mv. microvillus: cm. cell membrane. (Photograph supplied by Norton B. Gilula, Department of Physiology-Anatomy, University of California, Berkeley.) × 148,000.

Permeability of the membrane is determined by a combination of several factors:

1. The overall *size* of the substance attempting to pass through the membrane as compared to the size of the membrane pores and the mesh size of the protein coverings is important. In general, if the sum of the atomic weights of individual atoms in a

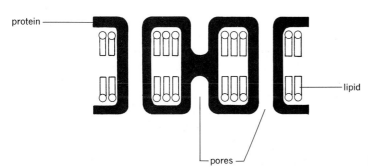

protein

lipid

pores

FIGURE 2–5. Diagram of the cell membrane.

molecule (molecular weight) is above 100, the size will be such as to restrict passage through the membrane.

2. The proteins of the membrane are capable of losing electrons or protons and may thus assume an *electrical charge.* They have *ionized.* If a substance attempting to cross the membrane is itself ionized, its passage may be accelerated if its charge is opposite to that of the membrane, or denied passage if the charge is the same.

3. If a substance *dissolves easily in fat* (high fat solubility) its passage may be more rapid inasmuch as a large part of the membrane consists of fat.

4. The presence in the membrane of specialized molecules known as *carriers,* capable of attaching to and transporting a substance, can insure passage of that substance without regard to the operation of physical factors such as those listed above.

The cytoplasm. This term refers to all of the cell substance within the plasma membrane excluding the nucleus. It is the factory area of the cell. It *receives* raw materials, *reduces* them to usable forms of

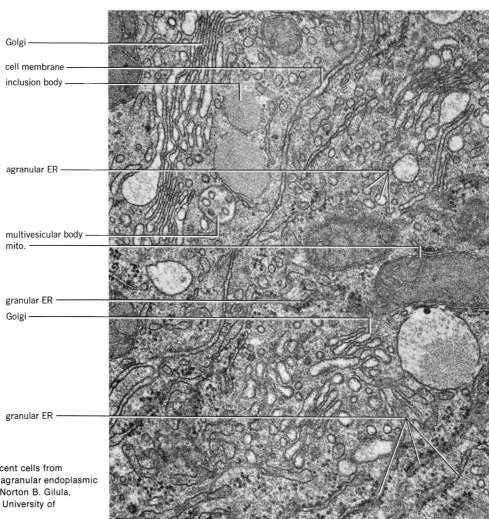

Golgi

cell membrane

inclusion body

agranular ER

multivesicular body

mito.

granular ER

Golgi

granular ER

FIGURE 2–6. Cytoplasm of two adjacent cells from trachea of lab mouse. Granular and agranular endoplasmic reticulum. (Photograph supplied by Norton B. Gilula, Department of Physiology-Anatomy, University of California, Berkeley.) × 38,000.

energy, *manufactures* new materials, *packages* materials for delivery elsewhere in the cell or body, and *disposes* of waste material. Activities in the cytoplasm are "organized chaos." Interference between different groups of chemical reactions is largely prevented by compartmentalization of each group of activities inside of formed structures within the cytoplasm. These formed structures are collectively referred to as the *organelles,* and include:

1. *The endoplasmic reticulum (ER)* (Fig. 2–6). The ER is a series of open channels permeating the entire cytoplasm, and reaching from membrane to nucleus. Its main function is that of *transport* of materials from one part of the cell to another. Microsomes are fragments of the ER.
2. *Ribosomes* (Fig. 2–7). Ribosomes are tiny granules of nucleic acid and protein. They are found either free within the cytoplasm or, more commonly, attached to the outer surface of the ER. Ribosomes are sites of *protein synthesis.*
3. *Mitochondria* (Fig. 2–8). Mitochondria are elongated or spherical

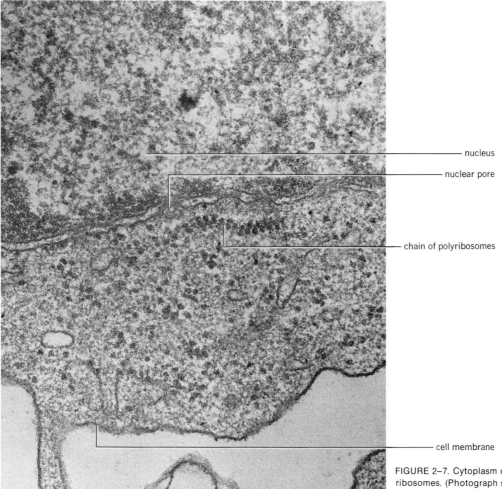

nucleus

nuclear pore

chain of polyribosomes

cell membrane

FIGURE 2–7. Cytoplasm of human lymphocyte. Polyribosomes. (Photograph supplied by Norton B. Gilula, Department of Physiology-Anatomy, University of California, Berkeley.) × 66,000.

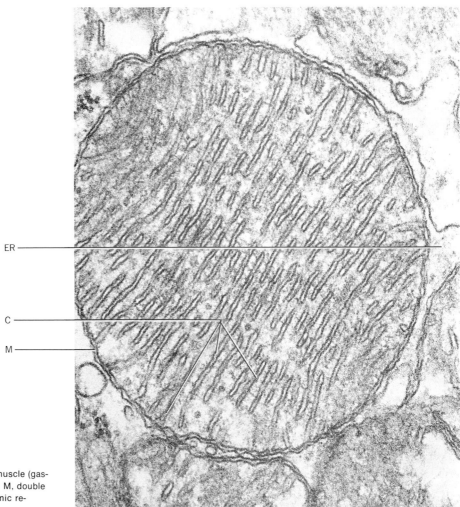

ER ———

C ———

M ———

FIGURE 2–8. Mitochondrion from skeletal muscle (gastrocnemius) of laboratory mouse. C, cristae; M, double membrane of mitochondrion; ER, endoplasmic reticulum. × 83,000.

organelles having a double wall. The outer wall is not folded, while the inner wall is folded into shelves or *cristae*. The organelle contains the materials necessary for the production of an energy-rich compound designated *adenosine triphosphate* or ATP. ATP is the chemical currency the cell spends to run its operations. This organelle is sometimes called the "powerhouse" of the cell due to its production of the fuel which runs cell machinery.

4. *The Golgi apparatus* (Fig. 2–9). The Golgi apparatus is a series of flattened channels *(Golgi membranes)* with expanded areas *(Golgi vacuoles)* within it. It is localized within the cell, usually close to the nucleus. It acts as an area where chemicals may accumulate and react with one another *(condensation membrane)* and functions in the packaging of materials produced in the Golgi membranes. The organelle is especially well developed in externally secreting glands, such as the pancreas and salivary glands. The secretions of such glands are rich in protein, in the

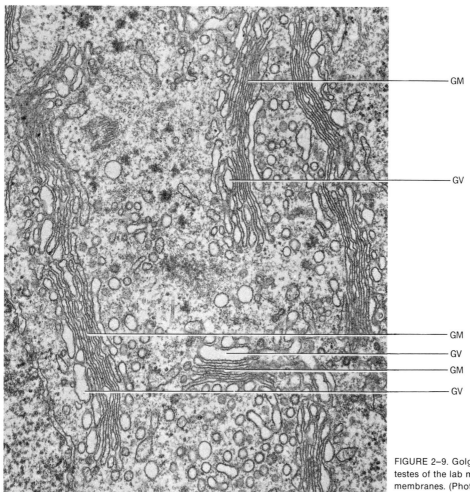

FIGURE 2–9. Golgi in the cytoplasm of a Sertoli cell in the testes of the lab mouse. GV, Golgi vesicles; GM, Golgi membranes. (Photograph supplied by Norton B. Gilula, Department of Physiology-Anatomy, University of California, Berkeley.) × 34,500.

form of enzymes, and so the Golgi apparatus appears to be important in the formation and release of such products.

5. *The central body* (Fig. 2–10). This organelle consists of two centrioles and a surrounding area of differentiated cytoplasm, the centrosome. All cells possess this organelle at some time during their life. Mature nerve cells lack it. The central body is instrumental in the process of cell division, and those animal cells which lack it are incapable of division. This is why destroyed nerve cells cannot be replaced by surviving cells.

6. *Lysosomes* (Fig. 2–11). Lysosomes contain powerful *hydrolytic enzymes* capable of breaking almost any large molecule into smaller units. They would destroy the cell itself if not contained by a limiting membrane. Because of this capability, lysosomes are sometimes referred to as "suicide packets." Lysosomes are useful in "digesting" foreign material taken into the cells. Products of lysosome action are utilized by other organelles.

32

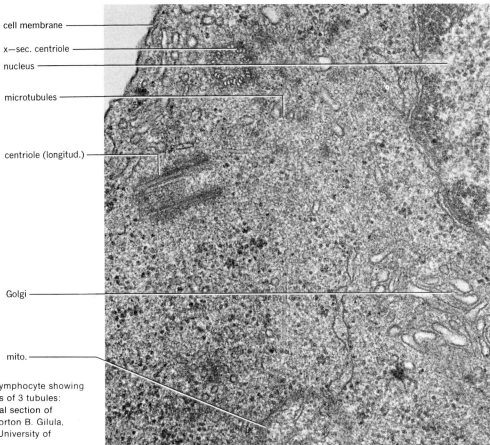

cell membrane

x—sec. centriole

nucleus

microtubules

centriole (longitud.)

Golgi

mito.

FIGURE 2–10. Cytoplasm of human lymphocyte showing centrioles of central body. Note 9 sets of 3 tubules: cross section of centriole; longitudinal section of centriole. (Photograph supplied by Norton B. Gilula, Department of Physiology-Anatomy, University of California, Berkeley.) × 49,500.

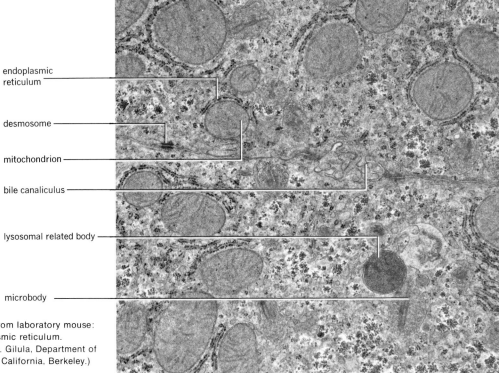

endoplasmic reticulum

desmosome

mitochondrion

bile canaliculus

lysosomal related body

microbody

FIGURE 2–11. Liver hepatocytes from laboratory mouse: lysosome; mitochondria; endoplasmic reticulum. (Photograph supplied by Norton B. Gilula, Department of Physiology-Anatomy, University of California, Berkeley.) × 16,500.

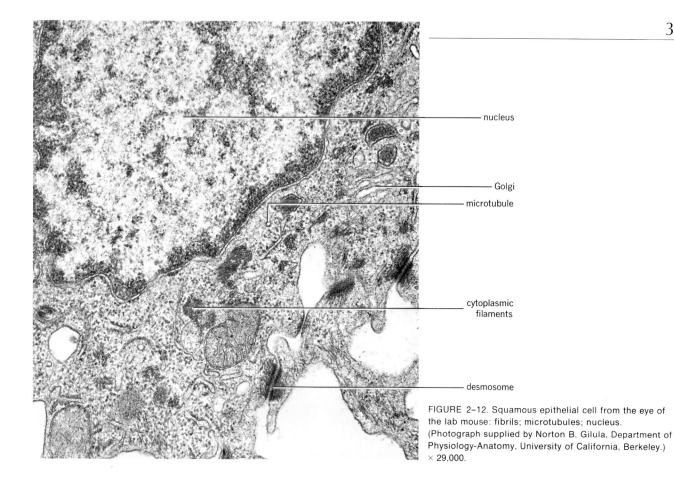

— nucleus

— Golgi
— microtubule

cytoplasmic filaments

— desmosome

FIGURE 2–12. Squamous epithelial cell from the eye of the lab mouse: fibrils; microtubules; nucleus. (Photograph supplied by Norton B. Gilula, Department of Physiology-Anatomy, University of California, Berkeley.) × 29,000.

7. *Vacuoles, fibrils, and tubules* (Fig. 2–12). Vacuoles are membrane-bounded sacs of fluid used for storage or excretion. Fibrils are strands of protein running through the cytoplasm. They may support the cell or give it the power of contractility. Tubules are of wide occurrence in cells and may give support or serve as an additional means of distributing materials throughout the cell.

The cytoplasm also contains droplets of fat, crystals of sugar and other raw materials, and wastes. Such materials are designated *inclusions*. They do not take an active part in the machinery of the cell, except as they are metabolized, or are the result of metabolism.

All organelles are in a constant state of flux. ER, mitochondria, and Golgi apparatus fragment and rejoin. The cell grows, may divide, and may transmit disturbances through itself. Organization is imparted to all these activities by the nucleus.

The nucleus (Fig. 2–13). This spherical structure is separated from the cytoplasm by a double-walled *membrane* similar in structure to the plasma membrane. The outer wall is derived from the ER, and thus the nucleus has direct access to materials in the ER cavities. The inner wall belongs to the nucleus proper. The two membranes come together at intervals forming thin areas designated as "nuclear pores" (Fig. 2–14). As in the plasma membrane, such areas are regions of greater permeability. Materials produced by the nucleus are believed to leave through these pores. The membrane encloses a *nuclear fluid* in which *nucleolus* and the *chromatin material* float. The nucleolus is composed chiefly

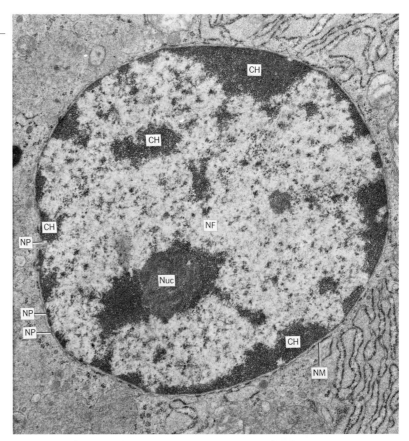

FIGURE 2–13. Nucleus from laboratory mouse trachea (mucus cell). NM, nuclear membrane; Nuc, nucleolus; NP, nuclear pores; CH, chromatin; NF, space containing nuclear fluid. (Photograph supplied by Norton B. Gilula, Department of Physiology-Anatomy, University of California, Berkeley.) × 15,000.

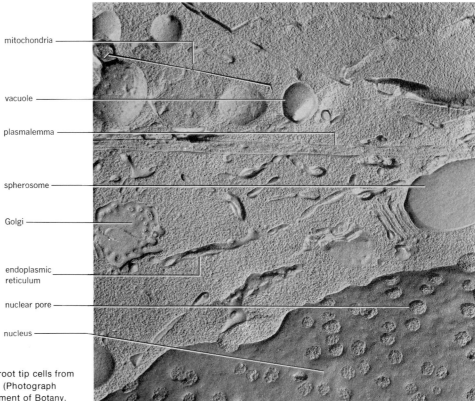

FIGURE 2–14. Nuclear pores in onion root tip cells from freeze-etch preparation: nuclear pores. (Photograph supplied by Dr. Daniel Branton, Department of Botany, University of California, Berkeley.) × 23,000.

of ribonucleic acid, or RNA, while the chromatin is composed of deoxyribonucleic acid, or DNA. These nucleic acids direct cellular activity, particularly the synthetic reactions of metabolism.

The typical unit which has been described reflects the activities occurring in the entire human body, and the body reflects the activities of individual cells. By the activities shown, the cell qualifies as living. When the activities described no longer occur, the cell is dead.

Table 2–1 summarizes the structure and function of the various parts of the cell.

TABLE 2–1
A Summary of Cell Structure and Function

Cell Part	Structure	Function	Comments
Plasma membrane	Three-layered sandwich of lipid and protein	Protects, limits, controls entry and exit of materials	Selectively permeable (semi-permeable)
Cytoplasm	Contains many organelles and inclusions	Factory area; synthesizes, metabolizes, packages	Bulk of cell
ER	Hollow channels in whole cytoplasm	Transport	"Circulatory system" of cell
Ribosome	Granules of nucleic acid and protein	Protein synthesis	Free or attached to ER
Mitochondria	Double layered with cristae	Production of ATP	"Powerhouse" of cell
Golgi apparatus	Flattened sacs with vacuoles	Secretion; condensation membrane	Packaging of materials
Central body	Centrioles and centrosome	Cell division	Lacking in mature nerve cells
Lysosome	Membrane surrounded sacs of enzymes	Reduces large molecules to smaller units	"Digestive system" of cell
Vacuoles	Membranous sacs of fluid	Storage and excretion	Universal occurrence
Fibrils	Protein strands	Support and movement	Extreme development in muscle cells
Tubules	Tiny hollow tubes	Support and conduction	Universal occurrence
Inclusions	Fat, sugar, wastes	Fuels and wastes	"Nonliving"
Nucleus	Membrane, nucleolus, chromatin and fluid	Directs cell activity	"Brains" of cell

Cell division

No cell lives forever. The actual length of time from "birth" to "death" of a cell varies from the few hours of life of a white blood cell to the hundred years of a nerve cell's existence. Replacement of cells which continually die is essential in such organs as the skin. The healing of wounds and the replacement of blood cells are other examples of continual need for replacement of dying units. In such cases new cells must be created which are like the old in both structure and function.

Mitosis.   The process of mitosis results in the formation of new cells genetically and morphologically like the old, and insures the continuance of a particular cell line. The process also enables the organism to increase in size. Mitosis is a continuous process, but for the purposes of description it is usually divided into five stages:

1. *Interphase.* The cell presents the appearance typical for its type. All cell structures are visible and the nuclear chromatin is present in the form of fine granules. The cell appears to be at rest, exhibiting no activity connected with division.
2. *Prophase.* The chromatin of the nucleus forms fine threads which then fragment and shorten into *chromosomes.* The chromosomes divide longitudinally into identical halves known as *chromatids.* The nuclear membrane disappears, the centrioles move to opposite poles of the cell, and a spindle of microtubules appears between the centrioles.
3. *Metaphase.* The chromosomes align themselves upon the center of the spindle.
4. *Anaphase.* The chromatids separate and move towards the centrioles, each member of a chromatid passing to a daughter cell. The cell membrane begins to constrict and divide the cytoplasm into daughter cells. In the human, the original diploid number of chromosomes, 46, is maintained.
5. *Telophase.* Nuclei reorganize within the two new cells, and the cytoplasm division is completed.

These events are diagrammed in Fig. 2–15. Since the chromosomes distributed to the daughter cells are identical, and are equal to the original number in the cell before division, the chromosome number and makeup characteristic for the species is maintained.

Meiosis.   A second type of cell division occurs within the ovaries and testes of the human. It is designated meiosis, and results in the production of daughter cells which contain one-half (haploid) the normal species number of chromosomes, 23 for the human. The cells are not like the parent in terms of identical genetic makeup. Cells undergoing meiosis pass through the same steps as were described in connection with mitosis, but pass through an additional division which reduces the chromosome number in half. The changes shown in meiosis are presented diagrammatically in Fig. 2–16. Further explanations of the details of this process will be found in the chapter on development of the individual.

The chemical constituents of the cell

Introduction.   Living material is composed of chemical substances in the form of free elements, ions, radicals, molecules, and compounds. Elements are substances in atomic form and are the simplest

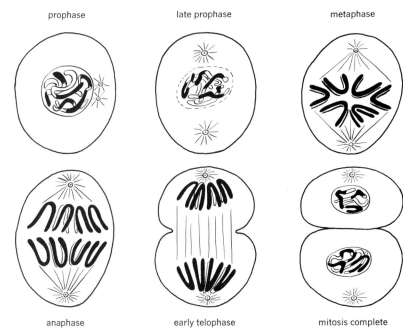

prophase   late prophase   metaphase

anaphase   early telophase   mitosis complete   FIGURE 2–15. Mitosis.

chemical units which retain the properties ascribed to them. Elements, either alone or in combinations known as radicals, are usually capable of gaining or losing electrons and may thus assume an electric charge. Such charged units are known as ions. A radical is a group of two or more atoms that act as a single atom, and which usually goes through a reaction without change; for example, bicarbonate (HCO_3^-), or phosphate ($PO_4^=$). A molecule is composed of two or more atoms or ions to create a substance having properties different from those of the materials forming it. The term compound conventionally refers to larger complex molecules or to the combination of molecules into a unit having different properties from the constituent parts. These materials may exist in two states within the cell. If they exist in the dissolved and unaggregated state in the cell, they form a *solution.* By aggregation, larger particles may be formed and are considered to be suspended rather than dissolved. Such suspensions are said to form a *colloidal suspension,* or colloid. Colloids are formed by proteins and other large molecules in the body. They do not diffuse easily, can "trap" and hold large amounts of water and can concentrate other materials by adsorption at their surfaces. The living material is thus a heterogeneous mixture of chemical substances exhibiting many different properties and existing in a variety of states.

The materials composing the cell. The chemicals found in the cell, from which the cell manufactures new materials, are the ones that are most plentiful in the environment.

1. *Water.* Composed of hydrogen and oxygen, water constitutes, on the average, 55 to 60 percent of the cell substance. The large amount of water present suggests that it has properties which fit it especially well to serve the cell.

   (a) Water is a good *solvent,* and what it cannot dissolve it may suspend. Water is not toxic to the cell. In large amounts, water may dilute cellular chemicals and interfere with cellular reactions.

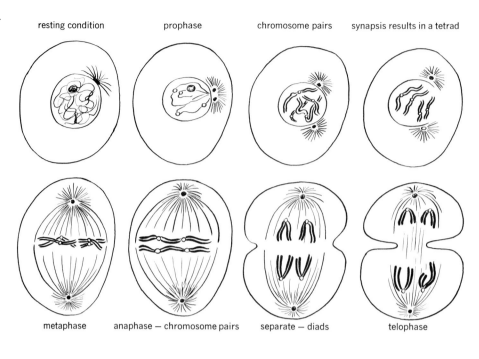

resting condition   prophase   chromosome pairs   synapsis results in a tetrad

metaphase   anaphase — chromosome pairs   separate — diads   telophase

First meiotic division

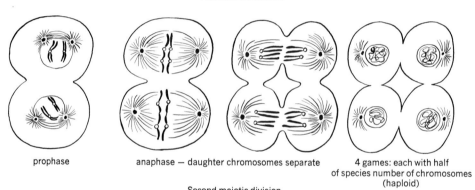

prophase   anaphase — daughter chromosomes separate   4 games: each with half of species number of chromosomes (haploid)

Second meiotic division

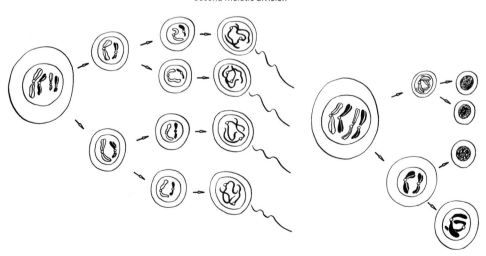

FIGURE 2–16. Meiosis.   spermatogenesis   oogenesis

(b) Water can *absorb much heat* energy without great changes in its own temperature. Cell metabolism produces much heat which must be eliminated. Water not only absorbs this heat within the cell, but, by means of the blood stream, can carry it to appropriate sites for elimination.

(c) Water causes many materials to *ionize.* Ionized substances enter chemical reactions more easily.

(d) Water is a *dipole molecule* and is polarized. The molecules can form "magnet-like" attachments to other charged particles and thus form hydration layers around these particles. For example:

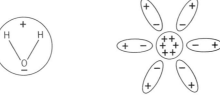

(A)   (B)

The dipole nature of the water molecule (A)
and the formation of a hydration shell around
a positively charged atom (B)

The "binding" of water is of importance in maintaining the water content of membranes and in the colloidal solutions of the cell itself, and can influence the passage of materials through membranes by altering the size of the material.

2. *Inorganic materials.* Usually referred to as "salts," these materials create the proper environment for cellular activity. They are instrumental in the creation of *osmotic gradients,* in *regulation of the reaction* (pH) of the cell, and confer an ability to *resist changes in pH.* Sodium, potassium, chloride, calcium, magnesium, and iron are of nearly universal occurrence in cells; phosphate, bicarbonate, and sulfate are common radicals; zinc and manganese are present only in some body cells.

3. *Proteins.* The name means "of first importance" and, in terms of basic structure and activity of the cell, proteins play a vital role. Proteins always contain carbon, hydrogen, oxygen, and nitrogen compounded into the building block of proteins, the *amino acid.* The acid always contains an amine ($-NH_2$) and carboxyl ($-COOH$) group. Amino acids are held together in the larger protein molecules by peptide bonds (Fig. 2–17a, b). Proteins are found in

FIGURE 2–17a. The linking of two amino acids by a peptide bond.

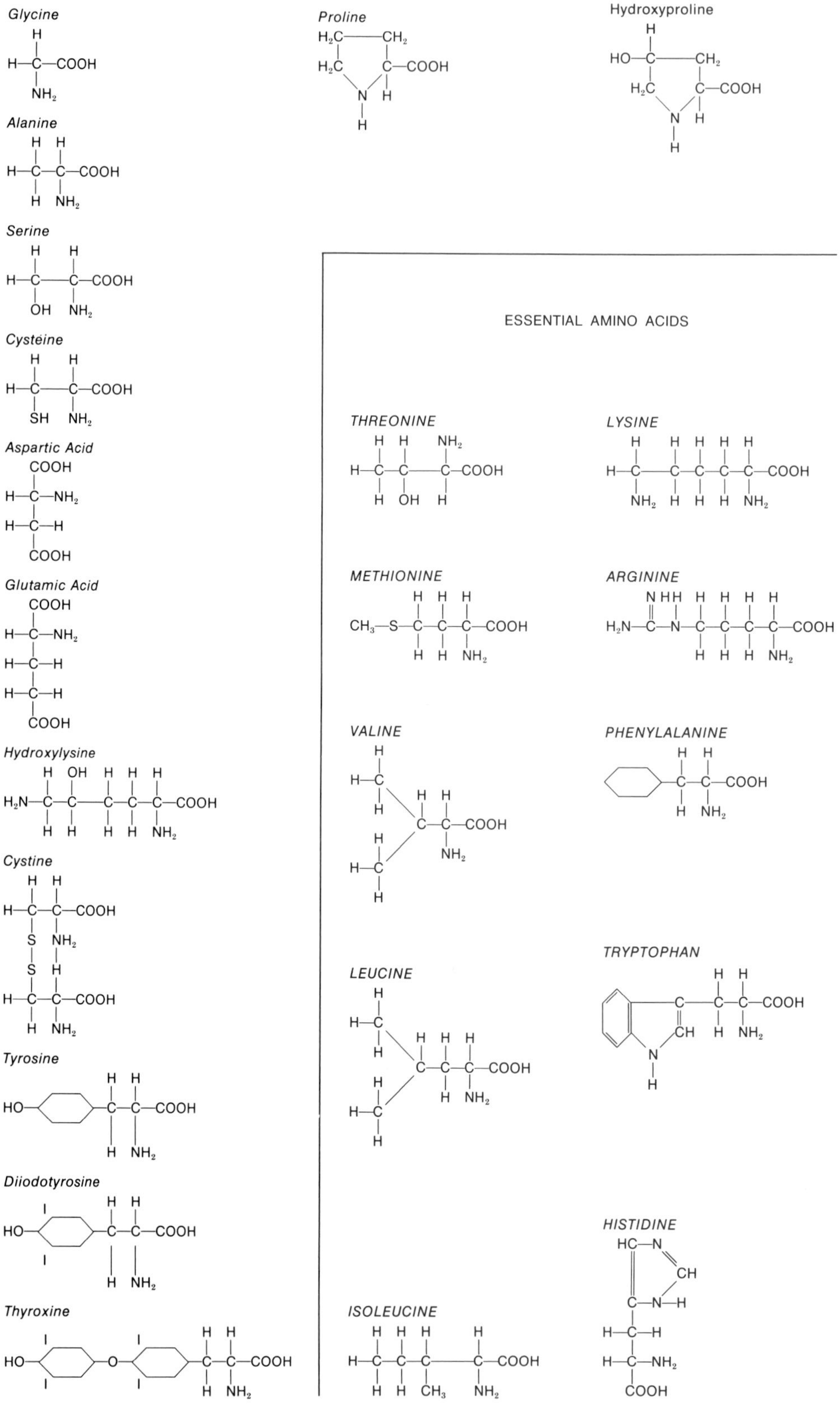

FIGURE 2–17b. Amino acids.

the *cell membrane;* they are found as a *cytoskeleton* within the cell; they form the *contractile elements* of muscle; they are the major components of connective tissue fibers; and they form *enzymes.*

Enzymes are *organic catalysts* which alter the velocity of chemical reactions. They are *specific* in the sense that they can affect only a single reaction, but can usually influence that reaction in both directions. They *require certain conditions* in terms of temperature and pH in order to exhibit their optimum activity, and work only on specific materials or substrates. Enzymes may be classified by substrate, type of activity, and where they act.

Intracellular enzymes operate within one cell, while *extracellular* enzymes are produced within a cell and are emptied into a hollow organ to exert their function. Table 2–2 summarizes some of the more important enzymes of the body.

TABLE 2–2
A Partial Classification of Enzymes

Name	Substrate (Material Attacked)	Action	Examples
Hydrolases	None specific. Fats, phosphate containing compounds, nucleic acids, many others	By adding water (hydrolysis), splits larger molecules into smaller units	Lipases, phosphatases, nucleases
Carbohydrases	Carbohydrates	Fragment carbohydrates by hydrolysis to smaller units	Salivary amylase, pancreatic amylase, maltase, lactase, sucrase
Proteases	Proteins, smaller units containing fewer amino acids	Fragment proteins by hydrolysis to smaller units	Pepsin, trypsin, erepsin
Phosphorylases	Many molecules	Split or activate molecule by addition of phosphate group	Muscle phosphorylase, glucokinase fructokinase
Dehydrogenases	Any compound capable of losing 2 H	Oxidizes compound by removal of 2 H	Succinic dehydrogenase, fumaric dehydrogenase
Oxidases	Any compound which may add oxygen	Oxidizes compound by addition of oxygen	Peroxidase
Aminases	Amine-containing compounds and keto acids	Addition or removal of amine groups	Transaminase, deaminase
Decarboxylases	Acids containing carboxyl group	Remove CO_2 from carboxyl group	Ketoglutaric decarboxylase
Isomerases	Chiefly carbohydrates	Move a radical to different part of molecule	Isomerase

4. *Lipids.* The term lipid refers to all substances soluble in fat solvents such as ether and chloroform, and insoluble in water. *Simple lipids* include the *fats,* waxes, and oils. *Compound lipids* are a combination of a lipid with some nonlipid fraction, such as *phospholipids. Derived lipids* include the *sterols,* and intermediates in the production or breakdown of the other lipid categories. The triglycerides are one of the most important lipids. They are one of the storage forms of fats in the body, and are the form in which many fats are ingested. The sterols and their derivatives are important in the formation of hormones from the gonads and adrenal gland. Figure 2–18 shows some of the more important lipids found in the body. Lipids are thus important structural elements in the body; they give form, shape, and insulation to the body and, as hormones, control a wide variety of body processes.

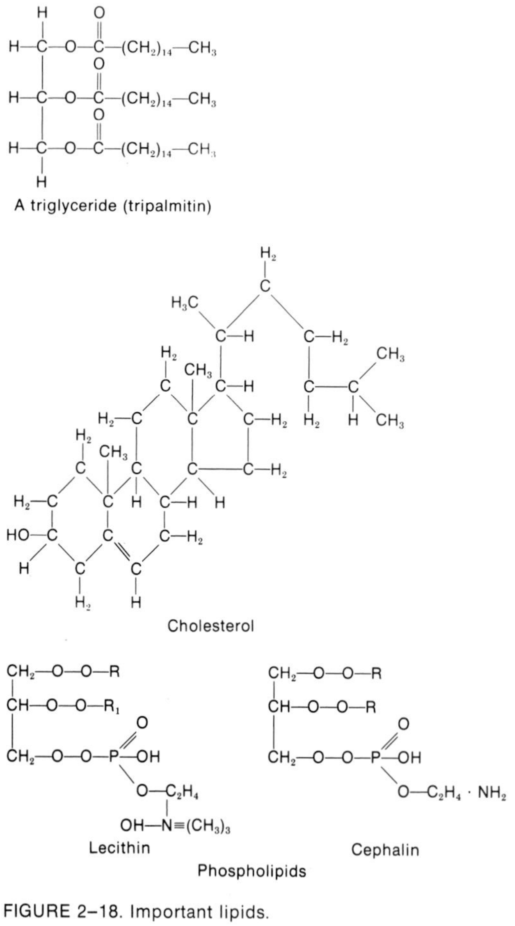

A triglyceride (tripalmitin)

Cholesterol

Lecithin   Cephalin

Phospholipids

FIGURE 2–18. Important lipids.

5. *Carbohydrates.* These compounds contain carbon, hydrogen, and oxygen. The hydrogen and oxygen is always in the same ratio as in water. *Monosaccharides,* also known as simple sugars, may be regarded as the building blocks of more complex carbohydrates.

The 5 and 6 carbon sugars are the most important physiologically. *Glucose* and *ribose* form large carbohydrates and nucleic acids respectively. Important monosaccharides are presented in Fig. 2–19.

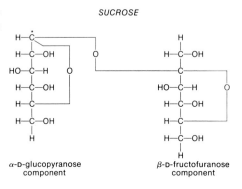

SUCROSE

α-D-glucopyranose
component

β-D-fructofuranose
component

D-Glucose
($C_6H_{12}O_6$)

D-Fructose
($C_6H_{12}O_6$)

D-Galactose
($C_6H_{12}O_6$)

D-Ribose
($C_5H_{10}O_5$)

D-Deoxyribose
($C_5H_{10}O_4$)

FIGURE 2-19. Some important monosaccharides.

MALTOSE (α form)

two α-D-glucopyranose components

Disaccharides are formed from two simple sugar units by the splitting out of water. *Maltose, lactose,* and *sucrose* are common examples of disaccharides, and are shown in Fig. 2–20.

Polysaccharides have many simple sugar units in them. *Glycogen* (Fig. 2–21) is a polysaccharide employed in the human body as a storage form of glucose. Carbohydrates form the preferred and primary *source of fuel* for cellular activity. They also serve structural functions when combined with other materials as in mucopolysaccharides.

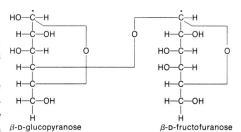

LACTOSE (β form)

β-D-glucopyranose
component

β-D-fructofuranose
component

FIGURE 2–20. Disaccharides

6. *Nucleic acids.* Nucleic acids are complex materials found in the nucleus and cytoplasm of cells. They are composed of *nucleotides* (Fig. 2–22) which in turn contain a *nitrogenous base,* a *sugar,* and *phosphoric acid.* Nucleic acids are of two general types, deoxyribonucleic acid (*DNA*) and ribonucleic acid (*RNA*). DNA is restricted to the nucleus, and RNA is found in both the nucleus and cytoplasm. The details of the involvement of the nucleic acids will be considered in Chapter 16 in connection with the metabolism of proteins.

7. *Trace materials.* Some substances essential to normal body function are required in only minute amounts. *Vitamins,* some *hormones,* and certain *metals* (Cu, Mo, Mg) are classed as trace materials. The functions of these substances will be considered in appropriate chapters later.

Table 2–3 summarizes the chemicals present in the living material, and their locations and functions.

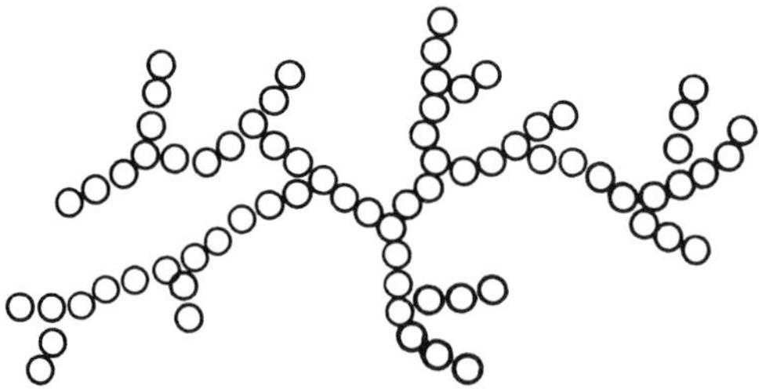

FIGURE 2-21. Schematic representation of the glycogen molecule. Each circle represents one simple sugar unit.

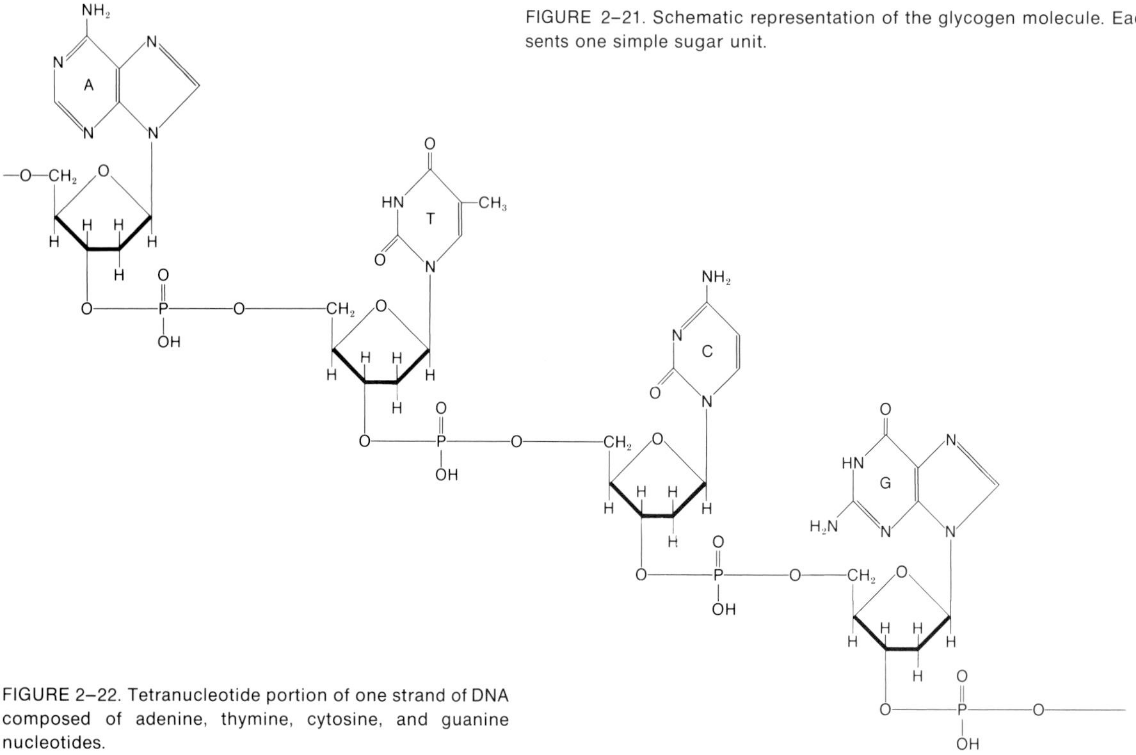

FIGURE 2-22. Tetranucleotide portion of one strand of DNA composed of adenine, thymine, cytosine, and guanine nucleotides.

Acquisition of materials by the cell

Introduction. All parts of the cell require chemical materials for continued activity and maintenance of structure. The required materials are in the cell's environment and must pass through at least one membrane in order to be utilized.

Two broad categories of processes by which materials cross membranes are recognized:

1. *Passive processes.* Materials moving by a passive process do so *without the cell contributing any work or energy to the process.* The only requirement for the process to proceed is a difference or gradient in concentration or pressure on the two sides of the membrane. The direction of movement will always be from the area of higher pressure or concentration to the area of lower pressure or concentration, and will tend to result in a condition of equilibrium on the two sides of the membrane.

2. *Active processes.* In an active process, the *cell contributes energy*

TABLE 2-3

A Summary of the Chemicals of the Living Material

Substance	Location in cell	Function
Water	Throughout	Dissolve, suspend, and ionize other materials; regulate temperature
Inorganic salts	Throughout	Establish osmotic gradients, pH, buffer capacity
Proteins	Membranes, cyto-skeleton, ribosomes, enzymes	Give form, strength, contractility, catalysts
Lipids	Membranes, Golgi apparatus, inclusions	Reserve energy source, form, shape, protection, insulation
Carbohydrates	Inclusions	Preferred fuel for activity
Nucleic Acids DNA	Nucleus, in chromo-somes and genes	Direct cell activity
RNA	Nucleolus, cytoplasm	Carry instructions, transport amino acids
Trace Materials Vitamins	Cytoplasm	Work with enzymes
Hormones	Cytoplasm	Work with enzymes
Metals	Cytoplasm	Specific functions in various synthetic schemes

and participates in the movement of material. By an active process, a material may be caused to pass through a membrane without regard to gradients.

Passive processes. *Diffusion* occurs as follows. Molecules are in constant motion. Assuming a greater concentration of a given type of molecule on one side of a membrane as compared to the other, and assuming that the membrane is permeable to the molecule, a gradient will be established from the side of greater concentration toward the side with a lesser concentration.

In a given unit of time, more collisions with the membrane will occur on the side with the greater concentration of material. There will therefore be a greater movement of molecules from the side with the greater concentration, and a net flow from that side will occur. As the numbers of molecules on the two sides approach equality, the rate of diffusion will decrease.

Dialysis is a term often used to refer to the diffusion of the soluble (as opposed to suspended) constituents in a solution. Dialysis is used to separate diffusible and nondiffusible molecules, for example, in the artificial kidney. Suppose that one has a fluid in a semipermeable tube and wishes to separate the smaller molecules from the larger ones. Assuming that the tube is permeable only to the smaller particles, placing the tube in a solution having no solutes would create a gradient for the smaller particles. They would then diffuse out of the tube, leaving the larger particles behind. Changing the solution around the tube would maintain the gradient, and a nearly complete separation of the particles will occur.

Filtration is the movement of materials across a membrane due to a *difference in pressure* on the two sides. This is a relatively non-selective process, since anything small enough to pass through the pores of the membrane will be forced through by the pressure. The membrane acts as a sieve. The rate of filtration will be determined by the pressure differential, and will continue unchanged if the pressure differential stays the same. Examples of filtration are seen in the passage of materials through capillary walls under the influence of the blood pressure.

Osmosis refers to the diffusion of solvent (water) only, through a membrane. Movement is brought about as the result of a concentration gradient for water. In terms of concentration (amount of solute per volume of solvent) the more solute there is in a given volume of solvent, the lower is the relative concentration of solvent. In effect, the amount of solute determines the water concentration. Therefore, a flow of water may be said to occur from the area of greater water concentration (or lower solute) towards the area of lesser water concentration (or more solute). The higher the solute concentration is in a given area, the greater will be the "pull" of these solutes on the water. This "pull" establishes an osmotic pressure whose magnitude depends on the total numbers rather than the kinds of solute molecules that are present. In theory, if the membrane is permeable only to water, a condition of equality can never be attained and the process will continue indefinitely. In practice, however, there are several factors which cause the process to cease.

1. As water moves across the membrane, the area receiving it will increase in volume and the amount of water it contains. When the weight of the accumulating water creates a hydrostatic pressure equal to that of the osmotic pressure causing the movement, the movement will cease.
2. If a cell or tissue offers resistance to being inflated by the incoming water, it may develop a "recoil pressure." This can be compared to the increasing difficulty that one might encounter in blowing up a large balloon. This will again create a "back pressure" which will stop osmosis.

Obviously, then, the concentrations of solutes and water on the two sides of a cell membrane will determine the direction and degree of water movement. By way of illustrating this principle, assume that an animal cell is placed into a solution and that concentrations of solute are equal inside and outside the cell. Osmosis into and out of the cell will proceed at equal rates, and the cell will not change its size. The cell may be said to be *isotonic* with respect to the solution.

If the cell is placed in a solution which has a solute concentration less than that within the cell, there will be a net flow of water into the cell, and the cell will swell. In this situation there is a greater concentration of water outside the cell, and the cell is in a *hypotonic* (hypo = less) *solution.*

If the cell is placed in a solution which has a solute concentration greater than that within the cell, there will be a net flow of water out of the cell and the cell will shrink. Such a solution is said to be *hypertonic* since it has a greater (hyper = more) solute concentration than the cell. These relationships are presented in Fig. 2–23.

The water content of cells is therefore extremely dependent upon the relative concentrations of solutes inside and outside of it. Recall that the inorganic salts play a major role in the determination of osmotic gradients.

The ability of a solution to create an osmotic pressure may be conveniently measured by the indirect method of determining the freezing point. For a given solvent (in this case water) a 1-molar solution (1 gram molecular weight of a substance per liter of water) of a non-ionizing substance will freeze at $-1.86°$ C. Such a solution is said to have a pressure of one *osmol* (Os). For ionizing materials, osmolarity is equal to molarity multiplied by the number of particles per mole resulting from ionization, or

$$Os = M \times \text{number of particles from ionization.}$$

The freezing point is depressed in proportion to the number of particles resulting from ionization. For sodium chloride (common table salt), a 1-molar solution is 86.3 percent ionized, and the osmolar concentration would be 1×0.863. It would therefore freeze at $-3.2°$ C. One may compare the numbers of osmotically active particles in solutions by the degree to which they depress the freezing point of water. In the body, concentrations of solutes are relatively low, so that the osmol becomes too large a unit to work with conveniently. The term

FIGURE 2–23. Osmosis. Black arrows indicate greatest direction of water flow.

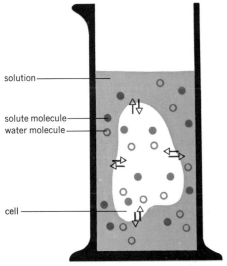
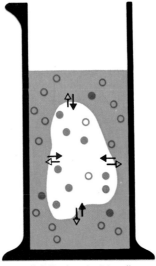
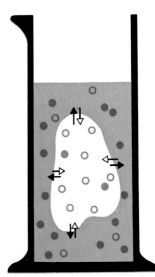

solution

solute molecule
water molecule

cell

Solution isotonic to cell   Solution hypotonic to cell   Solution hypertonic to cell

milliosomol (mOs) is usually employed. One milliosmol is 1/1000 osmol, and would depress the freezing point 0.00186° C.

Active processes.   Active processes involve the expenditure of energy by the cell and can move materials at a faster rate than the passive processes. More important, active processes can move a material against a concentration gradient.

Active transport. Three things are required for active transport:

1. A large *carrier* molecule, located in the cell membrane. The carrier is capable of attaching to and transporting some substance.
2. A *source of energy.* This activates the carrier and renders it capable of attaching to the material to be transported. ATP (adenosine triphosphate) is the source of energy contributed by the cell.
3. *Enzymes,* responsible for attaching and detaching the energizing molecule to the carrier. These are also contributed by the cell.

In general terms, the process proceeds as follows. The unactivated carrier is located on one side of the membrane. It is activated by an enzyme splitting the end phosphate group from an ATP molecule and attaching it to the carrier. The carrier attaches to the material to be transported, moves across the membrane, and encounters another enzyme which removes the phosphate group. The carrier is no longer capable of holding the transported material and discharges it to the outside of the membrane. The carrier then returns to the original side of the membrane, ready to repeat the process. In only a few cases has the chemical nature of the carrier been established. A specific illustration of the process is presented in Fig. 2–24, showing the "diglyceride transport system" operating in the transport of sodium.

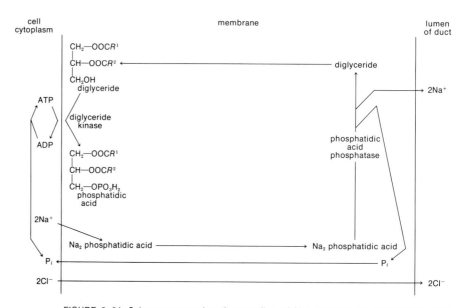

FIGURE 2–24. Scheme concerning the coupling of Na+ transport to phosphatidic acid metabolism. (After Hokin and Hokin, *J. Gen. Physiol.*, 44, 1960.)

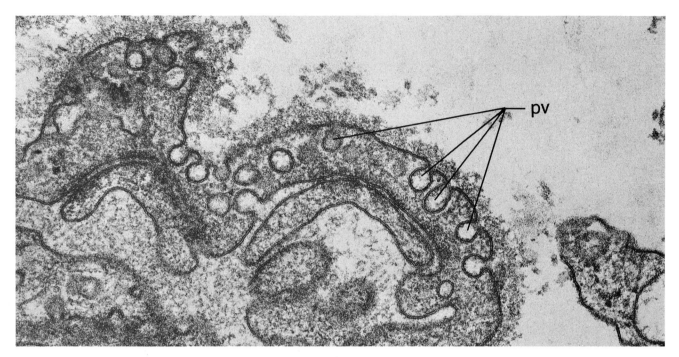

FIGURE 2–25A. Capillary from skeletal muscle of lab mouse demonstrates the structural basis for capillary permeability. PV, pinocytic vesicles. (Photograph supplied by Norton B. Gilula, Department of Physiology-Anatomy, University of California, Berkeley.) × 22,000.

If the movement of a substance by active transport is in the same direction as it would normally have occurred by a passive process, the process is termed *facilitated diffusion.*

A given transport system may be capable of carrying only a single material, or a single carrier may transport several substances. If a carrier can carry more than one material, and is faced with both at the same time, it will transport one material in preference to the other. The material which suffers a decreased transport rate is said to have undergone a *competitive inhibition.*

Pinocytosis (Fig. 2–25) is a process usually classed with the active processes, since the cell participates in it. The energy expenditure, however, is far less than that involved in active transport. Pinocytosis is an invagination or "sinking" of the cell membrane that ultimately forms a vacuole containing whatever was in the vicinity of the membrane where it invaginated. The cell thus brings in a part of its environment. Large molecules that have no transport system may thus be brought into the cell. A "reverse pinocytosis" (*emeiocytosis*) may operate; this enables vacuoles to empty their contents outside the cell and thus excrete waste material. *Phagocytosis,* considered by some to be an active process, is the engulfing of particulate matter by a cell. No membrane is crossed in this process.

By the active processes, the cell is capable of exerting considerable control over materials which enter and leave the cell.

Table 2–4 summarizes the major methods by which materials pass through membranes.

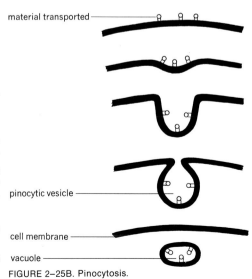

material transported

pinocytic vesicle

cell membrane

vacuole

FIGURE 2–25B. Pinocytosis.

TABLE 2–4

A Summary of Methods of Acquisition of Materials by the Cell

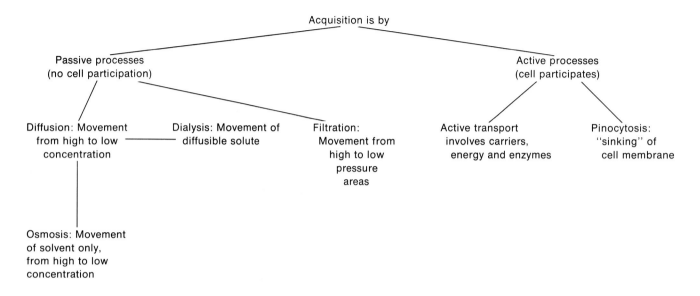

TISSUES

Cells that are similar in structure and function, together with all their associated intercellular material, constitute a tissue. Four tissue groups are recognized:

1. *Epithelial tissues.* Epithelia *cover and line* body surfaces and cavities. They are adapted to serve a wide variety of functions, such as protection, secretion, absorption, excretion, and reception of stimuli.
2. *Connective tissues.* These tissues *connect, support,* and *protect* other tissues of the body. Cartilage, bone, tendons, and ligaments are obvious examples of connective tissues. The blood is also included in the connective tissue group.
3. *Muscular tissues.* Muscle is specialized to *contract* or shorten, and thus cause the movement of materials through the body as well as the movement of the body through space.
4. *Nervous tissue.* This tissue is specialized to *react* to stimuli (irritability) and to *transmit* these impulses (conductivity) over the body.

ORGANS

Two or more tissues that are put together in a larger unit having specific functions constitute an organ. Organs are so numerous in the body as to preclude a listing of them. The reader should be able to cite many examples of organs.

SYSTEMS

Several organs working toward some larger goal constitute a system. A partial listing of systems follows:

1. *Integumentary system.* The skin and its appendages.
2. *Skeletal system.* The bones and joints.
3. *Muscular system.* The skeletal muscles of the body.
4. *Circulatory system.* The heart, blood, and blood vessels.
5. *Lymphatic system.* The lymph, lymph vessels, and the lymphoid organs such as thymus, spleen, and tonsils.
6. *Respiratory system.* The lungs and their associated structures.
7. *Digestive system.* The alimentary tract and the accessory organs of digestion, liver and pancreas.
8. *Urinary system.* The kidneys and their associated organs.
9. *Reproductive systems.* Testes, ovaries, and their associated organs.
10. *Nervous system.* The brain, spinal cord, and peripheral nerves.
11. *Endocrine system.* The ductless glands producing hormones.

GENERAL SUMMARY

From the discussion of the cell and its formation into larger levels of organization, we can perhaps better appreciate the statement that the cell is not only the structural unit of the body, but the functional unit as well. It is improbable that any process occurs in the organism without the direct or indirect intervention of a cell.

Since cells represent the point of ultimate responsibility for body activity, whatever problems the cell faces become problems for the body as a whole.

In the following chapters, we shall examine how the body as a whole solves the problems of acquisition of materials, metabolism, and elimination of wastes created by the cells themselves.

QUESTIONS

1. In what ways, both anatomically and functionally, does a single "typical cell" resemble the body as a whole?
2. If someone placed a piece of substance before you and asked you to determine if it was living, what would you test for, and how would you test for it?
3. What effect upon cellular activity would be seen if the lysosomes released their contents into a cell? Explain.
4. A water soluble, positively charged, large particle approaches a cell membrane. By what methods might its movement through the membrane be hindered or accelerated?
5. Establish some sort of hierarchy among the chemicals found in the cell, placing at the top of the list the one you consider most important. Justify the placement of each.
6. Contrast and compare mitosis and meiosis in terms of the similarity, structurally and functionally, of daughter cells to the parent.
7. What are some of the modern tools and techniques which have advanced our knowledge of the structure and function of the cell?

CHAPTER 3

Tissues

Over the structure of the cell rises the structure of plants and animals, which exhibit the yet more complicated elaborate combinations of millions and billions of cells coordinated and differentiated in the most extremely different ways. HERTWIG

EPITHELIUM

Introduction and general characteristics

Epithelia *cover* and *line* actual or potential free surfaces of the body. They are therefore in direct contact with both the external and internal environments of the body, and are important in aiding maintenance of the homeostasis of the internal environment. The free surfaces of the skin, digestive system, respiratory system, excretory system, reproductive systems, and body cavities have surfaces which normally do not have another tissue in direct contact with them, and these represent actual free surfaces. In a blood vessel, which is normally filled with blood, the free surface becomes apparent when the blood is drained from the vessel. Such a surface is an example of a potential free surface. Since epithelia exist upon surfaces and form barriers between the environment and the underlying tissues, they must form as dense a covering as possible. They consist therefore, of very *closely packed cells* with minimal amounts of material between the cells (intercellular material). Epithelia have *no blood vessels* of their own; the cells are nourished by diffusion from vessels in the underlying connective tissue. It would be impractical to have blood vessels so close to the surface that they are exposed constantly to danger of damage. Epithelia are *provided with nerves*, which pass into and between the cells. Epithelia usually rest upon a *basement membrane*, 55 a product of the epithelium and the underlying connective tissue. The membrane aids in affixing the epithelium to the connective tissue. Epithelia may originate from all three embryonic germ layers: ectoderm, endoderm, and mesoderm. Those developing from ectoderm and endoderm develop from germ layers which themselves have free surfaces in the embryo. Such epithelia are known as true epithelia. Epithelia developing from mesoderm, which does not have a free surface in the embryo, are termed false epithelia. One type of epithelium may develop from more than one germ layer, or one layer may give rise to several types.

Classification of epithelia

Most epithelia may be named according to three general criteria: (1) the *number of layers of cells* in the epithelium, (2) the *shape* of the surface cells in the epithelium, and (3) the presence of certain *modifications* upon the free surfaces of the cells. The epithelia which cannot be named by applying these criteria are sometimes designated as aberrant epithelia.

1. *Number of layers of cells.* A *simple* epithelium has only one layer of cells, all of which reach both the free surface and the basement membrane. The nuclei in the cells of a simple epithelium are all at the same level. A *stratified* epithelium consists of two or more layers of cells, of which only the top layer reaches the free surface and only the lowest layer reaches the basement membrane. In such a tissue, nuclei may be seen at many different levels, and the deeper-lying cells tend to become more round regardless of the shape of the surface layers. A stratified epithelium is therefore named according to the shape of the topmost layer of cells.

2. *Shape of cells in the epithelium.* There are three basic shapes in which epithelial cells may occur:

   (a) *Squamous.* The cells are flattened and length and width greatly exceed height. Nuclei are flattened parallel to the surface, and may form a bulge at the free surface.
   (b) *Cuboidal.* The cells are approximately equal in length, width, and height, and may be said to be isodiametric (equal). Nuclei are round, or nearly so, and the cell will appear square (if on a flat surface) or pyramidal (if lining a tubular structure).
   (c) *Columnar.* The cells are tall, and height exceeds length and width. The nuclei are compressed in a direction perpendicular to the free surface.

Intermediate forms between the three basic shapes are common. A rule of thumb that may be helpful when in doubt about the shape of any given cell is to look at the nucleus and determine its orientation and shape.

By combining these first two criteria, six different types of epithelia may be defined (Fig. 3–1A, B):

   (a) *Simple squamous.* One row of flattened cells, nuclei flattened parallel to the surface. Simple squamous epithelium lines the alveoli of the lungs, glomerular capsule, and eardrum. The thinness of the tissue permits easy passage of materials through it.

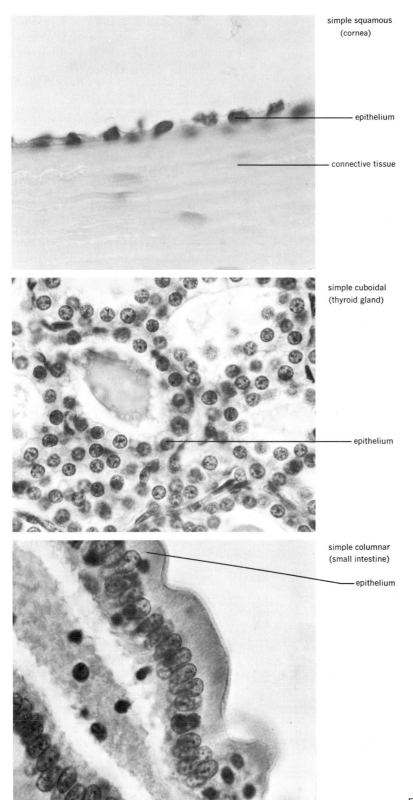

simple squamous
(cornea)

epithelium

connective tissue

simple cuboidal
(thyroid gland)

epithelium

simple columnar
(small intestine)

epithelium

FIGURE 3–1A. Simple epithelia.

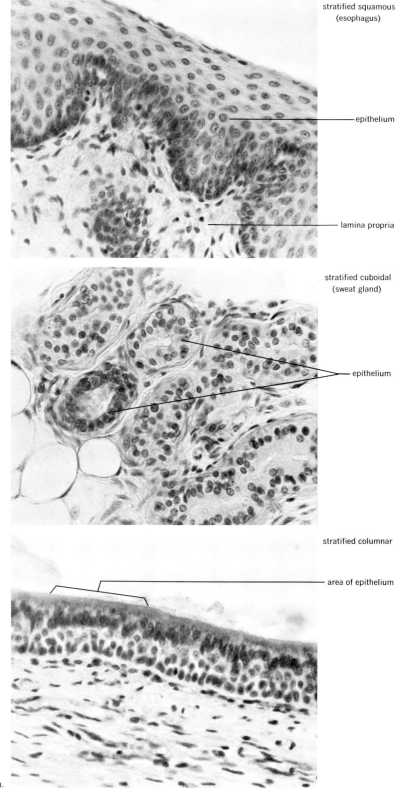

stratified squamous
(esophagus)

epithelium

lamina propria

stratified cuboidal
(sweat gland)

epithelium

stratified columnar

area of epithelium

FIGURE 3–1B. Stratified epithelia.

[1]■ Three tissues having an appearance identical to that of simple squamous epithelium, but differing in origin and potency are:

1. *Endothelium.* Mesodermal in origin, endothelium lines the internal surfaces of the heart, blood vessels, and lymph vessels.
2. *Mesothelium.* Mesodermal in origin, mesothelium lines body cavities not opening upon the body surface. Specifically it lines the abdominal, pleural, and pericardial cavities. (See Chapter 1.)
3. *Mesenchymal epithelium.* Originating from mesoderm, this epithelium lines the subdural and subarachnoid spaces of the central nervous system and the chambers of the eye and inner ear. ■

   (b) *Simple cuboidal.* One row of isodiametric cells, nuclei round. It covers the ovary, lines kidney tubules and the smaller ducts of many glands. It is very active in secretion and absorption.
   (c) *Simple columnar.* One row of tall cells, with nuclei elongated perpendicularly to the surface. The linings of the stomach, small intestine, gall bladder, and larger ducts are of this type of epithelium. It is also active in absorptive processes.
   (d) *Stratified squamous.* Several rows of cells, top layer(s) flattened. It is found in the epidermis of the skin, and lining the mouth, esophagus, anal canal, vagina and cornea. It may be cornified, and is the toughest of all epithelia.
   (e) *Stratified cuboidal.* Several rows of cells, top layer cuboidal. A rare tissue, it is found in the secretory portions of sweat glands.
   (f) *Stratified columnar.* Several rows of cells, top layer columnar. This tissue is found in the larynx.

Aberrant epithelia (Fig. 3–2). Five types of epithelium cannot be named according to the criteria given above.

1. *Pseudostratified epithelium* appears to be stratified and has nuclei at many different levels. However, all cells may be shown to reach the basement membrane, but not all reach the surface. The epithelium is therefore simple. It is the typical epithelium of the respiratory system and portions of the male reproductive system.
2. *Transitional* or *uroepithelium* is a stratified tissue, but the top layer of cells is not uniform in shape. The tissue has the ability to stretch greatly under tension, and the cells may thus show many "transitional" forms. It lines the renal pelvis, ureters, and urinary bladder.
3. *Syncytial epithelium* lacks membranes between cells and is one continuous multinucleated mass. It covers the villi of the placenta.
4. *Germinal epithelium* is found in the testis and is a stratified tissue containing specialized cells leading to the formation of spermatozoa.
5. *Neuroepithelium* is found in the retina, olfactory area, and cochlea and is highly specialized for sensory reception. The epithelium is composed of nerve cells and supporting elements. Such epithelium is usually named according to its function, for example, "olfactory epithelium."

---

[1] *Note:* Black squares indicate advanced matter.

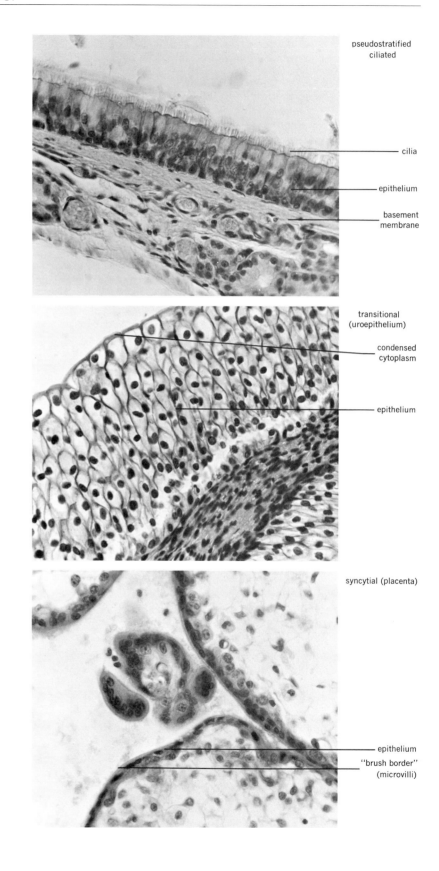

pseudostratified
ciliated

cilia

epithelium

basement
membrane

transitional
(uroepithelium)

condensed
cytoplasm

epithelium

syncytial (placenta)

epithelium

''brush border''
(microvilli)

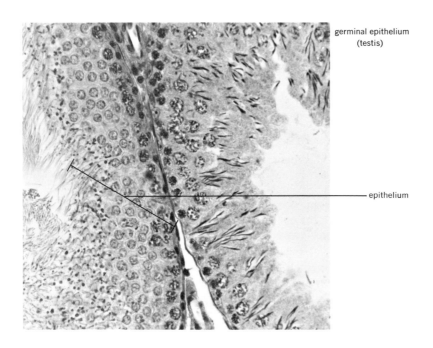

germinal epithelium
(testis)

epithelium

FIGURE 3–2. Aberrant epithelia.

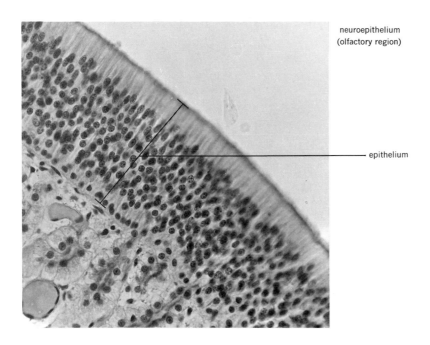

neuroepithelium
(olfactory region)

epithelium

Special surface modifications (Fig. 3–3A to E)

Since one side of the epithelium has a free surface, special structures or modifications of that surface may be present. Such modifications help to enable the epithelium to carry out its particular function more efficiently.

1. *Cilia and flagellae.* Cilia are multiple tiny hairlike projections arising from small granules located in the free ends of epithelial cells. According to some investigators, cilia are derived from centrioles. Flagellae are single extremely long processes conferring independent movement to the cell which possesses it, as in sperm. Both cilia and flagellae have a distinctive ultrastructure of smaller fibrils. Cilia may be motile (*kinocilia*) and aid the movement of materials over the cell surface. Motile cilia are the only surface modification normally included as part of the name of an epithelium. The adjective ciliated is used; we may speak, for example, of a simple columnar ciliated epithelium.

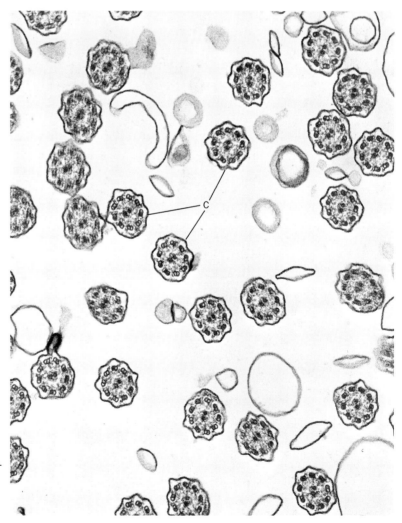

FIGURE 3–3A. Cross section of cilia from trachea of laboratory mouse. Note 9 pairs of peripheral tubules, one central pair. C, cilia. (Photograph supplied by Norton B. Gilula, Department of Physiology-Anatomy, University of California, Berkeley.) × 41,000.

2. *Microvilli.* Microvilli are tiny fingerlike projections of the cell cytoplasm at the free surface. They are commonly found on those epithelia active in absorptive and secretory processes. They serve to increase the surface area through which such transport may occur. On the intestinal epithelium, the microvilli are short and very regular in arrangement and form the "striated border" of that organ. In the kidney tubules, gall bladder, placenta, and lining of the abdominal cavity, the microvilli are much longer, forming "brush borders." *Stereocilia* are microvilli that appear to be attached at their free ends. They are found in the epididymis and serve to channel nutrients to the sperm stored in that organ.

3. *Condensed cytoplasm.* The cells of transitional epithelium have a thicker, denser layer of cytoplasm at the free surface of the top layer of cells, designated as condensed cytoplasm. The layer protects the cells from the acidic urine.

4. *Cuticles.* Layers of material that are produced at the free surface by cellular activity and may be separated from that surface with-

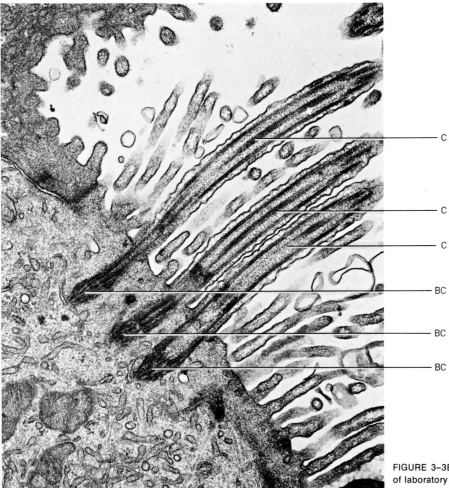

C

C

C

BC

BC

BC

FIGURE 3–3B. Longitudinal section of cilia from trachea of laboratory mouse. BC, basal corpuscle; C, cilia. (Photograph supplied by Norton B. Gilula, Department of Physiology-Anatomy, University of California, Berkeley.) × 29,000.

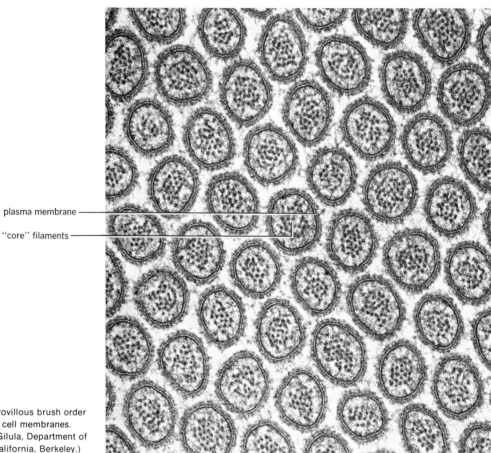

plasma membrane ⎯⎯

"core" filaments ⎯⎯

FIGURE 3–3C. Cross section of microvillous brush order of laboratory mouse. Note trilaminar cell membranes. (Photograph supplied by Norton B. Gilula, Department of Physiology-Anatomy, University of California, Berkeley.) × 102,000.

out damage to the cells form cuticles. True cuticles are rare in the body. The enamel of the tooth and the lens capsule are examples.

5. *Modifications of the intercellular surface.* Various devices (Fig. 3–4) may be employed to hold cells together and to seal the free surface between cells. The *zonula occludens* (the "terminal bar" of older literature) occurs at the free surface and involves an apparent fusion of membranes. The *zonula adherens* is found between cells and involves a separation of membranes with filaments obvious in the adjacent cytoplasm. The *macula adherens (desmosome)* also involves separated membranes, but with a dense plaque of cytoplasmic substance adjacent to the area of attachment. ■

Table 3–1 summarizes some basic facts about epithelia.

Epithelial membranes (Fig. 3–5A and B)

The combination of an epithelium and an underlying connective tissue layer forms an epithelial membrane. Two types of such membranes are recognized in the body.

Mucous membranes form the linings of the body cavities which open

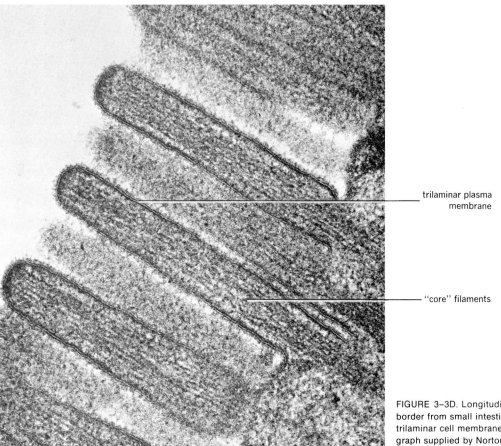

trilaminar plasma
membrane

"core" filaments

FIGURE 3–3D. Longitudinal section of microvillous brush border from small intestine of laboratory mouse. Note trilaminar cell membrane and "core" filaments. (Photograph supplied by Norton B. Gilula, Department of Physiology-Anatomy, University of California, Berkeley.) × 82,000.

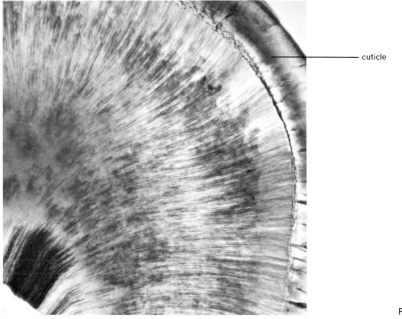

cuticle

FIGURE 3–3E. Cuticle (tooth enamel).

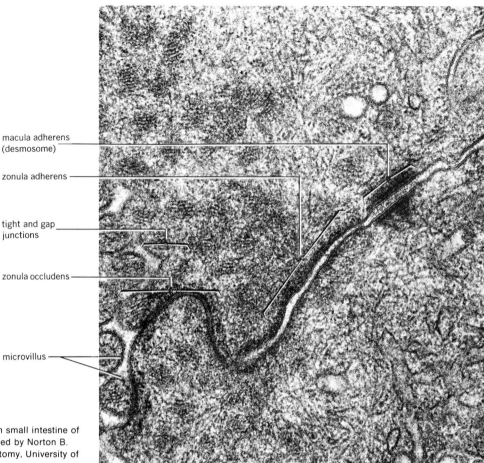

macula adherens
(desmosome)

zonula adherens

tight and gap
junctions

zonula occludens

microvillus

FIGURE 3–4. Junctional complex from small intestine of laboratory mouse. (Photograph supplied by Norton B. Gilula, Department of Physiology-Anatomy, University of California, Berkeley.) × 65,000.

on a skin surface. They include the membranes lining the cavities of the mouth, digestive tract, and the reproductive, respiratory, and urinary systems. The epithelial type may vary, but the epithelium is always moistened by mucus. The underlying connective tissue is known as the *lamina (tunica) propria.*

Serous membranes form the outer coat of visceral organs; line the true body cavities; and form omenta and mesenteries. The epithelial component is a simple squamous mesothelium applied to a sub-mesothelial connective tissue layer. Serous membranes are moistened by secretion of a watery fluid. Because they are always moist, they allow the free movement of materials over their surfaces or the movement between an organ and the lining of its cavity. In the chest cavity, this movement is of particular importance because of the change in shape and size of the heart and lungs as they work. The serous membranes forming the omenta and mesenteries may become a reservoir for storage of fat.

Glands

Secretion forms one of the primary activities of epithelia. Glands are specialized epithelial cells or groups of epithelial cells producing sub-

TABLE 3–1
Summary of Epithelium

Epithelial type	Characteristics	Examples of locations	Surface modification commonly present	Functions and comments
Simple squamous	One row flat cells, nuclei flattened and parallel to surface	Glomerular capsule, endothelium, mesothelium, Henle's loop of kidney	Microvilli (on meso-thelium)	Exchange, because of thinness
Simple cuboidal	One row isodia-metric cells, nuclei rounded	Kidney tubules, thyroid, sur-face of ovary	Microvilli (in kidney)	Secretion and absorption
Simple columnar	One row tall cells nuclei elongated perpendicular to surface	Stomach, small and large in-testines, gall bladder, ducts bronchioles, uterus	Microvilli (in gut) Cilia (in respiratory system)	Secretion and absorption (gut) movement of substances across sur-face if ciliated
Stratified squamous	Several layers of cells, top layer flattened	Skin (fully corni-fied), vagina (partially cornified), mouth, esoph-agus (slight or no cornifica-tion)	None	Protection, since cells are easily removed from the sur-face and replaced rapidly from below
Stratified cuboidal	Several layers of cells, top layer cuboidal	Sweat glands	None	Secretory
Stratified columnar	Several layers of cells, top layer columnar	Larynx, upper pharynx	Cilia	Transition form between stratified squamous and pseudo-stratified
Pseudo-stratified	All cells reach basement mem-brane, not all reach surface. Nuclei at many levels	Nasal cavity trachea, bronchi, male and female reproductive systems	Cilia	Movement of materials across surface
Transitional	Stratified, top layer not uni-form in shape	Kidney pelvis, ureter, bladder	Condensed cytoplasm	Stretches and protects cells from acidic urine
Syncytial	Simple, no mem-branes between cells	Placental villus	Microvilli	Secretion and protects
Germinal	Several layers showing stages of sperm forma-tion	Tubules of testis	None	Produces sperm
Neuro-epithelium	Nerve cells form part of epithelium	Taste buds, olfactory area, retina, cochlea	None	Sensory, usually as receptors of stimuli

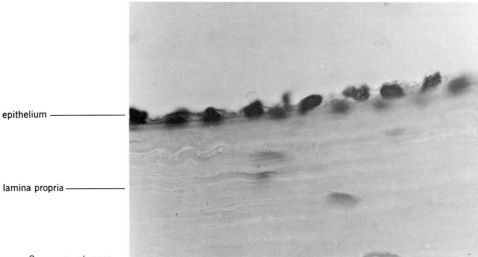

epithelium ————

lamina propria ————

FIGURE 3–5A. Epithelial membranes. Serous membranes.

stances differing in nature from the fluids otherwise available in the body. To produce a secretion different from the blood or tissue fluid, the cells must engage in metabolic activity over and above that required for their own maintenance. In short, production of a secretion involves the utilization of synthetic processes and active methods of transport.

The cells of a gland may remain in the epithelium *(intraepithelial)* or push into the underlying connective tissue, the latter forming a support for the gland. If a connection is retained to the epithelial surface, the gland is said to be an *externally secreting* or exocrine gland, and is possessed of ducts to convey the secretion to the exterior. If the surface connection is lost, the gland is ductless, and empties its secretion into the only other available channel, the blood stream. Such a gland is an *internally secreting* or endocrine gland. The structure of the endocrine glands is discussed in Chapter 22.

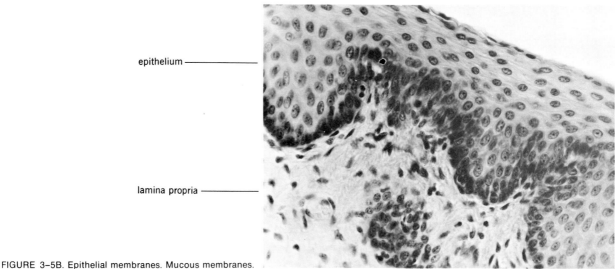

epithelium ————

lamina propria ————

FIGURE 3–5B. Epithelial membranes. Mucous membranes.

Exocrine glands may consist of only one cell *(unicellular)* or of many cells *(multicellular).* Mucus-secreting goblet cells of the intestine and respiratory passageways are examples of unicellular glands. All other glands are multicellular. Multicellular exocrine glands may be further characterized by the number of ducts conveying secretions from the secretory portion of the gland, and by the shape of the secretory portion.

A *simple gland* has but one duct in the entire gland. A *compound gland* has many ducts within it, although only one may empty on the epithelial surface. The secretory portion may be *tubular* or *alveolar (saccular)* in shape. Combining these criteria leads to the designation of a variety of glands. Further descriptive terms, such as branched or coiled may be applied to the glands.

The exocrine glands may further be characterized by the manner in which they produce their secretions. A *merocrine* gland produces a secretion in which no part of the cell itself is found. An *apocrine* gland loses some portion of its cells in producing the secretion. A *holocrine* gland either produces a cell as its secretion (testis) or the entire cell is shed into the secretion. Tables 3–2 and 3–3 summarize facts about exocrine glands.

CONNECTIVE TISSUES

Introduction and general characteristics

With the exception of the tissues lining joints (synovial membranes) and the tissues forming the bursae and tendon sheaths, connective tissues do not possess free surfaces. Unlike epithelia, they contain *widely spaced cells* and *large amounts of intercellular material.* With one exception (mesenchyme), the *intercellular material is fibrous.* Connective tissues are, as a rule, *vascular.* They transmit blood vessels if they have none of their own. As their name suggests, connective tissues *support* and *protect* other tissues or organs within the body. They form the primary structural tissue of the organism.

Classification and description of connective tissues

The many types of connective tissues may be grouped, and named individually, by using four basic criteria.

1. If the tissue is present only in embryo or fetus it is designated as an *embryonal tissue.* If the tissue is present without change after birth, it is designated as an *adult tissue.*
2. Adult tissues are grossly separated into three groups according to the *overall consistency of the intercellular material,* which may be more or less fluid, semisolid, or hard. They are further classed as *special* connective tissues if they exhibit a unique morphology, have a unique staining reaction, or contain some special product in their cells.
3. Specific types of nonspecialized (or general) adult connective tissues are classed according to the *type of fiber* which predominates in the intercellular material.
4. Adult connective tissues are further characterized by the pres-

TABLE 3-2

A Summary of
Exocrine Glands

Type of gland	Diagram	Characteristics	Examples
unicellular		one celled, mucus secreting	goblet cells of resp. and diges. system
multicellular simple tubular		one duct—secretory portion straight tube	crypts of Lieberkuhn of intestines
simple branched tubular		one duct—secretory portion branched tube	gastric glands, uterine glands
simple coiled tubular		one duct—secretory portion coiled	sweat glands
simple alveolar		one duct—secretory portion saclike	sebaceous glands
simple branched alveolar		one duct—secretory portion branched and saclike	sebaceous glands
compound tubular		system of ducts—secretory portion tubular	testes, liver
compound alveolar		system of ducts—secretory portion alveolar	pancreas, salivary glands mammary
compound tubulo—alveolar		system of ducts—secretory portion both tubular and alveolar	salivary glands

TABLE 3–3

Manner of Production of Secretion by Exocrine Glands

Type of gland	Examples
Merocrine—synthesized product independent of basic cell structure	Pancreas, salivary glands
Apocrine—some portion of the cell issued as part of the secretion	Mammary gland
Holocrine—product of gland is a cell, or cell as a whole is shed in the secretion	Testis, sebaceous glands

ence of a *cell that is characteristic for the group,* or a cell that is primarily responsible for the maintenance of the tissue.

Utilizing these criteria, we may construct the following "classification tree." The criteria used are shown in red. Major groups are shown in blue. Individual types are shown in black.

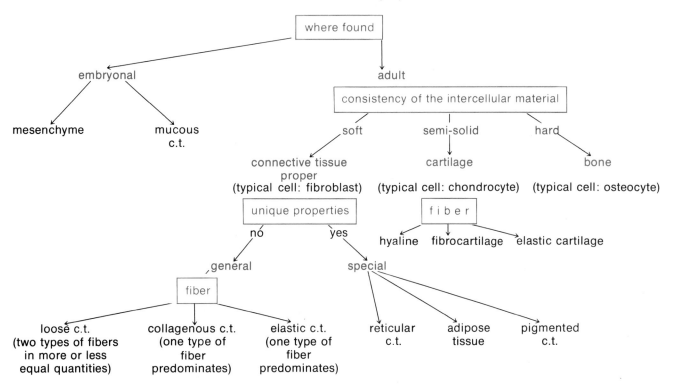

Embryonal connective tissues (Fig. 3–6).

A. *Mesenchyme.* Mesenchyme is the primitive or undifferentiated tissue from which all other connective tissues arise. The cells are

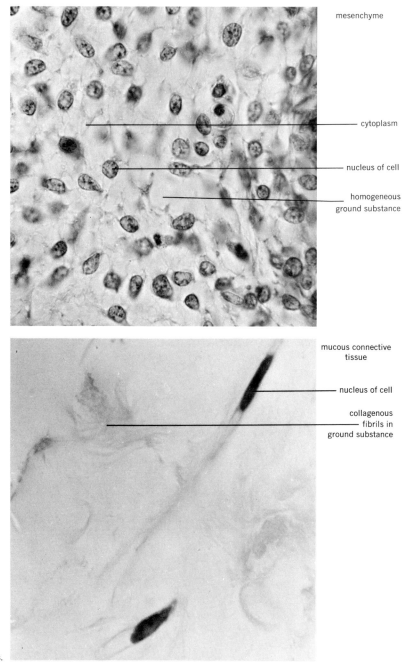

mesenchyme

cytoplasm

nucleus of cell

homogeneous
ground substance

mucous connective
tissue

nucleus of cell

collagenous
fibrils in
ground substance

FIGURE 3–6. Embryonal connective tissues.

stellate in shape and, typically, form networks. While the cells may appear to be connected, they behave independently, and may show ameboid motion. The cells lie in a histologically homogeneous ground substance composed of water, inorganic ions, and protein polysaccharides (mucopolysaccharides). The viscosity of the ground substance varies with the amount of an enzyme (hyaluronidase) present in the ground substance, becoming more liquid as amount of enzyme increases. Mesenchyme is seen in sections of vertebrate embryos, under the skin and along developing bones.

B. *Mucous connective tissue.* Mucous tissue is a differentiated tissue found only in the fetus. The cells are flattened or spindle-shaped, and are set in a mucoid (mucus-like) ground substance. The ground substance has many fine fibrils in it (collagenous fibrils; see below). Also known as Wharton's jelly, mucous tissue is found in the umbilical cord of the fetus.

Adult connective tissues *Connective tissues proper* (Fig. 3–7A, B) are characterized by having more or less fluid intercellular material, with the fibroblast as the typical cell. Five types are distinguished.

A. *Loose connective tissue* contains all of the structural elements found in other members of the group. Loose connective tissue is a nonoriented tissue. Two types of fibers, collagenous and elastic, are distributed in roughly equal quantities in the form of a three-dimensional "feltwork."

At least nine types of cells are sprinkled randomly through the fibrillar mass, and all elements are suspended in a homogeneous ground substance.

Collagenous fibers are 1 to 12 microns in diameter; they consist of subunits termed fibrils, and appear as wavy refringent bundles in the tissue. The fibers may branch, but the fibrils do not. The fibers are composed of a protein called collagen, which in turn is made up of tropocollagen molecules secreted by the fibroblasts. The molecules are apparently polymerized into collagen outside the fibroblast. Collagenous fibers are extremely strong, but show little or no elasticity.

Elastic fibers are also 1 to 12 microns in diameter, but they are always smaller in caliber than the collagenous fibers when the two occur together in any given tissue. They are highly refractile, show no subunit of structure (are therefore homogeneous), and freely branch and rejoin with one another. They consist of a protein called elastin and have little strength but much elasticity.

The three most common cellular elements of loose connective tissue are:

1. *Fibroblasts.* The cell cytoplasm is usually not visible in the typical slide, but gives the whole cell a flat or spindle shape outline if visible. Nuclei are large, oval, and typically possess 2 to 3 prominent nucleoli. The cells are thought to give rise to the fibers of the tissue.
2. *Histiocytes or macrophages.* These are cells capable of engulfing (phagocytosis) particulate matter, and thus form an important line of defense against bacterial invasion. They also aid in "cleaning up" the tissue if it has been damaged. The cells are of irregular outline, have a dense nucleus with a heavy nuclear membrane, and typically show irregular engulfed particles in the cytoplasm.
3. *Blood cells.* Because of their ameboid capacity, white blood cells are able to pass (*diapedesis*) through the walls of capillaries and enter the loose connective tissue. Lymphocytes, eosinophils, and tissue basophilis (mast cells) may be found. These cells may also

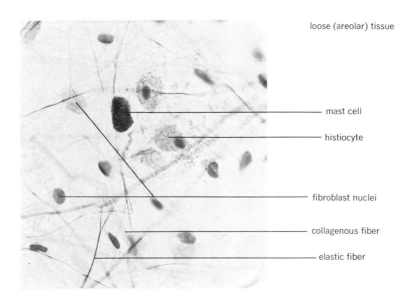

loose (areolar) tissue

mast cell

histiocyte

fibroblast nuclei

collagenous fiber

elastic fiber

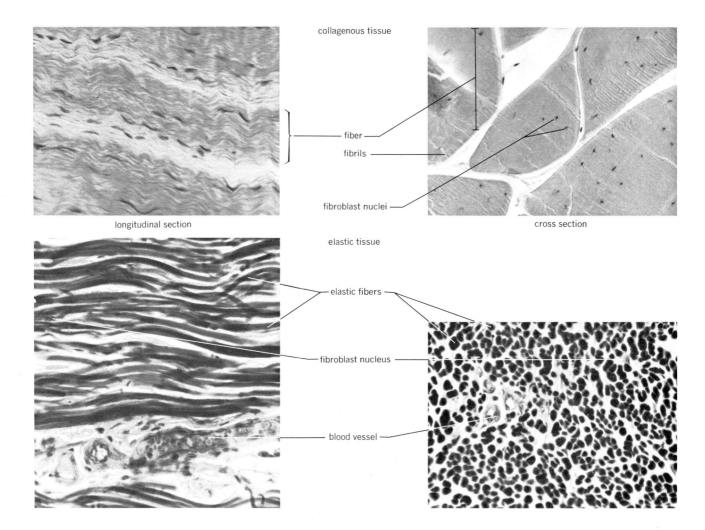

collagenous tissue

fiber

fibrils

fibroblast nuclei

longitudinal section

cross section

elastic tissue

elastic fibers

fibroblast nucleus

blood vessel

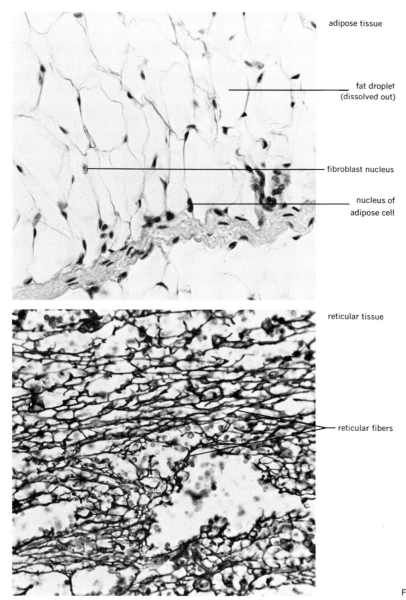

adipose tissue

fat droplet
(dissolved out)

fibroblast nucleus

nucleus of
adipose cell

reticular tissue

reticular fibers

FIGURE 3–7B. Special adult connective tissues.

be important in the defense against disease. The tissue basophil has granules which contain a heparin-like substance which may function to control the viscosity of the ground material. When tissue inflammation occurs, many neutrophils may also be found.

Other cell types found in the tissue include mesenchymal cells, typically located around blood vessels (pericytes), fat cells, pigment cells (melanocytes), and plasma cells. Plasma cells are an example of cells capable of forming antibodies. ■

The ground substance of loose connective tissue is a complex mix-

FIGURE 3–7A. *(left)* Adult connective tissues.

ture of water, collagen, glycoproteins, and lipids, whose viscosity may change under conditions of inflammation and injury. Such changes in viscosity may aid the penetration of blood cells or wall off an area and prevent spread of the abnormal process.

Loose connective tissue forms the subcutaneous layer of the skin (injections may be made into this tissue). It occurs as a "packing material" around body organs in the nooks and crannies of its cavities, and forms layers around organs such as the large arteries. Its fibers confer both strength and elasticity while permitting relatively free movement.

A variant of loose connective tissue, termed dense irregularly arranged connective tissue, consists essentially of compacted loose connective tissue. The fibrillar elements are somewhat larger than in loose connective tissue. The tissue forms the dermis of the skin and submucosa of the digestive tract.

Loose connective tissue may contain large numbers of pigment cells, forming pigmented connective tissue. Such tissue is found in the retina, choroid, and iris of the eye.

B. *Collagenous connective tissue* has a predominance of collagenous fibers arranged in an oriented manner. Flattened fibroblasts are found in rows between the fibers. The tissue forms round structures in ligaments and tendons, and flattened membranes in the sclera of the eye and in aponeuroses.

C. *Elastic connective tissue* has a predominance of elastic fibers also arranged in an oriented fashion. Fibroblasts are present in the spaces between the meshes of the tissue. Elastic connective tissue forms the middle layer (media) of the aorta and pulmonary artery, and is found in the ligamentum nuchae and ligamenta flava of the vertebral column.

D. *Reticular connective tissue* is composed of small irregular-diameter interlacing fibers, and forms the framework of many body organs, such as the spleen, liver, and lymph nodes. The tissue also holds smooth muscle cells into sheets. Reticular fibers are antecedent to collagenous fibers and may be demonstrated by silver containing stains.

E. *Adipose tissue* consists essentially of loose connective tissue, wherein the primitive fibroblasts have become specialized for the storage of fat. The adipose cells typically show a "signet ring" shape, with the cytoplasm pushed peripherally in the cell by a large fat droplet. (In most slides, the fat has been dissolved out of the cell by the alcohols used in preparing the slide, and so the cell presents an open cavity.) Adipose tissue insulates, acts as a shock absorber, is a storage form of energy, and gives form and strength to the body as a whole. The subcutaneous layer of the skin is a good place to find it.

Cartilages (Fig. 3–8).   Cartilages have a semisolid intercellular substance termed the *interterritorial matrix.* The characteristic cells, *chondrocytes,* typically lie singly or in groups in cavities in the matrix known as *lacunae.* Matrix substance between cells in a lacuna forms the *territorial matrix,* and a darker staining region of newly formed matrix, the *capsule,* surrounds the lacuna. The surface of a cartilage is covered by a membranous *perichondrium.* Three varieties of cartilage are recognized.

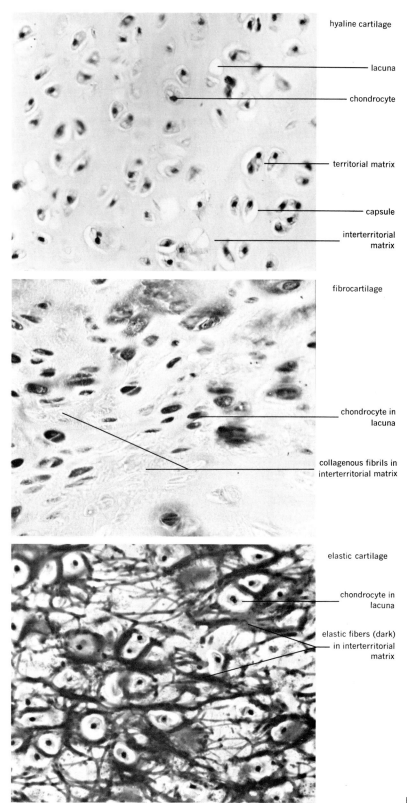

hyaline cartilage

lacuna

chondrocyte

territorial matrix

capsule

interterritorial matrix

fibrocartilage

chondrocyte in lacuna

collagenous fibrils in interterritorial matrix

elastic cartilage

chondrocyte in lacuna

elastic fibers (dark) in interterritorial matrix

FIGURE 3-8. Cartilages.

A. *Hyaline cartilage* (gristle) is, in fresh section, a translucent firm mass. On a slide, in addition to the features described above, the interterritorial matrix *appears* devoid of fibers. Fibers are present, but not visible. Hyaline cartilage forms most of the embryonic skeleton, covers the end of bones forming freely movable joints, and is found in the nasal septum, costal, tracheal, and laryngeal cartilages.

B. *Fibrocartilage* has large numbers of visible collagenous fibers in the matrix. Other features are the same as previously described. Fibrocartilage forms the intervertebral discs and the symphysis pubis.

C. *Elastic cartilage* has many visible elastic fibers in the matrix as the differentiating characteristic. It may be found in the epiglottis and external ear.

Growth of cartilage occurs by two methods. *Appositional growth* is the transformation of primitive cells in the perichondrium into chondrocytes and the formation of new cartilage at the surface of the mass. *Interstitial growth* is formation of new cartilage on the walls of the lacunae within the mass. Obviously the greatest amount of cartilage growth is appositional growth, for to reduce the size of the lacuna is to sow the seeds of self-destruction. If damaged, cartilage heals slowly. Most cartilages have a poor blood supply and regeneration is both slow and limited.

Bone. Bone has a hard intercellular material, due to the deposition of *inorganic salts* in the matrix. The typical cells, *osteocytes,* lie in *lacunae* in the hard intercellular material. *Canaliculi* (tiny canals) connect lacunae with one another and give access to nearby blood vessels. Two types of bone are recognized. They differ, not in chemical composition, but in arrangement. Both types are about 35 percent organic substance, consisting of bone collagen fibers, and cells; and 65 percent inorganic substance, chiefly $[Ca_3(PO_4)_2]_3 \cdot Ca(OH)_2$. *Cancellous* or *spongy bone* consists of interlacing plates and bars of bony material with many spaces between. This type of bone is found in the ends of long bones and inside flat and irregularly shaped bones. *Compact bone* (Fig. 3–9) is extremely dense and consists of longitudinally oriented subunits of structure, the *osteons (Haversian systems).* A canal in the center of the system carries blood vessels which communicate with the outside of the bone through obliquely oriented *Volkmann canals.* Bone is constantly remodeled, particularly during youth, and it is possible to find incomplete systems in compact bone. These represent remodeled osteons and are known as *interstitial lamellae.* The outer surface of the bone is covered by *circumferential lamellae,* lacking osteons.

A membranous *periosteum* covers bone except on articular surfaces. It is two-layered. The inner or *osteogenic layer* contains cells capable of differentiating into osteoblasts and forming new bone. The outer or *vascular layer* contains fewer cells and more blood vessels. The *endosteum,* a condensed layer of bone marrow stroma, lines the internal surface (marrow cavity) of long bones.

Other cells of bone. Osteoblasts, recognizable by their position in rows on the surface of newly forming bone and by their deeply basophilic cytoplasm, are bone-forming cells. *Osteoclasts,* giant multinucleated cells, also may be found on bone surfaces, and the functions of bone destruction and bone remodeling are attributed to them. Both

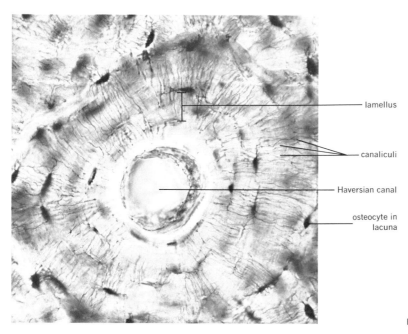

lamellus

canaliculi

Haversian canal

osteocyte in
lacuna

FIGURE 3–9. Compact bone (ground preparation).

cells may be formed from mesenchyme or fibroblast cells. Bone marrow cells fill the spaces between spicules of spongy bone. As bone is remodeled, particularly in young people, the spongy component tends to orient itself to resist the direction of force imposed upon the bone. Figure 3–10 shows the vaults and bridgelike pattern of the spongy bone. This arrangement confers much strength with lightness upon the bone as a whole.

Table 3–4 summarizes facts about the connective tissues.

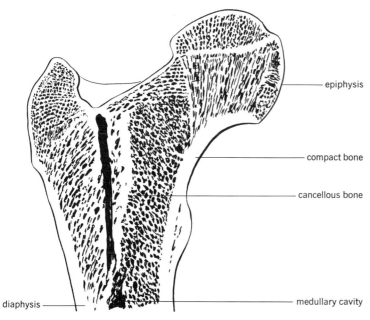

epiphysis

compact bone

cancellous bone

diaphysis

medullary cavity

FIGURE 3–10. Longitudinal section of proximal end of femur.

TABLE 3–4

A Summary of the Connective Tissues

Adult or embryonic	Tissue group; consistency of intercellular material (ICM)	Typical cell	Tissue type	Other cells commonly found	Characteristics	Examples of locations
Embryonic	—	Mesenchyme	Mesenchyme	None	Stellate ameboid cells in networks	In embryo around forming bones; under skin
Embryonic	—	Modified mesenchyme	Mucous c.t.	None	Cells flattened collagenous fibrils in ground substance	Umbilical cord
Adult	ICM Fluid c.t. proper	Fibroblast				
			Loose c.t.	Histiocyte white blood cells, plasma cells, fat cells	Feltwork of collag. and elastic fibers	Around organs, subcutaneous tissue
			Dense irregular c.t.	Histiocyte white blood cells, plasma cells, fat cells	Compacted loose c.t.	Submucosa of digestive tract, dermis of skin
	General		Collagenous	None	Wavy bundle of fibers, fibrils present	Ligaments, tendons, sclera of eye
			Elastic c.t.	None	Branching, homogeneous fibers	Aorta, ligamentum nuchae ligamenta flava
	Special		Reticular c.t.	Reticular cell	Fine, irregular branching fibers	Interior of liver, spleen, lymph nodes
			Adipose tissue	Fibroblasts	"signet ring" cells with fat drops inside	Anywhere (almost); subcutaneous tissue of skin
			Pigmented	—	Masses of cells containing pigment	Iris, retina of eye; choroid of eye

TABLE 3–4 *con't.*

A Summary of the Connective Tissues

Adult or embryonic	Tissue group; consistency of intercellular material (ICM)	Typical cell	Tissue type	Other cells commonly found	Characteristics	Examples of locations
	Cartilage, ICM semisolid	Chondrocyte				
			Hyaline	None	Clear matrix (fibers not visible) cells in lacunae; capsule around lacunae	Embryonic skeleton, costal cartilages, nasal cartilages, articular surfaces
			Fibrocartilage	None	Visible collagenous fibers in matrix; other features as in hyaline	Intervertebral discs, symphyses
			Elastic cartilage	None	Visible elastic fibers in matrix; other features same as in hyaline	Epiglottis, external ear
	Bone, ICM hard	Osteocyte	Spongy bone	None *in* tissue	Interlacing plates and bars	Ends and interiors of bones
			Compact bone	None *in* tissue	Osteons present	Shafts of bones

BONE FORMATION

Two different processes lead to the formation of bone. *Intramembranous formation* occurs in the flat bones of the cranium, part of the clavicle and on the surface of all bones. It may be characterized as a direct formation of bone from connective tissue. *Intracartilagenous (endochondral) formation* gives rise to all other bones and is indirect in the sense that a cartilage model is first formed which is replaced by bone.

Intramembranous formation (Fig. 3–11)

In an area where intramembranous formation is to occur, mesenchymal cells increase in number (hyperplasia) and size (hypertrophy) and there

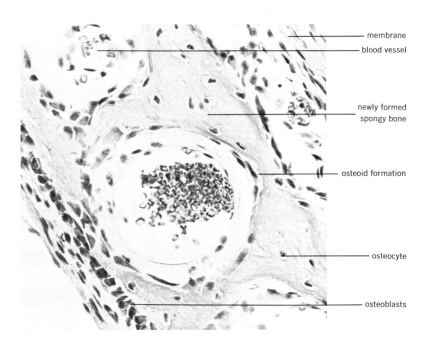

membrane
blood vessel

newly formed
spongy bone

osteoid formation

osteocyte

osteoblasts

FIGURE 3–11. Intramembranous bone formation.

is an increase in vascularity of the region. Some mesenchyme cells assume the typical appearance of osteoblasts. Collagenous fibrils are laid down in the spaces between cells, along with a semisolid ground substance. The entire mass (sometimes called osteoid) is calcified by the deposition of inorganic material in it.

Intracartilagenous formation (Fig. 3–12)

In intracartilageous formation, the outline of the bone is first formed in hyaline cartilage. The processes then proceed in continuous fashion, leading to the formation of new bone. For purposes of clarity of description and understanding, the process will be described as though it occurred in discrete steps.

1. Cartilage cells in the center of the model enlarge and increase in number, squeezing the matrix into thin bars and plates between the cells.
2. As these changes occur in the center of the model, primitive cells of the perichondrium differentiate into osteoblasts, and form, intramembranously, a bone collar around the weakened central area. The part of the membrane overlying the collar thus becomes a periosteum.
3. The cartilage matrix is calcified by deposition of inorganic material in it. This constitutes a provisional calcification.
4. Simultaneously with step 3, vascular mesenchyme (periosteal buds or sprouts) penetrates the bone collar and invades the calcified cartilage.

   The bud follows the lines established by the cartilage cells. As the bud penetrates and destroys the cartilage cells, it drops off behind it undifferentiated cells. Some of these cells form bone marrow cells; others may be recognized as osteoblasts. Osseous tissue (osteoid) is applied to the surfaces of the surviving calcified matrix and this is subsequently calcified to form new bone. Thus, two calcifications take place, and the original cartilage model is replaced by bone tissue.

5. The bud sends lateral branches towards the ends of the model, and ahead of the advancing bud a series of changes occurs which essentially recapitulate the changes which first occurred in the center of the model.

Five zones may be recognized (Fig. 3–12).

(a) Far towards the ends of the bone is unchanged cartilage the zone of reserve cartilage.
(b) Next there is the zone of multiplication, where the cartilage cells are undergoing hyperplasia.
(c) A zone of cell and lacunar enlargement is next towards the center, wherein the cartilage cells undergo hypertrophy and alignment into rough rows.
(d) A zone of calcification is next seen, where the cartilage is provisionally calcified.
(e) Finally, there is a zone of cartilage removal and bone deposition occasioned by the advancing bud.

6. As these changes are taking place in the central part of the bone, the bone collar is being extended, and changes begin to occur in the ends of the bone similar to those taking place in the center. The central changes constitute a primary ossification center and the changes in the ends of the bone, secondary ossification centers. ■

It should be emphasized that the type of bone first formed by either process is of the cancellous variety. Formation of osteons occurs as the lateral branches of the bud fill in from its walls toward the center by the application of concentric lamellae of bony tissue. The bud therefore becomes the blood vessel of an Haversian canal. The original point of invasion of the bud through the bone collar may remain as a Volkmann canal. Volkmann canals also reach the marrow cavity.

The regions of bone formation in the center and ends of the bone do not meet if the intervening cartilage replaces itself faster than it is destroyed by the processes described. It forms an *epiphyseal cartilage,* permitting continued growth in length of the bone. The rate of replacement of this cartilage appears to be directly influenced by hormones, chiefly growth hormone of the pituitary gland. At about 25 years of age (again due to hormonal influences, chiefly those designated as sex hormones), all epiphysial cartilages become closed, that is, replaced by bone. The bone can thus no longer grow in length. Growth at the surface (intramembranously) can still occur, however, which increases the diameter of the bone.

The physiology of calcification

Although it is basically unknown exactly how calcification occurs, there is general agreement that two sets of conditions must be met in order for calcification to occur. Humoral factors, concerned with the accumulation of organic and inorganic materials in the body fluids must op-

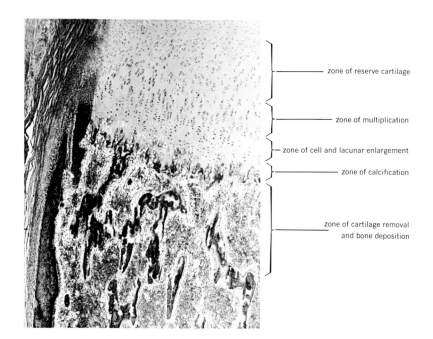

zone of reserve cartilage

zone of multiplication

zone of cell and lacunar enlargement

zone of calcification

zone of cartilage removal
and bone deposition

FIGURE 3–12. Intercartilaginous bone formation.

Legend key for FIGURE 3–12 *(bottom).*
Diagram of the development of a typical long bone as shown in longitudinal
sections. *Green,* bone, *blue,* calcified cartilage, *red,* arteries. *a', b', c',
d', e',* cross sections through the centers of *a, b, c, d, e,* respectively.
a, cartilage model, appearance of the periosteal bone collar; *b,* before
the development of calcified cartilage; *c,* or after it, *d; e,* vascular
mesenchyme has entered the calcified cartilage matrix and divided it
into two zones of ossification, *f; g,* blood vessels and mesenchyme enter
upper epiphyseal cartilage; *h,* epiphyseal ossification center develops
and grows larger; *i,* ossification center develops in lower epiphyseal
cartilage; *j,* the lower and, *k,* the upper epiphyseal cartilages disappear
as the bone ceases to grow in length, and the bone marrow cavity is
continuous throughout the length of the bone. After the disappearance
of the cartilage plates at the zones of ossification, the blood vessels of
the diaphysis, metaphysis, and epiphysis intercommunicate. (Bloom and
Fawcett, *A Textbook of Histology,* courtesy of W. B. Saunders Co.)

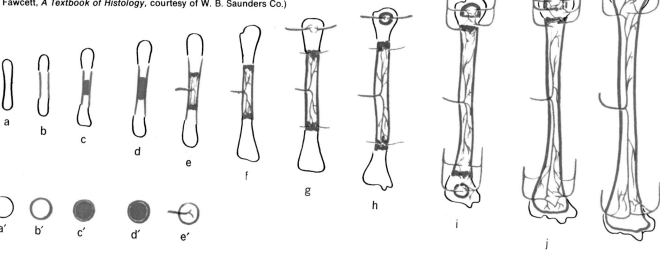

a   b   c   d   e   f   g   h   i   j   k

a'   b'   c'   d'   e'

erate first. These factors involve absorption of adequate amounts of protein, calcium, and phosphate from the gut or from the placenta until some critical concentration in the body fluids is reached. Second, a local mechanism that renders a tissue calcifiable must operate. The local mechanism may involve enzymatic activity which changes a tissue from noncalcifiable to calcifiable. Once this is done, calcification may apparently proceed without enzymes, if a nucleus or seed for crystallization is provided. It has been suggested that perhaps the bone collagen acts as a template or surface to seed the reaction. Crystal formation always occurs in the matrix, never in the cells, and will continue only if adequate supplies of materials are provided. Resorbtion of bone may take place through chelation, triggered by the osteoclasts. Chelation involves the formation of poorly dissociated ring structures of salts with metallic ions. Osteoclasts may supply a material having a stronger affinity for calcium than the bone matrix. The substance supplied is not known, but is probably an organic material.

Other influences on calcification. *Vitamin A deficiency* retards maturation, growth, and degeneration of cartilage cells within the bone. The normal sequences of changes cease. Surface bone formation continues and thus the bones may assume bizarre shapes. *Excess Vitamin A* results in brittleness through stimulation of excess osteoclastic activity.

Vitamin C deficiency results in failure to produce normal intercellular material anywhere in the body. In bone, the intercellular material produced is not calcifiable and soft bones result.

Vitamin D is concerned primarily with absorption of necessary bone minerals from the gut. Deficiency of this substance therefore results in poor calcification (rickets) because of insufficient minerals in the body fluids.

■ Disorders in connective tissue formation in general

In view of the many requisites for formation of ground substance and fibers of connective tissue and of the many things which may influence the course of development of the tissue, perhaps it is not surprising that the possibility of malfunctions of formation exist. It is perhaps more surprising that disorders do not occur more frequently.

The term collagen diseases is applied to those conditions which result in pathological alteration in the connective tissue elements of the body, chiefly the fibrous portions of these tissues.

Collagen diseases appear to have some hereditary basis in the sense that in many cases the production of collagen proceeds at a faster rate than it can be utilized. This occurs without parallel increases in the numbers of fibroblasts, the supposed producers of collagen. It is therefore suspected that there is a basic metabolic defect of the synthetic processes within the fibroblast which may be attributed to altered gene controlled enzymatic mechanisms. A second factor in collagen diseases appears to be the formation of defective collagen, which is chemically different enough from normal collagen as to not be recognized by the antibody producing sites of the body as "self." The collagen is thus treated as a foreign protein, antibodies are manufactured against it,

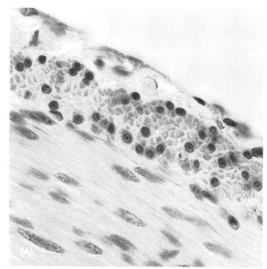

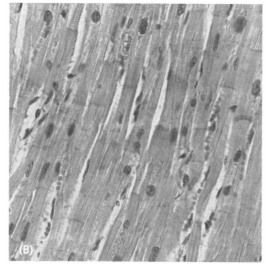

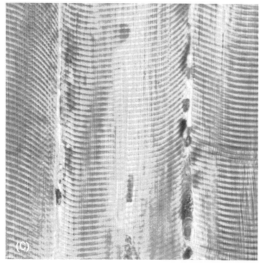

and inflammation and degeneration result. This pattern is one of auto-immunity, and is more common in diseases states than formerly recognized.

The collagen diseases are more common in the elements of connective tissue proper. Cartilages are more susceptible to disorders affecting the production of calcifiable matrix in the steps leading to bone formation. Achondroplasia results from "mutation" or abnormalities in the chromosomes and leads to defective endochondral bone formation. Intramembranous formation is usually unaffected or may be increased. As a result, the bones are short and wide, may be deformed (particularly in the appendages), and the general appearance presented is that of a normal head and trunk with dwarfing of the appendages. Mental ability usually is not affected, and achondroplastics may survive until old age. ■

The bones themselves are subject to a variety of disorders. *Osteoporosis* results from a decrease in bone matrix formation as a result of osteoblastic failure, or deficiency of necessary building blocks in bone formation. The bones show a "moth-eaten" appearance. Demineralization and collapse often follows. Supplying missing nutrients may suffice to ameliorate the condition. Several hormones, notably sex hormones, aid in maintaining bone matrix; therefore, treatment with appropriate hormones may prevent osteoporosis such as that following the menopause.

Osteomalacia is characterized by failure to calcify newly formed bone matrix. This is usually directly traceable to inadequate saturation of the body fluids with calcium and phosphate. Vitamin D may be deficient, leading to inadequate absorption of calcium, or kidney excretion may be excessive. Supplying the missing nutrient or increasing blood calcium levels by intravenous or oral intake of calcium usually aids in the treatment of the condition.

Osteomyelitis is an acute infection of the bone and bone marrow, and is most frequently caused by the organism *Staphylococcus aureus*. An inflammatory reaction with softening and necrosis of hard tissue results. The disease occurs most often in young people, whose bones are nourished by terminal capillary loops. These loops favor the settling of bacteria in this area. The shaft (diaphysis) of the bone is involved almost to the exclusion of the ends (epiphyses).

MUSCULAR TISSUES (Fig. 3–13)

Muscular tissues are the *contractile* tissues of the body. By shortening, they cause the movement of the body through space, and the movement of materials through the body. Three types of muscular tissue are found in the body. Detailed descriptions of each are provided in later chapters.

Smooth muscle (Chapter 12) occurs in the body viscera and around blood vessels.

Cardiac muscle (Chapter 11) is found only in the heart.

Skeletal muscle (Chapter 9) attaches to and moves the skeleton.

FIGURE 3–13. Three types of muscle. A. Smooth muscle. B. Cardiac muscle. C. Skeletal muscle.

NERVOUS TISSUE (Fig. 3–14)

Nervous tissue is the *irritable* (excitable) and *conductile* tissue of the body, originating and transmitting nerve impulses to body organs. Detailed descriptions of the nervous tissue is found in Chapter 20.

QUESTIONS

1. Compare and contrast epithelia and connective tissues as to general morphology, locations and functions.
2. What relationships exist between the epithelium found lining an organ and the functions that organ serves? Give examples.
3. In the development of bone by the intracartilagenous method, what substances are required in order to insure normal growth, and at what stages in the development do possibilities for malformation exist?
4. What varieties of glands are found in the body? Does a correlation exist between the manner in which the secretion is produced and the function the secretion serves?

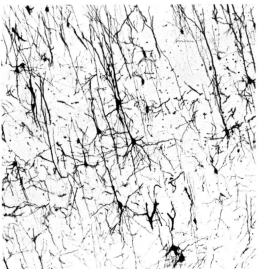

FIGURE 3–14. Nervous tissue, cerebrum.

CHAPTER 4

Body Fluids and
Their Regulation

All the vital mechanisms, however varied they may be, have only one object, that of preserving constant the conditions of life in the internal environment. CLAUDE BERNARD

THE BODY FLUIDS

Introduction

Life is said to have originated in the sea, that is, in a watery environment. The original cells thus existed in an isotonic medium which provided an immediately available source of nutrients, and a means to carry away wastes of cellular activity, both chemical and thermal. The body possesses fluids which act in the same manner. Since the body's fluids are enclosed within the covering of the body they are often said to constitute an *internal environment* for the cells. These body fluids bathe the cells directly and therefore their osmotic pressure and reaction (acidity or alkalinity) must be maintained within rather narrow limits if normal cellular function is to be maintained.

Fluid compartments

The fluids contained within the blood and lymph vessels of the body and within and around the cells constitute the total body water. On the average, the volume of such fluids amounts to 65 percent of the body weight in adult males, and 55 percent in adult females. Actual values for individuals are quite variable depending principally upon the amount of fat present in the body and upon age. Lean individuals have a higher percent of body water than obese individuals, and older individuals

91

have a smaller relative amount of water. The various body fluids are separated from each other by capillary and cell membranes, and have different compositions. For these reasons, total body water is usually subdivided into "compartments."

The term *extracellular fluid* (ECF) refers to all body fluids not found within cells. Two major and several minor subdivisions in this compartment are recognized:

Major subdivisions:
1. The *plasma,* or liquid portion of the blood.
2. *Interstitial fluid,* or the fluids outside of blood vessels and cells which bathe the cells and fill the lymphatic vessels.

Minor subdivisions:
1. *Synovial fluid,* the fluid filling the cavities of freely movable joints in the body. Total volume is about 100 ml.
2. *Cerebrospinal fluid,* the fluid filling (a) the ventricles of the brain, (b) the spinal canal, and (c) the subarachnoid spaces. Its volume is approximately 200 ml.
3. *Ocular fluids,* the fluids filling the chambers of the eye. Its volume is 50 ml.

The characteristics of these small fluid compartments are considered in appropriate chapters elsewhere in the text.

The *intracellular fluid* (ICF) is the fluid contained within the cells themselves.

The subdivisions or compartments are summarized in Table 4–1.

TABLE 4–1

Major Subdivisions of the Body Water

Compartment	Approximate % of body weight	Approximate % of body fluid	Amount in a 70-Kg man (liters)
Extracellular fluid	25	35	17
Plasma	5	7	3
Interstitial fluid	20	28	14
Intracellular fluid	40	65	29
Total body water	65	100	46

Sources of water for the various compartments; water balance

The primary source of fluid for all compartments is fluid which is ingested into the alimentary tract. On the average, 2300 ml of water is taken in food and drink each day. About 200 ml of water is produced metabolically each day by combustion of fuels within the cells. The body water thus gains about 2500 ml per day. If total fluid volume is not to alter drastically, intake must be balanced by output. Routes of fluid loss include: the lungs, through which about 300 ml per day is lost as air is exhaled; the skin, which accounts for a loss of about 500 ml per day, via perspiration, and the kidney, through which about 1500 ml per day is lost as urine. An additional 200 ml is lost through the feces and mouth cavities. These relationships are summarized in Table 4–2.

TABLE 4–2

Water Balance in the Body

Input		Output	
Source	Amount/Day (ml)	Source	Amount/Day (ml)
Ingestion of food and drink	2300	Lungs	300
		Skin	500
		Urine	1500
Metabolic water	200	Feces and mouth	200
	Total 2500		Total 2500

Water which is absorbed from the digestive tract passes for the most part into the blood vessels of the tract and thus becomes a part of the plasma compartment. The plasma compartment has four routes of exit for fluid. The lungs, skin, and kidneys represent routes by which water is irretrievably lost. The fourth route of exit from the plasma is passage into the interstitial compartment. Fluid in the interstitial compartment may move back into the blood vessels, into the cells, or pass into the lymphatic vessels. The latter vessels ultimately deliver their fluid to the veins in the shoulder region. Water within cells themselves has really only one place to go, into the interstitial compartment. The relationships of these compartments, their routes of entry and exit, and their volumes are indicated in Fig. 4–1.

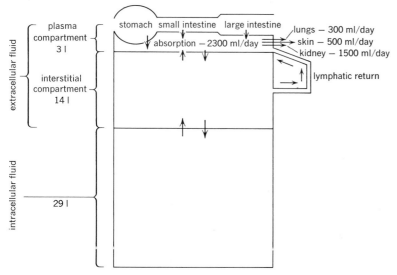

FIGURE 4–1. Relationship of body fluid compartment and routes of loss and exchange.

Composition of the compartments

Fig. 4–2 indicates the types and concentrations of materials in the three major compartments. These substances are in part responsible for exchange of fluid between compartments. Several important conclusions may be drawn from the figure:

1. The height of the columns indicates the total amount of solute in each compartment, and in general gives a clue to the osmotic pressure within each area.
2. The cell appears to be hypertonic to the other two areas, and would thus have a tendency to accumulate water by osmosis.
3. Interstitial fluid has the lowest solute concentration and highest fluid content and would tend to lose water to the other compartments.
4. The composition of plasma and interstitial fluid is quite similar as to kinds of ions present. The chief differences is a greater protein content in the plasma.
5. The ionic types present in extracellular fluid and intracellular fluid are quite different. Sodium and chloride are the chief extracellular ions, while potassium, phosphate and ionized protein are the chief intracellular ions.

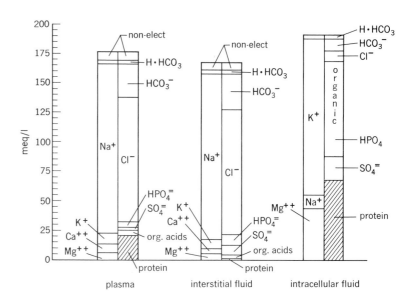

FIGURE 4–2. A comparison of the constituents of the three major compartments of the body water.

Exchange between compartments

Solutes in the digestive tract are, for the most part, actively transported from the fluids of the tract into the blood vessels of the tract. Removal of solute creates an osmotic gradient for water, and it moves passively into the plasma compartment. Once into the plasma compartment, the fluids are subjected to the pressure created by heart action. This hydrostatic pressure not only circulates the fluid through the vessels, but creates a filtration pressure which can cause materials to move through capillary walls. The plasma constituents filter through the capillaries, but most plasma proteins are too large to filter and remain in the capillaries. These proteins exert an osmotic pressure *(oncotic pressure)* tending to draw fluids back into the capillaries. The filtration pressure created by the heart action amounts to about 30 mm Hg, while the oncotic pressure amounts to about 20 mm Hg. There is, therefore, a net pressure of 10 mm Hg favoring fluid passage out of the plasma compartment. The interstitial compartment thus has a tendency to become inflated with fluid. Removal of fluid from this compartment is achieved by osmotic flow into the cells, since the cells contain more solute than the interstitial fluid, and by movement into the lymphatic vessels. Movement of fluid into the lymphatics is accomplished by the hydrostatic pressure of the tissues and by the massaging action of the muscles as they contract and relax. The cell has a tendency to accumulate fluid and must constantly return water into the interstitial compartment. This return is accomplished by a reverse pinocytosis (emeiocytosis) or movement of fluid-filled vacuoles to the cell membrane and emptying of contents to the outside. The main forces responsible for exchange between compartments are summarized in Fig. 4–3.

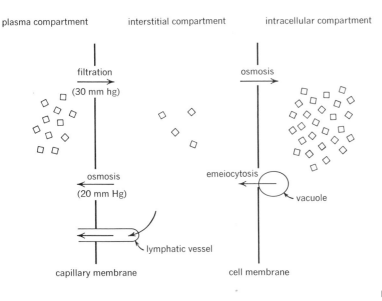

FIGURE 4–3. Forces responsible for fluid exchange between compartments.

Other factors which may influence movement between compartments include:

1. The permeability of the capillary walls. Capillaries are generally conceded to have pores in their walls. Increase in diameter of the vessels (for example, with increased blood pressure) can increase the passage of larger molecules through the vessel walls.
2. Hormones. Antidiuretic hormone and aldosterone govern the reabsorption of water and sodium ion from the fluids filtered in the kidney. The water content of the plasma compartment is therefore controlled by these substances.

We thus see that the amount of water present in any given compartment is determined primarily by passive processes. Hormones and blood vessel permeability govern chiefly the passage of solutes, which may in turn influence water passage.

Edema is a condition caused by accumulation of fluid in the interstitial compartment. Recalling the forces which control exchange between the compartments, several conditions which predispose to the creation of edema may be listed:

1. Increase of flltration of fluid from the plasma or an increased capillary permeability permitting loss of plasma protein. Hypertension or venous obstructions will raise filtration pressure, while increased permeability is a common consequence of inflammation.
2. Loss of plasma protein, reducing return of fluid to the plasma compartment. This is commonly seen in kidney disease which permits protein loss in the urine.
3. Decrease in the pumping efficiency of the heart, for example, in congestive heart failure. This may raise venous pressure and retard absorption of tissue water into the capillaries.
4. Retention of salts. This will cause retention of sufficient water to keep the solution isotonic. Kidney disease associated with low blood pressure prevents efficient filtration of solutes.
5. Blockage of the lymphatic drainage from the interstitial compartment. Parasitic invasion of lymph channels, as in elephantiasis, may block the channels and prevent return of fluid to the veins.

Excessive water intake in relation to loss may produce *overhydration* and symptoms of water intoxication. Excessive water intake tends first to dilute the extracellular fluid in relation to the cell. Water then moves into the cell and dilutes its contents. Severe disturbances, particularly of central nervous system function, including disoriented behavior, convulsions, coma, and death, may result unless water loss is promoted. In infants and children, relatively small amounts of water administered as enemas or intravenously may bring on the symptoms of intoxication, because of a lower total body water volume. *Dehydration,* or a loss of water greater than intake, causes a concentration of salts in the extracellular fluid. Water moves osmotically out of the cells. Drying of mucous membranes, a sensation of thirst, and increase in body temperature (fever) result. Normally, intake of water by drinking returns the situation to normal. If the individual is unconscious, intravenous administration of fluids is often necessary.

■ Measurement of compartment volumes

The volumes of fluid in the various compartments may be measured if a material can be found which will confine itself chiefly to one area. This material is injected in known concentration into a blood vessel, time is allowed for thorough mixing of the substance in the compartment, and then a sample is withdrawn and the concentration of the substance is determined. The urine is also analyzed for the material inasmuch as some of the substances is usually excreted. This method is termed the "dilution method," and the volume in which the substance is distributed is calculated by the following equation:

$$\text{volume of distribution} = \frac{\text{quantity administered minus quantity excreted}}{\text{concentration of sample}}$$

Only total body water, plasma volume, and extracellular volume can be measured directly. Interstitial volume is calculated by subtracting plasma volume from total extracellular volume, and intracellular volume is calculated by subtracting extracellular volume from total body water. Table 4–3 indicates some of the substances used to measure the volumes of the various compartments and the values obtained:

TABLE 4–3

Measurement of Water Compartment Volumes

Compartment to be measured/calculated	Substances employed	Volume in a 70-Kg man. (in liters; approximate)
Total body water, measured	Heavy water (D_2O) Antipyrine	46
ECF, measured	Polysaccharides Disaccharides (sucrose)	17
Plasma, measured	Evans blue dye Radioactive serum albumin	3
Interstitial, calculated (ECF minus plasma)	–	14
Intracellular, calculated (total minus ECF)	–	29

REGULATION OF REACTION OF BODY FLUIDS

Introduction

The body fluids normally have the ability to resist changes in their reaction, or degree of acidity and alkalinity. The term "buffering" is used to describe this ability to resist reaction changes. The primary force tending to alter fluid reaction is the addition of hydrogen ion (H^+) to the fluids.

Hydrogen ion is derived from the hydrogen atom by a loss of the single orbiting electron the atom possesses. The ion is extremely reactive and combines with water to form a hydronium ion (H_3O^+). The H_3O^+ is the form in which the H^+ actually exists in the body. Conventionally, however, it is referred to as H^+, and this terminology will be followed in the ensuing discussion.

Acids and bases

A material capable of losing H^+ is termed an "acid," while a substance capable of accepting H^+ is a "base." "Strong acids" are those substances which give up H^+ easily, and which usually have a high degree of ionization or disassociation. Hydrochloric acid, for example, is a strong acid because it is 100 percent dissociated in dilute solutions:

$$HCl \longrightarrow H^+ + Cl^-$$

A "weak acid" is a substance which has a lesser tendency to release its H^+, and consequently has a low degree of dissociation. Acetic acid is a weak acid, being only slightly dissociated in solution:

$$HAc \underset{\longleftarrow}{\longrightarrow} H^+ + Ac^-$$

The concentration of hydrogen ion, represented by $[H^+]$, is conveniently described by the symbol pH. Strictly defined, pH, or degree of acidity of a solution, is given by the equation:

$$pH = \log \left(\frac{1}{[H^+]} \right) \text{ or } pH = -\log [H^+]$$

The pH scale runs from 0 to 14, or from strongly acid to weakly acid solutions. Conversely it may be said that this scale represents a change from weakly basic to strongly basic solutions. At a pH of 7, a solution is neither acidic nor basic; it is neutral. A change of one pH unit (for example, from 6 to 5) represents approximately a ten-fold change in $[H^+]$. These basic facts indicate the severity of the problem faced by the body in its attempts to remove H^+.

Buffer systems

If a weak acid and its completely ionized salt are present in a solution, the addition of H^+ to the solution will result in little change of pH. Such a pair of substances constitutes a "buffer pair" or "buffer system," and it is capable of buffering or removing free H^+ from the solution. The operation of such a system may be illustrated by the buffer system of acetic acid (HAc) and sodium acetate (NaAc):

$$HAc \longrightarrow H^+ + Ac^-$$
$$NaAc \longrightarrow Na^+ + Ac^-$$

If H^+ is added, it reacts with Ac^- supplied by NaAc to form the slightly dissociated HAc.

$$H^+ + Na^+ + Ac^- \longrightarrow HAc + Na^+$$

The body has several buffer systems which operate in the fashion indicated for example:

1. Carbonic acid and sodium bicarbonate.

$$H_2CO_3 \longrightarrow H^+ + HCO_3^-$$

$$NaHCO_3 \longrightarrow Na^+ + HCO_3^-$$

2. Proteins with H^+ and proteins with Na^+ or K^+. The carboxyl group of the amino acid can lose H^+ and acts as a weak acid.

$$R - COOH \longrightarrow R - COO^- + H^+$$

$$R - COONA \longrightarrow R - COO^- + Na^+$$

3. Phosphate systems, with K^+ or Na^+.

$$NaH_2PO_4 \longrightarrow NaHPO_4^- + H^+$$

$$Na_2HPO_4 \longrightarrow NaHPO_4^- + Na^+$$

The first system forms the primary buffer system of the extracellular fluid. The body normally maintains a ratio of 1 molecule of H_2CO_3 to 20 molecules of $NaHCO_3$. The extra HCO_3^- constitutes the "alkali reserve" of the body and represents the extra buffering capacity of the fluid.

Sources of hydrogen ion

Most of the H^+ added to the body fluids arises from three primary sources:

1. The complete combustion of carbon compounds. Complete combustion of any fuel produces carbon dioxide. This reacts with water to produce carbonic acid, which though classed as a weak acid, is produced in quantity in the body.

$$CO_2 + H_2O \longrightarrow H_2CO_3 \longrightarrow H^+ + HCO_3^-$$

To remove this source of H^+, one must eliminate the CO_2 leading to its formation.
2. Production of organic acids from the incomplete combustion of carbohydrates, fats, and protein. Lactic acid, pyruvic acid, and fatty acids are examples of such materials, all of which can liberate H^+ from their carboxyl groups.
3. From medication with ammonium salts or other acidic materials. For instance, ammonium chloride is a common constituent of materials which increase mucus secretion in the respiratory system and therefore act as expectorants facilitating expulsion of sputum. After absorption, the ammonium portion of the molecule is converted to urea in the liver with production of H^+. This H^+ is normally buffered by the bicarbonate system.

Methods of removing hydrogen ion

Several methods of minimizing H^+ added to the body fluids are utilized by the body.
1. *Dilution.* As H^+ are produced by cells, they are dissipated through the extracellular fluid. Although it does not remove H^+, this device prevents local buildup of H^+.
2. *Buffering* by the use of systems such as those described in the previous section.
3. The *excretion of CO_2* by the lungs. In the lungs, an enzyme designated carbonic anhydrase increases the rate of decomposition of carbonic acid into CO_2 and H_2O. The CO_2 is then exhaled.

$$H_2CO_3 \xrightarrow{\frac{carbonic}{anhydrase}} H_2O + CO_2$$

To replenish the H_2CO_3, the bicarbonate ion recombines with H^+, thus removing H^+.

$$H^+ + HCO_3^- \longrightarrow H_2CO_3$$

4. The *secretion of H^+* by the kidney. By active transport, the kidney tubule cells are able to move H^+ into the tubule lumen where they will ultimately be excreted in the urine.
5. *Uptake by hydrogen acceptors.* In several metabolic schemes, there are special compounds known as hydrogen acceptors. (NAD or DPN, NADP or TPN, FAD.) These compounds accept hydrogen ion as they are produced during catabolism of foodstuffs. Such compounds ultimately contribute these hydrogens to the formation of water.

Acid-base disturbances

The mechanisms described are able to maintain body fluid pH within the range of 7.4 ± 0.04. If these mechanisms allow a greater fluctuation of pH, acidosis or alkalosis will occur. Acidosis occurs if the pH falls below 7.36, and alkalosis occurs if pH rises to more than 7.44. Four major types of acid-base disturbances are recognized.

1. *Respiratory acidosis* occurs if passage of CO_2 through the lung epithelium is decreased. Retention of CO_2 with consequent increase of H_2CO_3 will result, and the pH will fall. The normal compensatory reactions of increased rate and depth of breathing may not suffice to remove excess CO_2. In lung diseases of various types, the diffusing surface may be decreased or thickened, thereby decreasing diffusion of CO_2 from the bloodstream.
2. *Metabolic acidosis* results from excess production of H^+ during abnormal or exaggerated metabolic processes. More H^+ and organic acids may be produced than the buffers can handle, decreasing total buffer base. In diabetes, for example, there is an increased combustion of fats with production of vast amounts of H^+, which is buffered by normal mechanisms. Continued production of CO_2 leads to exhaustion of buffer, and the pH decreases.
3. *Respiratory alkalosis* results from excessive removal of CO_2 through the lungs, upsetting the normal balance of H_2CO_3 and

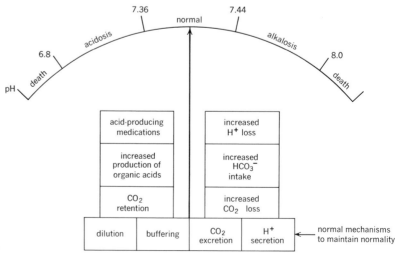

FIGURE 4-4. Mechanisms affecting body fluid pH.

$NaHCO_3$. The pH rises. Hysteria and salicylate (aspirin) poisoning are common examples of conditions resulting in this type of alkalosis.

4. *Metabolic alkalosis* occurs if there is excessive intake of alkalies, as in the use of antacids, or excessive loss of H^+ from the body, as in vomiting. As H^+ falls, total buffer base, HCO_3^-, rises and the pH rises. A simple method of summarizing the mechanisms responsible for maintenance or alteration of pH is shown in Fig. 4–4.

The principles described in the foregoing sections will be applied in later chapters as the functions of specific body systems are discussed. The student should make sure he understands the methods the body possesses for exchanging fluids and solutes between the various compartments, and for resisting changes in pH. These mechanisms are of vital importance in maintaining overall body homeostasis.

QUESTIONS

1. Assuming the intake of a large volume of distilled water, what would happen, and why, to the volumes of the plasma, interstitial, and intracellular fluid compartments?
2. After reviewing Table 4–1, predict the effects each of the following would have on water balance. Which compartment(s) would suffer a change?
   (a) Excessive sweating.
   (b) A burn, involving only loss of fluid.
   (c) Kidney disease, permitting loss of plasma protein.
3. What conclusions may be drawn as to direction of water movement from the figure illustrating the ionic composition of the fluid compartments?
4. Compare and contrast the composition of the three major fluid components.
5. What is buffering? Illustrate its operation with two systems normally found in the body.

CHAPTER 5

The Body as a Whole

Every man is a volume if you know how to read him.
WILLIAM ELLERY CHANNING

SURFACE ANATOMY

Introduction

The study of anatomy should not be limited to the dissection of lifeless organisms or to the study of brightly colored samples of tissue under the microscope. To retain the image of the body as a living, functioning and dynamic organism is essential to complete and satisfying understanding of both structure and function. Having studied the individual components of the body, let us now consider the body as a unit.

When viewed from the external surface, each of us presents many anatomical *landmarks*. These landmarks may be used in their own right, and as reference points for locating deeper lying structures. The purpose of this chapter is to present the body as a whole, in the hope that the student may achieve a better "feel" for his own structure. To this end, as many of the landmarks as possible should be located on his own body.

Terms of direction

By convention, all descriptive terms conveying information as to the location of structures relative to one another are defined assuming that the body is in the *anatomical position.* In this position, the body is standing erect with the eyes level and directed forward, the arms at the sides with the palms forward, and the feet parallel with the heels approximated (Fig. 5–1). The terms employed are:

105

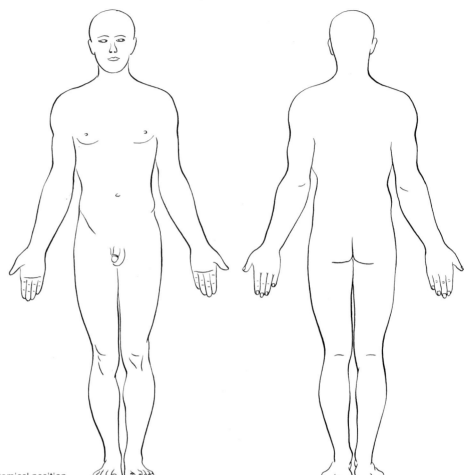

FIGURE 5–1. Anatomical position.

1. *Anterior or ventral.* The front or belly side of the body.
2. *Posterior or dorsal.* The back of the body.
3. *Superior.* Above, or something on a higher portion of the body than the original point of reference.
4. *Inferior.* Below, or something on a lower portion of the body than the original point of reference.
5. *Medial.* A line running from the center of the forehead to between the feet defines the center line of the body. Medial implies a position closer to the center line. The term mediad is used to denote movement *toward* the midline.
6. *Lateral.* A structure is removed in a sidewards direction from the center line. Laterad denotes movement *away from* the midline.
7. *External.* The general connotation of this term is "outside of." It is most commonly employed in referring to the outer surface of the body as a whole or to a position further removed from the interior of a hollow organ.
8. *Internal.* The general connotation of this term is "inside." It is used to refer to structures lying within the body or closer to the interior of a hollow organ.

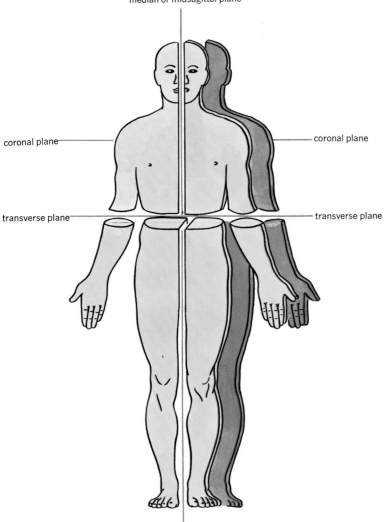

median or midsagittal plane

coronal plane — — coronal plane

transverse plane — — transverse plane

FIGURE 5–2. Planes of section.

9. *Superficial.* This term refers to something closer to the external body surface. It carries a similar connotation to external, but is a preferable term when referring to a structure which approaches, but does not reach, the surface.

10. *Deep.* This term implies removed from the surface. It suggests that a structure lies covered by other structures.

11. *Proximal.* Strictly defined, proximal means closest to the point of attachment of a part to the body or to the midline. For example, the shoulder is proximal to the elbow; the wrist is proximal to the fingers.

12. *Distal.* Strictly defined, distal means further from the point of attachment or the midline. The wrist would be distal to the elbow. Proximal and distal are most commonly employed to designate opposite ends of bones or appendages. The terms lose their significance if applied to the trunk, head, or neck.

Planes of section (fig. 5–2)

Further information as to positioning of body structures may be gained through the study of sections cut through various planes of the body. A section which divides the body into equal right and left halves is a *median* or *midsagittal plane.* Any plane parallel to the median plane is *parasagittal.* A section which divides the body into front and back halves is a *frontal* or *coronal plane.*

Transverse or *horizontal planes (cross sections)* are at right angles to the others and divide the body into superior and inferior portions. The terms longitudinal and cross sections are employed to describe sections of organs themselves. A longitudinal section parallels the longest dimension of the organ. A cross section follows the shortest dimension of the organ and is perpendicular to a longitudinal section.

General surface anatomy

The external appearance of the body varies according to age, sex, and state of nutrition. Premature infants typically lack the thick subcutaneous layer of adipose tissue found in full term infants, and the skin appears to hang loosely upon the body. Full term infants have a chubby appearance due to overall deposition of fat. The buccal fat pad in the cheeks gives the face the typical infant appearance. As growth occurs, the distribution of fat becomes more even on the body and is nowhere excessively abundant. At the time of adolescence and puberty, a sex difference becomes apparent. The female accumulates adipose tissue in restricted areas of the body (fat pads), including the mammary glands, over the shoulders, buttocks, inner and outer sides of the thighs, lower abdomen, and over the symphysis pubis. In the male, the pads are thinner and more evenly distributed over the body, unless the individual is grossly overweight. In the latter case, the abdomen assumes the greatest role as a fat storage area. In middle age, obesity again becomes more predominant, while in old age, disappearance of fat and loss of elasticity of the skin causes the skin to hang loosely upon the body. Figure 5–3 shows adult male and female bodies. The male figure may be described as being triangular in shape with broad shoulders and narrow hips. The body is angular and lacks the rounded contours seen in the female. Muscular development is greater, and the general posture is more erect. The female body is more diamond-shaped. The shoulders are narrower, the hips broader. The body is rounded and softer in appearance due to the lesser development of muscles and a thicker layer of adipose tissue under the skin. The posture appears less erect, due primarily to a more pronounced curvature of the lower back.

Certain descriptive terms may be applied to the grosser body area (Fig. 5–4). These terms should be learned as an introduction to the regional description which follows.

REGIONAL SURFACE ANATOMY

The head

Cranium (Fig. 5–5). The upper or cranial portion of the head presents an outline which adheres closely to the structure of the bony parts. The tissues covering the bones are deficient in subcutaneous fat, and thus the bony prominences may be appreciated easily. The *frontal eminences*

FIGURE 5–3. *(right)* Male and female adults.

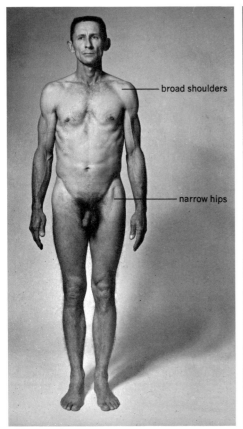
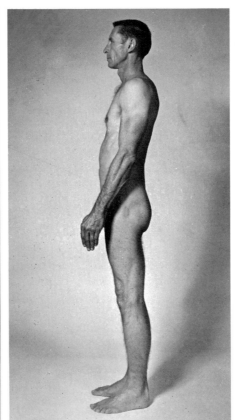
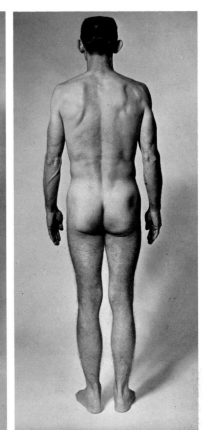

broad shoulders

narrow hips

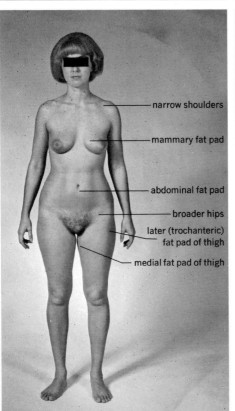
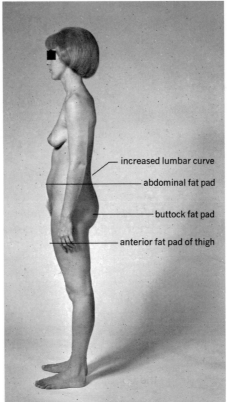
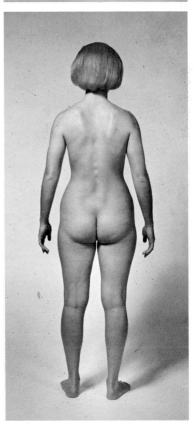

narrow shoulders

mammary fat pad

abdominal fat pad

broader hips

later (trochanteric) fat pad of thigh

medial fat pad of thigh

increased lumbar curve

abdominal fat pad

buttock fat pad

anterior fat pad of thigh

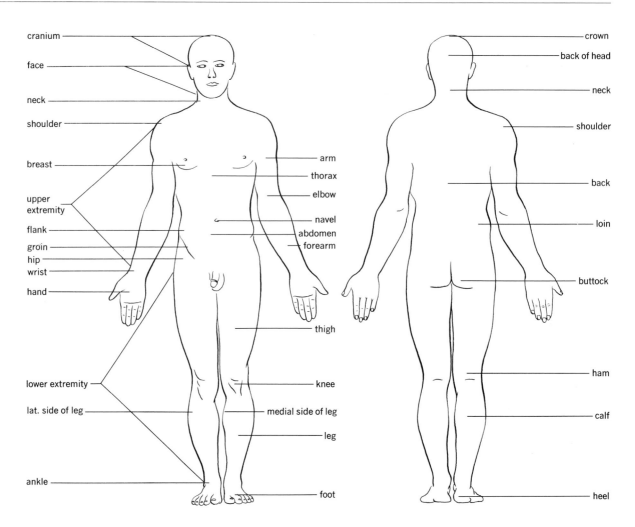

FIGURE 5-4. General descriptive areas of the body.

are slight projections on either side of the midline of the upper forehead. The *external auditory meatus* marks the opening of the ear on the side of the head. *Parietal eminences* occur about two-thirds of the way back on the upper cranium. The *external occipital protuberance* lies at the back of the head where it joins the neck. The *mastoid process* lies medial and slightly posterior to the ear lobe. On the sides of the cranium, in the area commonly known as the temple, the pulsations of the *superficial temporal artery* may be felt. It lies directly over bone, is easily compressed, and thus this location may serve as a pressure point for controlling bleeding on the cranium.

Face (Fig. 5-6). Bony landmarks on the face are generally somewhat obscured by the flesh of the face. Variable amounts of fat and muscle overlie the bony parts. Just superior to the medial margins of the eyebrows, the *superciliary ridges* may be located. Internal and deep to the ridges are located the *frontal sinuses.* These are cavities within the bone which communicate with the nasal cavities. The eye resides within the *orbit,* the latter delimited from the rest of the face by the *orbital margin.* The "cheekbones" are formed by the zygomatic and maxillary bones, and lie inferiorly and laterally to the orbits. The *max-*

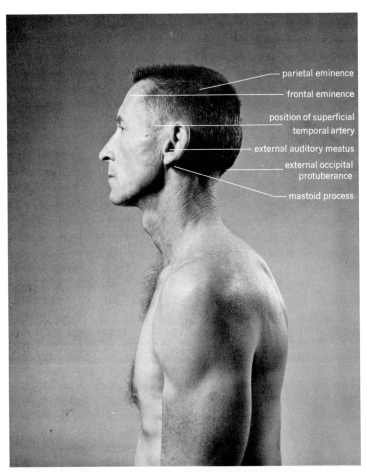

parietal eminence

frontal eminence

position of superficial
temporal artery

external auditory meatus

external occipital
protuberance

mastoid process

FIGURE 5–5. Side view of cranium.

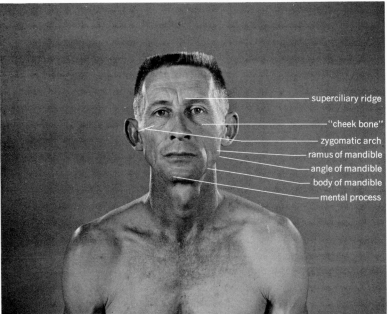

superciliary ridge

"cheek bone"

zygomatic arch

ramus of mandible

angle of mandible

body of mandible

mental process

FIGURE 5–6. Front view of face.

illary sinus lies within the maxillary bone below the orbit. Extending posteriorly from the chin are the lower borders of the *body of the mandible,* which bend at the *angle,* and connect with the superiorly direct *rami* (singular, ramus). Opening the mouth widely protrudes the mandible and the rounded *condyloid process* may be felt crowning the ramus. Several nerves exit from foramina or holes in the skull and pass over the face beneath the skin. The *infraorbital nerve* exits below the orbit and supplies the face, anterior teeth, and nose with sensory fibers. The *mental nerve* emerges from the sides of the mandible about an inch posterior to the chin. The *external maxillary artery* traces a course from the inner margin of the orbit along the nose and over the side of the face and mandible. Its pulsations may be appreciated about an inch anterior to the mandibular angle on the lower margin of the mandibular body. This area constitutes a pressure point for controlling facial bleeding.

The *masseter* muscle overlies the ramus of the mandible. The *parotid salivary gland* covers the anterior part of this muscle. The *temporalis* occupies most of the lateral side of the cranium above the *zygomatic arch.* Both muscles are easily felt when the jaw is clenched. The facial nerve emerges from the substance of the parotid gland and supplies the facial musculature. If damaged at this point, a unilateral facial paralysis may result.

The infant skull (Fig. 5–7) presents several landmarks in addition to those mentioned above. The *anterior fontanel* (soft spot) is open from birth until a year or so of age, and represents an area where the cranial bones have not yet knit together. Several large scalp veins are usually visible. The fontanel is sometimes used as a site for withdrawal of blood, and the scalp veins are used for intravenous administration of fluid if other veins cannot be utilized.

The neck

The bony parts of the neck are well covered by muscle and fascia, and are difficult to palpate. In the posterior midline of the neck, at its base, a large protuberance may be felt. It is the spinous process of the seventh cervical vertebra. About one-half inch below and in front of the mastoid process, the transverse processes of the first cervical vertebra (atlas) may be located. The *hyoid bone* may be located approximately at the level of the lower border of the mandible, in the anterior part of the neck. Inferior to the hyoid bone is the prominent *thyroid cartilage* of the larynx. The *cricoid cartilage* lies inferior to the thyroid cartilage. Below the cricoid cartilage may be felt the *cartilagenous rings* of the trachea. The muscles of the floor of the mouth and neck outline a number of anatomical triangles which contain within them important vascular and nervous structures.

The *posterior cervical triangle* (Fig. 5–8) lies between the posteriorly positioned trapezius muscle, the posterior border of the sternocleidomastoid muscle, and the upper border of the clavicle. The omohyoid muscle passes obliquely through the triangle and divides it into an upper, larger *occipital triangle,* and a lower, smaller *omoclavicular* (subclavian) triangle. The occipital triangle passes the 11th cranial nerve (accessory) on its way to innervate the trapezius muscle. The omoclavicular triangle contains the roots of the subclavian artery and vein, and the trunks of the brachial nerve plexus to the upper appendage.

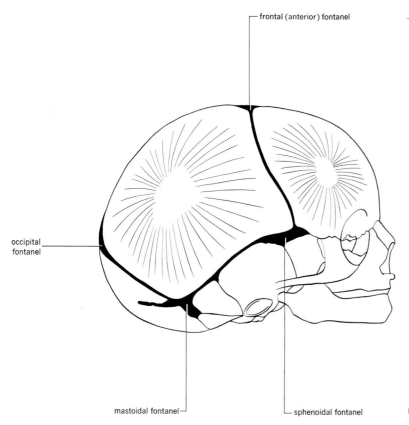

frontal (anterior) fontanel

occipital fontanel

mastoidal fontanel

sphenoidal fontanel

FIGURE 5-7. Skull of a newborn.

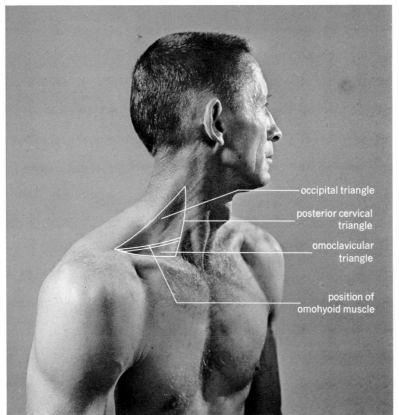

occipital triangle

posterior cervical triangle

omoclavicular triangle

position of omohyoid muscle

FIGURE 5-8. Posterior cervical triangle.

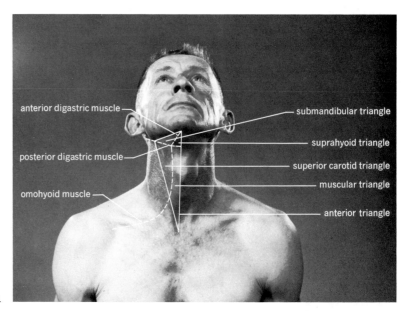

anterior digastric muscle —

posterior digastric muscle—

omohyoid muscle —

— submandibular triangle

— suprahyoid triangle

— superior carotid triangle

— muscular triangle

— anterior triangle

FIGURE 5–9. Anterior triangle.

The space behind the clavicle is termed the supraclavicular notch, and by pressing here, the pulsations of the subclavian artery may be felt. This is a poor pressure point, inasmuch as the artery is difficult to compress against a bone and cannot be totally blocked.

The *anterior triangle* (Fig. 5–9) is bounded anteriorly by the midline of the neck, posteriorly by the sternocleidomastoid muscle, and superiorly by the body of the mandible. It is subdivided into four smaller triangles:

1. The *inferior carotid (muscular) triangle* lies between the anterior midline of the neck and the omohyoid and sternocleidomastoid muscles. It contains the thyroid gland, larynx, upper portion of the trachea, common carotid artery, internal jugular vein, and vagus nerve.

2. The *superior carotid triangle* is bounded by omohyoid, sternocleidomastoid, and posterior digastricus muscles. The bifurcation of the common carotid artery into internal and external carotid arteries, and portions of the internal jugular vein and vagus nerve lie within this triangle.

3. The *submandibular (digastric) triangle* is bounded by the anterior digastricus and stylohyoideus muscles and by the mandibular body. The facial nerve, parotid, and submandibular salivary glands lie in this triangle.

4. The *suprahyoid (submental) triangle* is bounded by the anterior digastric, the midline of the neck and body of the hyoid bone. Several large lymph nodes and the anterior jugular vein occupy the triangle. ■

The internal jugular veins are often used to draw blood samples on infants.

The trunk

The chest or thorax. The upper portion of the thorax is nearly covered by the large muscle of the breast, the *pectoralis major.* Laterally, this muscle forms the anterior boundary *(anterior axillary fold)* of the axilla or armpit. The posterior boundary *(posterior axillary fold)* is formed by the lateral margin of the *latissimus dorsi.*

The major bony landmarks of the thorax are provided by the sternum, ribs, clavicles, and scapulae. The upper border of the sternum is marked by the *suprasternal (jugular) notch.* The junction between the upper portion of the sternum *(manubrium)* and the central portion *(body)* occurs at an *angle.* A ridge is thus formed, about 2 inches below the notch, which is designated as the sternal angle. It marks the point of attachment of the cartilage of the second rib. The third and lowest portion of the sternum, the *xiphoid process,* may be felt in the midline below the body. The process is cartilaginous in younger individuals and becomes ossified in the adult.

Several vertical lines may be drawn for reference points on the thorax (Fig. 5–10). The *anterior and posterior axillary lines* are drawn along the corresponding folds. The *midaxillary line* lies halfway between the preceding two. The *midsternal line* lies through center of the suprasternal notch and sternum. The *midclavicular line* is drawn halfway between the midsternal line and the point of the shoulder. The *lateral sternal line* lies along the lateral margin of the sternum, while the *parasternal line* lies halfway between lateral sternal and mid-clavicular lines.

The heart is included between the left midclavicular and right parasternal lines. The anterior border of the right lung approaches the midsternal line. The spaces between the ribs are closed by the inter-costal muscles, and are known as interspaces. In the left fifth inter-space, the apex beat of the heart may be felt.

The back (Fig. 5–11). A furrow down the middle of the back marks the position of the *spinous processes of the vertebrae.* The ridges of muscle on either side of this furrow are caused by the *erector spinae muscles.* The number of any particular vertebral spine may be deter-mined by counting from the prominent seventh cervical spine at the base of the neck. There are 7 cervical, 12 thoracic, and 5 lumbar vertebrae. In the lower back, the highest points of the iliac bones are at the level of the fourth lumbar spinous process. Lumbar punctures are commonly done at the fourth lumbar level. The puncture cannot harm the spinal cord, for it ends at the level of the second lumbar vertebra. A single line of reference may be drawn vertically on the back, passing through the inferior angle of the scapula. The major muscles visible on the back are the superiorly placed *trapezius* and the more inferior *latissimus dorsi.* On the infant, the skin of the back is rather loose and affords an easy site for subcutaneous injections.

The abdomen. The abdomen is delimited largely by muscles, so that the contour established is determined by the tone of these muscles and the amount of fat in the subcutaneous tissues. The upper border of the abdomen is established by the curve of the ribs and costal cartilages. The lower border is set by a curved line running from the iliac crests through the pubis. The midline of the abdomen is formed by a single layer of muscle, the rectus abdominus. The oblique muscles and trans-verse abdominal form the remaining portion of the abdominal wall. Lines drawn vertically and horizontally through the umbilicus divide the abdomen into four *quadrants.* These are designated as the right and left upper, and right and left lower quadrants. *Smaller subdivi-sions* may be created by drawing the lines shown in Fig. 5–12. In the male, the inguinal canal passes the spermatic cord from the

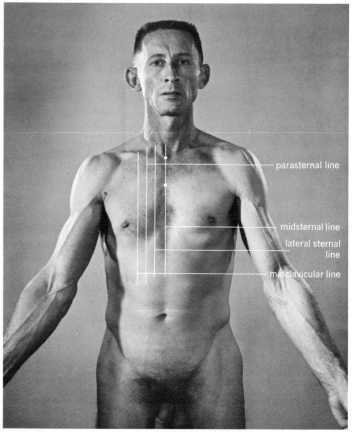

parasternal line

midsternal line

lateral sternal line

midclavicular line

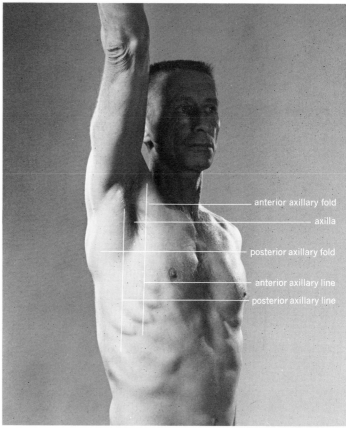

anterior axillary fold

axilla

posterior axillary fold

anterior axillary line

posterior axillary line

FIGURE 5–10. Thoracic reference lines. *Above*, Thoracic. *Below*, Axillary.

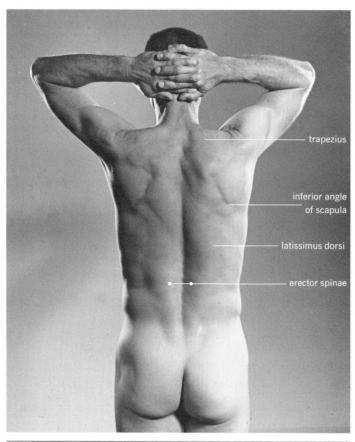

trapezius

inferior angle
of scapula

latissimus dorsi

erector spinae

FIGURE 5–11. The back.

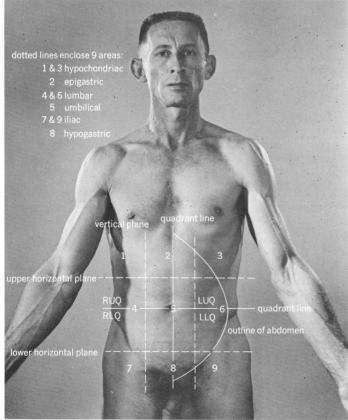

dotted lines enclose 9 areas:
  1 & 3 hypochondriac
    2   epigastric
  4 & 6 lumbar
    5   umbilical
  7 & 9 iliac
    8   hypogastric

vertical plane

quadrant line

1   2   3

upper horizontal plane

RUQ   LUQ
RLQ 4 5 6 LLQ   quadrant line

outline of abdomen

lower horizontal plane

7   8   9

FIGURE 5–12. Abdominal reference lines.

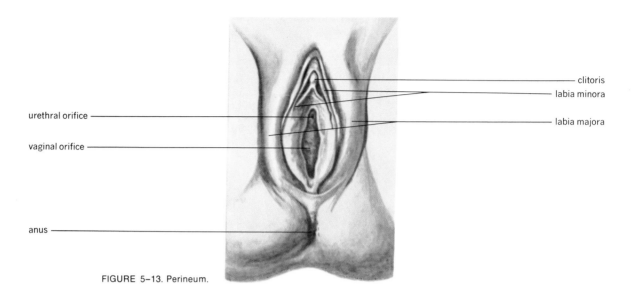

urethral orifice

vaginal orifice

anus

clitoris

labia minora

labia majora

FIGURE 5–13. Perineum.

scrotum. The canal is a common site of hernia.

The *perineum* or floor of the abdominal cavity (Fig. 5–13) is formed by the *levator ani muscle* and contains the openings of the digestive system *(anus),* and urinary and reproductive systems *(vagina* and *urethra* of the female; *urethra* of the male). In the female, the labia surround the urethral and vaginal openings.

The upper limb (Fig. 5–14)

Bones. The *clavicles* and *scapulae* form the pectoral girdle giving support and attachment for the upper limbs to the thorax. The clavicle is subcutaneous throughout most of its length. The scapula lies buried beneath muscle, except for the point of the shoulder *(acromion process)* and the *scapular spine.* The latter may be felt as a ridge on the upper back, running obliquely upwards and outwards toward the shoulder.

The bone of the upper arm, the *humerus,* has its shaft surrounded by muscles, and only the features on the proximal and distal ends can be easily palpated. The *greater tuberosity* lies on the lateral aspect of the proximal end, while the *medial and lateral epicondyles* form prominences on the distal end.

The forearm is composed of two bones, a medial *ulna* and a lateral *radius.* The *olecranon process* of the ulna forms the point of the elbow posteriorly. The posterior border of the ulna may be felt throughout its length, and it terminates in a distal *styloid process.* The process forms a prominent bulge just above the wrist. The radius is covered by muscle except at the distal end, where it too forms a *styloid process.*

On the anterior aspect of the wrist, the lateral elevations are produced by the *scaphoid* and *trapezoid* carpal bones. A prominent medial elevation marks the position of the *pisiform* bone of the carpals. The *metacarpals* are appreciated best by feeling the posterior surface of the hand. These are numbered from one to five, beginning with the thumb. The fifth metacarpal is subcutaneous. The knuckles are formed by the distal ends of the metacarpals. There are fourteen *phalanges* forming the digits. There are four fingers, each with three phalanges. The thumb contains only two phalanges.

Muscles. The *deltoid* covers the shoulder and is composed of anterior, middle, and posterior fiber groups. These parts may be palpated easily if the arm is elevated.

The anterior aspect of the humerus is covered by the superficial *biceps* and deep *brachialis.* The separation of the two biceps heads may be appreciated by pushing a finger between them about two-thirds of the way up the length of the muscle. The lateral, medial, and posterior aspects are covered by the three heads of the *triceps muscle.*

The anterior aspect of the forearm is occupied by muscles which flex the wrist and fingers, as in making a fist. These are difficult to separate. At the wrist, when it is strongly flexed, may be seen the *tendon of the palmaris longus.* On the anterior side of the elbow is located the *antecubital fossa.* It lies between the medial biceps tendon and the laterally placed brachioradialis muscle.

Arteries. The *subclavian artery,* already described, becomes the *axillary artery* across the armpit, and may be located medial to the upper end of the biceps. The axillary artery becomes the *brachial artery* on the upper arm, and may be traced along the humerus medial to the biceps. The artery divides at the elbow to form a *radial* and *ulnar artery* following the respective bones. Both arteries are deeply placed through most of their course. The radial artery becomes superficial at the wrist and lies lateral to the palmaris tendon. The pulse is commonly taken at this point.

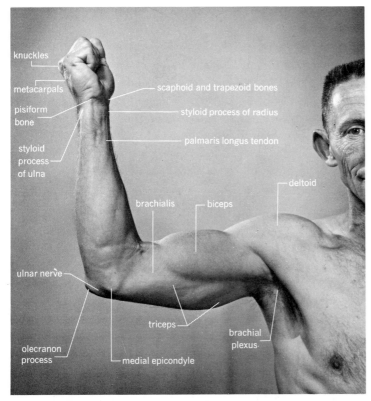

FIGURE 5–14. Upper limb.

Veins. The deep veins follow the bones of the upper limb and have the same names as the arteries. They cannot be recognized because of their deep placement and because they collapse when pressure is applied. The superficial veins are easily recognized, particularly in thin individuals. The veins on the back of the hand form the *dorsal venous network* and connect at the wrist to form two main trunks. The *cephalic vein* lies laterally on the anterior aspect of the forearm, and the *basilic,* medially. In the antecubital fossa, the two are connected by the *median antecubital vein.* The latter vessel is the one most commonly employed as a site of venipuncture. The cephalic continues superficially to the level of the clavicle lateral to the biceps. The basilic disappears beneath connective tissue fasciae about 2 inches above the elbow.

Nerves. At only two points on the upper limb may nerves be palpated. Members of the *brachial plexus* may be felt on the lateral axillary wall. The *ulnar nerve* is located in the groove between the medial epicondyle of the humerus and the olecranon process of the ulna. The nerve at this point constitutes the "crazy bone."

The lower limb (Fig. 5–15)

Bones. The pelvic girdle, formed by the two *os coxae* ("hip bones") supports the lower limbs. These bones approach the surface anteriorly

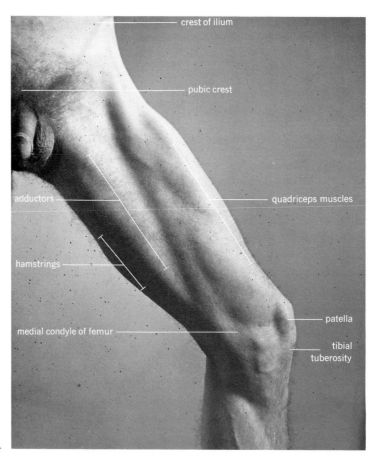

FIGURE 5–15. Lower limb.

to form the *anterior superior iliac spine* and *crest of the ilium.* The *pubic crest* is located in the anterior midline at the inferior aspect of the abdomen. The *ischial tuberosities* are located beneath the buttock muscles. The *greater trochanter* of the femur or thigh bone appears laterally on the proximal end of the thigh. The shaft of the femur is covered by muscle on all aspects. Distally, the femur presents *medial* and *lateral condyles* and *epicondyles.* Anteriorly the *patella* (knee cap) lies over the knee joint. The leg is formed by two bones, the larger, medially placed *tibia,* and the smaller, laterally placed *fibula.* Just below the patella, the tibia shows the *tibial tuberosity.* The anterior medial aspect of the tibia lies subcutaneously along its entire length, forming the "shin." Distally the tibia terminates in a large bulge, the *medial malleolus.* The *fibular head* may be felt proximally on the lateral leg, and distally it terminates in the *lateral malleolus.*

The heel is formed by the *calcaneous bone.* The dorsal aspect of the foot is formed by the cuboid, navicular, and cuneiform bones. These may be appreciated in thin individuals. The five *metatarsals* and fourteen *phalanges* form the remainder of the foot.

Muscles. The *quadriceps* group of muscles occupies the anterior and lateral aspect of the thigh. Posteriorly, the *hamstrings* cover the femur, above which may be recognized the large rounded *gluteus maximus* or buttock muscles. The *gluteal fold* lies at the junction of these two muscle groups. Symmetry of these folds indicates proper

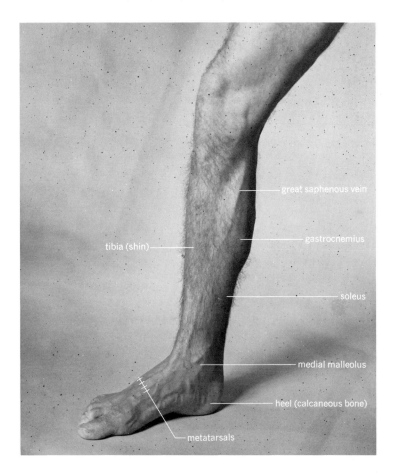

tibia (shin)

great saphenous vein

gastrocnemius

soleus

medial malleolus

heel (calcaneous bone)

metatarsals

hip-joint anatomy. Medially, the thigh is covered by the *adductor muscles.*

The *tibialis anterior* muscle lies lateral to the anterior border of the tibia. Its tendon becomes prominent across the anterior aspect of the ankle when the foot is raised. The fibula is covered by the *peroneal muscles.* The two-headed *gastrocnemius* forms the bulge of the calf and leads into the *Achilles tendon* (tendo calcaneus) to the heel. The *soleus* may be seen on either side of this tendon on the lower half of the leg.

Blood vessels. The pulsations of the *femoral artery* may be felt at the groin. The artery may be palpated again on the posterior aspect of the knee, as the *popliteal artery.* On the distal aspect of the anterior tibia, the *anterior tibial artery* may be traced into the *dorsis pedis artery* of the foot. The superficial veins consist of a medial-lying *great saphenous,* and a posterior *small saphenous.* These veins are especially prominent at the ankle. They are common sites for intravenous injection in infants.

Nerves. Only one nerve may be palpated on the lower limb. The *common peroneal nerve* winds around the lateral side of the neck of the fibula.

QUESTIONS

1. Describe, using correct anatomical terminology, the placement of the following organs on or in the body, and relate each to the other:
    (a) The nose
    (b) The larynx
    (c) The anterior triangle
    (d) The umbilicus
    (e) The pubic crest
    (f) The knee
    (g) The liver
    (h) The kidney
    (i) The urinary bladder
    (j) The brain
2. What organs are included (a) within the 4 quadrants of the abdomen, and (b) within the 9 smaller abdominal regions?

CHAPTER 6

Integumentary System

THE BODY COVERING

Introduction

The external surface of the body is formed by the integumentary system. The integumentary system includes the *skin,* derivatives of the skin such as *hair, nails,* and *glandular* structures, and several specialized types of *receptors.*

The skin itself is the largest organ of the body. It forms a pliable protective covering over the external body surface. The term protective here includes not only resistance to bacterial invasion or attack from the outside, but also protection against large changes in the internal environment. Control of body temperature, prevention of excessive water loss, and prevention of excessive loss of organic and inorganic materials are necessary to the maintenance of internal homeostasis and continued normal activity of individual cells. In addition, the skin acts as an important area of storage, receives a variety of sensations, and synthesizes several important substances used in the overall body economy.

An adult has an average skin surface area of 1.75 square meters (about 3000 square inches). Grossly. the skin presents several unique features. *Flexion creases* appear where the skin folds during the movement of joints. The hand shows many such creases. *Flexion lines* occur where the skin must stretch, but to limited degrees. The back of the

125

hand and fingers show such lines well. *Friction ridges* occur on those parts of the body involved in grasping. These ridges occur on the finger and toe tips and on the sole of the foot and palm. On the fingers, they are used for identification purposes because the pattern of the ridges in these areas is characteristic for the individual. Firmness of attachment varies from loose (as over the elbow) to tight (as on the scalp). Generally, the greater the degree of movement required, the looser the attachment will be. Skin varies in thickness in different areas of the body. In general, the thickness depends upon the degree of mechanical attrition to which the area is subjected. The thickest skin (5 to 6 mm) is found on the hands and feet. Callouses and bunions represent even greater thickening in areas of extreme pressure or wear and tear. The eyelids, eardrum, and penis have the thinnest ($\frac{1}{2}$ mm) skin covering. Other areas of the body have skin of thickness intermediate (1 to 2 mm) between that of the areas mentioned. Skin may be pigmented to various degrees due to the presence of pigment or pigment cells. Various shades of redness are dependent upon the calibre of the blood vessels in the skin and upon the degree of oxygenation of the blood in these vessels.

Structure of the skin

Histologically, skin is classed as being thick or thin. The terms not only refer to total depth of the covering, but also imply different microscopic structure. Both types of skin possess a superficial epidermis, an underlying dermis (corium), and a subcutaneous layer (superficial fascia) which affixes the skin to underlying tissues or organs.

Thick skin (Fig. 6–1 A and B). *Epidermis.* As a whole, the epidermis may be classified as a stratified squamous epithelium resting upon a basement membrane. Five major layers of cells are found in the epidermis of thick skin. From the deepest layer outwards, these layers are:

1. *Stratum basale (stratum cylindricum, stratum germinativum).* The basale is formed of a single layer of columnar or cylindrical cells whose lower ends have processes fitting into reciprocally shaped "pockets" in the basement membrane. A firm fixation of the epidermis to the underlying tissue is thus assured. The epithelium renews itself by mitosis from this layer.

2. *Stratum spinosum.* The spinosum is composed of 8 to 10 layers of polygonal cells. In life, the cells fit closely together and are attached to one another by desmosomes. The techniques employed in preparing a tissue section for microscopic viewing cause these cells to shrink. They pull away from one another, except at desmosomal attachments. The cells thus show what appear to be processes extending from one to another. These processes or "spines" are responsible for the name of the layer.

3. *Stratum granulosum.* The granulosum is composed of 2 to 5 layers of rhombic cells containing dark staining granules of keratohyaline. This material represents the first step in the formation of keratin, a proteinaceous substance which will fill the surface cells. Granulosum cells usually show nuclei in various stages of degeneration. The cells are so far removed from the dermal blood vessels that they cannot be maintained by diffusion of nutrients.

4. *Stratum lucidum.* The lucidum consists of 3 to 4 layers of flat

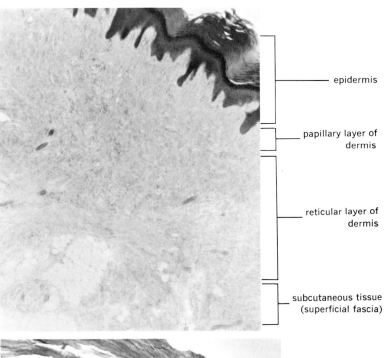

FIGURE 6–1A. Thick skin, general view.

epidermis

papillary layer of dermis

reticular layer of dermis

subcutaneous tissue (superficial fascia)

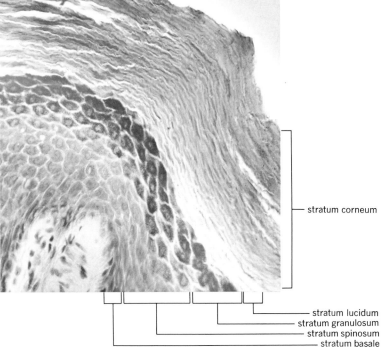

stratum corneum

stratum lucidum
stratum granulosum
stratum spinosum
stratum basale

FIGURE 6–1B. Epidermis of thick skin.

clear refractile cells containing semiliquid droplets of eleidin. Eleidin represents the second stage in keratin formation. No cellular structure is evident in this layer.

5. *Stratum corneum.* The corneum consists of 25 to 30 layers of flat dead scale-like cells filled with keratin. The most superficial layers are constantly being shed and form the *stratum disjunction.*

The lower surface of the entire epidermis is folded. Downwards projections of epidermis are termed *rete pegs.*

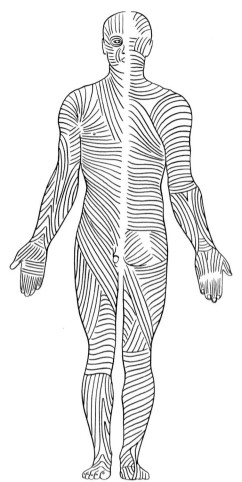

FIGURE 6–2. Cleavage lines of the skin.

The dermis. Normal dermis is a tough, flexible, and elastic layer of dense irregularly arranged connective tissue. Its elasticity creates *cleavage lines* (Fig. 6–2) for the skin as a whole. Skin, if cut, draws away from the wound in certain well-defined directions. These are the cleavage lines. (Surgeons generally make incisions parallel to these lines for more rapid healing.) The upper layer of the dermis is termed the *papillary layer.* It is about one-fifth of the total dermis and possesses *papillary pegs,* upfolds which interdigitate with the downfolds of the epidermis. Connective tissue fibers in the papillary layer are smaller than in the remainder of the dermis. The *reticular layer* forms the remainder of the dermis. The dermis contains many blood vessels arranged in two layers. In the deep portion of the reticular layer, many sinuous arteries are found. Branches ascend to the lower border of the papillary layer to form a subpapillary arterial plexus. From this plexus, capillaries form loops within the papillary pegs (Fig. 6–3). The arrangement of the capillary loops and their connecting vessels allows more or less blood to flow through the dermis as a whole, a fact of importance in temperature regulation.

Subcutaneous layer. The subcutaneous layer is composed of loose connective tissue typically containing much adipose tissue *(panniculus adiposus).* There is no clear line of demarcation between it and the dermis, other than fatty infiltration. The tissue is loose enough to accommodate significant volumes of fluid, and it is into this layer that subcutaneous injections are made.

Thin skin (Fig. 6–4). In thin skin, all layers are reduced in depth. In addition, a granulosum and lucidum are usually lacking. If subjected to increased attrition, thin skin may assume the appearance of thick skin.

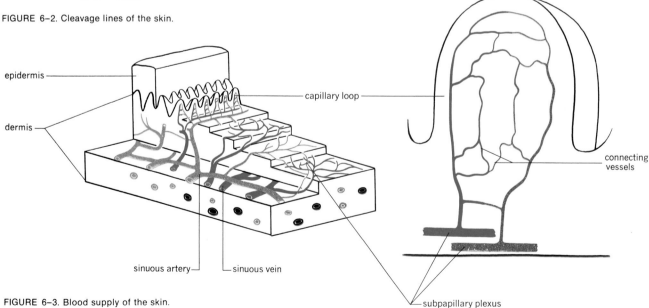

FIGURE 6–3. Blood supply of the skin.

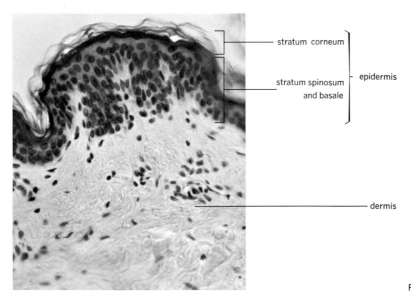

— stratum corneum

— stratum spinosum
and basale
} epidermis

— dermis

FIGURE 6-4. Thin skin, general view.

Appendages of the skin

Hair (Fig. 6–5). Hair is widely distributed over the human body. Only the anterior surface of the hands and fingers, soles of the feet, posterior aspect of the last two phalanges of the fingers, lips, nipples, umbilicus, and the skin-covered portions of the male and female genitalia are truly hairless. In lower animals, the hair performs important temperature regulating and protective functions. In the human, the total hair covering is not dense enough to be of any importance in these respects. Hair color is provided by melanins located within the hair, or by the presence of air within the hair shaft. "Gray hair has air." Lanugo or down hair appears first on the fetal body, and is replaced by terminal hairs.

A downgrowth of epidermal cells into the dermis forms a *hair follicle.* Germinal cells at the base of the follicle give rise to the hair itself. These cells are nourished by vascular dermal papillae. The hair itself has a central *medulla,* surrounding *cortex,* and a covering *cuticle.* The follicle consists of *inner and outer root sheaths* and a *connective tissue sheath.* The hair has a *bulb* on its lower end. The *shaft* is the visible portion of the hair. Hairs always have *sebaceous glands* associated with them. These glands secrete an oily sebum which keeps the hairs pliable and the skin moist. Hairs are set obliquely into the skin, and in the obtuse angle between the follicle and the surface is found the *arrector pili muscle.* When contracted, the muscle pulls the hair into a more vertical position. In hair-covered animals, this action increases the insulative capacity of the fur and makes the animal look larger. In man, "goose flesh" is the only result; this occurs as the erected hair pushes a mound of skin to one side.

Nails (Fig. 6–6). Nails are modified corneal and lucidal layers of the digits. The nail has a *root, body,* and *free edge,* and rests upon the *nail bed.* The *eponychium* and *hyponychium* are skin layers at the base of the nail and under the free edge, respectively. The *lunula* is a whitish

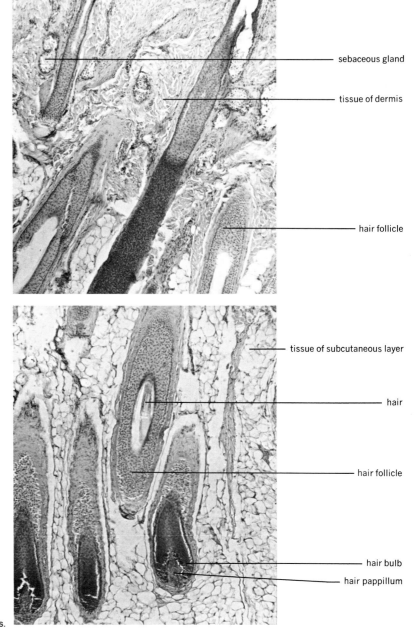

sebaceous gland

tissue of dermis

hair follicle

tissue of subcutaneous layer

hair

hair follicle

hair bulb
hair pappillum

FIGURE 6–5. Hairs.

half-moon shaped area under the base of the nail, and it represents the active growing region of the nail. The nail is bordered proximally and laterally by *nail folds.* In both longitudinal and cross sections the nails are curved, giving much strength for the thinness.

Glands. Two important types of glands are associated with and derived from the epidermis. *Sebaceous glands* (Fig. 6–7) are simple alveolar holocrine glands. Indifferent cells line the gland and transform into sebaceous cells, which then disintegrate to form the secretion (sebum).

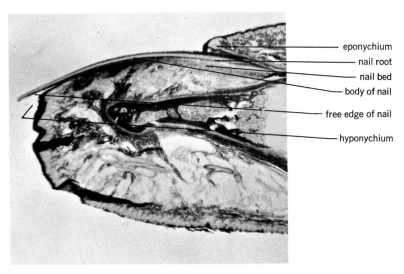

eponychium
nail root
nail bed
body of nail
free edge of nail
hyponychium

A. longitudinal section

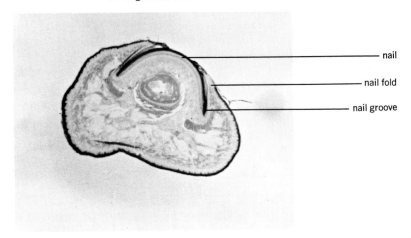

nail
nail fold
nail groove

B. cross section

FIGURE 6–6. Nails.

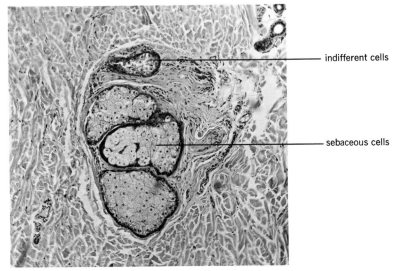

indifferent cells

sebaceous cells

FIGURE 6–7. Sebaceous gland.

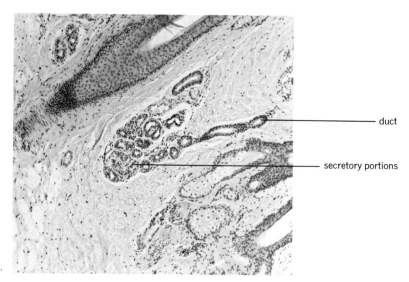

FIGURE 6-8. Sweat gland.

Sudoriferous (sweat) glands (Fig. 6–8) are simple coiled tubular structures. The so called *eccrine* sweat glands are widely distributed over the body and produce a watery merocrine secretion important in temperature regulation. In the axilla, anal, genital, and ear lobe regions of the body are found *apocrine* sweat glands. Their secretion is whitish and cloudy and contains much organic substance.

CHARACTERISTICS AND FUNCTIONS OF THE SKIN

Mechanical properties

Skin on the body is normally under tension and will retract if it is cut. This phenomenon indicates that the elastic elements of the dermis are normally slightly stretched. With loss of elastic fibers (for example in aging) the skin loses its elasticity and tends to sag. *Resiliency* of the skin may be defined as the tendency or ability to return to its original shape when stretched. This property varies with age and again depends upon the presence of elastic fibers in the dermis. Young skin averages a 92 percent resiliency; aged skin, 81 percent. It is of interest to note that application of estrogenic hormone preparations to nonresilient skin can lead to redevelopment of elastic fibers and an increase in blood supply of the dermis. Some gain in resiliency may be realized. The *tensile strength* of the skin refers to the force required to elongate it. In general, the younger the skin, the more easily it stretches. Older skin is tougher.

Electrical behavior

The skin behaves as though it were a negatively charged membrane. It may absorb anions, but does not pass them. Basic dyes penetrate easily, acidic ones do not. An "electrical double layer," consisting of H^+ externally and OH^- internally exists at the line of junction between the cornified and noncornified layers of the skin. This layer tends to limit the passage of charged substances.

Percutaneous absorption

This term refers to the penetration of substances through the skin and into the blood stream. Historically, the skin was considered to be impermeable to all substances. We know today that gases and lipid soluble substances pass relatively easily through the skin, but that the skin remains essentially impermeable to electrolytes and water. Such results imply a barrier to the penetration of materials. The barrier does not lie in the corneum. The corneum behaves like a gross sieve. An initial barrier is imposed by the electrical double layer at the cornified–noncornified epidermal junction. The electrical properties of this area probably limits the passage of electrolytes. A second barrier exists at the basement membrane. Most noncharged substances appear to be held up at this level. If a substance passes these two barriers, no restriction is imposed upon its entry into the blood stream.

Fat solubility increases penetration, chiefly because the skin surface is covered by a waxy layer composed of cholesterol, cholesterol esters, and waxes. This layer hinders the penetration of aqueous solutions, as well as facilitating penetration of solutions of oils. Substances which dissolve keratin (*keratolytics*), including salycylic acid (aspirin), enhance penetration of fat soluble substances.

A second possible route of passage through the skin is afforded by the hair follicles and sweat glands. These might be suspected of affording an easier route of entry, since no penetration of a primary barrier is involved. Some absorption takes place through hair follicles, but apparently none occurs through sweat glands.

Some specific substances and their routes and manner of passage through the skin are summarized below:

Substance	Route	Reason for penetration
Phenols	Skin surface	Coagulates skin proteins and destroys barrier
Hormones (testosterone, estrogen)	Skin surface	Fat soluble
Vitamins (A,D,K)	Skin surface	Fat soluble
Organic bases	Skin surface	Fat soluble
Aspirin	Skin surface	Keratolytic
Gases (O_2, CO_2)	Skin surface	Are not changed; small molecules which diffuse rapidly
Animal and vegetable fats	Hair follicle	

These data suggest that application of substances directly to the skin has practical value in the treatment of certain conditions. If the object is to achieve a higher local concentration of substance, it is of value.

Remember, however, that the vehicle employed to dissolve the substance is important, as is the depth to which any material may penetrate the skin.

Sensory functions

The skin mediates four main modalities of sensation:

1. Pain, including itch and tickle.
2. Touch, light and heavy.
3. Cold.
4. Warmth.

The distribution of the "sensory spots" serving the four major modalities are in the form of a mosaic, and differ in density in different areas of the body. The sense of light touch, attributed to the *Meissner's corpuscles* (Fig. 6–9) are irregular areas about 5 mm in diameter. They are most numerous on the finger tips and on the lips. The sensation of touch may also be aroused by bending of hairs. Heavy touch is assigned to the *Pacinian corpuscle* (Fig. 6–10). These receptors lie deeper in the dermis than the Meissner's corpuscles, and require a heavier pressure to be stimulated.

Pain elicited from the skin has two qualities. It may be bright and pricking (temporary and easily localized), or burning and longer lasting.

Thermal sensations are attributed to *Ruffini corpuscles* (warmth) and *Krause corpuscles* (cold) (Fig. 6–11). Excitation of thermal receptors occurs when the normal skin temperature gradients are changed by thermal stimuli.

Thermoregulation

The body loses heat by four methods.

1. *Radiation* involves the transfer of heat from a warmer to a cooler area by electromagnetic waves (infrared). Under normal conditions, about 60 percent of the body's heat loss is by this method.
2. *Conduction* is transfer of heat from a warmer to a cooler area if they are in contact. Normal clothing plays little role in changing physical heat loss. Heat stasis may result, with change of total body temperature if insulating materials are employed.
3. *Convection* is transfer of heat to a moving medium, such as air or water. Convection carries off a great deal of heat as long as the moving medium has a lower than body temperature.
4. *Evaporation* causes heat loss because of the heat required to vaporize water on the skin surface.

The role of the skin in heat loss revolves primarily around its ability to act as a radiant surface and transfer heat from the cutaneous blood vessels to the exterior. Obviously, the amount of heat available for transfer depends on the diameter of the cutaneous vessels and the amount of blood therefore circulated through the skin. Control of the vessel diameter is served nervously (Fig. 6–12). Within the hypothalamus are "heat gain" and "heat loss" centers. These centers receive input information from superficial and deep-lying thermal

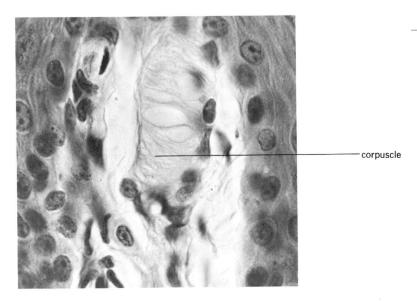

FIGURE 6–9. Meissner's corpuscle.

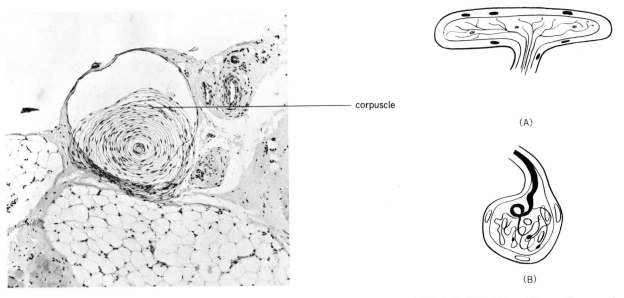

FIGURE 6–10. Pacinian corpuscle.

FIGURE 6–11. Ruffini (A) and Krause (B) corpuscles.

receptors, and they also monitor the temperature of the blood. The heat gain center sends nerve fibers to cutaneous vessels which constrict them, and to skeletal muscles to increase their tone. Heat radiation through the skin is thus decreased, while heat production is increased. The heat loss center sends fibers to cutaneous vessels to dilate them, and to sweat glands to stimulate secretion. Radiation and evaporation are increased.

Sweating forms a second mechanism of thermoregulation which directly involves the skin. Eccrine sweat glands produce the clear sweat responsible for heat regulation. Eccrine sweat is composed of 99 to 99.5 percent water and 0.5 to 1 percent solids. Half of the solids are inorganic, chiefly NaCl: the other half are organic, including urea, lactic acid, amino acids, glucose, and B vitamins. Freshly secreted

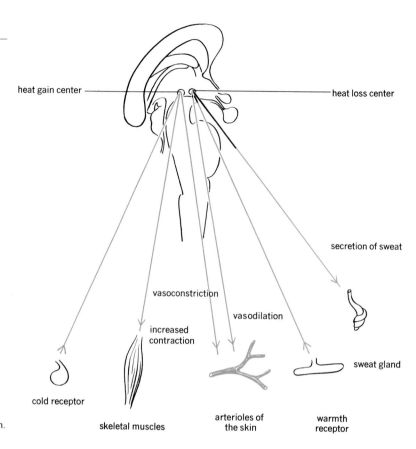

heat gain center — heat loss center

secretion of sweat

vasoconstriction

vasodilation

increased
contraction

sweat gland

cold receptor

FIGURE 6–12. Nervous control of temperature regulation.

skeletal muscles

arterioles of
the skin

warmth
receptor

eccrine sweat has a pH of 5.73 to 6.49. Control of eccrine secretion is nervous to those glands located on the palms, soles, head, and trunk. Glands in other areas appear to respond directly to local application of heat. The term *hyperidrosis* means excessive sweating. Apocrine gland secretion contains protein, sugars, ammonia, ferric ion, and fatty acids. The secretion is cloudy, may contain odoriferous substances (sex attractants in some animals), or be colored (hippopotamus, "sweating blood"). Bacterial action upon the constituents of apocrine sweat may increase odor, but is not necessarily required to create an odor. Apocrine glands receive a dual innervation, and respond primarily to mental rather than thermal changes.

Synthetic functions

Although the skin manufactures many substances used in its own economy, for example, keratin, there is one substance manufactured which exerts effects upon areas far removed from the skin. *Vitamin D* is manufactured from dehydrocholesterol, a sterol found in skin. Through the action of ultraviolet light on the above compound, vitamin D_3 is formed, absorbed into the blood stream, and becomes involved in the metabolism of calcium and phosphate in the body.

Burns

Exposed as it is, the skin is very susceptible to damage by flame, scalding, or contact with hot objects. A *first degree burn* occurs when there is reddening of the skin surface. The reddening is due either to direct damage to capillaries, causing their dilation, or to the release of chemical substances (for example, histamine) which dilate the vessels. A sunburn is a good example of a first degree burn.

Second degree burns result in blistering involving usually only the epidermis or at most the papillary dermal layer.

Third degree burns destroy the deep tissues of the skin, dermis, and fascia.

Extensive burns are serious primarily because they remove the covering which prevents water and protein loss from the body fluids. Changes in fluid concentration of solute, loss of blood volume, pH changes and increased susceptibility to infection in the denuded areas may lead to fatal results.

Rashes

It is beyond the scope of this text to delve deeply into the various and multitudinous causes of skin rashes. Suffice it to say that the skin is a very sensitive indicator of the presence of foreign materials on or in the body. The larger number of skin rashes develop as a result of some variety of antigen-antibody reaction involving the presence of a sensitizing agent. Dermatitis and skin edema are common results of this type of reaction. "Diaper rash" results from other causes, chiefly the plugging of sweat ducts by long exposure to water and swelling of the surface cells of the skin. The presence of ammonia, derived from urea in the urine by bacterial action, may increase the severity of the rash.

Pigmentation of the skin

The epidermis and dermis normally contain a variety of pigments or pigment cells which impart a color to the skin other than that due to vascular factors. *Melanin,* a yellow to black pigment, is located in the stratum basale in Caucasian races. Melanin is found in all epidermal layers in the Negro race. Exposure to ultraviolet radiation increases the amount and darkens the color of melanin, leading to tanning and protection against the radiation. Melanin is formed in melanocytes, by the action of an enzyme (tyrosinase) on the amino acid tyrosine. Lack of the enzyme results in no pigment formation (albinism). *Melanoid,* similar to melanin, develops from keratin, and is found chiefly in the stratum corneum. *Carotene,* a yellow-orange pigment is found in the stratum corneum and in the fatty areas of the dermis. It accounts for the yellowish color of the skin of Oriental races.

QUESTIONS

1. Discuss the structure of the skin as related to the functions it serves in the maintenace of homeostasis.
2. Contrast and compare the two types of skin.
3. What are the various factors which combine to create skin color?
4. What detrimental results occur when skin is removed, as in an abrasion or a burn? Relate each to the functions the skin serves.

Articulations

INTRODUCTION

A joint or articulation is formed where a bone joins another bone, or where a cartilage joins a bone. The structure of the joint depends largely upon the function it must serve. Accordingly, the union may be rigid, or may permit a variable degree of motion. If permitted, motion may occur in one, two, or three planes of movement, and the joint may be termed uniaxial, biaxial, or triaxial. Joints depend for their security upon closely fitting bony parts, ligaments, or muscles. The closer the fit of the bones, the stronger the joint, but the greater the restriction on axes of movement. The joints which permit the greatest range of motion are generally held together by muscles and are most liable to dislocation.

Motions permitted in freely movable joints are of four general types (Fig. 7–1):

1. *Gliding motions* occur in a side-to-side and back-and-forth direction. A twisting or rotational motion is generally not permitted because of ligaments or the proximity of other bones. Such a joint would be classed as biaxial.
2. *Rotation* occurs as a turning about a central point or around the long axis of a bone without any other motion being permitted. If permitting only one type of motion, a joint would be termed uniaxial.

141

3. *Angular movements* alter the angle between two bones. Flexion decreases the angle. Extension increases it. Abduction moves a part away from the midline of the body or midline of the appendage. Adduction moves the part towards the midline of body or appendage. Joints usually permit the corresponding pairs of actions, singly or together. Uniaxial or biaxial joints may permit angular movements.

4. *Circumduction* occurs when the distal end of an appendage is caused to describe a circle, while the proximal end remains essentially stable within the joint capsule. Circumduction involves an orderly progression of flexion, abduction, adduction, and extension, and generally involves rotation as well. Since movement occurs in all three planes, a joint permitting circumduction is usually triaxial.

CLASSIFICATION OF JOINTS

Utilizing the criteria of structure, and degree and type of motion permitted, three categories of joints are recognized.

1. *Fibrous joints* (Fig. 7–2) *(juncturae fibrosae, immovable joints),* No joint cavity is present and the bones involved are held together by thin fibrous membranes or by surrounding connective tissue. The category includes several subgroups.
   (a) *Syndesmosis.* Two bones are held rigidly by connective tissue between them or surrounding the area. The junction between the distal ends of the tibia and fibula is a syndesmosis.
   (b) *Sutures.* Found only in the skull, sutures involve a nearly bone to bone union with only minimal amounts of connective tissue between the bones.

■ *True sutures (sutura vera)* involve interlocking between the bones. *Dentate sutures (sutura dentata)* are formed by rather long toothlike interdigitating processes. The suture between the two parietal bones is a dentate type. *Serrate sutures (sutura serrata)* involve fine sawlike processes which interlock. The joint between the two portions of the frontal bone is a serrate suture. A *limbose suture (sutura limbosa)* involves both interlocking and beveling, for example, in the suture between the parietal and frontal bones.

False sutures (sutura notha) involve no interlocking. A *squamous suture (sutura squamosa)* is beveled, as in the suture between the parietal and temporal bones. A *plane suture (sutura plana, sutura harmonia)* involves two flat surfaces in approximation. Plane sutures are extremely weak. The suture between the nasal bones is of this type. A *schindylesis* occurs where a blade like portion on one bone fits into a cleft on the other, as between the sphenoid and ethmoid. A *gomphosis* occurs where a conical projection fits into a socket. The only example of a gomphosis in the body is the teeth articulating in the sockets of the jaw bones. ■

2. *Cartilagenous joints (juncturae cartilaginea, amphiarthroses, slightly movable joints)* (Fig. 7–3). No joint cavity is present and a pad of cartilage joins the two bones. Fibrocartilage or hyaline cartilage is the joining material. Since both types of cartilage

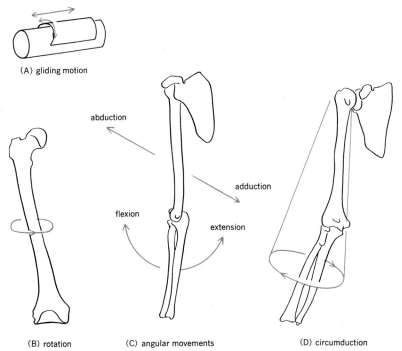

(A) gliding motion

(B) rotation

(C) angular movements

abduction

adduction

flexion

extension

(D) circumduction

FIGURE 7-1. Motions permitted in freely movable joints.

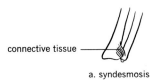

connective tissue

a. syndesmosis

b. true sutures

dentate suture

serrate suture

limbose suture

c. false sutures

squamous suture

plane suture

schindylesis

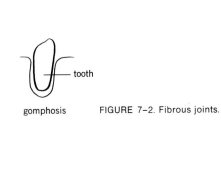

tooth

gomphosis

FIGURE 7-2. Fibrous joints.

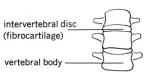

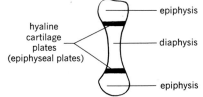

intervertebral disc
(fibrocartilage)

vertebral body

hyaline
cartilage
plates
(epiphyseal plates)

epiphysis

diaphysis

epiphysis

(A) synchondrosis

(B) symphysis

FIGURE 7-3. Cartilagenous joints.

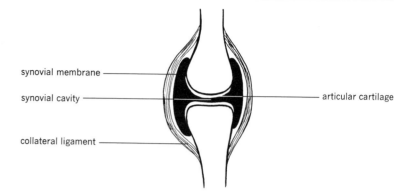

FIGURE 7–4. Typical synovial joint.

synovial membrane

synovial cavity

collateral ligament

articular cartilage

have only a little flexibility, the joints formed have but slight movement. According to the type of cartilage present, two sub-types of cartilagenous joints are recognized.

(a) *Synchondrosis.* The connecting material is hyaline cartilage. Such joints occur between the ends (epiphyses) of a growing bone, and its shaft (diaphysis). The cartilage will ultimately be replaced by bone; the joint is therefore only temporary.

(b) *Symphysis.* The connecting material is fibrocartilage, as in the intervertebral joints, or symphysis pubis.

3. *Synovial joints (diarthroses, juncturae synoviale, freely movable joints).* The typical diarthroid joint (Fig. 7–4) has a joint cavity *(synovial cavity),* a *synovial membrane* lining the cavity and secreting a *synovial fluid* to lubricate the joint, and surrounding *capsular (collateral) ligaments* to hold the joint together. The two bones involved in the joint have their articular surfaces covered with hyaline cartilage for smooth action. Such joints may have a pad of cartilage between the two bones for shock absorbing or security purposes. Six varieties (Fig. 7–5) of diarthroid joints are recognized, based upon the structure of the bones involved and the degrees or numbers of direction of movement permitted.

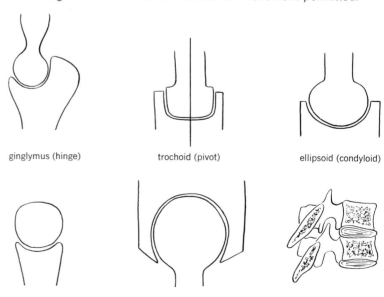

ginglymus (hinge)

trochoid (pivot)

ellipsoid (condyloid)

saddle

spheroidal (ball and socket)

arthrodial (gliding)

FIGURE 7–5. Types of diarthroid joints.

(a) *Hinge (ginglymus).* A spool-like surface of one bone fits into a half-moon surface on the other bone. Only a single plane of motion is permitted, usually flexion-extension, and the joint is uniaxial. The elbow and knee joints are examples of this type.

(b) *Pivot (trochoid).* A conical, pointed, or rounded surface articulates with a shallow depression on the other bone. Only rotation is permitted and the joint is again uniaxial. The joint between the radius and humerus is a pivot.

(c) *Ovoid (condyloid).* An egg-shaped surface on one bone fits into a reciprocally shaped surface on the other. The joint permits side-to-side and back-and-forth motions and is biaxial. The wrist joint is of this type.

(d) *Saddle.* Imagine a saddle on the back of a horse, the saddle representing one bone, the horse the other. Such a joint permits two directions of movement, side-to-side and back-and-forth, and is biaxial. The joint between the carpal and the metacarpal of the thumb is a saddle joint.

(e) *Ball-and-socket (spheroidal).* A ball-like surface fits into a cup-like depression on the other bone. All planes of motion are permitted and the joint is triaxial. The shoulder and hip joints are of this type.

(f) *Gliding (arthrodial).* The bony surfaces are nearly flat, and side-to-side and back-and-forth motion is permitted. The joint is biaxial. Intercarpal joints are of this type.

A summary of the articulations is presented in Table 7–1.

Synovial fluid is the fluid filling synovial joints. It is secreted by the synovial membrane and has a total volume in man of about 100 ml. Its function is to provide a thin film of fluid over the articulating surfaces for smooth action as the bones move upon one another. Its chemical and physical properties resemble those of interstitial fluid, from which it is derived. Analysis of knee joint synovial fluid in man shows it to have the following characteristics:

Total solids (percent)	3.4
pH	7.39
Viscosity (relative to water)	235
Specific gravity (average)	1.012
Na (meq)	135
Protein (g/100 ml)	2.8
Mucin (g/100 ml)	0.85
Uric acid (g/100 ml)	3.9
Hyaluronic acid (mg/100 ml)	155

The high viscosity of the fluid may be attributed to its mucin and protein content.

If excess amounts of fluid are produced (for example, after injury to the joint), movement of the joint may become restricted.

Arthritis is a term employed to designate inflammation of a joint. Arthritis may occur after infection or injury to a joint.

Gout commonly affects joints and is characterized by deposition of sodium ureate crystals in the joint. The disease is regarded as an inherited disorder of nitrogenous base (purine) metabolism.

TABLE 7–1

Summary of Articulations

Major Category	Type	Characteristics	Axes of motion permitted	Examples
Fibrous joints		No cavity, immovable	None	Sutures, syndesmoses
	Syndesmosis	Bones held together by connective tissue	None	Tibia-fibula
	Suture	Bone-bone joint, minimal c.t. between	None	Skull
	True sutures (sutura vera)	Interlocking		
	Dentate	Toothed interlocking parts	None	Parietal bones
	Serrate	Finely toothed inter-locking parts	None	Frontal
	Limbose	Interlocking and beveled	None	Frontal-parietal
	False sutures (sutura notha)	No interlocking		
	Squamous	Beveled	None	Parietal-temporal
	Plane	Flat surfaces opposed	None	Nasal bones
	Schindylesis	Blade in cleft	None	Spenoid-ethmoid
	Gomphosis	Cone in socket	None	Teeth in jawbones
Cartilagenous joints		No cavity; slightly movable; a bone, cartilage, bone joint	Compression	Synchondroses, symphyses
	Synchondrosis	Bone, hyaline cartilage bone	None	Shaft and ends of long bones Temporary
	Symphysis	Bone, fibrocartilage, bone	None	Symphysis pubis, vertebral column
Synovial joints		Cavity, freely movable	One to three	Six types
	Hinge	Spool in halfmoon	Uniaxial; flexion and extension	Knee, elbow, ankle
	Pivot	Cone in depression	Uniaxial; rotation	Radio-humeral, atlas-axis.

TABLE 7–1 *con't.*

Summary of Articulations

Major Category	Type	Characteristics	Axes of motion permitted	Examples
Synovial joints		Cavity, freely movable	One to three	Six types
	Ovoid	Egg in depression	Biaxial; flexion-extension, adduction	Wrist
	Saddle	"Saddle on horse"	Biaxial; flexion-extension, abduction-adduction	Carpo-metacarpal joint of thumb
	Ball and Socket	Ball in cup	Triaxial; flexion-extension, adduction-abduction, rotation (circum-duction)	Hip, shoulder
	Gliding	Nearly flat surfaces opposed	Biaxial; flex-ion-extension, abduction-adduction	Intercarpal joints, intertarsal joints

As the student studies the skeleton, he should identify and classify the joints encountered. Further detailed descriptions of selected joints will be provided in the appropriate sections.

QUESTIONS

1. Compare fibrous, cartilagenous, and synovial joints as to structure and freedom of movement.
2. Synovial joints may be described as uniaxial, biaxial, or triaxial. Pick 6 synovial joints in the body, and classify them in terms of degrees of movement permitted, type of movement permitted, and type of joint.
3. Look at a skull and determine what type of suture is found in each joint you can find. Does the skull have only sutures? Explain.

CHAPTER 8

The Skeletal System

His bones are as strong as pieces of brass; his bones are like bars of iron.   OLD TESTAMENT

151

THE SKELETAL SYSTEM

Definition

The skeletal system of man is made up predominately of organs called bones and of minor components of cartilage. The bones are joined at the articulations or joints, enabling them to be moved in meaningful relationship one to another. The skeletal muscles provide the energy source and are capable of converting this stored chemical energy into mechanical energy, that is, energy of action.

Kinds of Skeletons

The skeleton of man, as of other vertebrates, is a living internal or *endoskeleton*. As such, it grows as the body grows; it adapts itself to the condition of life of the individual. It is capable of self repair following disease or injury. This is in contrast to the external or exoskeleton so well demonstrated in the insects and other Arthropods. The *exoskeleton* is nonliving, but is produced by underlying living tissues. The organism, to grow, must shed the exoskeleton and then replace it after growth is accomplished. It does not have the adaptability of an endoskeleton. It is, however, a good protection for the animal. The exoskeleton in vertebrates is limited to scales, shells (as in turtles), feathers, and (in man) to nails and hair. In this chapter we are concerned only with the endoskeleton.

Functions

The skeletal system is a framework of support for the soft tissues of the body. It is basic to the form of the body, and its erect posture. The skeletal muscles attach to the bones and, in the appendicular skeleton especially, the bones are used as levers, with the articulations acting as fulcra around which movement takes place. The skeleton therefore plays a passive, but essential, role in movement.

The skeleton provides protection for vital organs, such as the central nervous system, which is housed in the cranial cavity (formed by the bones of the skull) and vertebral canal (formed by the vertebrae). The heart, lungs, and major blood vessels are enclosed in the thoracic cavity with its protective framework of vertebrae, ribs, sternum, and costal cartilages.

The skeleton is a reservoir of minerals such as calcium and phosphorus, and the bones become involved in the metabolism of these substances. Certain bones also serve as centers for blood cell formation or hemopoiesis. Blood formation takes place mostly in red marrow in the proximal epiphyses of the femur and humerus, in the ribs, sternum, clavicle, os coxae, vertebrae, and in the diploe of skull bones. Yellow marrow, found mostly in the shafts of long bones, may also become active in the formation of red cells, granulocytes, and platelets. It is an emergency reserve for blood cell formation.

From another point of view, bones, because they are among the most stable of body organs and are the ones most likely to be preserved in the earth's crust, are a valuable source of information in evolutionary studies. They have been the objects of search for centuries by archaeologists and paleontologists in all parts of the world, and have contributed much to our understanding of the history of life and of man.

Classification of bones

Bones, as organs, are usually classified according to shape and fall logically into four groups; long, short, flat, and irregular.

Long bones are those whose length is greater than their width. The humerus, femur, tibia, and radius are obvious examples. Not so obvious are the bones of the fingers and toes, the phalanges, which one might be inclined to call short.

Short bones are considered to be those of the carpus (wrist) and tarsus (ankle); blocky bones closely joined; and those in which differences in length and width are not significant.

Flat bones usually are composed of two, more or less parallel plates of compact bone separated by spongy bone. They present large surfaces for muscle attachment. The scapulae and bones of the skull, particularly those enclosing the cranial cavity, are good examples.

Irregular bones are those of complex shape and structure and require individual description. The vertebrae and some of the bones of the face fall into this category.

Surface features of bones

The bones present many surface markings which suggest the function in which the bones are involved. The ends of weight-bearing long bones are enlarged, hard, smooth, and covered with a thin layer of hyaline cartilage. Articulating surfaces of bones vary in shape and partly determine the kind of action possible at joints. Bones have roughened areas and prominences of different forms for muscle, tendon, and ligament attachment. Some bones have grooves for the passage of blood vessels over their surfaces, and perforations through the bone for passage of blood vessels or nerves. Depressions on the bone surface may serve also for articulation or for attachment of ligaments or muscles.

Surface features of bone may be viewed as follows:

Articular surfaces:

Head.   A rounded surface often set off from the rest of the bone by a neck, as in the femur or humerus.

Condyle.   A relatively large, convex prominence; for example, the occipital condyles or the prominences seen at the ends of long bones such as the femur.

Facet.   A smooth, flat or shallow surface as seen on vertebrae for the attachment of ribs.

Nonarticular surfaces:

Process.   The generic term for any prominent, roughened projection from a bone.

Trochanter.   A large, blunt process found only on the femur.

Tuberosity.   A large, often rough, eminence as seen on the ischium.

Tubercle.   A smaller, rounded eminence as seen on the humerus.

Spine.   An abrupt projection from the bone surface which may be blunt or sharply pointed, such as the spine of the scapula.

Crest.   A prominent, often roughened, border or ridge, such as the crest of the ilium.

Line.   A slight ridge, as seen in the linea aspera of the femur.

Epicondyle.   A prominence above or upon a condyle, such as the epicondyles of the femur.

Fovea.   A shallow depression, such as the fovea capitis of the head of the femur.

Fossa.   Generally, a deeper depression; for example, the olecranon and coranoid fossae at the distal end of the humerus.

Sulcus.   A groove, such as the sagittal sulcus of the skull.

Fissure.   A narrow, cleftlike opening, as in the orbital fissure of the sphenoid bone.

Incisura.   A notch, usually in the border of a bone, such as the sciatic incisura or notch.

Meatus.   A canal running *within* a bone, such as the external auditory meatus.

Foramen.   An opening *through* a bone for the passage of nerves and/or blood vessels.

Organization of the skeleton

The skeleton is made up of 206 bones (Fig. 8–1). They are generally organized in the following logical manner for purposes of learning and

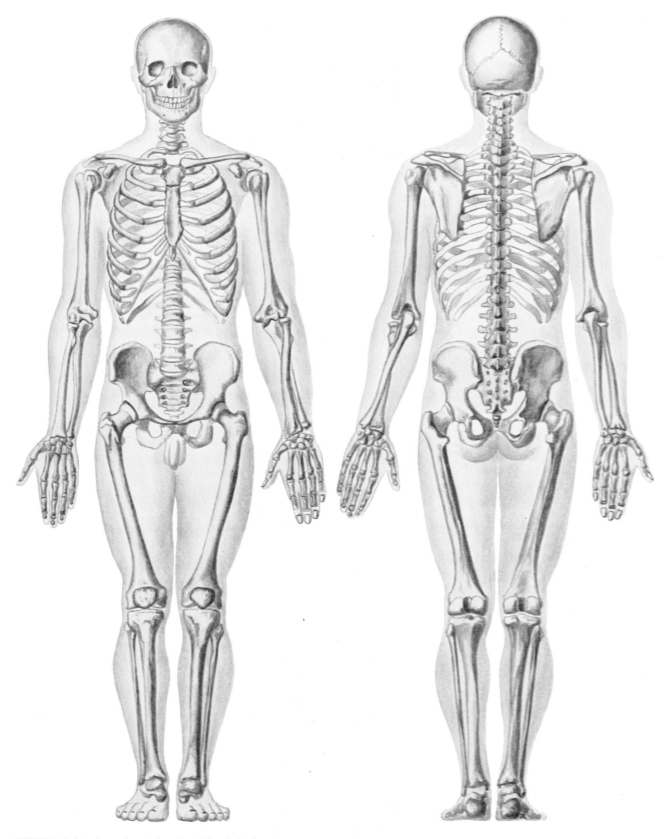
FIGURE 8-1. Anterior and posterior view of the skeletal system.

understanding. This organization also takes into consideration factors relative to function and to evolutionary history. The axial and the appendicular skeletons, for example, show some measure of independence in their origin and evolution. It is only in the tetrapod (four-footed) vertebrates that their components become closely involved.

Axial skeleton	80 bones
Skull	29 bones
cranium	8 bones
face	14 bones
hyoid	1 bone
ossicles (ear bones)	
malleus	2 bones
incus	2 bones
stapes	2 bones
Vertebral column	26 bones
cervical vertebrae	7 bones
thoracic vertebrae	12 bones
lumbar vertebrae	5 bones
sacrum	1 bone (5 fused)
coccyx	1 bone (3 to 5 fused)
Thorax	25 bones
sternum	1 bone
ribs	24 bones
Appendicular skeleton	126 bones
Shoulder girdle	4 bones
clavicle	2 bones
scapula	2 bones
Upper extremity	60 bones
humerus	2 bones
ulna	2 bones
radius	2 bones
carpals	16 bones
metacarpals	10 bones
phalanges	28 bones
Pelvic girdle	2 bones
os coxae	2 bones
Lower extremity	60 bones
femur	2 bones
fibula	2 bones
tibia	2 bones
patella	2 bones
tarsus	14 bones
metatarsus	10 bones
phalanges	28 bones
	Total 206 bones

If one were to count the two *sesamoid bones* which occur under the head of the first metatarsal bone of each foot, the total number of bones would be 210

THE SKULL

The skull rests on the superior end of the vertebral column which, because of its special modifications, allows freedom of movement of the skull and hence of the head. This versatility of movement of the skull increases the effectiveness of special sense organs, such as the eyes, ears, and nose, which are housed in and protected by the skull bones. The skull forms the cranial cavity, housing and protecting that crowning glory of vertebrate evolution, the human brain. Embryologically, morphologically, and physiologically, the skull is a double structure. It is composed of two sets of bone embryologically: an outer set which develops within the skin, and an inner set which is performed in cartilage—*intramembranous* and *intracartilaginous* bones. As development progresses, these become so closely joined that they lose their identity and form a unified whole, the neurocranium. Morphologically, one set of skull bones, those of the *neurocranium,* surround the brain end of the neural tube; the other set encircles the superior end of the digestive system, the *splanchnocranium.* Physiologically, the skull serves not only for support and protection, but through the freely movable mandible contributes to mastication and to speech.

Classification of skull bones

For purposes of description and of organized learning, the bones of the skull may be divided into two groups: cranial bones and facial bones.

> Cranial bones (8)
> Single bones:
> frontal
> occipital
> ethmoid
> sphenoid
> Paired bones:
> temporal
> parietal
> Facial bones (14)
> Single bones:
> mandible
> vomer
> Paired bones:
> maxillae
> zygomatics
> nasals
> lacrimals
> palatines
> inferior nasal conchae

In addition to the 22 bones listed above, there are bony ossicles in the middle ear cavity of each temporal bone, the malleus, incus, and stapes. There is also a hyoid bone in the base of the tongue which rightfully belongs to the skull. These bones bring the total number in the skull to 29.

General description

The bones of the cranium and of the face are immovably joined by fibrous articulations, with the exception of the mandible. The mandible forms with the temporal bone a synovial, free moving joint, enabling this bone to function in chewing, talking, and other related actions. The fibrous articulations of the skull are mostly of the suture type with interdigitating, overlapping, or abutting articular surfaces or edges.

Cranial vault sutures.   The cranial vault of the skull (Fig. 8–2) shows clearly some of the sutures, namely the *coronal suture,* which joins the frontal and parietal; the *sagittal suture* between the two parietal bones; and the *lambdoidal suture* between the parietals and the occipital. Laterally, the parietals are joined with the temporal bone at the *squamosal suture.*

Fontanels.   In the skull of the newborn (Fig. 8–3) membranous areas, the fontanels, persist at the points of joining of the above named sutures. It should be recalled that the bones of the cranial vault are intramembranous bones and the fontanels are fibrous membranes in which ossification has not yet taken place. Also, the two frontal bones have not yet joined on the midline and one sees here at a slightly later stage a frontal (metopic) suture which in the adult has grown together and is obliterated. Between parietals and frontal bones at the midline of the skull is the largest fontanel, the *frontal or anterior fontanel.* This is used as a landmark in obstetrical diagnosis. It closes at about 18 months of

FIGURE 8–2. Internal surface of ''cap'' of skull.

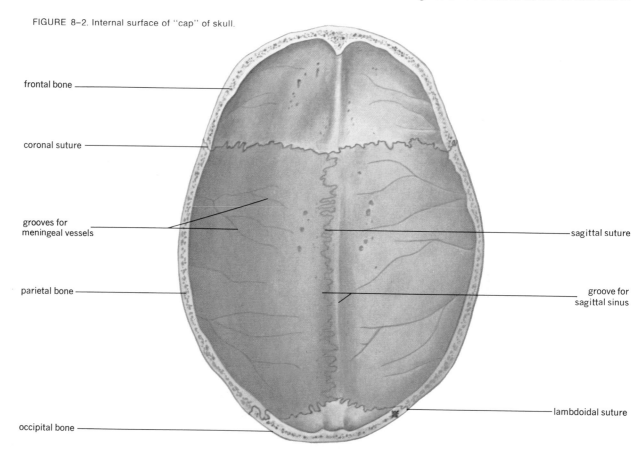

frontal bone

coronal suture

grooves for
meningeal vessels

parietal bone

occipital bone

sagittal suture

groove for
sagittal sinus

lambdoidal suture

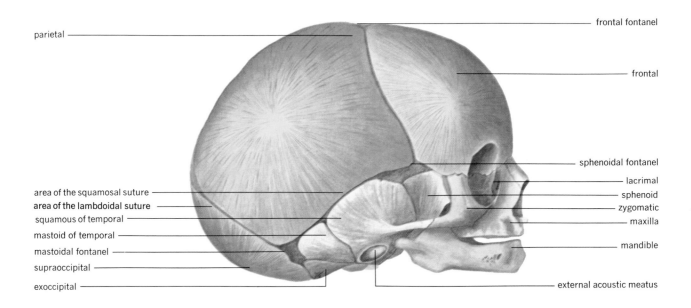

parietal

frontal fontanel

frontal

sphenoidal fontanel

lacrimal

sphenoid

zygomatic

maxilla

mandible

area of the squamosal suture

area of the lambdoidal suture

squamous of temporal

mastoid of temporal

mastoidal fontanel

supraoccipital

exoccipital

external acoustic meatus

FIGURE 8–3. Lateral and basal view of the skull of a newborn.

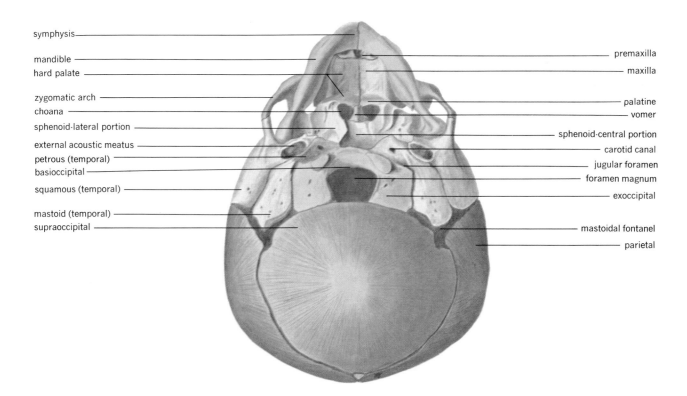

symphysis

mandible

hard palate

zygomatic arch

choana

sphenoid-lateral portion

external acoustic meatus

petrous (temporal)

basioccipital

squamous (temporal)

mastoid (temporal)

supraoccipital

premaxilla

maxilla

palatine

vomer

sphenoid-central portion

carotid canal

jugular foramen

foramen magnum

exoccipital

mastoidal fontanel

parietal

age. The *occipital or posterior fontanel* is at the midline where the lambdoidal and sagittal sutures meet. It closes about two months after birth. Two pairs of lateral fontanels are also found: (1) the *anterolateral or sphenoidal fontanels,* which lie at the juncture of parietal, frontal, sphenoid, and temporal bones, and (2) the *posterolateral or mastoidal fontanels,* which are where occipital, parietal, and temporal bones join. These fontanels and the fibrous membranes which persist between the bones of the cranial vault ease the process of childbirth because they allow the bones to override each other to accommodate to the size of the birth canal. Small bones of irregular occurrence may appear along the sutures and hence are called sutural or Wormian bones.

Numerous other sutures appear between the bones of the face and in the base of the skull, and are usually named according to the bones involved.

Anterior view (Fig. 8–4).   From this view, the bones which underlie the forehead and the face are shown. Conspicuous features are the orbits (orbital fossae) housing the eye; the nasal fossae housing the nose; and the jaws, which mark the entrance to the digestive system and support the teeth.

The *squama* or *vertical plate* of the frontal bone underlies the forehead and forms the anterior part of the cranial vault. It slopes gradually forward from the coronal suture and then turns abruptly downward. To either side of the midline it bulges slightly to form the frontal eminences. Below each eminence, a flattened ridge, the *superciliary arch,* curves transversely across the bone at about the level of the eyebrow. Between the frontal eminences and the superciliary arches is a flattened area, the *glabella,* which extends downward to where the frontal and nasal bones articulate. The frontal bone thickens below the superciliary arches and turns under and backward at the *supraorbital margins* as a *horizontal or orbital plate* to form the roof of the orbits and part of the floor of the cranial cavity. In the supraorbital margin is a notch or foramen for passage of the supraorbital nerves and blood vessels, the *supraorbital notch* or *foramen.* The *frontal sinuses* lie medially behind the superciliary arches and empty into the nasal cavities (Fig. 8–5).

The *orbit* houses the bulb or eyeball, and the muscles, nerves, and blood vessels associated with it, as well as nerves and blood vessels passing to the face. The orbit is pyramidal in shape with the apex pointing inward. The base of the pyramid forms the orbital aperture. Its superior margin is formed as indicated above, by the supraorbital margin of the frontal bone. The medial margin of the orbit is formed mostly by the frontal process of the maxillary bone, while the inferior margin is formed by the body of the maxilla and a part of the zygomatic bone. The zygomatic also forms most of the lateral margin of the orbit. The medial wall of the orbit is formed by the frontal process of the maxilla, the lacrimal, and the lamina orbitalis of the ethmoid bone (Figs. 8–6 and 8–9). The roof is formed by the orbital plate of the frontal bone; the floor is formed by the maxilla and zygomatic bones; and the lateral wall is formed anteriorly by the zygomatic and posteriorly by the great wing of the sphenoid.

Running along the floor of the orbit toward the apex and between the maxilla and great wing of the sphenoid is an irregular crevice, the *inferior orbital fissure,* which in life carries nerves and blood vessels to the deep structures of the face in the *infratemporal fossa.* At the

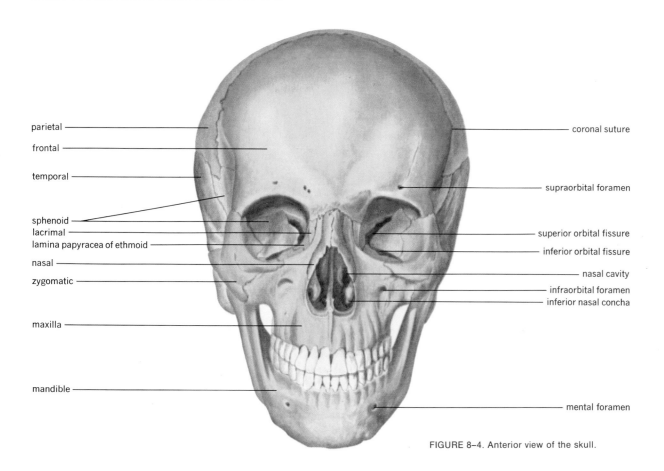

parietal

frontal

temporal

sphenoid
lacrimal
lamina papyracea of ethmoid

nasal

zygomatic

maxilla

mandible

coronal suture

supraorbital foramen

superior orbital fissure
inferior orbital fissure

nasal cavity

infraorbital foramen
inferior nasal concha

mental foramen

FIGURE 8–4. Anterior view of the skull.

apex of the orbit is seen the *optic foramen and canal* which carries the optic nerve and ophthalmic artery between the cranial cavity and the orbit. Just lateral to the optic canal is the *superior orbital fissure* which transmits the oculomotor, trochlear, ophthalmic division of the trigeminal, the abducens nerves, a branch of the middle meningeal artery, and the ophthalmic veins into the orbit.

The *lacrimal bone,* in the anterior medial wall of the orbit, is the smallest bone of the face (Fig. 8–7). It is best identified by a vertical ridge, the *posterior lacrimal crest,* which divides its lateral surface into two areas. In front of the crest is a groove which joins with a similar groove on the posterior surface of the frontal process of the maxilla to form the *lacrimal sulcus,* which houses the lacrimal sac, and part of the *nasolacrimal duct,* which empties into the lateral wall of the nasal fossa.

The *nasal fossa* is triangular in form (Fig. 8–4). Its margins consist of the two nasal bones superiorly and of the maxillae laterally and inferiorly. The nasal bones together form the bridge of the nose. Internally, the fossa is divided into two cavities by the perpendicular plate of the ethmoid and the vomer (Figs. 8–5, 8–9, and 8–10). These constitute the bony nasal septum which in the living subject is extended forward by a septal cartilage which gives flexibility to the external nose. The nasal fossae are further subdivided internally by the shelf-like *inferior nasal conchae* which project medially from the lateral bony

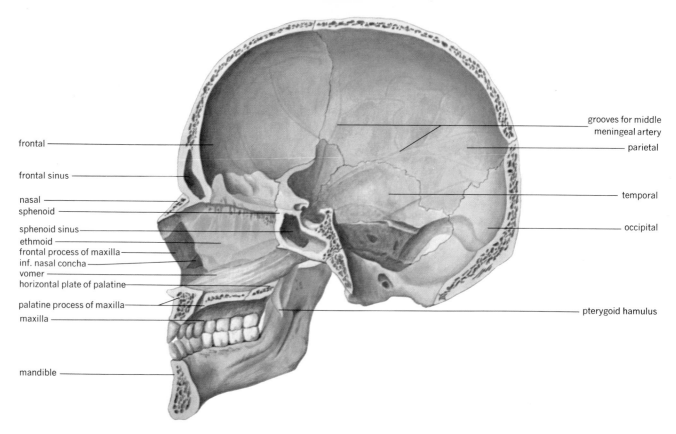

frontal

frontal sinus

nasal
sphenoid

sphenoid sinus
ethmoid
frontal process of maxilla
inf. nasal concha
vomer
horizontal plate of palatine

palatine process of maxilla
maxilla

mandible

grooves for middle meningeal artery

parietal

temporal

occipital

pterygoid hamulus

FIGURE 8–5. View of a median section of the skull.

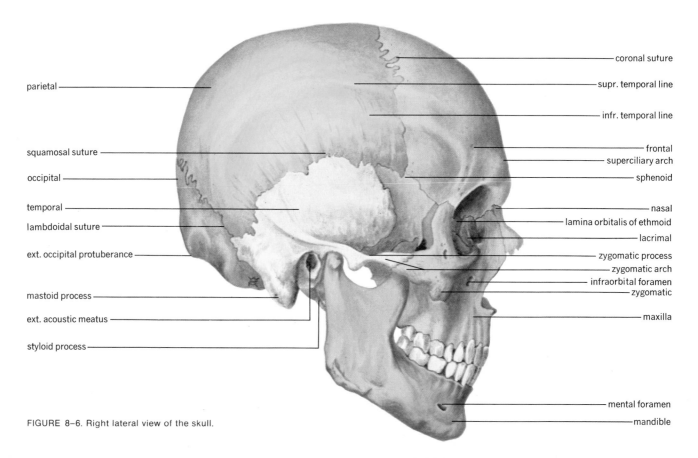

parietal

squamosal suture

occipital

temporal

lambdoidal suture

ext. occipital protuberance

mastoid process

ext. acoustic meatus

styloid process

coronal suture

supr. temporal line

infr. temporal line

frontal
superciliary arch

sphenoid

nasal
lamina orbitalis of ethmoid
lacrimal
zygomatic process
zygomatic arch
infraorbital foramen
zygomatic

maxilla

mental foramen
mandible

FIGURE 8–6. Right lateral view of the skull.

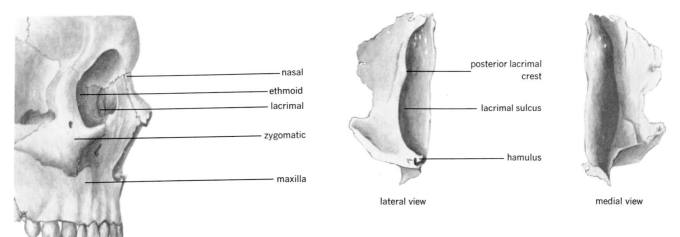

nasal
ethmoid
lacrimal
zygomatic
maxilla

posterior lacrimal crest
lacrimal sulcus
hamulus

lateral view

medial view

FIGURE 8-7. The right lacrimal bone in situ.

walls of the fossae (Figs. 8–8, 8–11). The *middle and superior nasal conchae,* which are a part of the lateral mass of the ethmoid, further subdivide and add surface area to the walls of the nasal fossae or cavities. The ethmoid also, through its cribriform plate, forms the roof of the nasal fossae and transmits the olfactory nerves to the cranial cavity through its foramina. The horizontal portions of the palatine

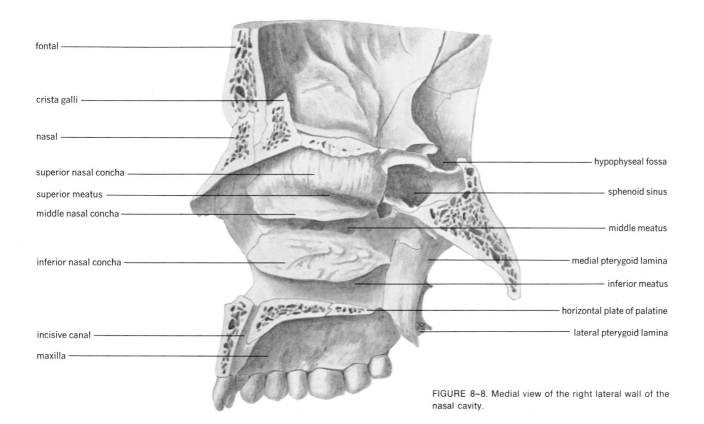

fontal
crista galli
nasal
superior nasal concha
superior meatus
middle nasal concha
inferior nasal concha
incisive canal
maxilla

hypophyseal fossa
sphenoid sinus
middle meatus
medial pterygoid lamina
inferior meatus
horizontal plate of palatine
lateral pterygoid lamina

FIGURE 8-8. Medial view of the right lateral wall of the nasal cavity.

bones and the palatine processes of the maxillae form the floor of the nasal fossae, while the vertical portions of the palatine bones contribute to their lateral walls (Fig. 8–12).

The *maxillary bones* (Fig. 8–13) have large sinuses, the *maxillary sinuses* or antra of Highmore, within their bodies. These communicate with the nasal cavities. The maxillaries also bear teeth in sockets on their alveolar processes. Below the orbit is the *infraorbital foramen* for the passage of infraorbital vessels and nerves. The *zygomatic process* projects laterally from the body of the maxilla to join the zygomatic bone.

The *mandible* (Fig. 8–14), which is separate and free moving, completes the skeleton of the face. It articulates by means of the only synovial joint in the skull with the temporal bone to form the *temporomandibular joint.* It consists of a horseshoe-shaped body which arches backward, and rami which rise vertically on each side to articulate with the skull. The body has an alveolar process superiorly with sixteen sockets to house the lower teeth. The inferior border of the body is rounded and supports the musculo-membranous structures which form the floor of the mouth. The two halves of the mandible develop separately and join on the anterior medial line to form the *mental or mandibular symphysis.* A *mental foramen* appears on each side of the symphysis, transmitting mental nerves and vessels. To either side of the symphysis is a depression for the sublingual salivary gland, the *sublingual fossa,* and posterior to that and extending to the ramus is a submandibular fossa for the salivary gland of the same name. A prominent ridge, the *mylohyoid line,* extends upward and backward from near the symphysis to the ramus. At about the center of the medial side of the ramus is a *mandibular foramen* which carries inferior dental (alveolar) nerves and vessels into the lower jaw. In front and above the foramen is a flap of bone, the *lingula.*

Lateral view (Fig. 8–6).   The study of this aspect of the skull allows a review of the bones of the anterior aspect as seen in profile and clarifies their relationships. The frontal bone is seen as a part of the cranial vault. It articulates at the coronal suture with the parietal bones; with the great wing of the sphenoid; and, by its zygomatic process, to the frontal process of the zygomatic bone. The zygomatic bone in turn articulates, as described above, with the zygomatic process of the maxilla anteriorly and forms a part of the margin and lateral wall of the orbit. Posteriorly, by its temporal process, it articulates with a long process of the temporal bone, the *zygomatic process.* Together, these structures form the *zygomatic arch* which bridges a depressed area on the lateral surface of the skull, the *temporal fossa.* The temporal fossa houses the temporalis muscle, an important muscle of the mandible. This aspect also reveals to better advantage the relationship of the body to the ramus of the mandible and the angle. The superior side of the ramus has two processes; the thin, triangular *coronoid process* in front and the thick *condyloid process* in back. These are separated from each other by the prominent *mandibular notch.* The head of the condyloid process fits transversely into the mandibular fossa of the temporal bone to form the *temporomandibular joint* (Fig. 8–6). This joint contains an articular disc and has a synovial membrane which forms separate cavities on each side of the disc. A loose, common, fibrous, articular capsule encloses the whole joint, and added support and strength are given on the anterior and lateral

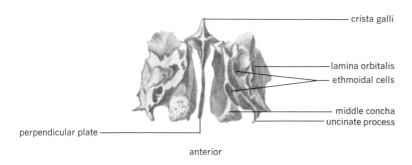

crista galli

lamina orbitalis
ethmoidal cells

middle concha
uncinate process

perpendicular plate

anterior

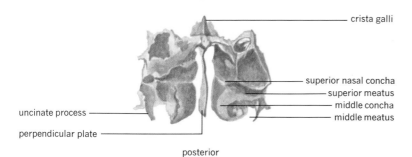

crista galli

superior nasal concha
superior meatus
middle concha
middle meatus

uncinate process

perpendicular plate

posterior

FIGURE 8–9. Anterior and posterior views of the ethmoid.

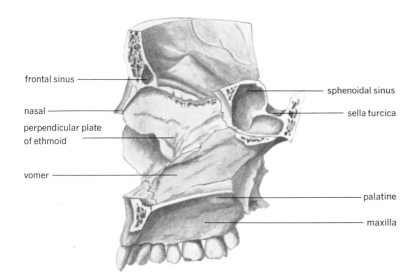

frontal sinus

sphenoidal sinus

nasal

sella turcica

perpendicular plate
of ethmoid

vomer

palatine

maxilla

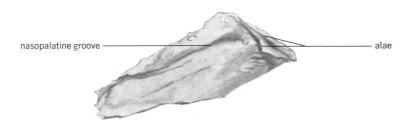

nasopalatine groove

alae

FIGURE 8–10. The right vomer in situ, medial view.

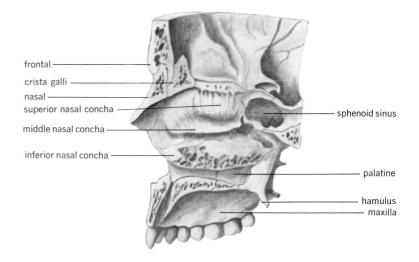

frontal

crista galli

nasal

superior nasal concha

middle nasal concha

inferior nasal concha

sphenoid sinus

palatine

hamulus

maxilla

lacrimal process

maxillary process

ethmoid process

ethmoid process

lacrimal process

FIGURE 8–11. The right inferior nasal concha in situ and medial and lateral view.

surfaces by a temporomandibular ligament. Opening and closing the jaws is primarily a hinge movement, the head of the mandible moving on the concave inferior surface of the articular disc. When the mandible is protruded, the disc and mandibular head move as one onto the articular eminence which lies in front of the mandibular fossa. When moved backward (retraction), the disc and mandibular head slip back into the mandibular fossa. In side-to-side movements, the disc and head move as a unit across the mandibular fossa. Chewing, biting, and grinding may combine all of the above movements.

The remaining bones of the lateral aspect of the skull are the parietal, the temporal, and the occipital.

Note the *superior temporal line* which arises from the zygomatic process of the frontal bone, arches backward over the side of the frontal and parietal, and then curves forward along the supramastoid crest to the base of the zygomatic process. A less conspicuous inferior temporal line parallels it. Inferior to the temporal lines the *squamous suture* can be seen, which articulates the *squamous portion* of the temporal with the parietal bone.

The *temporal fossa,* more clearly shown here, is shallow posteriorly, but dips anteriorly to its deepest point behind the frontal process of the zygomatic bone. It houses the temporal muscle which inserts on the coronoid process of the mandible and, when contracted, elevates the mandible to close the mouth.

The other part of the temporal bone seen in the lateral aspect is the auditory-mastoid area (Fig. 8–15). The *external auditory meatus* is the

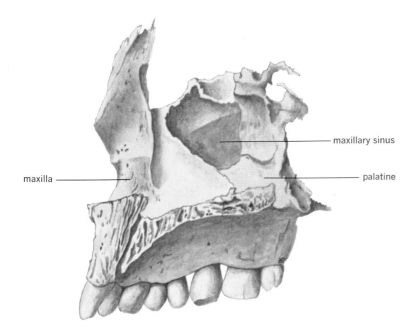

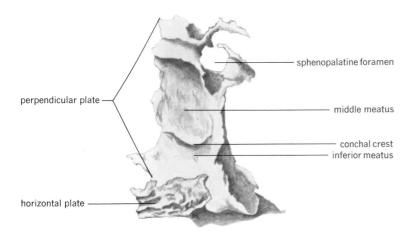

FIGURE 8–12. The right palatine bone in situ, medial view.

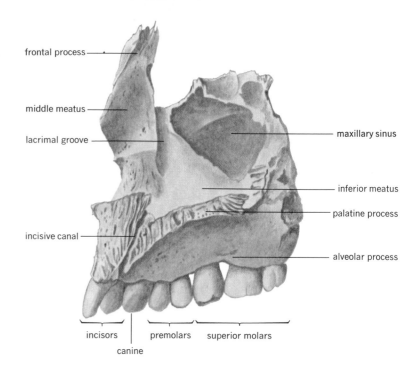

frontal process

middle meatus

lacrimal groove

incisive canal

maxillary sinus

inferior meatus

palatine process

alveolar process

incisors

canine

premolars

superior molars

FIGURE 8–13. Medial view of the right maxilla.

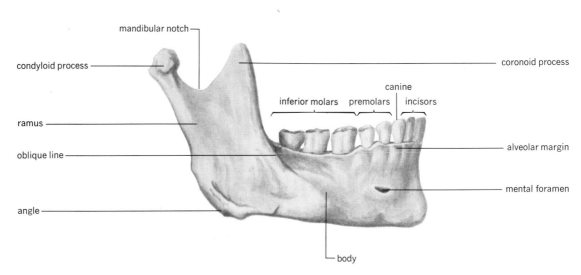

mandibular notch

condyloid process

ramus

oblique line

angle

coronoid process

canine

inferior molars   premolars   incisors

alveolar margin

mental foramen

body

FIGURE 8–14. The lateral and medial view of the mandible.

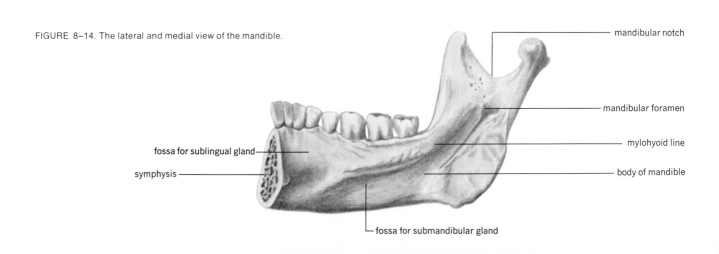

mandibular notch

mandibular foramen

mylohyoid line

body of mandible

fossa for sublingual gland

symphysis

fossa for submandibular gland

conspicuous opening which leads into the middle ear cavity. It lies just behind the mandibular fossa. Projecting forward and downward from the external auditory meatus is the slender, long *styloid process* which serves for the attachment of certain tongue, pharynx, and neck muscles. Posterior to the meatus is the conspicuous, blunt, triangular *mastoid process,* the interior of which is hollowed out by the mastoid air cells which communicate with the middle ear cavity. At the posterior side of the lateral aspect, the confluence of the squamosal and lambdoidal sutures is seen, and the occipital bone. The latter is better described from the basal aspect of the skull.

The base of the skull. The base of the skull is best seen after removal of the mandible (Fig. 8–16). As seen in this aspect, it is oval in outline. The maxillary bones are seen anteriorly bearing the teeth in the sockets of their alveolar margins. Within the horseshoe formed by the teeth is the *hard palate,* which is composed of the horizontal palatine processes of the maxillae and the horizontal portion of the palatine bones. In the midline anteriorly is the *incisive fossa;* posteriorly, the hard palate forms part of the border of the *choanae or internal nares.* The vomer divides the choanae into right and left portions and forms a part of the bony nasal septum. The body of the sphenoid and its medial pterygoid lamina complete the walls of the choanae. Lateral to the medial pterygoid lamina is the *pterygoid fossa,* bordered laterally by the lateral pterygoid lamina. The infratemporal fossa lies between the lateral pterygoid lamina and the zygomatic arch and is occupied in life by muscles of mastication. Lateral and superior to the infratemporal fossa is the *temporal fossa.* The great wing of the sphenoid spreads laterally and superiorly to form a part of the wall of the temporal fossa and of the orbit. Posteriorly, the body of the sphenoid articulates with the body of the occipital bone. In the living subject the area between the choanae and in front of the occipital bone is occupied by the pharynx.

Laterally, the sphenoid and occipital bones articulate with the temporal bone. The mandibular fossa can be seen at the base of the zygomatic process and in front of it the articular eminence. The squamous portion of the temporal is best studied in the lateral view of the skull, but the *petrous portion* pushes mediad in between the sphenoid and the occipital. It is the hardest bone in the body and houses the essential parts of the ear (Fig. 8–17). The *carotid canal* opens on its inferior surface. A long slender spine, the styloid process, projects downward from the petrous temporal. It is frequently broken off in preserved skulls. Posterior and lateral to the styloid process is the *mastoid portion* of the temporal with its mastoid process which contains the *mastoid air cells.*

The *occipital bone* forms the most posterior part of the base of the skull (Fig. 8–18). From its articulation with the sphenoid it spreads posteriorly and laterally, articulating with the petrous and mastoid portions of the temporal bones. It has a large opening, the *foramen magnum,* through which the brain stem passes to continue into the spinal cord. Two large *occipital condyles* lie to either side of the anterior border of the foramen magnum. From the foramen magnum the occipital turns gradually and then abruptly upward, articulating, as noted previously, along the lambdoidal suture with the parietal bones. Medially this portion has an *external occipital crest* which ends superiorly in the external occipital protuberance. The external surface of the occipital carries several *nuchal lines.*

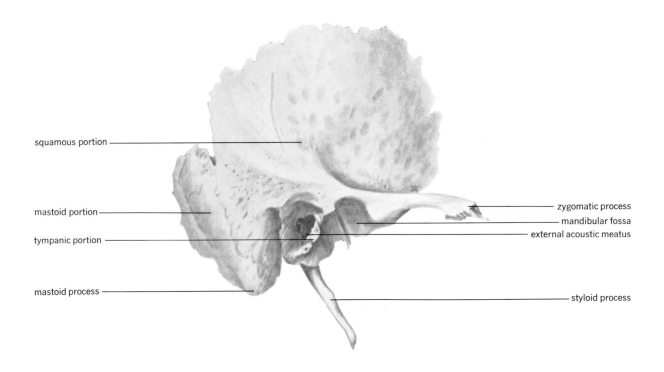

squamous portion

mastoid portion

tympanic portion

mastoid process

zygomatic process

mandibular fossa

external acoustic meatus

styloid process

FIGURE 8–15. The right temporal bone. (*above*) Lateral view. (*below*) Medial view.

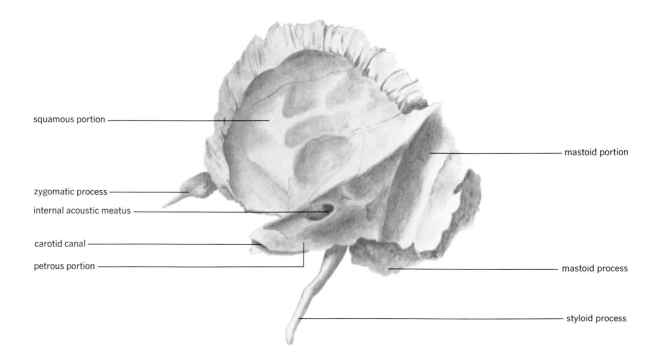

squamous portion

zygomatic process

internal acoustic meatus

carotid canal

petrous portion

mastoid portion

mastoid process

styloid process

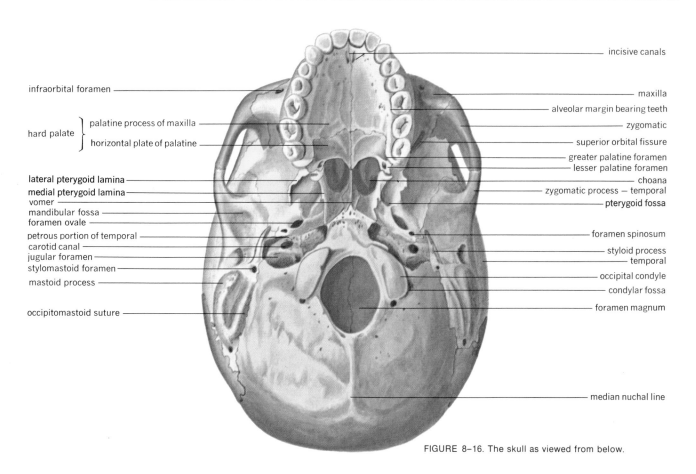

incisive canals

infraorbital foramen

maxilla

alveolar margin bearing teeth

hard palate {
  palatine process of maxilla
  horizontal plate of palatine
}

zygomatic

superior orbital fissure

greater palatine foramen

lesser palatine foramen

lateral pterygoid lamina

medial pterygoid lamina

vomer

mandibular fossa

foramen ovale

petrous portion of temporal

carotid canal

jugular foramen

stylomastoid foramen

mastoid process

occipitomastoid suture

choana

zygomatic process – temporal

pterygoid fossa

foramen spinosum

styloid process

temporal

occipital condyle

condylar fossa

foramen magnum

median nuchal line

FIGURE 8–16. The skull as viewed from below.

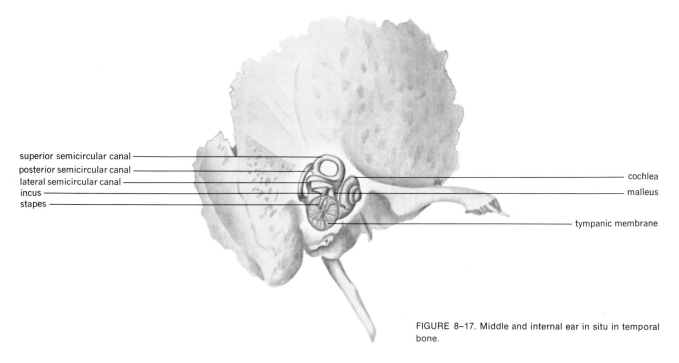

superior semicircular canal

posterior semicircular canal

lateral semicircular canal

incus

stapes

cochlea

malleus

tympanic membrane

FIGURE 8–17. Middle and internal ear in situ in temporal bone.

TABLE 8–1

Foramina of the Skull of Man

Foramina of facial bones	Location	Structures passing through
1. Incisive (Stensen; Scarpa, sometimes)	Posterior to incisor teeth in hard palate	Stensen, anterior branches of descending palatine vessels; Scarpa, nasopalatine nerves
2. Greater palatine	Palatine bones at posterior-lateral angle of hard palate	Greater palatine nerve
3. Lesser palatine	Palatine bones at posterior-lateral angle of hard palate	Lesser palatine nerves
4. Supraorbital (or notch)	Frontal bone, above orbit	Supraorbital nerve and vessels
5. Infraorbital	Maxilla, body of	Infraorbital nerve and vessels
6. Zygomaticofacial	Zygomatic bone, malar surface	Zygomaticofacial nerve and vessels
7. Zygomaticotemporal	Zygomatic bone, temporal surface	Zygomaticotemporal nerve
8. Zygomaticoorbital	Zygomatic bone, orbital surface	Zygomaticotemporal and zygomaticoorbital nerves
9. Mental	Mandible, lateral anterior surface, inferior to second premolar	Mental nerve and vessels
10. Mandibular	Mandible, about center of medial side of ramus	Inferior alveolar vessels and nerves

Foramina of cranial bones	Location	Structures passing through
1. Olfactory	Ethmoid, cribriform plate	Olfactory nerves (I)
2. Optic	Sphenoid, superior surface	Optic nerves (II); ophthalmic arteries
3. Superior orbital fissure	Sphenoid, between small and great wings	Oculomotor (III); trochlear (IV); trigeminal (V), ophthalmic branch; abducens (VI); orbital branches of middle meningeal artery; superior ophthalmic vein; branch of lacrimal artery
4. Inferior orbital fissure	Sphenoid; zygomatic; maxilla; palatine	Trigeminal (V), maxillary nerve; infraorbital vessels
5. Rotundum	Sphenoid, greater wing	Maxillary nerve (V)
6. Ovale	Sphenoid, greater wing	Mandibular nerve (V); accessory meningeal artery; lesser petrosal nerve
7. Spinosum	Sphenoid, greater wing	Mandibular nerve, recurrent branch (V); middle meningeal vessels
8. Lacerum	Sphenoid, greater wing and body; apex of petrous temporal; base of occipital	Meningeal branch of the ascending pharyngeal artery; internal carotid artery
9. Internal acoustic meatus	Temporal, petrous portion	Facial nerve (VII); acoustic nerve (VIII); internal auditory artery; nervus intermedius
10. Jugular	Temporal, petrous portion; occipital	Glossopharyngeal (IX), vagus (X); accessory (XI); internal jugular vein
11. Hypoglossal canal	Occipital	Hypoglossal nerve (XII); meningeal artery
12. Condyloid canal	Occipital	Vein from transverse sinus
13. Carotid canal	Temporal, petrous portion	Internal carotid artery
14. Stylomastoid	Temporal, between mastoid and styloid processes	Facial nerve (VII)
15. Mastoid	Temporal, mastoid portion	An emissary vein
16. Foramen magnum	Occipital	Medulla oblongata and its meninges; accessory nerves (XI); vertebral arteries

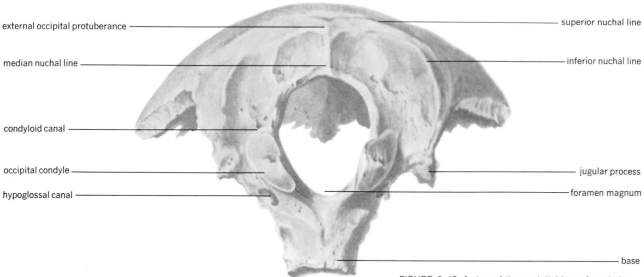

external occipital protuberance

median nuchal line

condyloid canal

occipital condyle

hypoglossal canal

superior nuchal line

inferior nuchal line

jugular process

foramen magnum

base

FIGURE 8-18. A view of the occipital bone from below.

A number of foramina are easily recognized in this view of the base of the skull. They provide for the transmission of blood vessels and nerves. These should be noted on the drawings of the skull and studied in Table 8–1.

The cranial cavity (Figs. 8–2 and 8–19).   The internal surface of the cranial vault is concave in all directions and its markings reflect structures on the surface of the brain. The *superior sagittal sinus* leaves a groove along the midline. *Middle meningeal arteries* leave an impression of themselves as they course upward across the parietal bones. Impressions of lesser vessels may also be observed as can the coronal and sagittal sutures.

Looking into the floor of the cranial cavity one sees three large fossae arranged like steps. The highest step, the anterior cranial fossa, houses the large frontal lobes of the cerebrum and lies above the orbits. The central step, the middle cranial fossa, receives the temporal lobes of the cerebrum. The lowest step, the posterior cranial fossa, contains in life the medulla oblongata and the cerebellum. Each of these fossae reflects by the markings on its walls the form of the brain and the blood vessels which in life rest against these walls.

The walls of the *anterior cranial fossa* are composed largely of the frontal bone, its orbital plates forming a large part of the floor of the fossa. Between the orbital plates and fitting into the ethmoid notch of the frontal bone is the *cribriform plate* of the ethmoid. These plates are perforated by numerous foramina which carry the olfactory nerve filaments from the olfactory cavities. Arising between the troughs of the cribriform plates is a sharp upward projection, the *crista galli,* which serves as a point of attachment for the dura mater of the brain (Fig. 8–9). On the midline of the anterior vertical wall of the anterior cranial fossa is a sharp frontal crest which affords attachment for the falx cerebri, a dural membrane which lies between right and left cerebral hemispheres of the brain.

The posterior part of the floor of the anterior cranial fossa and its

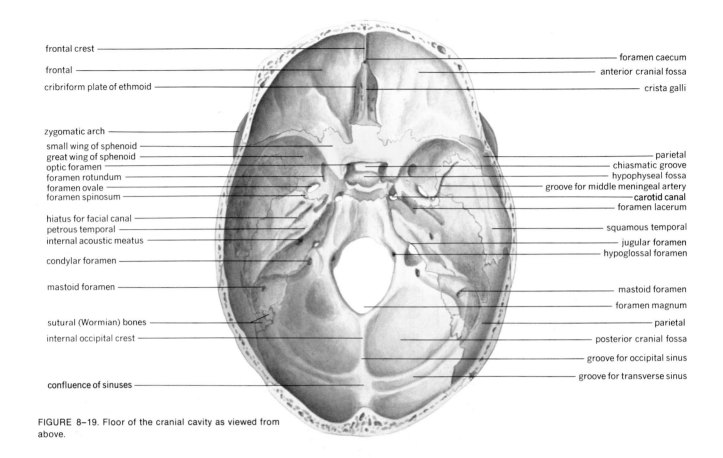

frontal crest
frontal
cribriform plate of ethmoid
zygomatic arch
small wing of sphenoid
great wing of sphenoid
optic foramen
foramen rotundum
foramen ovale
foramen spinosum
hiatus for facial canal
petrous temporal
internal acoustic meatus
condylar foramen
mastoid foramen
sutural (Wormian) bones
internal occipital crest
confluence of sinuses

foramen caecum
anterior cranial fossa
crista galli
parietal
chiasmatic groove
hypophyseal fossa
groove for middle meningeal artery
carotid canal
foramen lacerum
squamous temporal
jugular foramen
hypoglossal foramen
mastoid foramen
foramen magnum
parietal
posterior cranial fossa
groove for occipital sinus
groove for transverse sinus

FIGURE 8–19. Floor of the cranial cavity as viewed from above.

posterior sharp margin are formed by the body and lesser wings of the sphenoid bones. To either side of the midline this margin of the sphenoid projects backward, overhanging the middle cranial fossa as the *anterior clinoid processes* (Fig. 8–20). The optic canals which transmit the optic nerves emerge from under these processes. They lead into the chiasmatic groove where a partial crossing over of the optic nerves takes place.

The anterior cranial fossa terminates abruptly at its posterior margin. Behind and below it is the *middle cranial fossa.* The medial portion of the middle cranial fossa is formed by the elevated *body* of the sphenoid bone (Fig. 8–20). This is often called the *sella turcica* or *Turk's saddle.* A deep fossa in its superior surface is the *hypophyseal (pituitary) fossa* and houses in life the hypophysis or pituitary gland. The hypophyseal fossa is bounded anteriorly by the *tuberculum sellae,* from the posterior border of which project small *medial clinoid processes.* The posterior wall of the fossa is the *dorsum sellae* bearing *posterior clinoid processes.*

The medial part of the fossa drops off laterally into deep cups which are bounded anteriorly by the *great wings of the sphenoid,* laterally by the squamous portion of the temporal, and posteriorly by the petrous temporal. Five prominent foramina are found in the floor and walls of the middle cranial fossa. Anteriorly and medially and under the anterior clinoid processes are the *superior orbital fissures* which transmit the oculomotor (III), trochlear (IV), ophthalmic division of the trigeminal (V), and the abducens (VI) nerves into the orbits. Below the orbital fissures are the *foramina rotunda* which transmit the maxillary (V) nerves and, posterior and lateral to these, the *foramina ovale* for the passage of the mandibular (V) nerve. The last foramina in this row are the small *foramina spinosae* which carry the middle meningeal arteries and the recurrent branches of the mandibular nerves. Medial to this row of foramina and to each side of the sella turcica are shallow carotid grooves and, below them, the *foramina lacerum* which transmit the internal carotid arteries and the cavernous sinuses. (Table 8–1.)

The *posterior cranial fossa* is the largest and deepest of the three cranial fossae. It is limited anteriorly by the dorsum sellae of the sphenoid bone, the base of the occipital, and the petrous portions of the temporal bones. The mastoid portions of the temporals, the mastoid angles of the parietals, and the occipitals limit the fossa laterally while the occipital completes its posterior wall. The large foramen magnum

FIGURE 8–20. Superior and posterior view of sphenoid.

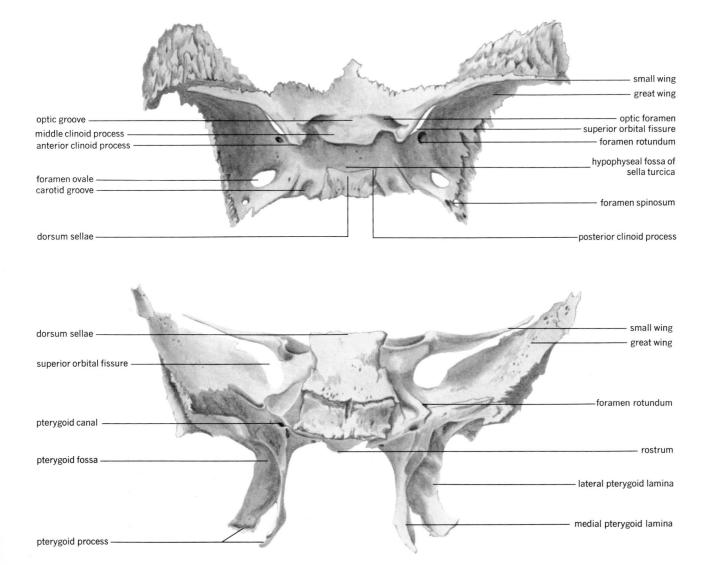

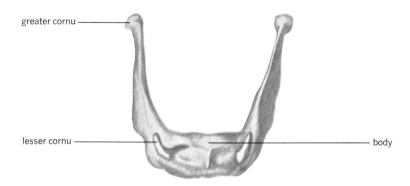

greater cornu

lesser cornu

body

FIGURE 8–21. Anterior view of the hyoid bone.

occupies the central portion of the floor of the fossa. From its posterior border an *internal occipital crest* extends along the midline of the occipital bone and gives attachment to the falx cerebri. At the end of the occipital crest is the *internal occipital protuberance.* Grooves for venous sinuses are prominent in the walls of the posterior cerebral fossa. The groove for the *occipital sinuses* follows the internal occipital crest. At the internal occipital protuberance the grooves for occipital, sagittal, and transverse sinuses come together. The *transverse sinuses* go laterally from the protuberance and then downward and medially to end at the *jugular foramina* which lie between the base of the occipital and the petrous portion of the temporal. *Mastoid foramina* and *condyloid canals* may be seen entering the lower part of the grooves for the transverse sinuses. A large opening is seen in the posterior part of the petrous temporal, the *internal acoustic meatus,* which transmits the facial and vestibulocochlear nerves and auditory arteries.

The *hyoid* (Fig. 8–21) or tongue bone lies above the larynx and is suspended from the tips of the styloid processes of the temporal bone by the stylohyoid ligament. It is composed of a transverse *body* and *greater* and *lesser cornua.* It serves for the attachment of a number of muscles of the tongue, the neck, and pharyngeal regions.

THE VERTEBRAL COLUMN

Definition and function (Fig. 8-22)

The vertebral column or spine is a part of the axial skeleton. It is a strong, flexible rod which protects the spinal cord, gives support to the head, to the thorax, and through it to the superior limb, and to the pelvic girdle and inferior limb. It provides protection for the vital organs in the thorax and for structures within the pelvic cavity. It serves as a base for muscle attachment and is a key to the posture of the individual. Its *intervertebral foramina* transmit the spinal nerves.

The vertebral column is composed of a series of 33, or sometimes 34, vertebrae which are arranged in the following groups: 7 cervical, 12 thoracic, 5 lumbar, 5 sacral, and 4 to 5 coccygeal. In the adult, the sacral vertebrae are fused into one bone, the sacrum, and the coccygeal into one bone, the coccyx. Counted on this basis the vertebral column contains 26 individual bones. Also an important part of the column are

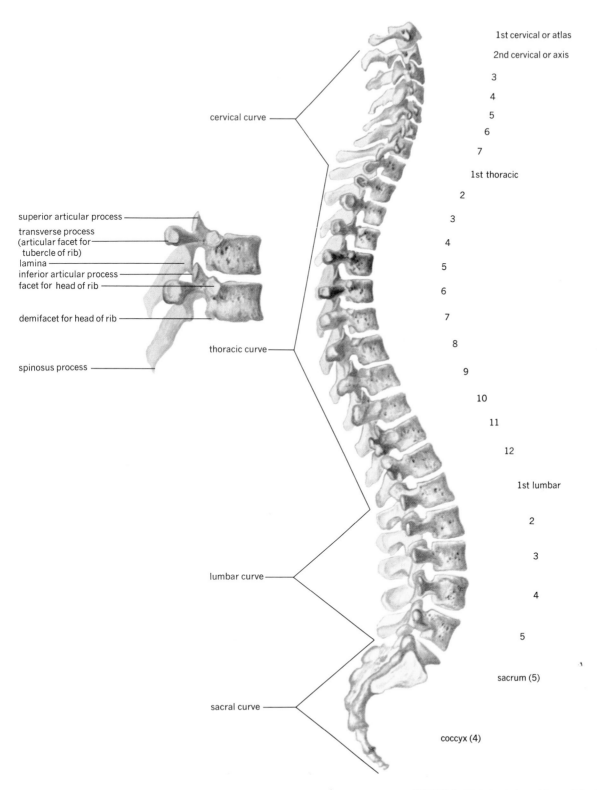

1st cervical or atlas

2nd cervical or axis

3

4

5

6

7

1st thoracic

2

3

4

5

6

7

8

9

10

11

12

1st lumbar

2

3

4

5

sacrum (5)

coccyx (4)

cervical curve

superior articular process

transverse process
(articular facet for
tubercle of rib)

lamina

inferior articular process

facet for head of rib

demifacet for head of rib

spinosus process

thoracic curve

lumbar curve

sacral curve

FIGURE 8–22. Lateral view of the vertebral column.

the *fibro-cartilaginous intervertebral discs* which alternate with the vertebrae and lie between their bodies.

Curvatures of the spine (Fig. 8-22)

Viewed from the lateral side, the vertebral column shows four curves which are alternately convex and concave anteriorly. In the fetus these curves do not exist. Instead, there is one large curve which is concave anteriorly. As a child learns to hold his head erect (at about the third month), the cervical vertebrae are lifted upward and a *cervical curve* gradually develops. As the child pulls itself up to the erect position and learns to walk, an anterior convexity develops in the lumbar region, the *lumbar curve.* In the thoracic and sacrococcygeal regions the curvatures retain the anterior concavity of the fetus. The *thoracic and sacral curves* are for this reason referred to as *primary curves;* the *cervical and lumbar curves,* having been modified from the fetal condition, are *secondary curves.* If the vertebral column is viewed from the anterior or posterior aspect it is normally without curves although slight curvatures to the right are quite common.

Abnormal curvatures of the spine are sometimes seen. A marked curvature to the right or left is called *scoliosis.* It may be caused by muscular paralysis on one side of the spine, by poor posture, or by disease processes which deteriorate the bone. An excessive posterior convexity of the spine is called *kyphosis* and usually involves the thoracic region giving a "hump-backed" appearance. There may be an excessive lumbar curvature or sway-back condition known as *lordosis.* These abnormal curves may be caused by tuberculosis of the vertebrae, causing a softening of the bodies of the vertebrae so that they crush under the influence of the weight of the body. The column then tends to telescope or fold, exaggerating otherwise normal curves. Other disease processes affecting bone may cause the same problems; poor posture, malnutrition (rickets), and wearing of improper shoes may be causative factors.

Components of a typical vertebra (Fig. 8–22)

The vertebrae have a common pattern of structure, although they differ in size, shape, and in finer details. Each has evolved adaptive features which determine its function as a part of the vertebral column.

The parts of a typical vertebra are the body (centrum), vertebral arch, and a variety of processes, articulating surfaces, and foramina. The body and vertebral arch form the walls of the *vertebral foramen* through which the spinal cord passes in the living subject.

The *body* forms the anterior and thickest part of the vertebra. Its flattened, rough superior and inferior surfaces give attachment to the fibrocartilaginous intervertebral discs. The circumferences of these surfaces are slightly raised to form a rim. Nutrient foramina are common in the anterior and lateral surfaces, and on the posterior surface facing the vertebral canal are one or two irregular apertures for the passage of the basivertebral veins.

The *vertebral arch* extends posteriorly from the body of the vertebra. It is composed of a stout *pedicle* on each side and flat *laminae* which extend from the pedicles mediad, where they meet on the midline to complete the arch.

Seven processes arise from the vertebral arch. At the junction of the pedicle and lamina on each side a *transverse process* extends laterally. A single *spinous process* extends posteriorly and inferiorly from the posterior midline of the vertebral arch. Two pairs of articular processes extend from the junction of pedicles and laminae, one pair superiorly the other inferiorly. A layer of hyaline cartilage covers their articular surfaces in the living subject.

Regional variations in the vertebral column

While all vertebrae have a common pattern of structure, the vertebrae of each region have their own identifying features. The sizes of the bodies vary, getting larger and stronger in progressing from the superior end of the column to the lower lumbar region. This is a logical arrangement, since the more inferior vertebrae carry more of the body weight.

The intervertebral discs also become gradually thicker and are thickest in the lumbar region. The *vertebral canal,* though fairly uniform in size throughout its length, enlarges slightly and becomes more triangular in the cervical and lumbar regions where it accommodates the cervical and lumbar enlargements of the spinal cord. In the thoracic region, it is rounder and smaller.

The *spinous processes* are short and bifid in most of the cervical region. They become longer and quite narrow in the thoracic region and slant downward so that each process overlaps the one inferior to it like shingles on a roof. In the lower thoracic, and especially in the lumbar region, they become more and more massive and are square as seen from the lateral aspect.

The *transverse processes* of the cervical region are most distinctive in that they have a transverse foramen for the passage of vertebral arteries and veins. In the thoracic region they possess shallow articular facets or demifacets for the articulation of the tubercles of the ribs. In the lumbar region neither facets nor transverse foramina are found and the processes come straight out to the sides.

The *articular processes* vary greatly from the superior to the inferior ends of the spine. They are farthest apart in the cervical region, come quite close together in the thoracic region, and, in the lumbar region, the articulating surfaces of the superior processes turn inward and grasp the outward-turning inferior processes. This prevents rotation of the spine in this region.

Specialization of vertebrae (Fig. 8–23)

Some vertebrae have become highly adapted and serve special functions. Notable among these are the first and second cervical vertebrae, called the atlas and axis, respectively. The *atlas* is a ring of bone with anterior and posterior arches and large lateral masses. The body has been lost and has become attached to the body of the axis as the dens, where it serves as a pivot around which the atlas can rotate and allows the movements involved in shaking the head to mean "no." The lateral masses have on their superior surfaces large concave facets for articulation with the occipital condyles of the occipital bone (the atlantooccipital joint). At this articulation the head can be flexed and extended as

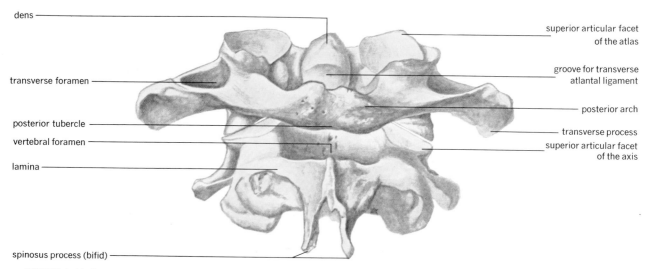

dens

transverse foramen

posterior tubercle

vertebral foramen

lamina

superior articular facet of the atlas

groove for transverse atlantal ligament

posterior arch

transverse process

superior articular facet of the axis

spinosus process (bifid)

FIGURE 8–23. Posterior view of the atlas and axis.

in shaking the head to mean "yes." Some lateral movement is also allowed. Circular, flat (or slightly concave) articular facets are found on the inferior surfaces of the lateral masses which articulate with the axis. The transverse processes of the atlas are wide and each contains a large transverse foramen.

The *axis* is characterized by the prominent *dens* mentioned above, and by a prominent bifid spinous process and small transverse processes. Its body is prolonged downward anteriorly where it overlaps the body of the third cervical vertebra.

The *seventh cervical vertebra,* or *vertebra prominens,* is marked by its large, nonbifid spinous process which serves for attachment of the ligamentum nuchae.

The *sacrum* is composed of five modified and fused sacral vertebrae and serves as a strong foundation for the attachment of the pelvic girdle (Fig. 8–24). It is triangular in shape, and on its concave pelvic surface are four transverse lines that mark the joining of the bodies of the vertebrae. Four pairs of large foramina, the *anterior sacral foramina,* are seen here. The posterior surface is convex and has along its midline the *middle sacral crest* and three or four tubercles representing reduced spinous processes of the vertebrae. Where the laminae and spinous processes do not form, a *sacral hiatus* or opening is left. This occurs, for example, in the region of the fourth and fifth sacral vertebrae. Four *posterior sacral foramina* are seen and near the apex of the sacrum are the *sacral cornua.*

Laterally, the sacrum is broad above where it has a large *auricular surface* for articulation with the ilium to form the sacroiliac joint. Below this, it narrows into little more than a ridge.

The triangular *coccyx* is the most rudimentary part of the vertebral column, and is a vestige of a tail (Fig. 8–24). It is made up of four vertebral components, all of which are hardly recognizable as vertebrae. Only the first one has small transverse processes and coccygeal cornua which project upward to join the sacral cornua. It also has an intervertebral disc and is joined to the sacrum. The remaining coccygeal vertebrae, each smaller than the one above, are little more than reduced bodies or centra.

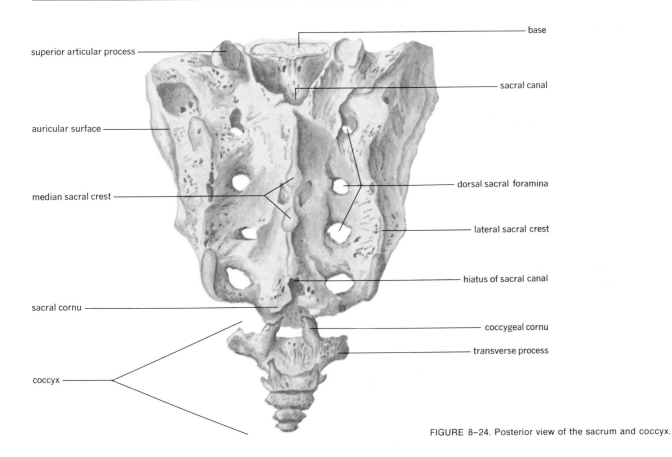

superior articular process

auricular surface

median sacral crest

sacral cornu

coccyx

base

sacral canal

dorsal sacral foramina

lateral sacral crest

hiatus of sacral canal

coccygeal cornu

transverse process

FIGURE 8–24. Posterior view of the sacrum and coccyx.

Articulations and movements of the vertebral column

The articulations within the vertebral column fall into two categories, cartilaginous and synovial. The cartilaginous articulations are of the symphysis type. They consist of pads of fibrocartilage, the *intervertebral discs,* interposed between the bodies of adjacent vertebrae. The cartilages vary in thickness and therefore in resilience; they are thickest in the lumbar region, thinnest in the thoracic. This is one of the factors which influences movement in the vertebral column; the greatest movement being in the lumbar region, the least in the thoracic.

The *synovial articulations* lie between the superior and inferior articulating surfaces of the vertebrae and, as indicated earlier, the relationships of these processes is another determining factor in movement. Since these processes are directed more or less upward and downward in the cervical region they allow, along with the symphyses, a variety of movements: flexion, extension, rotation, abduction, and adduction. In the thoracic region the articular processes face anteriorly and posteriorly and all of the above movements are possible, but more restricted. In the lumbar region the articular processes face outward and inward, one grasping the other in such a way as to prevent rotation. Yet, because the intervertebral discs are so thick and compressible, flexion and extension are greatest in the lumbar region and some abduction and adduction are possible.

The joint between the atlas and axis, the *atlantoaxial joint,* is a special

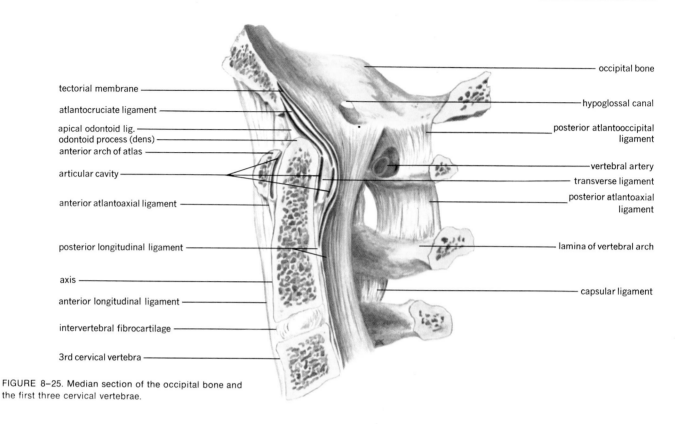

tectorial membrane

atlantocruciate ligament

apical odontoid lig.
odontoid process (dens)
anterior arch of atlas

articular cavity

anterior atlantoaxial ligament

posterior longitudinal ligament

axis

anterior longitudinal ligament

intervertebral fibrocartilage

3rd cervical vertebra

occipital bone

hypoglossal canal

posterior atlantooccipital
ligament

vertebral artery
transverse ligament
posterior atlantoaxial
ligament

lamina of vertebral arch

capsular ligament

FIGURE 8–25. Median section of the occipital bone and the first three cervical vertebrae.

case based upon the modifications of these vertebrae described above (Fig. 8–25).

Articulations between the vertebral column and other skeletal structures also concern us here. The relationship of the atlas to the skull at the *atlantooccipital joint* has been described. Some of the ligaments which maintain it are shown in Fig. 8–25.

The sacrum joins the ilium to form the *sacroiliac joint.* This is partly a fibrous joint and partly a synovial joint; the former giving strength and the latter slight movement. Since the whole weight of the trunk, head, and neck rests on this joint, it is under stress at all times, except when the individual is lying down. Strong ligaments support it as seen in Fig. 8–43.

The thoracic vertebrae give attachment to the ribs at articular facets on their bodies and transverse processes. These are synovial joints allowing the ribs to swing forward and upward.

THE THORAX (Fig. 8-26)

The thorax consists of twelve pairs of ribs, a sternum, and the costal cartilages. The thoracic vertebrae complete the thorax posteriorly. The narrow inlet of the conical-shaped thorax is superior; its broad outlet, which is closed by the diaphragm, is inferior. It is flattened from front to back and in cross section is kidney-shaped since the ribs swing posteriorly from their attachment to the thoracic vertebrae and then forward.

Ribs and costal cartilages (Figs. 8–26 to 8–29)

The first seven, or *true ribs,* attach through their costal cartilages to the sternum. The eighth through the twelfth are *false ribs,* so-called because their costal cartilages do not attach directly to the sternum. The costal cartilages of the eighth through the tenth ribs attach to each other and then to the cartilage of the seventh rib. The eleventh and twelfth ribs are called "floating ribs" because their anterior ends terminate freely in the body wall.

From number one to seven the ribs increase in length; they then taper off to the small twelfth rib. A typical rib, such as number six, consists of a *head* which has two articular facets separated by an interarticular crest. The facets articulate with the demifacets of the bodies of adjacent thoracic vertebrae. Next to the head is a short *neck* which ter-

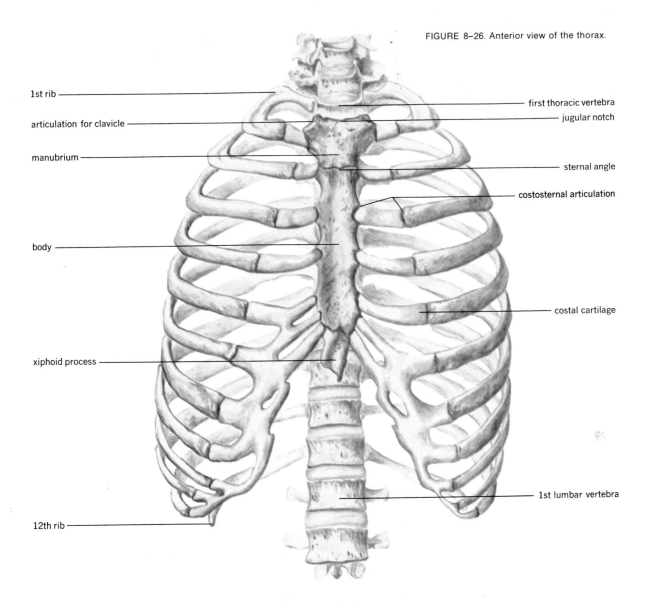

FIGURE 8–26. Anterior view of the thorax.

1st rib

articulation for clavicle

manubrium

body

xiphoid process

12th rib

first thoracic vertebra

jugular notch

sternal angle

costosternal articulation

costal cartilage

1st lumbar vertebra

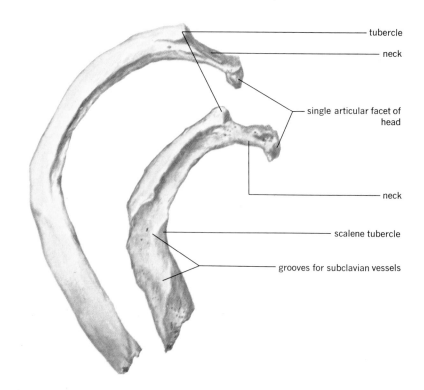

FIGURE 8–27. Superior view of the right first and second ribs.

minates at the *tubercle*. The tubercle is divided into articulating and nonarticulating portions. The lower and more medial of the tubercles articulates with the facet on the transverse process of a thoracic vertebra. The nonarticular tubercle serves for the attachment of a ligament. Beyond the tubercle the rest of the rib constitutes the *body or shaft.* A short distance from the tubercle the shaft bends abruptly to form the *angle.* The superior border of the shaft of the rib is rounded; the inferior border is sharper. The external surface of the rib is convex; the inner surface has a *costal groove* for the passage of the intercostal vessels and nerves. The anterior end of the rib is flattened and has a depression in it for attachment of the costal cartilage.

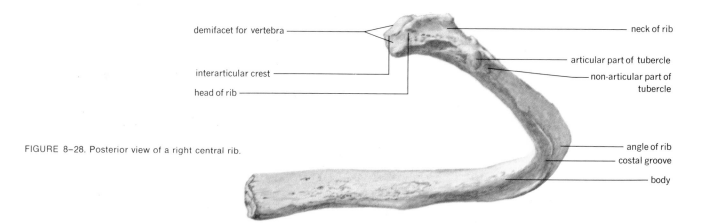

FIGURE 8–28. Posterior view of a right central rib.

Some ribs deviate from the above description. The first rib is usually the shortest and is flattened in the horizontal plane. It has the greatest curvature, but no angle, and the head has a single facet for articulation with the first thoracic vertebra. The tenth, eleventh, and twelfth ribs also have only one articular facet on their heads, and the eleventh and twelfth have no neck or tubercle. The twelfth rib may have no angle and is sometimes shorter than the first rib.

Sternum (Fig. 8–26)

The sternum is located in the midline of the anterior thoracic wall to which it contributes. It is made up of three parts: a superior manubrium, an elongated body, and an inferior xiphoid process. The *manubrium* is triangular in shape and has on its superior surface a depression, the *suprasternal notch,* to each side of which are oval articular surfaces for the proximal ends of the clavicles. On the lateral border, superiorly, there is a facet for articulation of the costal cartilage of the first rib; inferiorly, there is a demifacet for the cartilage of the second rib. Where the manubrium joins the body there is a raised transverse line called the *sternal angle.* The *body* of the sternum is flat, broadest at its lower end, and laterally has demifacets at the superior and inferior ends with four full facets between them. These are for the attachment of ribs two to seven. The xiphoid varies in form and is cartilaginous in early age. It gradually ossifies with accumulating years. It has a demifacet laterally at its superior end which, with the demifacet of the body, serves for attachment of the seventh costal cartilage.

Functions

The thorax provides protection for vital organs such as the heart and lungs. It gives support to the pectoral (shoulder) girdle. It is essential in the breathing process because the inspiration of air is dependent upon the enlargement of the thorax by muscle action. During inspiration, the ribs are lifted forward and upward, the movement taking place at the synovial joints by which the ribs articulate to the thoracic vertebrae.

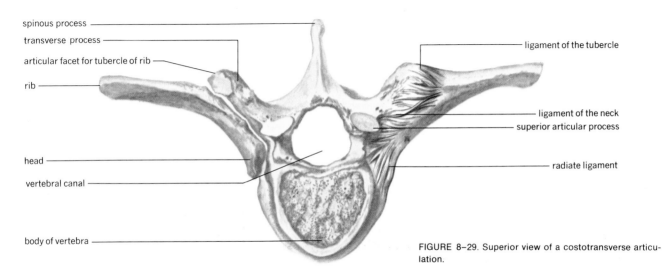

FIGURE 8–29. Superior view of a costotransverse articulation.

The contraction of the diaphragm increases the vertical diameter of the thorax. These movements, by increasing the volume of a closed cavity, reduce intrathoracic pressure and air rushes into the expanded lungs. The expiration of air takes place as muscle action ceases. It is a passive process in normal quiet breathing. The chest wall falls back into its original position; the lungs (since they have elastic tissue) recoil, and air is expelled. Expiration can be made active and more complete by contraction of abdominal muscles which put pressure on the viscera; the viscera, in turn, put pressure on the diaphragm and push it upward.

APPENDICULAR SKELETON

Definition and functions

The appendicular skeleton, composed of pectoral and pelvic girdles and upper and lower extremities, serves the locomotor and manipulatory needs of the body. When vertebrates evolved from the aquatic to the terrestrial environment, paired limbs changed from balancing organs in an environment of water which gave support to the body to organs of support and locomotion. It became necessary to lift the body above the substratum in order to move effectively. Many terrestrial vertebrates, such as salamanders and alligators, still have their paired limbs out to the side; in higher vertebrates, like mammals, they are placed under the torso and so lift it well above the ground for more efficient locomotion. In man, the body became erect and the upper extremities were made available for a variety of uses in the manipulation and modification of the environment. Man's hand is his best tool and, coupled with a superior brain, has enabled him to dominate his world.

As this diversity of functions of the girdles and the upper and lower extremities suggest, there is a corresponding variation in their structure. The *pectoral* or *shoulder girdle* has no articulation with the vertebral column. Through its anterior structure, the clavicle, it articulates with the sternum. Its posterior component, the scapula, rides freely in a complex musculature. The shoulder joint itself is freely movable. The whole complex is relatively weak, but highly versatile in movement. This freedom of movement serves as a good basis for the upper extremity, enabling it to be used in a multitude of ways and positions to serve man's needs.

The *pelvic* (hip) *girdle,* in contrast, is articulated securely to a special fused part of the vertebral column, the sacrum. It gives a strong and stable support to the lower extremity on which the weight of the body is carried and balanced in locomotion (walking, running, and jumping) or just in standing. The hip joint where the lower extremity joins the pelvic girdle, although it is a ball-and-socket or universal joint, is still a strong one, the head of the femur fitting the deep bony acetabulum of the girdle.

Pectoral girdle (Figs. 8–30 to 8–33)

The pectoral girdle consists of two pairs of bones, the posterior scapulae and the anterior clavicles (collar bones). This girdle rests upon the

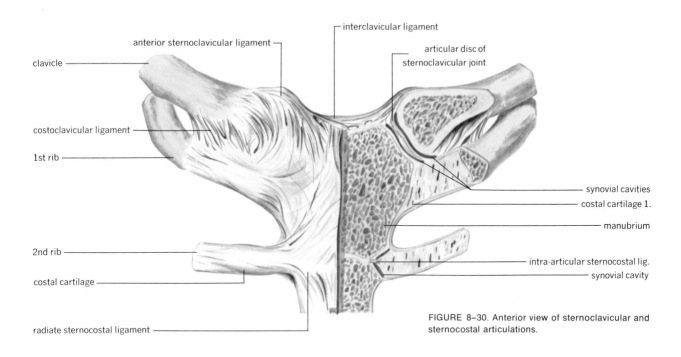

FIGURE 8–30. Anterior view of sternoclavicular and sternocostal articulations.

Labels (clockwise from top):
interclavicular ligament
anterior sternoclavicular ligament
articular disc of sternoclavicular joint
clavicle
costoclavicular ligament
1st rib
2nd rib
costal cartilage
radiate sternocostal ligament
synovial cavities
costal cartilage 1.
manubrium
intra-articular sternocostal lig.
synovial cavity

sternal extremity
acromial extremity
coracoid tuberosity

FIGURE 8–31. Anterior view of the right clavicle.

thorax, but articulates with it only anteriorly where the clavicle joins the manubrium of the sternum (the *sternoclavicular joint*) (Fig. 8–30).

Clavicles (Figs. 8–30 and 8–31). As seen from the superior or inferior aspect the clavicles are shaped like a shallow *S*. Their medial ends are rounded and articulate with the sternum. Their lateral ends are flattened and broad, and articulate with the acromion processes of the scapulae—the *acromioclavicular joints* (Fig. 8–32). On the inferior surface of the bone laterally is a *coracoid tuberosity* for the attachment of the coracoclavicular ligament which crosses over to the coracoid process of the scapula to form a fibrous joint. The clavicle holds the shoulder joint out to the side so that the arm can swing freely; when it is broken the shoulder collapses. Since the clavicle is subcutaneous, it is subject to fracture by blows to the shoulder. Also, because it is the only means by which the pectoral girdle–upper extremity complex is attached to the axial skeleton, it is often broken by falling on the outstretched arm, for example, when one uses the arm to break a fall.

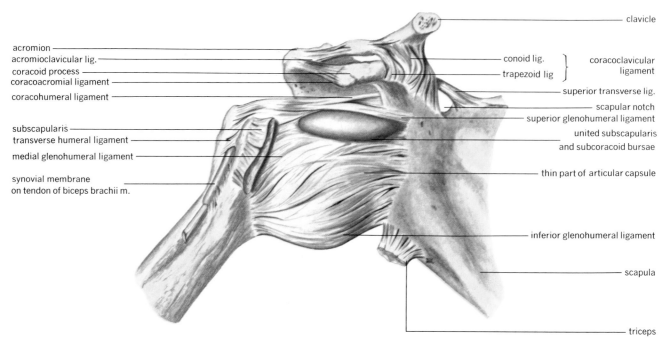

acromion
acromioclavicular lig.
coracoid process
coracoacromial ligament
coracohumeral ligament

subscapularis
transverse humeral ligament
medial glenohumeral ligament

synovial membrane
on tendon of biceps brachii m.

clavicle

conoid lig.
trapezoid lig

coracoclavicular
ligament

superior transverse lig.
scapular notch
superior glenohumeral ligament
united subscapularis
and subcoracoid bursae

thin part of articular capsule

inferior glenohumeral ligament

scapula

triceps

FIGURE 8-32. Anterior view of the joints of the right shoulder.

FIGURE 8-33. Anterior and posterior views of the right scapula.

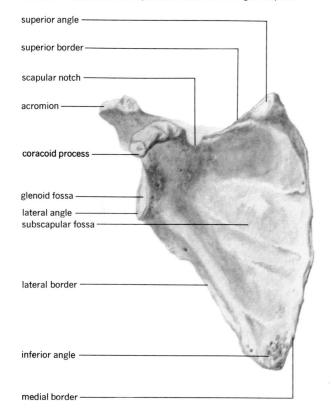

superior angle

superior border

scapular notch

acromion

coracoid process

glenoid fossa
lateral angle
subscapular fossa

lateral border

inferior angle

medial border

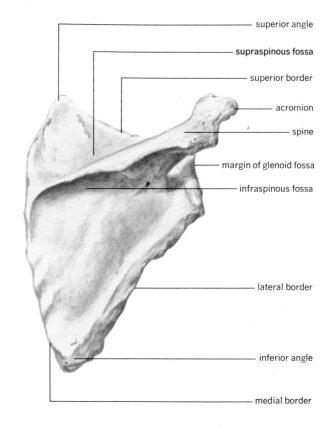

superior angle

supraspinous fossa

superior border

acromion

spine

margin of glenoid fossa

infraspinous fossa

lateral border

inferior angle

medial border

Scapulae (Figs. 8–32 and 8–33). The scapulae lie over the posterior wall of the rib cage between ribs two and seven and about 2 inches from the vertebral column. They consist of a triangular, flattened *body* from which arises posteriorly a *spine,* the lateral end of which is expanded and flattened to form the *acromion.* The body has a thin *medial* or *vertebral border,* a thick *lateral* or *axillary border,* and a *superior border.* It has three angles: superior, inferior, and lateral. At the lateral angle is a smooth, ovoid-shaped depression, the *glenoid fossa,* which in the living subject is deepened by a rim of cartilage, the *glenoid labrum,* and which serves for the articulation of the humerus. A *supraglenoid tubercle* is found above the rim of the fossa, and an *infraglenoid tubercle* lies below. At the beginning of the superior border is a strong, curved *coracoid process.* Just beyond the coracoid process on the superior border is the scapular notch. The superior border, the shortest of the three borders, ends at the superior angle. The anterior or costal surface of the body of the scapula is concave, the *subscapular fossa,* and lies over the rib cage. The posterior surface is convex and is divided by the spine into two unequal parts: a smaller *supraspinous fossa* above the spine and a large *infraspinous fossa* below the spine. These fossae are the surfaces of origin for some of the major muscles of the shoulder.

Upper extremity (Figs. 8–34 to 8–36)

The bones of the upper extremity consist of: the humerus in the arm; the radius and ulna in the forearm; and, in the hand, the carpals (wrist), metacarpals (palm), and phalanges (fingers or digits)—a total of 30 bones.

Humerus (Fig. 8–34).   The humerus is the longest and largest bone of the upper extremity. It articulates at its proximal end with the glenoid fossa of the scapula to form the shoulder joint. At its distal end it joins the radius and ulna at the elbow joint. Its proximal end consists of a head, neck, greater and lesser tubercles, and the intertubercular sulcus or groove. The *head* fits into the glenoid fossa. The neck separates the head of the humerus from the rest of the bone. This is the *anatomical neck.* A second neck, the *surgical neck,* is the narrow part of the humerus just below the tubercles which is frequently the site of fracture. The *intertubercular groove (bicipital groove)* is fairly deep and lies on the upper anterior surface of the humerus between the two tubercles. In the intact body, the tendon of the long head of the biceps brachii muscle lies in this groove.

The *shaft or body* of the humerus is almost cylindrical at its upper end. It gradually becomes triangular, and at its lower end is flattened and broad where it is continuous with the distal extremity of the bone. At about the middle of the shaft, on the anterolateral surface, is a slightly raised, roughened area, the *deltoid tuberosity.* It serves for the attachment or insertion of the deltoideus muscle.

The distal extremity of the humerus is flattened anteroposteriorly and curved slightly forward. It has a double articulating surface consisting of the lateral capitulum and the adjoining medial trochlea. The radial, coronoid, and olecranon fossae and the lateral and medial epicondyles are the other features on the distal end of the humerus. The *capitulum* articulates with the head of the radius; the *trochlea* articulates with the

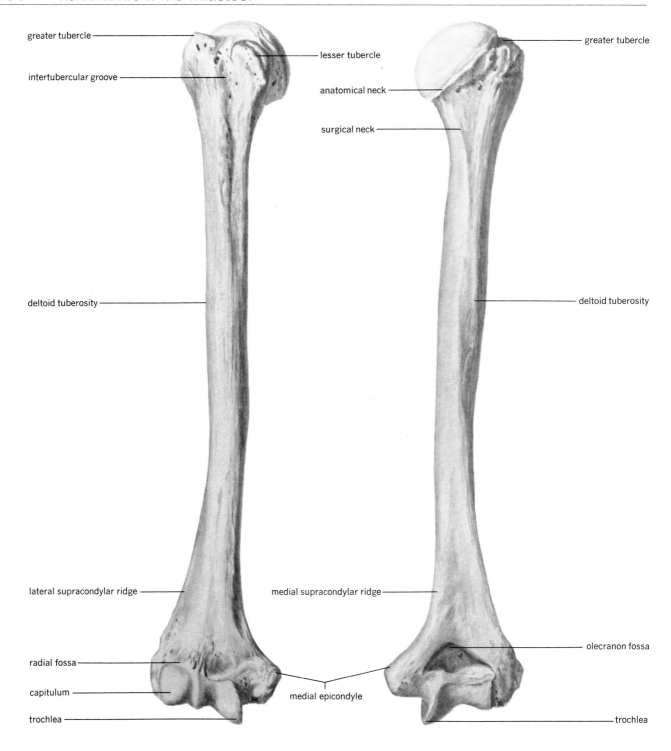

greater tubercle

lesser tubercle

intertubercular groove

greater tubercle

anatomical neck

surgical neck

deltoid tuberosity

deltoid tuberosity

lateral supracondylar ridge

medial supracondylar ridge

olecranon fossa

radial fossa

capitulum

trochlea

medial epicondyle

trochlea

FIGURE 8–34. Anterior and posterior view of the right humerus.

semilunar notch of the ulna, to form the hinge-type elbow joint. The *radial* and *coronoid fossae* are on the anterior surface of the bone; the former receives the head of the radius and the latter receives the coronoid process of the ulna when the forearm is flexed. The *olecranon fossa,* on the posterior side of the bone, receives the olecranon process

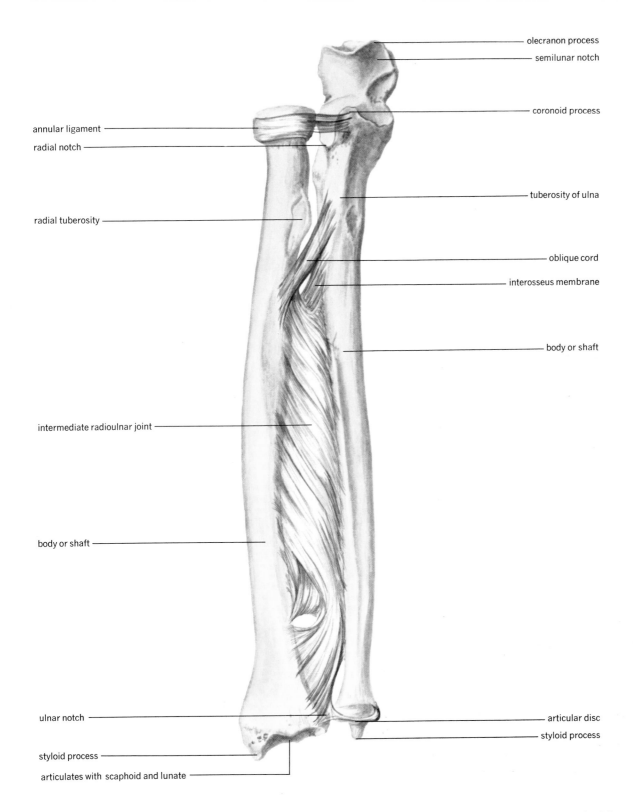

olecranon process

semilunar notch

coronoid process

annular ligament

radial notch

tuberosity of ulna

radial tuberosity

oblique cord

interosseus membrane

body or shaft

intermediate radioulnar joint

body or shaft

ulnar notch

articular disc

styloid process

styloid process

articulates with scaphoid and lunate

of the ulna when the forearm is extended and limits that movement. The *lateral and medial epicondyles* are eminences which lie beside and above the condyles.

Ulna (Fig. 8–35). The ulna is the medial bone of the forearm and is larger and longer than the radius, to which it lies parallel. Its proximal

FIGURE 8–35. Anterior view of proximal, intermediate, and distal articulations of the right forearm.

end is large and highly adapted to form a part of the elbow joint. The bone tapers off through its shaft to the smaller distal end which articulates with the radius.

The proximal end of the ulna is dominated by the *olecranon process* which forms the tip of the elbow. The deep *semilunar notch* serves to receive the trochlea of the humerus. Below the semilunar notch is the *coronoid process,* and in front of it is the *tuberosity* of the ulna, a roughened area for the insertion of the brachialis muscle. The *radial notch* is an articular depression on the upper part of the ulna lateral to the coronoid process. The round head of the radius fits into this notch. The shaft or body of the ulna has a sharp lateral *interosseous border* for the attachment of the *interosseous ligament* which connects with a similar border on the medial side of the shaft of the radius. The distal end of the ulna is small and round with a blunt projection, the *styloid process,* on its posterior side.

Radius (Fig. 8–35). The radius is the lateral bone of the forearm. Its proximal end is round and its circumference fits into the radial notch of the ulna; its superior surface, which is slightly concave, articulates with the capitulum of the humerus. On the medial side just below the proximal end is a raised, roughened area, the *radial tuberosity,* for the insertion of the biceps brachii muscle. The shaft of the radius, almost uniform in diameter through most of its length, widens distally to form a broad, concave inferior surface for articulation with the scaphoid and lunate of the carpus or wrist. The distal end has vertical grooves on its convex posterior surface for the passage of tendons. There is a *styloid process* on the lateral side and a concave *ulnar notch* medially for articulation with the distal end of the ulna.

Carpus (Fig. 8–36). The carpus consists of eight bones arranged in two transverse rows, each with four bones. In the proximal row, from lateral to medial side, are the scaphoid (navicular), lunate, triquetrum (triangular), and pisiform; in the distal row, lateral to medial, are the trapezium (greater multangular), trapezoid (lesser multangular), capitate and hamate. The names of the bones are somewhat suggestive of their shapes. They are bound closely together, but the joints between them are synovial. Little movement is allowed, however, the greatest being in the midcarpal or transverse intercarpal joint which lies between proximal and distal rows of carpals.

Metacarpus (Fig. 8–36). The metacarpus consists of five bones, which are numbered from one to five, starting with the lateral bone. Each has a proximal base, a shaft, and a distal head. The first metacarpal is more freely movable than the other four, which articulate with one another and with carpal bones. The heads of the metacarpals articulate with the digits.

Phalanges (Fig. 8–36). The phalanges make up the digits. There are two phalanges in digit one (the thumb), and three in each of the other four digits. The phalanges are called proximal, middle, and distal phalanges.

Articulations of the upper extremity. The principal joints to be considered here are the shoulder, elbow, and wrist joints. Others will be mentioned and some understanding of them can be gained by careful study of the illustrations.

The *shoulder joint* brings the upper extremity into functional relationship to the shoulder girdle. The rather expansive, rounded head of

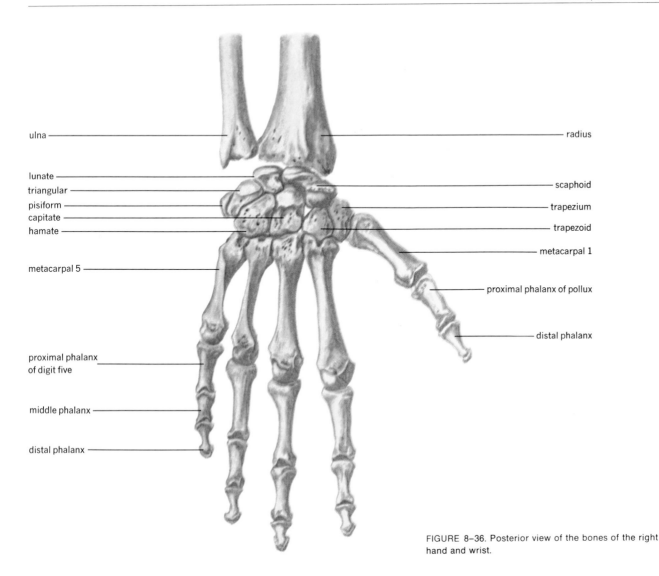

ulna

lunate
triangular
pisiform
capitate
hamate

metacarpal 5

proximal phalanx
of digit five

middle phalanx

distal phalanx

radius

scaphoid

trapezium

trapezoid

metacarpal 1

proximal phalanx of pollux

distal phalanx

FIGURE 8–36. Posterior view of the bones of the right hand and wrist.

the humerus fits into the shallow glenoid fossa of the scapula. Because of the great discrepancy in the surface areas of these two articulating surfaces, only a small part of the head of the humerus is at any time in contact with the glenoid fossa. Its surface is covered in part by the glenoid labrum of the cartilage, which deepens the glenoid fossa in the living subject, and by the *articular capsule* which encloses the joint. The articular capsule is also a loose fitting structure which attaches around the border of the bony glenoid fossa and around the anatomical neck of the humerus allowing a separation of as much as 2.5 cm between the bones. This laxity of the joint, along with the relatively loose attachment of the scapula, allows for the great freedom of movement which characterizes the shoulder and hence the arm. The fibrous articular capsule is lined internally by a *synovial membrane*. The fibrous capsule has one opening for the passage of the tendon of the long head of the biceps brachii. The tendon does not penetrate the synovial membrane however, but pushes against it and becomes surrounded by it.

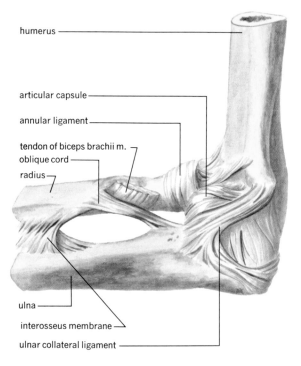

humerus

articular capsule

annular ligament

tendon of biceps brachii m.

oblique cord

radius

ulna

interosseus membrane

ulnar collateral ligament

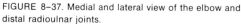

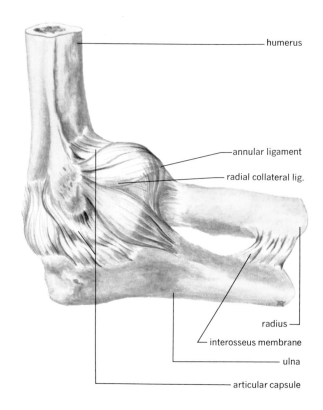

humerus

annular ligament

radial collateral lig.

radius

interosseus membrane

ulna

articular capsule

FIGURE 8–37. Medial and lateral view of the elbow and distal radioulnar joints.

The shoulder joint is further supported by the *coracohumeral* and *glenohumeral ligaments,* and by muscles such as the supraspinatus above; the tendons of the teres minor and infraspinatus posteriorly; the subscapularis anteriorly; and the long head of the triceps below. Flexion, extension, abduction, adduction, rotation, and circumduction are movements possible in the shoulder joint.

The *elbow joint* (Fig. 8–37), broadly considered, includes the joining of the semilunar notch of the ulna with the trochlea of the humerus; the relation of the head of the radius with the capitulum; and the relation of the circumference of the radial head with the radial notch of the ulna. All of these are enclosed under a common capsule and have a common synovial cavity. The radioulnar joint may also be considered separately. The capsule is reinforced by radial and ulnar collateral ligaments.

The *humeroulnar articulation* is a typical hinge joint. The semilunar notch fits closely over the trochlea, allowing only flexion and extension of the forearm. The humeroradial articulation, where the concave superior surface of the radius fits the humeral capitulum, follows the hinge action of the elbow. It also allows the rotation of the radius at the proximal *radioulnar articulation* where the circular head of the radius, held in position by an *annular ligament,* turns against the radial notch of the ulna. This rotation allows supination and pronation of the hand. An *intermediate radioulnar joint* consists of an interosseous membrane which extends between the adjacent borders of the shafts of the radius and ulna. It is a fibrous joint. The distal radioulnar joint is held by the dorsal and palmar radioulnar ligaments. The movement here is pivot-like, the distal end of the radius, at its ulnar notch, riding around the

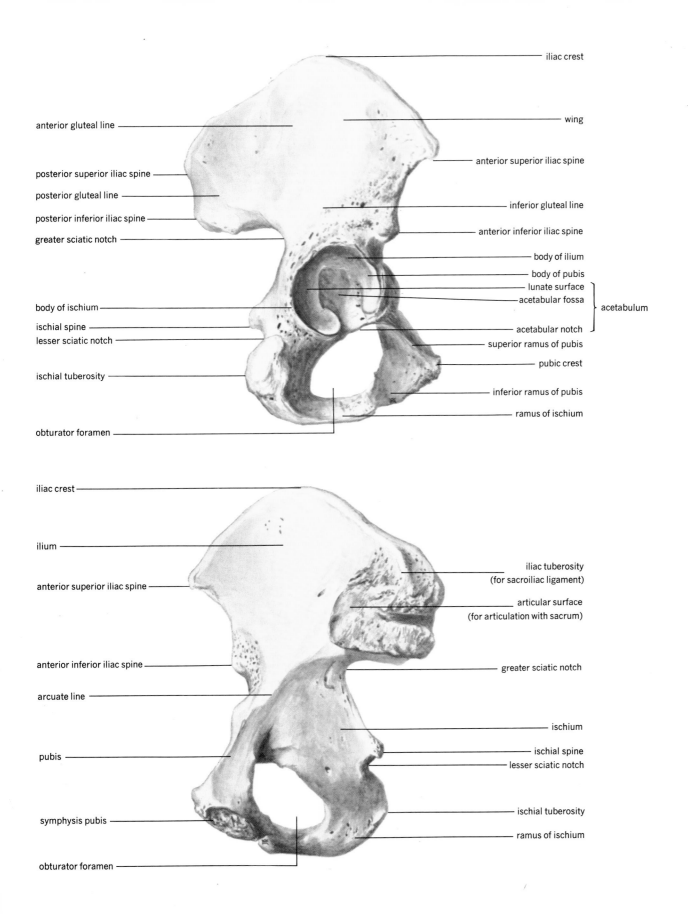

iliac crest

anterior gluteal line

wing

posterior superior iliac spine

anterior superior iliac spine

posterior gluteal line

inferior gluteal line

posterior inferior iliac spine

anterior inferior iliac spine

greater sciatic notch

body of ilium

body of pubis

lunate surface

acetabular fossa

body of ischium

acetabulum

ischial spine

acetabular notch

lesser sciatic notch

superior ramus of pubis

pubic crest

ischial tuberosity

inferior ramus of pubis

ramus of ischium

obturator foramen

iliac crest

ilium

iliac tuberosity
(for sacroiliac ligament)

anterior superior iliac spine

articular surface
(for articulation with sacrum)

anterior inferior iliac spine

greater sciatic notch

arcuate line

ischium

ischial spine

pubis

lesser sciatic notch

ischial tuberosity

symphysis pubis

ramus of ischium

obturator foramen

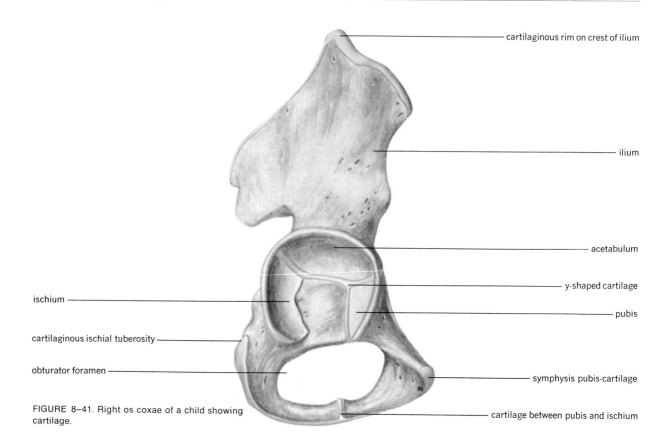

cartilaginous rim on crest of ilium

ilium

acetabulum

y-shaped cartilage

pubis

ischium

cartilaginous ischial tuberosity

obturator foramen

symphysis pubis-cartilage

cartilage between pubis and ischium

FIGURE 8–41. Right os coxae of a child showing cartilage.

The internal surface of the ilium has a smooth concave surface called the *iliac fossa* where the iliacus muscle has its origin. Also, posterior to the fossa is a divided roughened area, the superior portion of which is the *iliac tuberosity* and the inferior portion of which is the *auricular surface,* for articulation with the sacrum.

The external surface of the ilium shows three arched lines: the posterior, anterior, and inferior gluteal lines. Between them the three gluteal muscles have their origin.

Ischium (Figs. 8–40 and 8–41).   The ischium, the lower and posterior part of the pelvic girdle, contributes about two-fifths of the acetabular wall. Posteriorly, on the body of the ischium, there is a prominent *ischial spine* below which is the lesser sciatic notch and below it, in turn, the *ischial tuberosity*. The remaining portion of the ischium is the ramus. It extends anteriorly and slightly upward and joins the inferior ramus of the pubis and, with the pubis, forms the *obturator foramen.*

Pubis (Figs. 8–40 and 8–41).   The pubis is the anterior and inferior part of the pelvis. Its body forms the remaining one-fifth of the acetabular wall. From the body the *superior ramus* of the pubis extends downward to the *symphysis pubis* where the pubic bones of opposite sides are articulated. The *inferior ramus* extends downward and backward from the symphysis to join the ramus of the ischium.

The bony pelvis (Fig. 8–42).   As indicated above, the bony pelvis is composed of the two os coxae or hip bones anteriorly and laterally, and the sacrum and coccyx posteriorly. The pelvis is divided into a greater (false) pelvis above and a lesser (true) pelvis below by a plane which passes through the promontory of the sacrum, the *arcuate (iliopectineal) lines,* and the superior margin of the pubic bones and the symphysis pubis.

The *greater or false pelvis* is formed laterally by the expansive ilia; posteriorly it is formed by the upper portion of the sacrum; and anteriorly there is no bony component, only the abdominal wall. It contains parts of the intestines, gives them support, and also throws some of their weight onto the abdominal wall.

The *lesser* or *true pelvis* is well protected by its walls, which are composed of a part of the ilium, the ischium, pubis, sacrum, and coccyx. It has openings above and below, called the *inlet* and *outlet* of the pelvis, respectively. It contains the urinary bladder anteriorly; the rectum posteriorly; and between these organs, in the female, the vagina and part of the uterus.

Differences between male and female pelves (Fig. 8–42).   The female pelvis is characterized by adaptations to the childbearing function. It is wider, more spacious, and lighter in construction than that of the male. It is shallower than that of the male and its diameters are greater. The ilia flare out more to the sides than in the male, which makes for the broader hips of the female. The inlet of the lesser or true pelvis of the female is almost round, that of the male is heart-shaped. The cavity of the true pelvis is shallower and wider than in the male. The sacrum is shorter, wider, and less curved than in the male; the coccyx is more flexible; the ischial spines and tuberosities turn outward and hence are farther apart; the pubic arch forms an obtuse angle rather than the acute angle of the male. These features all contribute to the wider outlet of the true pelvis in the female. This is an accommodation to the birth of the child.

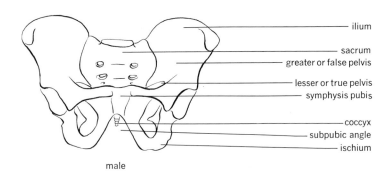

ilium
sacrum
greater or false pelvis
lesser or true pelvis
symphysis pubis
coccyx
subpubic angle
ischium

male

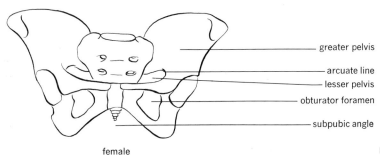

greater pelvis
arcuate line
lesser pelvis
obturator foramen
subpubic angle

female

FIGURE 8–42. Scheme of male and female pelvis.

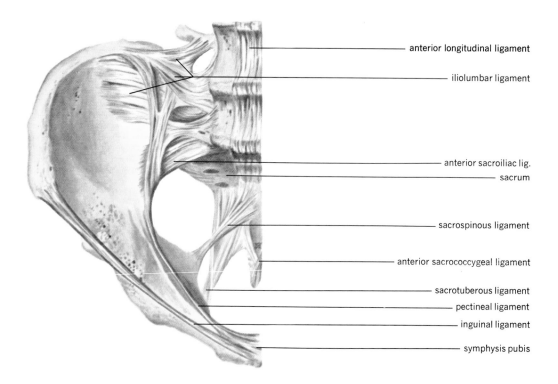

anterior longitudinal ligament

iliolumbar ligament

anterior sacroiliac lig.

sacrum

sacrospinous ligament

anterior sacrococcygeal ligament

sacrotuberous ligament

pectineal ligament

inguinal ligament

symphysis pubis

FIGURE 8–43. Anterior and posterior view of the articulations of the bony pelvis.

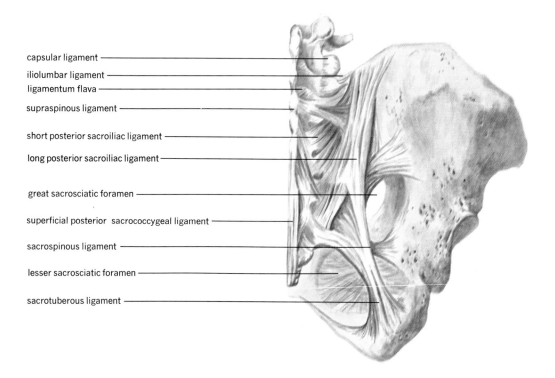

capsular ligament

iliolumbar ligament

ligamentum flava

supraspinous ligament

short posterior sacroiliac ligament

long posterior sacroiliac ligament

great sacrosciatic foramen

superficial posterior sacrococcygeal ligament

sacrospinous ligament

lesser sacrosciatic foramen

sacrotuberous ligament

Articulations of the pelvis (Fig. 8–43). The articulations of the pelvis are the sacroiliac joint, the symphysis pubis, and the sacrococcygeal symphysis. The *sacroiliac joint* is a double structure in that the upper part is a fibrous articulation, the lower part synovial. The joint, however, is almost immovable because of the roughened interlocking surfaces of the articulating bones. It does give resilience to this weight-bearing structure. Iliolumbar, anterior sacroiliac, long and short posterior sacroiliac ligaments support this articulation directly while the sacrospinous and sacrotuberous ligaments help by preventing any abnormal tipping of the pelvis.

The *symphysis pubis,* between the two pubic bones anteriorly, is a cartilaginous joint similar to those between the bodies of the vertebrae. It is held by the superior pubic ligament which lies between the pubic tubercles, and the arcuate pubic ligament which arches inferior to the cartilage and between the inferior pubic rami.

The *sacrococcygeal symphysis* is similar to those joints between the bodies of the vertebrae. Its ligaments are also similar such as the dorsal, ventral, and lateral sacrococcygeal ligaments and the interarticular ligament (Fig. 8–43).

All three of the articulations of the pelvis are influenced by hormones during pregnancy. The ligaments are softened, as are the fibrocartilages of the symphysis joints, resulting in slight looseness and movement of the joints which is an aid to child birth.

The lower extremity (Figs. 8–44 to 8–48)

The lower extremity consists of the femur in the thigh; the patella; the tibia and fibula in the leg; and, in the foot, the tarsals (ankle), metatarsals, and phalanges—a total of 30 bones.

Femur (Fig. 8–44).   The femur is the largest and longest bone of the skeleton. Its head articulates into the acetabulum and the shaft of the bone is directed downward and mediad to approach the femur of the opposite side. This convergence of the femurs brings the knee joints nearer the line of gravity of the body. The female pelvis, being broader than the male, causes the bones to converge even more and thus many women tend to be knock-kneed.

The femur consists of a body or shaft and proximal (upper) and distal (lower) ends. The proximal end consists of a rounded head, neck, and greater and lesser trochanters. The head is nearly spherical with a smooth surface, except for the oval *fovea capitis* near its center where the *ligamentum teres* is attached. The *neck* extends between the head and the shaft of the femur. It angles upward, mediad, and slightly forward. The angle between the neck and the shaft is slightly less acute in the male than in the female due to the greater width of the female pelvis. The greater and lesser trochanters are prominences at the junction of the neck and the shaft. Between them on the anterior surface is a narrow, roughened *intertrochanteric line,* while on the posterior surface of the bone they are connected by a raised *intertrochanteric crest.*

The *shaft* is cylindrical in cross section and smooth except for a vertical, roughened ridge on the posterior surface. The ridge is divided superiorly into three, and inferiorly into two, diverging lines. This is the *linea aspera* and serves for the attachment of a number of thigh muscles. One of the three upper diverging lines runs toward the greater

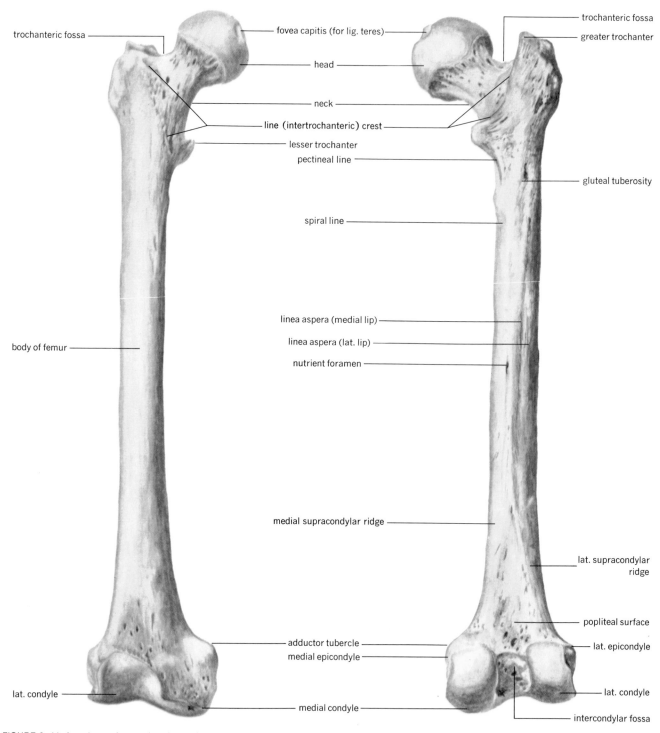

trochanteric fossa — fovea capitis (for lig. teres) — trochanteric fossa

trochanteric fossa — greater trochanter

head

neck

line (intertrochanteric) crest

lesser trochanter

pectineal line — gluteal tuberosity

spiral line

linea aspera (medial lip)

linea aspera (lat. lip)

body of femur — nutrient foramen

medial supracondylar ridge — lat. supracondylar ridge

popliteal surface

adductor tubercle — lat. epicondyle

medial epicondyle

lat. condyle — lat. condyle

medial condyle — intercondylar fossa

FIGURE 8–44. Anterior and posterior views of the right femur.

trochanter and is the *gluteal tuberosity.* The second, directed toward the lesser trochanter, is the *pectineal line.* The third line twists around beneath the lesser trochanter, and is the *spiral line.* The two lower diverging lines are the *lateral* and *medial supracondylar ridges,* which enclose a triangular space, the *popliteal surface.*

The distal or lower expanded end of the femur includes: the *medial* and *lateral condyles* for articulation with the tibia; a depressed area between them posteriorly, the *intercondyloid fossa,* which is limited above by the *intercondyloid line;* and the *lateral and medial epicondyles* lying above the condyles. A small tubercle, the *adductor tubercle,* is found where the medial supracondylar ridge joins the epicondyle. The femur articulates above with the acetabulum of the pelvis, below with the tibia, and anteriorly and distad with the patella.

Patella (Fig. 8–45).   The patella lies in the tendon of the quadriceps femoris muscle. Bones of this type, formed in tendons, are called *sesamoid bones.* It articulates with only one bone, the femur.

Tibia (Fig. 8–46).   The tibia is the medial and larger of the two bones of the leg. It is the weight-bearing bone of the leg, carrying weight to the ankle of the foot. Its entire length is subcutaneous and can be felt along the anterior and medial surface of the leg. It is often called the "shinbone."

The proximal end of the tibia is expanded into *lateral and medial condyles* to form a broad surface for articulation with the large condyles of the femur. The two shallow articular areas of the tibia are separated by an intercondylar eminence and are deepened by flat, semicircular wedges of cartilage, the *lateral and medial semilunar (menisci) cartilages.* On the posterolateral aspect of the lateral condyle is the fibular facet for articulation with the fibula. A *tibial tuberosity* is seen on the anterior surface of the upper end of the bone for attachment of the patellar ligaments.

The shaft or body of the tibia tapers away from the proximal end and widens again distally. It has a raised line on its lateral border, the *interosseous border,* for the attachment of the interosseous membrane which extends between the tibia and fibula.

The distal end of the tibia is shaped to articulate with the talus bone at the ankle joint. It is broad and has a downward projecting *medial malleolus* which grasps the medial side of the talus. On the lateral side of the lower end of the tibia is a fibular notch for articulation with the fibula.

Fibula (Fig. 8–46). The fibula is a slender bone, triangular in cross section, which articulates laterally and posteriorly with the tibia. Its upper end is rounded and has an articular surface medially for articulating with the tibia. Its lower end is triangular and, as the *lateral malleolus,* extends beyond the tibia to form the lateral part of the ankle joint. The lateral and medial malleoli can be easily palpated.

FIGURE 8–45. Anterior and posterior view of the right patella.

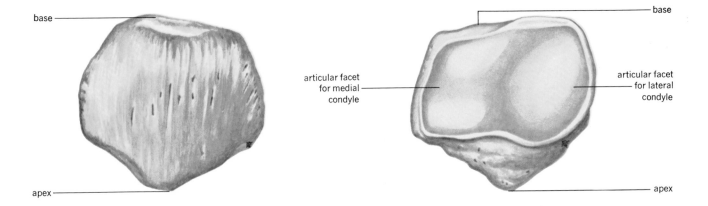

base

apex

base

articular facet for medial condyle

articular facet for lateral condyle

apex

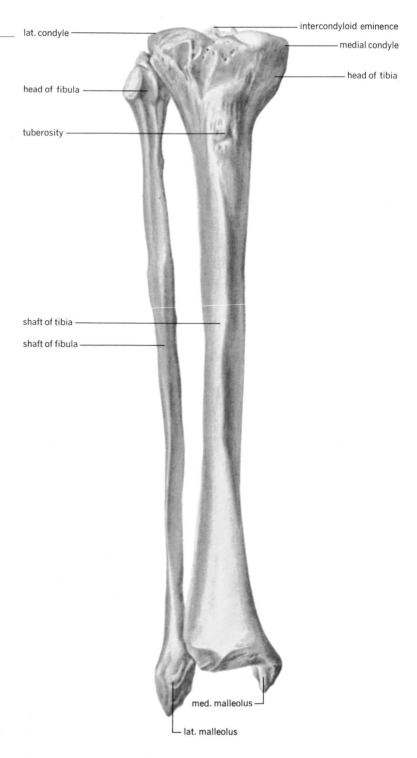

lat. condyle

intercondyloid eminence

medial condyle

head of tibia

head of fibula

tuberosity

shaft of tibia

shaft of fibula

med. malleolus

lat. malleolus

FIGURE 8–46. Anterior view of the right tibia and fibula.

An interosseous membrane connects the twisted shaft of the fibula with the tibia.

Tarsus (Figs. 8–47 and 8–48). The bones of the tarsus, seven in number, can be divided into two groups: (1) the talus and calcaneus on the posterior part of the foot, and (2) the cuboid, navicular, and three cuneiforms anteriorly. The *talus,* formerly called the astragalus, is the only bone of the foot to articulate with the tibia and fibula. These three bones form the hinge-like ankle joint at which the entire foot can be dorsiflexed or plantar flexed. The talus bears the entire weight carried by its extremity. One half of the weight is transferred downward from the talus to the heel, the other half forward to the tarsals which form the keystones of the arch of the foot. They increase the flexibility of the foot, especially its twisting movements.

Metatarsus (Figs. 8–47 and 8–48). The five metatarsals articulate proximally with the cuboid and with the middle, intermediate, and lateral cuneiforms. Their distal articulation is with the proximal phalanges of the toes. They form the forward pillar of the longitudinal arch. The first metatarsal is thicker and shorter than the rest and has two small sesamoid bones under its head.

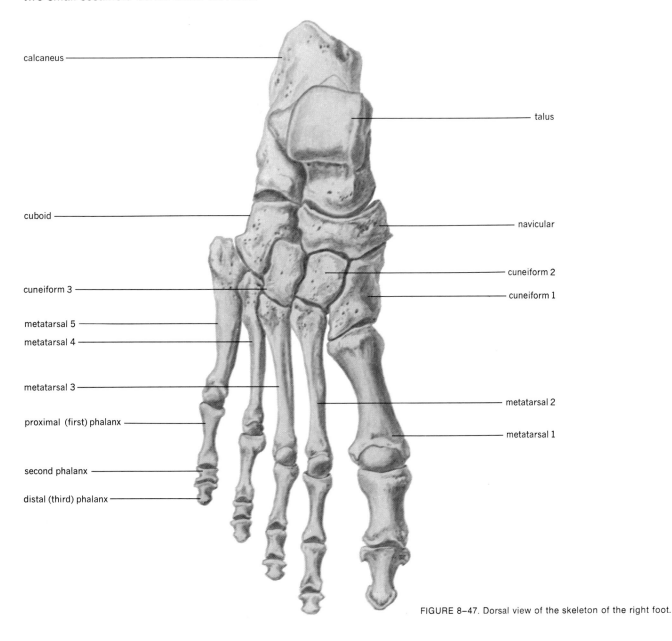

FIGURE 8–47. Dorsal view of the skeleton of the right foot.

FIGURE 8–48. Medial view of the skeleton of the right foot.

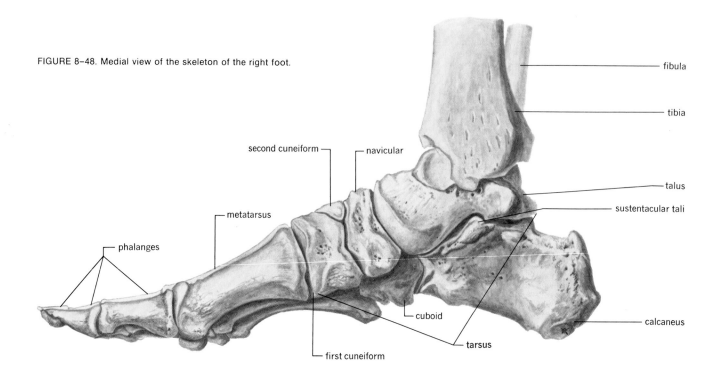

Phalanges (Figs. 8–47 and 8–48).   The arrangement of phalanges is essentially like that of the hand. The first or big toe has two phalanges, quite large and heavy, and the other four toes have three each. The proximal phalanges are the largest; the middle ones are short; and, except for the first toe, the distal ones are short and have expanded ends.

Articulations of the lower extremity (Figs. 8–49 to 8–53).   The *hip joint* is a good example of a synovial joint which serves three important functions. It supports one half of the body weight above the pelvis; it is involved in the transmission of weight; it is responsible for movement. It involves the os coxae and the femur; the former presenting a deep bony socket, the acetabulum, and the latter a rounded head which fits into the acetabulum. The acetabulum is deepened in the living subject by a rim of cartilage, the *acetabular labrum,* and a transverse ligament bridges the gap formed by the acetabular notch. The acetabulum, thus constituted, virtually grasps the head of the femur beyond its greatest diameter and thus serves to strengthen the joint. The fibrous capsule is attached to the rim of the acetabular labrum and encloses most of the neck of the femur. The synovial membrane follows the fibrous capsule and is extremely lax and is reflected upon the neck of the femur. It becomes an acetabular fat pad in the acetabular fossa.

The ligaments of the hip joint are thickenings of the fibrous capsule and are named for the parts they connect; they are the iliofemoral, pubofemoral, and ischiofemoral ligaments. These ligaments serve to check the range of movement of the pelvis on the femurs more than to strengthen a joint which is already sturdy because of its bony components. A *ligament of the femoral head* (ligamentem teres fe-

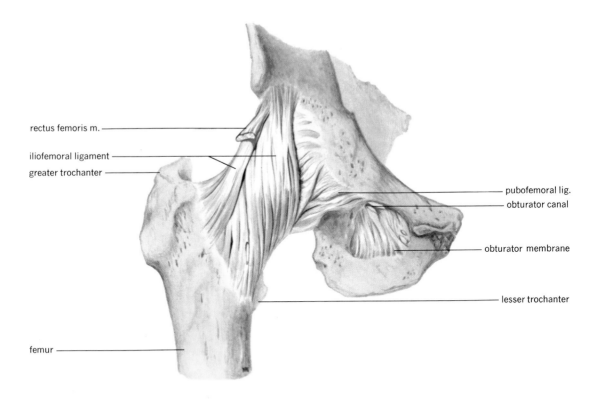

rectus femoris m.

iliofemoral ligament

greater trochanter

pubofemoral lig.

obturator canal

obturator membrane

lesser trochanter

femur

FIGURE 8–49. Anterior view and coronal section of the right hip joint.

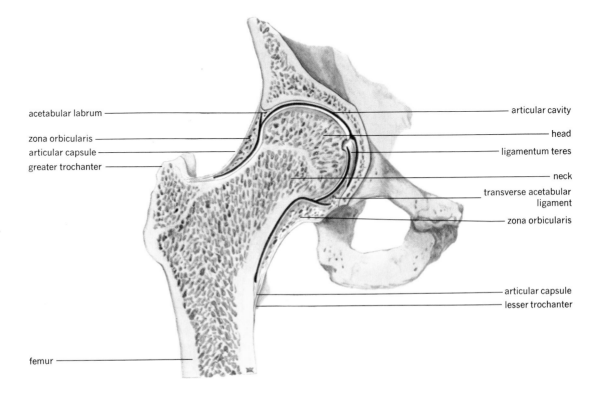

acetabular labrum

zona orbicularis

articular capsule

greater trochanter

articular cavity

head

ligamentum teres

neck

transverse acetabular ligament

zona orbicularis

articular capsule

lesser trochanter

femur

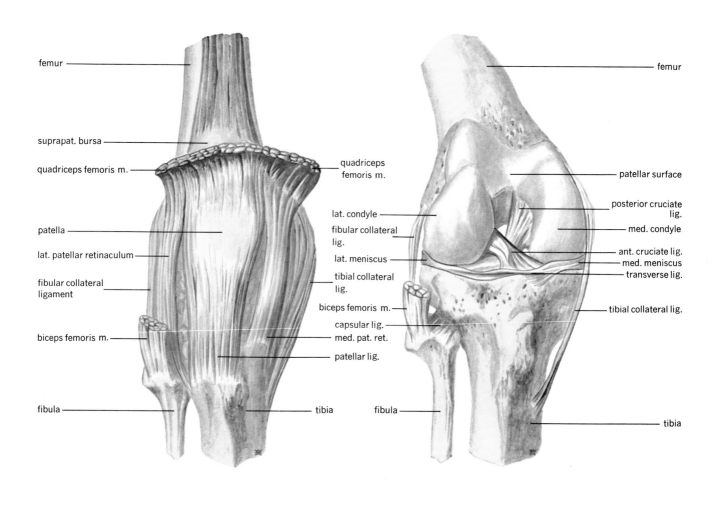

femur

suprapat. bursa

quadriceps femoris m.

patella

lat. patellar retinaculum

fibular collateral ligament

biceps femoris m.

fibula

quadriceps femoris m.

lat. condyle

fibular collateral lig.

lat. meniscus

tibial collateral lig.

biceps femoris m.

capsular lig.

med. pat. ret.

patellar lig.

tibia

femur

patellar surface

posterior cruciate lig.

med. condyle

ant. cruciate lig.

med. meniscus

transverse lig.

tibial collateral lig.

fibula

tibia

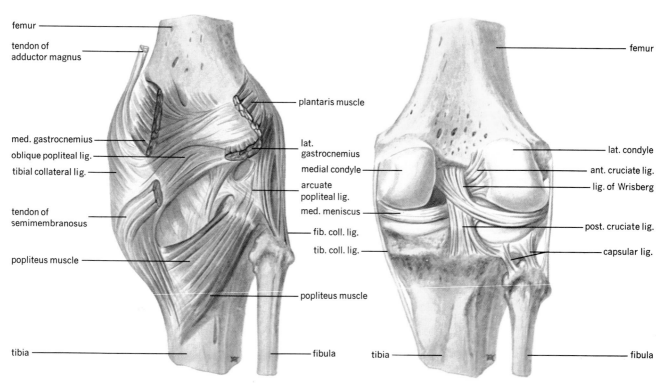

femur

tendon of adductor magnus

med. gastrocnemius

oblique popliteal lig.

tibial collateral lig.

tendon of semimembranosus

popliteus muscle

tibia

plantaris muscle

lat. gastrocnemius

medial condyle

arcuate popliteal lig.

med. meniscus

fib. coll. lig.

tib. coll. lig.

popliteus muscle

fibula

femur

lat. condyle

ant. cruciate lig.

lig. of Wrisberg

post. cruciate lig.

capsular lig.

tibia

fibula

moris) runs from the fovea of the femoral head into the acetabulum to carry blood vessels into the head.

The movements of the hip joint are flexion, extension, abduction, adduction, and medial and lateral rotation. While the movements at this joint are relatively free they do not equal the range of those of the shoulder joint where the scapula is also free to move.

The *knee joint* is the most complicated joint in the body and is inherently unstable. (Fig. 8–50). Yet the two knee joints together carry the total weight of the body superior to them and are involved in the important functions of walking, running, standing, and kicking. The large medial and lateral condyles of the femur rest upon the smaller and shallow corresponding condyles of the tibia. The condyles of the tibia are deepened by crescentic, wedge-shaped rims of fibrocartilage, the semilunar cartilages or menisci.

The fibrous articular capsule encloses not only the articulations between the femur and tibia, but also the one between patella and femur. The capsule is thin posteriorly and is strengthened laterally and medially by the *fibular and tibial collateral ligaments* and anteriorly is replaced by the tendon of the quadriceps femoris muscle and the patella. The collateral ligaments prevent side-to-side movements of the joint. As the joint becomes straight at full extension the collateral ligaments tighten, which adds stability to the joint. A thickening in the posterior part of the capsule, the *oblique popliteal ligament,* also prevents overextension. A transverse ligament runs between the menisci anteriorly.

The *cruciate ligaments* are intraarticular, arising from the non-articular portion of the superior surface of the tibia. One passes anterior and the other posterior to the intercondylar eminence. They cross one another as they rise through the intercondylar fossa. The *anterior cruciate ligament* passes backward and attaches to the internal surface of the lateral condyle of the femur. The *posterior cruciate ligament* crosses the anterior ligament and passes forward to attach to the internal surface of the medial condyle of the femur. These ligaments are stabilizers of the knee joint and prevent antero-posterior displacement.

The synovial cavity is large and has many ramifications due to the large intercondylar fossa and the many pecularities of this complex joint. It extends several inches upward under the quadriceps tendon and muscle, forming a synovial pouch filled with synovial fluid and acting as a bursa. The synovial membrane lines all of the synovial cavity except the bearing surfaces of the bones and the menisci.

The knee joint is a hinge allowing extensive flexion and extension and slight rotation when the knee is in the flexed condition.

The *tibiofibular joints* are three in number, the superior tibiofibular, the interosseous membrane, and the inferior tibiofibular (Figs. 8–51 and 8–52). At none of these joints is there any appreciable amount of movement. They allow the fibula to spring a bit as an accommodation to ankle movement without losing the stability necessary for foot action.

The *superior tibiofibular* joint is a synovial joint and allows some gliding movement between the tibia and fibula. It is held by anterior and posterior ligaments of the fibular head.

FIGURE 8–50. Anterior and posterior views of the superficial and deep structures of the right knee.

←

The *interosseous membrane* stretches between the interosseous borders of tibia and fibula to form a fibrous joint. The interosseous membrane receives muscle attachments which extend onto it from the adjacent bones.

The *inferior tibiofibular joint* is a fibrous joint, a syndesmosis. An interosseous tibiofibular ligament holds the fibula firmly in the fibular notch of the tibia while anterior and posterior tibiofibular ligaments strengthen the joint externally.

The *talocrural or ankle joint* is of the synovial hinge variety (Figs. 8–51 and 8–52). The convex upper surface of the talus fits into the concave inferior end of the tibia and is grasped medially by the medial malleolus of the tibia and laterally by the lateral malleolus of the fibula. This arrangement allows dorsiflexion and plantar flexion but lateral movements are largely prevented by the presence of reinforcing ligaments in the fibrous capsule of the joint.

There are a number of collateral ligaments which extend from the tibial and fibular malleoli to the talus and from the malleoli to other tarsal bones (Fig. 8–51). Among these are the anterior and posterior talofibular ligaments, and the calcaneofibular ligament on the lateral side. On the medial side are found the anterior and posterior tibiotalar ligaments, the tibiocalaneal ligament, and the tibionavicular ligaments. These medial ligaments are quite close together and spread out fanlike. They constitute what is commonly called the *"deltoid ligament."* The lateral ligaments are the ones injured when the foot is forcefully twisted inward; the deltoid ligament is injured upon violent eversion. In these instances we say we have sprained the ankle.

Reference to Fig. 8–52 will give a good idea of the complexity of joints in the foot. They fall roughly into five classifications: intertarsal, tarsometatarsal, intermetatarsal, metatarsophalangeal, and interphalangeal joints. All are synovial joints.

Among the intertarsal joints the *transverse tarsal or midtarsal joint* is of particular importance. It lies between the talus and navicular and the calaneus and cuboid. At this point movements of abduction, inversion, and eversion of the foot take place. The talocalcaneal joint is also involved with these movements.

The *tarsometatarsal joints* are between the three cuneiforms, the cuboid, and the bases of the metatarsals (Fig. 8–52). Note that the medial cuneiform is longer than the others and that the second metatarsal, articulating with the middle cuneiform, is wedged in between the medial and lateral cuneiforms and metatarsals one and two. Movement at these joints is limited to a slight gliding of one bone in relation to another because dorsal, plantar, and interosseous tarsometatarsal ligaments bind them closely together.

Intermetatarsal joints involve both proximal and distal ends of these bones. Dorsal, plantar, and interosseous metatarsal ligaments bind these bones proximally, allowing only slight gliding movements. Distally, the heads of all of the metatarsals are joined by a transverse metatarsal ligament. Recall that in the hand the corresponding transverse metacarpal ligament did not join the first metacarpal.

The *metatarsophalangeal joints* allow flexion, extension, abduction, and adduction. They are condyloid joints. They are held by plantar ligaments ventrally, while dorsally the tendons of extensor muscles take

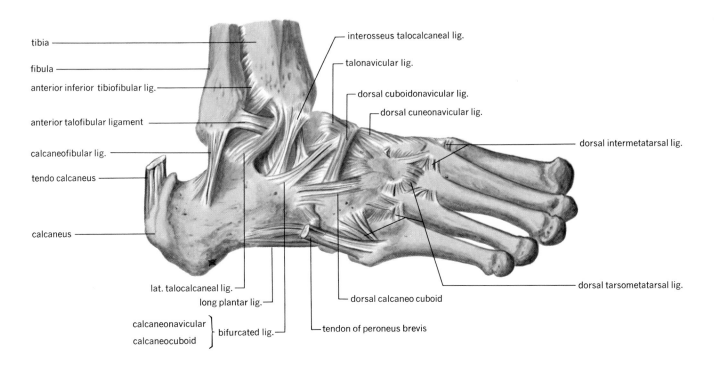

tibia

fibula

anterior inferior tibiofibular lig.

anterior talofibular ligament

calcaneofibular lig.

tendo calcaneus

calcaneus

interosseus talocalcaneal lig.

talonavicular lig.

dorsal cuboidonavicular lig.

dorsal cuneonavicular lig.

dorsal intermetatarsal lig.

dorsal tarsometatarsal lig.

lat. talocalcaneal lig.

long plantar lig.

calcaneonavicular
calcaneocuboid } bifurcated lig.

dorsal calcaneo cuboid

tendon of peroneus brevis

FIGURE 8–51. Ligaments and tendons of the lateral and medial views of the right foot.

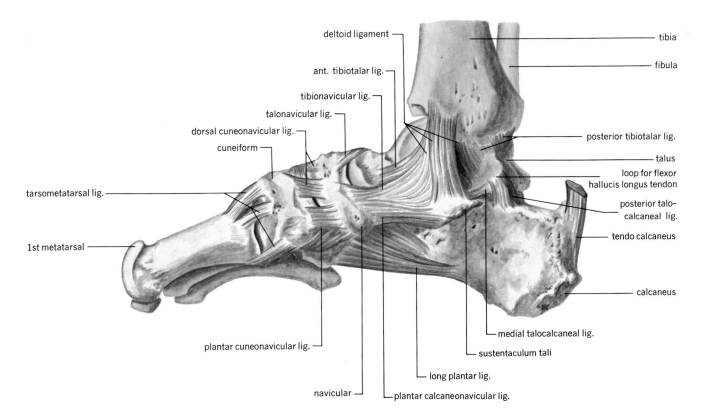

deltoid ligament

ant. tibiotalar lig.

tibionavicular lig.

talonavicular lig.

dorsal cuneonavicular lig.

cuneiform

tarsometatarsal lig.

1st metatarsal

tibia

fibula

posterior tibiotalar lig.

talus

loop for flexor
hallucis longus tendon

posterior talo-
calcaneal lig.

tendo calcaneus

calcaneus

medial talocalcaneal lig.

sustentaculum tali

plantar cuneonavicular lig.

navicular

long plantar lig.

plantar calcaneonavicular lig.

the place of ligaments. Strong, rounded collateral ligaments are found on the lateral and medial sides of each joint.

Interphalangeal joints are ginglymoid or hinge joints and hence allow only flexion and extension. Their ligaments are essentially like those of the metatarsophalangeal joints, namely, plantars and collaterals.

Arches of the foot. The arches of the foot are the lateral and medial longitudinal arches and the transverse arch. The former have been briefly described above. The *medial longitudinal arch* is higher than the lateral as can be seen by a footprint of a normal foot, in which the medial side does not leave a print. The medial arch starts in the calcaneus and rises to its sustentaculum tali and to the head of the talus. It then descends forward through the navicular, the three cuneiforms, and the first three metatarsals, whose distal ends or heads form three of the six contacts of the arched foot. The talus is the keystone of this arch; the calcaneus the posterior pillar; and the remaining bones, the anterior pillar. The lateral arch, like the medial one, begins at the calcaneus as its posterior pillar; it rises to its high point and keystone in the cuboid, and its anterior pillar consists of the fourth and fifth metatarsals whose

FIGURE 8–52. Synovial articulations of right foot as shown in section.

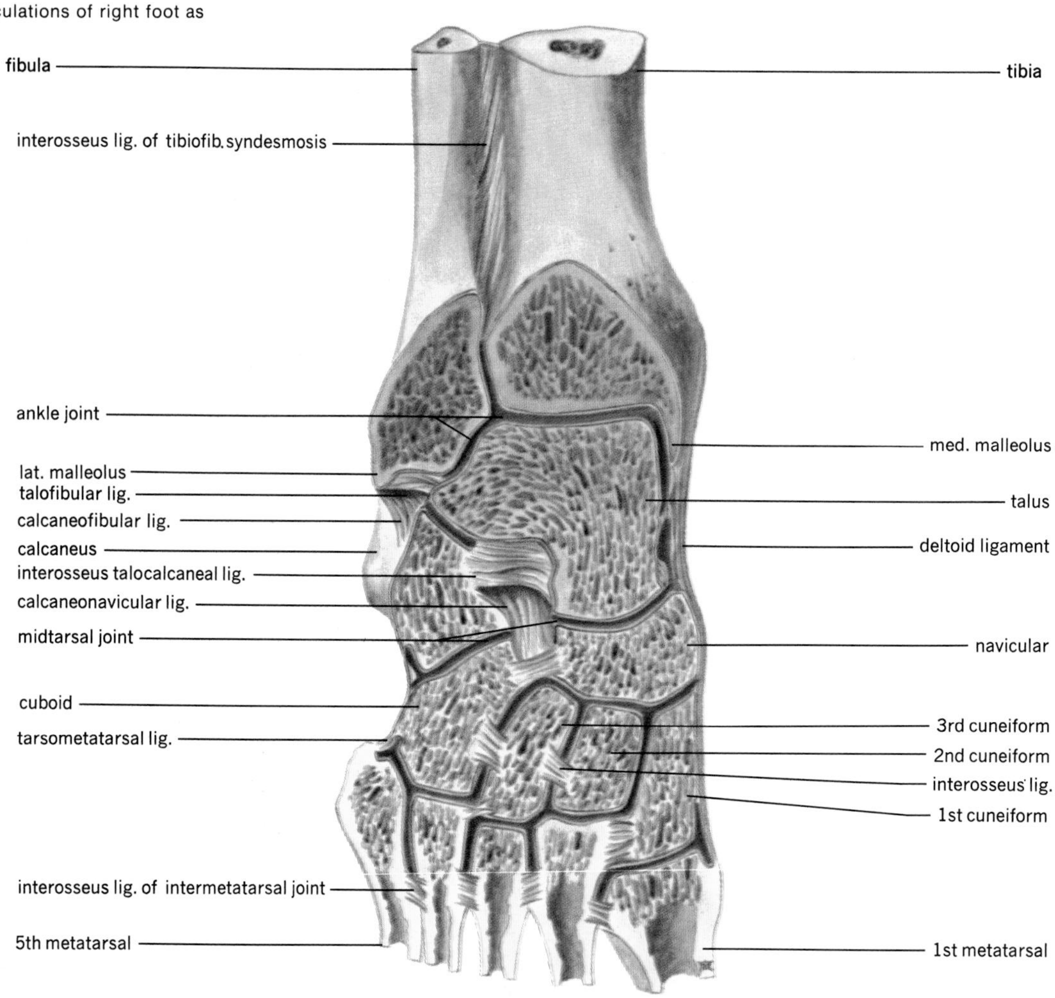

heads complete the six points of contact of the longitudinal arch. The points of contact, in summary, are the calcaneus and the heads of the five metatarsals.

The *transverse arch* involves the concave inferior surfaces of the navicular and cuboid and the interlocking cuneiforms. The latter hold the articulating metatarsals in the same domed position.

The arches are held primarily by ligaments and in part by muscles such as the peroneus longus and the tibialis posterior.

While many ligaments have been described in the account of the foot, all contributing to maintaining its integrity, three are of special importance in relationship to the arches of the foot. The medial longitudinal arch is supported by a strong, elastic ligament which runs longitudi-

FIGURE 8-53. Plantar ligaments and tendons of the right foot.

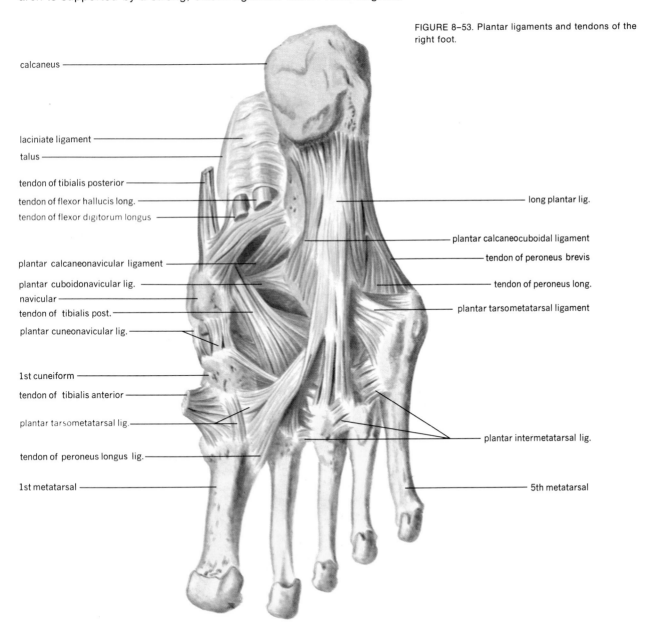

calcaneus

laciniate ligament

talus

tendon of tibialis posterior

tendon of flexor hallucis long.

tendon of flexor digitorum longus

plantar calcaneonavicular ligament

plantar cuboidonavicular lig.

navicular

tendon of tibialis post.

plantar cuneonavicular lig.

1st cuneiform

tendon of tibialis anterior

plantar tarsometatarsal lig.

tendon of peroneus longus lig.

1st metatarsal

long plantar lig.

plantar calcaneocuboidal ligament

tendon of peroneus brevis

tendon of peroneus long.

plantar tarsometatarsal ligament

plantar intermetatarsal lig.

5th metatarsal

nally under the talus from the calcaneus to the navicular, the *plantar calcaneonavicular ligament.* It is often called the "spring ligament" because of its elasticity and the resilience it gives to the foot as it prevents the flattening of the arch. This ligament is coated with hyaline cartilage which enables the bones to move more freely on it.

The much lower lateral longitudinal arch is supported and strengthened by the *long plantar* and the *plantar calcaneocuboid ligaments.*

These arches not only give resilience, strength and stability to the foot but spread the superimposed weight of the erect body about equally between the calcaneus and the heads of the metatarsals. In walking, one comes down on the calcaneus (heel) and then the weight is shifted forward along the lateral side of the foot to the heads of the metatarsals. The head of the first metatarsal takes about 50 percent of the weight as one comes up on the front of the foot and is stronger in construction to carry this greater burden.

BONE MARROW (MYELOID TISSUE) AND HEMOPOIESIS
(Fig. 10–1)

Bone marrow, found in the medullary cavities of long bones and in the spaces in spongy bone, is the site of hemopoiesis or the development of blood. Bone marrow consists of a supporting and sponge-like framework of reticular tissue, the stroma, in which are located blood vessels, the sinusoids, and various stages in the development of blood cells. Marrow is divided into active red marrow, and yellow marrow. The marrow in the embryo and newborn is all red, but as age progresses some of it is gradually changed to yellow marrow, until in the adult there are about equal amounts of each. Red marrow in the adult is found in the vertebrae, sternum, ribs, the diploe of skull bones and, usually, in the epiphyses of the femur and humerus. Yellow marrow is in the medullary cavities of long bones.

Contained in the stroma of the bone marrow are primitive and phagocytic reticular cells. These are attached to the reticular fibers. The sinusoids are lined with flattened, fixed macrophage cells like those found in the sinuses of lymph nodes. Through the walls of the sinusoids, cells from the marrow constantly pass into the blood stream. Fat cells are always present in small numbers in red bone marrow or myeloid tissue but are so numerous in yellow marrow that they obscure most other cells. Lymph nodules are found in bone marrow but no lymph vessels have been demonstrated.

Free cells in bone marrow are varied in form and consist of (1) the mature erythrocytes and granular leucocytes, just as found in the blood stream, and (2) immature cells, the hemocytoblasts, which, through a series of developmental stages, give rise to the various blood cells. Lymphoid tissues such as lymph nodes and spleen give rise to lymphocytes and monocytes. Since, as indicated above, mature blood cells are found in the myeloid or marrow tissue, it serves as a ready or emergency source of these cells as well as a place for their manufacture. The spleen and liver are also areas of blood cell formation in the embryo.

Destruction of blood cells

In man, red corpuscles have a life span of about 120 days. Red cell destruction is normally carried out by action of tissue macrophage cells. In the destruction, hemoglobin is split into hematin and globin. The hematin is then further reduced into iron and into bilirubin which is excreted by the liver as a part of the bile. The iron is reused or stored.

White cells, or leucocytes, vary in their life spans. The granular types may live in the circulation for as long as 14 days or as short as two days. The nongranular leucocytes, on the other hand, may survive in the circulation for 100 to 200 days. Aging leucocytes, as in the case of erythrocytes, are disposed of through the cells of the reticulo-endothelial system (the fixed phagocytes), especially in the liver and spleen.

QUESTIONS

1. What constitutes the skeleton of man?
2. What are the functions of the skeletal system?
3. Most of the articulations of the skull are of the fibrous type. Name one that is not.
4. Define suture, fontanel, endochondral bone, and Wormian bone.
5. Name the sutures which can be seen in a superior aspect of the cranium. What additional suture would be seen in the skull of the newborn?
6. Name the bones which form the walls of the orbit.
7. What bones make up the nasal septum?
8. What bones form the "bridge of the nose"?
9. What is the smallest bone of the face?
10. What bones contribute to the floor of the nasal fossa?
11. What constitutes the zygomatic arch?
12. What are the three portions of the temporal bone?
13. Where are the structures of the internal ear?
14. Name the sinuses of the skull.
15. Of what practical significance are the fontanels of the skull?
16. Describe the sella turcica (Turk's saddle).
17. Name the cranial fossae and indicate what is housed in each one.
18. How do you explain the presence of fontanels in the skull of the newborn?
19. What kinds of articulations are found in the human skull? Give examples.
20. What functions are served by the intervertebral discs? The intervertebral foramina?
21. What is the basis for calling some of the normal curvatures of the spine primary; others secondary?
22. Name the abnormal curvatures which occur in the human vertebral column and state some of the causes of such curvatures.
23. What type of articulation is found between the inferior and superior articular processes of adjacent vertebrae?
24. Describe the joints which enable the skull to be moved in relationship to the vertebral column.
25. Distinguish between true, false, and floating ribs.
26. How does a typical rib, such as number six, articulate with the vertebral column and sternum? What kinds of articulation are involved?
27. Name the components of the sternum.
28. Define suprasternal notch, sternal angle.
29. Describe the function of the thorax in breathing.
30. Contrast the structures and functions of the pectoral girdle and shoulder joint with the pelvic girdle and hip joint.
31. Describe the structural features at the distal end of the humerus and the proximal ends of the radius and ulna which form the elbow and proximal radioulnar joints.

32. Describe the distal radioulnar joint and the wrist joint in terms of their bony relationships and their supporting capsules and ligaments. What kinds of movement are allowed?
33. What are the basic structural differences between the first metacarpal and its digit (thumb) and the remaining four metacarpals and their digits?
34. In the carpal region, where do we find the greatest amount of movement?
35. In what classification do we place intercarpal articulations?
36. Name the components of the os coxae.
37. How would you distinguish the male from the female bony pelvis?
38. Describe (a) the sacroiliac joint; (b) the symphysis pubis.
39. The bones of the knee do little to maintain the integrity of this important joint. What structures do give it strength and how are they arranged?
40. Why do women have more of a tendency to be knock-kneed than men?
41. Describe the ankle joint and how it works.
42. Compare the articulations of the foot with those of the hand, especially in relationship to the first digits and their metacarpals or metatarsals.
43. Describe the longitudinal arches, naming their keystone bones and their anterior and posterior pillars.
44. What functions are served by the arches of the feet?
45. Define bone marrow.
46. How do hemopoietic tissues differ in the adult and in the fetus and newborn?
47. What happens to blood cells after they have degenerated and died?

CHAPTER 9

Muscular System

In life's small things be resolute and great,
To keep thy muscle trained; know'st thou when Fate
Thy measure takes, or when she'll say to thee,
'I find thee worthy; do this deed for me?'

JAMES RUSSELL LOWELL

219

STRUCTURE AND FUNCTION

Introduction

Movement is often said to be one of the prime characteristics of living things. All living forms exhibit some form of motion, ranging from the cytoplasmic streaming of plant cells to the movement of the body of a whale through the sea. The term movement should be understood to include not only the movement of a body through space, but also the transport of materials through the body. Accordingly, three types of muscle are found in the mammalian body, each possessing a structure and function particularly suiting it for the tasks it must perform. *Cardiac* (heart) and *smooth* (visceral) muscle are involuntary types, whose structure and functions are considered in appropriate chapters (Chapters 11 and 12). *Skeletal* or *striated muscle* is a voluntary type attaching to the skeleton and responsible for movement of the body through space. The muscular system of the body consists of these skeletal muscles.

The structure of skeletal muscle

Skeletal muscle (Fig. 9–1) is composed of long, cylindrical fibers (cells), 10 to 100 microns in diameter and up to 6 cm (about 2½ inches) in length. The fibers are surrounded by a membrane known as the *sarcolemma.* The sarcolemma encloses the cytoplasm of the fiber, the

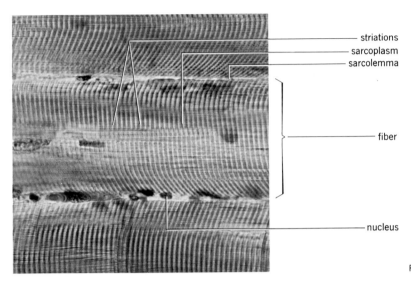

striations
sarcoplasm
sarcolemma

fiber

nucleus

FIGURE 9–1. Skeletal muscle structure.

sarcoplasm. Sarcoplasm in turn contains peripherally placed *multiple nuclei,* and longitudinally arranged *myofibrils.* Stained myofibrils exhibit banding or cross striation, which gives the entire fiber a banded or striated appearance. At the ultramicroscopic level, it may be demonstrated that the myofibrils consist of longitudinally arranged *myofilaments.* These are of two types, designated as thick and thin filaments, and are arranged as shown in Fig. 9–1. The thin filaments are composed of a protein designated *actin;* the thick filaments are composed of a protein designated as *myosin.* Other substances which may be isolated from skeletal muscle include creatine phosphate, adenosine triphosphate (ATP), glucose, and a variety of inorganic ions, including Ca^{++}, $PO_4^{=}$, Na^+, and K^+. Skeletal fibers also possess a system of intracellular tubules, designated as the *sarcoplasmic reticulum* (comparable to the endoplasmic reticulum of other cells) and the system known as the *"T tubules"* (Fig. 9–2). The sarcoplasmic reticulum usually lacks ribosomes and is intimately associated with the myofibrillae. The T tubules communicate with the extracellar fluid of the fiber and with the sarcoplasmic reticulum by way of the cisternae. A pathway for free movement of organic and inorganic material is thus provided, a fact of importance in muscular contraction.

Organization into muscles

Individual fibers are bound together by reticular fibers forming the *endomysium.* Groups of fibers are bound into fasciculi by connective tissue (chiefly collagenous fibers) known as the *perimysium.* Many fasciculi are bound together into the muscle by the *epimysium.* These connective tissue bindings are continuous with one another and with the tendons which are usually located on the ends of the muscle. In addition to providing a means of holding the muscle together, the connective tissue serves to transmit the blood vessels and nerves to the muscle fibers. Skeletal muscle has a nearly one to one ratio of capillaries and muscle fibers, and is provided with both sensory and motor nerves.

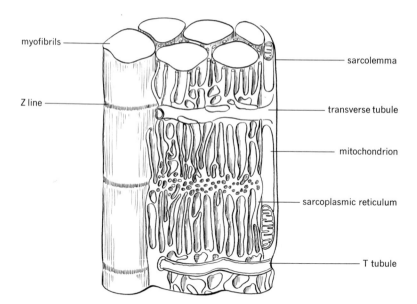

FIGURE 9–2. Sarcoplasmic reticulum and transverse (T) tubules.

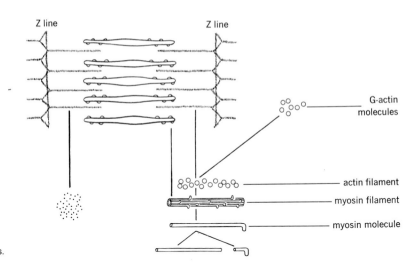

FIGURE 9–3. Myofilament arrangement in the myofibrils.

Physiology of muscular activity

Although the exact details of muscle contraction are not known, the mechanics of contraction are best explained by the so-called *interdigitating filament model*. According to this theory, the myofilaments are arranged in a definite pattern (Fig. 9–3), with their ends overlapping or interdigitating with one another. The myosin filaments are possessed of extensions referred to as "cross bridges" which, in the resting muscle, simply project toward, but are not chemically attached to, the actin filaments. The resting muscle is therefore a plastic, easily extended structure. Passive extension of a resting muscle causes the filaments to slide past one another. In addition, in the resting muscle, most of the muscle Ca^{++} is located in the T system and/or in the sarcoplasmic reticulum tubules, and there are ATP molecules attached to the tips of the cross bridges. Myosin possesses the ability to split a phosphate from the ATP, that is, it possesses "ATPase" activity.

It is presumed that upon stimulation of the muscle, the permeability of the tubule systems is increased, and Ca^{++} moves into the region of the myosin. The myosin is activated, splits a phosphate group from the ATP molecule on its cross bridges, and allows a cross link to be made between the myosin ADP molecule and a "receptor site" on the actin. The essential feature here is that the two proteins are chemically linked together into an actomyosin complex.

What happens next is a matter of conjecture. It has been suggested that the cross link is broken, and a new one established at a new receptor site further down the molecule. The process is repeated (Fig. 9–4), each new linkage drawing the actin filaments closer together. The fibers are thus shortened in a series of movements resembling the rowing of a boat.

In any event, the muscle as a whole undergoes a shortening. Relaxing is accomplished by removal of calcium (active transport?), breaking of the actomyosin complex, and a return to original length by elastic recoil of the protein strands.

Energy for activity

The immediate source of energy for contractile activity comes from the ATP molecules on the myosin. Essential to continued activity of the muscle is maintenance of the ATP supply. The necessary energy is provided in three ways:

1. Creatine phosphate breaks down (through enzymatic activity) to provide energy for ATP resynthesis.

$$\text{creatine phosphate} \xrightarrow[\text{phosphatase}]{\text{creatine}} \text{creatine} + \text{phosphate} + \text{energy}$$

$$\text{energy} + \text{ADP} + \text{phosphate} \longrightarrow \text{ATP}$$

2. The breakdown of glucose in the muscle also provides energy for ATP synthesis.

$$\text{glucose} \longrightarrow \text{pyruvic acid} + \text{energy}$$
$$\text{energy} + \text{ADP} + \text{phosphate} \longrightarrow \text{ATP}$$

3. If the muscle is not supplied with enough oxygen to aid in the further combustion of pyruvic acid (oxygen debt), pyruvic acid is changed to lactic acid. A small percentage of the lactic acid is oxidized to provide energy to resynthesize the remainder to glucose and thus maintain the second energy source.

$$\text{pyruvic acid} \xrightarrow{+ 2H} \text{lactic acid}$$

$$1/5 \text{ lactic acid} \longrightarrow \text{energy} + CO_2 + H_2O$$
$$\text{energy} + 4/5 \text{ lactic acid} \longrightarrow \text{glucose}$$

Types of contraction of skeletal muscle

If the foregoing series of reactions results in an actual shortening of the muscle, an *isotonic contraction* has occurred, and the muscle causes movement of a body part. If shortening is prevented, but tension develops and the body part is held but does not move, an *isometric*

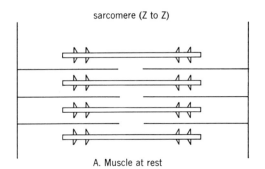

sarcomere (Z to Z)

A. Muscle at rest

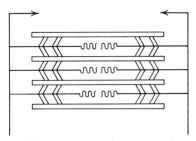

B. Contracted—cross bridges formed, Z lines approach each other

FIGURE 9–4. The events of muscular contraction.

contraction has occurred. Isometric contractions occur commonly to maintain body posture. In the laboratory, an isolated muscle may be stimulated with a single impulse to create a *twitch.* Twitches do not normally occur in the living body, but are useful to demonstrate the phases in muscular activity. Figure 9–5 shows a record of a single twitch with phases and times indicated. It may be noted that the total time required for a skeletal muscle to complete all phases of activity does not exceed 0.1 second, making it the fastest reacting type of muscle.

Skeletal muscle is capable of reacting to a second stimulus even before it has completed its response to a first stimulation. It is said to have an extremely *short refractory period.* The short refractory period enables the muscle to be *tetanized,* or thrown into a sustained contraction. Figure 9–6 shows the development of tetanus with increasing frequency of stimulation. It may be noted that tetanus is actually the result of the fusion of separate twitches, with relaxation not being allowed to occur.

FIGURE 9–5. A single muscle twitch as recorded on a kymograph.

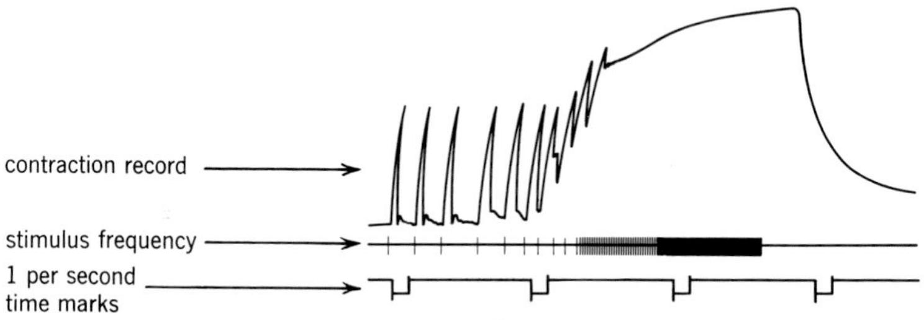

FIGURE 9–6. The development of tetanus as recorded on a kymograph. Increasing frequency of stimulation results in "fusion of twitches" to give smooth sustained contraction.

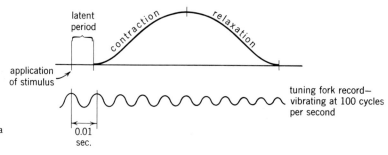

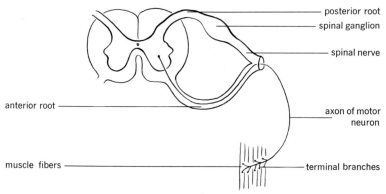

FIGURE 9–7. The motor unit.

The problem of *tonus,* or sustained partial contraction in skeletal muscle remains to be resolved. The nerves which supply and activate a skeletal muscle are arranged to form motor units (Fig. 9–7). These are composed of a single nerve fiber plus the muscle fibers it supplies. A typical muscle contains hundreds or thousands of motor units, and can *grade the strength* of its contraction according to the number of motor units that are active. Remembering that a skeletal muscle is normally activated through its nerve supply, any contraction, partial or complete, must be associated with recordable electrical (nerve) impulses. A muscle at complete rest shows no impulses, and therefore it is not contracted. If one interprets tone as a partial contraction associated with recordable impulses, then tone exists, but it is still dependent on stimulation from the brain.

Treppe. If a muscle receives repetitive stimuli of a strength designed to activate all its motor units, the first few stimuli may result in an increasing strength of contraction (Fig. 9–8). Since all fibers are active, the increasing strength of contraction cannot be attributed to more active fibers. It is suggested that "warming up," with decreased viscosity of the sarcoplasm, enables more energy to be directed towards shortening and less to overcoming internal resistance, thus increasing the degree of contraction.

The foregoing discussion has emphasized the structure and some of the basic physiological properties of skeletal muscles. These are summarized in Table 9–1.

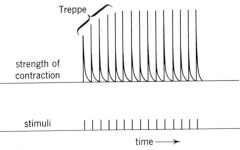

FIGURE 9–8. Treppe.

Muscular action

Muscles can only actively shorten, and they cause movement by pulling upon the skeleton. Since joints may allow several directions of motion, it follows that there must be separate muscles or groups of muscles to achieve each type of motion. A muscle causing a desired action is said to be an *agonist or prime mover.* Usually, a muscle having an opposite effect upon the bone must relax to permit the movement to occur. This second muscle is termed the *antagonist.* Most joints are operated by *antagonistic groups* of muscles. Still another type of muscle occasions little movement of a joint, but may steady the joint or eliminate undesirable movements. Such a muscle is termed a *synergist* (syn = together; ergos = work), or *fixator.* Fixators take no part in the movement, but steady the body to provide a stable platform against which motion may occur. A muscle attaches, usually via a tendon, to one bone or a connective tissue sheath, which moves to a lesser degree than the bone or connective tissue to which the other end of the muscle attaches. The lesser moving part serves as the *origin* of the muscle and the greater moving part serves as the *insertion.* The *action* is the particular type of motion which occurs as the muscle pulls upon the insertion.

The muscles utilize the bones and joints as levers to achieve movement.

Any lever has several basic parts:

1. The *fulcrum* (F). This is the point about which the lever moves. In the body, the fulcrum is provided by a joint.
2. The point of *effort* (E). This is where the muscle inserts; the point at which the pull of the muscle is applied.

TABLE 9-1

Summary of Structure and Properties of Skeletal Muscle

Item	Function or Structure	Comments
Fiber	The basic unit or "cell" of skeletal muscle	Unit of structure of all named muscles
Sarcolemma	Membrane around fiber	Provides limit to fiber, some control of entry
Sarcoplasm	The cytoplasm of the fiber	Contains typical organelles
Myofibrils	Longitudinally arranged units of the fiber	Cross banded
Myofilaments	Protein strands longitudinally arranged inside myofibrils	Contractile units of the muscle
ATP, CP, K^+, Ca^{++}, $PO_4^=$, glucose	Chemicals, within the fiber	Necessary for contraction and nutriment
Sarcoplasmic reticulum	The "endoplasmic reticulum" of the fiber	Houses Ca^{++} until required for contraction
T tubules	Tubules separate from sarcoplasmic reticulum which communicate to fiber exterior	Avenue for passage of substances into fiber
Endomysium Perimysium Epimysium	Connective tissue binding fibers into a muscle	Endo—binds fibers together Peri—surrounds fasciculi Epi—surrounds muscle
Interdigitating filament model	Theory of arrangement of filaments in myofibrils	Allows explanation of muscle contraction
Muscle contraction	Shortening of fibers	Ultimate force of movement
Creatine Phosphate	Compound in muscle releasing energy for ATP synthesis	Immediate source of energy for ATP synthesis
Lactic acid	Partially combusted to provide energy for ATP synthesis	Occurs only when O_2 deficient
Isometric contraction	Tension developed but no shortening	Posture
Isotonic contraction	Contraction with shortening	Movement
Twitch	Illustrates phases of muscular activity	0.1 second duration
Tetanus	Fusion of twitching; sustained maximal contraction	Result of short refractory period
Tonus	Fusion of twitches; sustained partial contraction	May not exist in resting muscle
Treppe	Increasing strength of contraction with repeated strong stimuli	"Warming up" allows greater contraction

3. The *resistance* (R). This is the weight which must be moved by the muscle's pull. It is usually considered to be concentrated at some point on the lever.
4. The *effort arm* (EA). This is the distance from fulcrum to point of effort. It is usually some part of the length of the bone being moved.
5. The *resistance arm* (RA). This is the distance from the fulcrum to the area where the resistance is concentrated. It also is a distance along the length of the bone.

According to the placement of fulcrum, effort, and resistance relative to one another, three classes of levers are recognized (Fig. 9–9).

Class I. The fulcrum is placed between effort and resistance. See-saws, post hole diggers, and pry bars are everyday examples of first class levers. Not too many levers of this class are found in the body, for it is not common to have a joint in the center of the bone. They are usually placed at the ends of a bone. The action of the triceps on the elbow is a good example.

Class II. The fulcrum is towards the end of the lever, and resistance is placed betweeen fulcrum and effort. A wheelbarrow is an example of a second class lever. This class requires less effort to move a given weight than the other classes. It is not agreed that any second class levers exist in the body. Raising oneself on ones toes, the fulcrum being the ball of the foot, is considered by some to be an example.

Class III. The fulcrum is towards the end, and effort is between fulcrum and resistance. This type is the most common in the body, for it provides a joint at the end of a bone, and much space for muscle attachment along the effort arm.

"Efficiency" of a given type of lever may be calculated by the simple formula:

$$E \times EA = R \times RA$$

For example, in the triceps action upon the elbow, typical measurements might be:

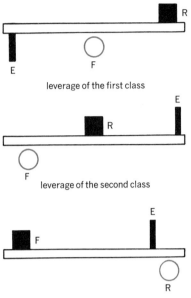

leverage of the first class

leverage of the second class

leverage of the third class

FIGURE 9–9. The three classes of levers.

E = ?   Thus E × 1 = 10 × 10
EA = 1 inch   or E = $\dfrac{10 \times 10}{1}$

R = 10 lbs   E = 100 lbs to move
RA = 10 inches   a 10-lb weight

For the second class

E = ?   Thus E = $\dfrac{100 \times 12}{10}$

EA = 10 inches
R = 100 lbs   E = 120 lbs
RA = 12 inches

For the third class

E = ?   E = $\dfrac{10 \times 12}{2}$

EA = 2 inches
R = 10 lbs   E = 60 lbs
RA = 12 inches

It becomes obvious that first and third class levers require much more power than is exerted by the resistance, to make them work. However, these classes achieve a great range of motion when power is applied.

Arrangement of fibers in a muscle

According to placement on the body and the area to which the effort of a muscle is applied, muscles may vary in the arrangement of fibers. Following is a simple classification of fiber arrangement.

name	external view	cross section	examples
fusiform	tendon / belly (fibers) / tendon		appendage muscles; bone central, muscles around it
pennate (feather form)			
unipennate	fibers / tendon		rhomboid, semimembranosus
bipennate			abdominal muscles
multipennate			deltoid, quadriceps, triceps
circumpennate			orbicularis muscles of eyes and mouth

Tendons and bursae (Fig. 9–10)

Tendons usually make the attachment of a muscle to the bone for several reasons:

1. The muscular portion may not be long enough to reach between the two bones. A tendon can be any length, and also is small in size.
2. Tendons may pass over bony prominences which would destroy a muscle fiber.
3. Because of their small size, many more tendons can pass over a joint than could accommodate the fleshy portion.

Tendons are subjected to wear where they pass over bony prominences. Small fluid-filled sacs known as bursae are placed between the tendon and the bone to cushion the tendon. *Bursitis* is an inflammation of the bursae.

Actions

As indicated previously, muscle contraction results in the motion of a bone, the motion caused being termed the action of the muscle. Table 9–2 defines the basic types of actions. Note that the actions are paired in antagonistic movements.

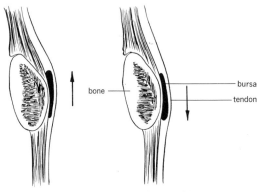

FIGURE 9–10. The relation of tendons and bursae. Bursae facilitate movement of tendons over bones.

TABLE 9–2

Actions

Action	Definition
Flexion	Decrease of angle between two bones
Extension	Increase of angle between two bones
Abduction	Movement away from the midline (of body or part)
Adduction	Movement toward midline (of body or part)
Elevation	Upward or superior movement
Depression	Downward or inferior movement
Rotation	Turning about the longitudinal axis of the bone
Medial	Toward midline of body
Lateral	Away from midline of body
Supination	To turn the palm up or anterior
Pronation	To turn the palm down or posterior
Inversion	To face the soles of the feet toward each other
Eversion	To face the soles of the feet away from each other
Dorsiflexion (= flexion)	At the ankle, to move the top of the foot toward the shin
Plantar flexion (= extension)	At the ankle, to move the sole of the foot downward, as in standing on the toes

THE MUSCLES

Classification of muscles

Description of individual body muscles will be carried out on a regional basis, with special attention being paid to the muscles operating as functional groups. The student may be aided in remembering the muscles by noting that muscles are named by *shape, origin* and *insertion, position* on the body, or by the *action* they cause. It might also be helpful at this time to review the chapter on articulations, and actions in the present chapter. Figure 9–11 presents a general view of the muscular system of man. Rather general descriptions of each muscle will be found in the following Section. Detailed descriptions of the muscles will be found in the table at the end of the chapter.

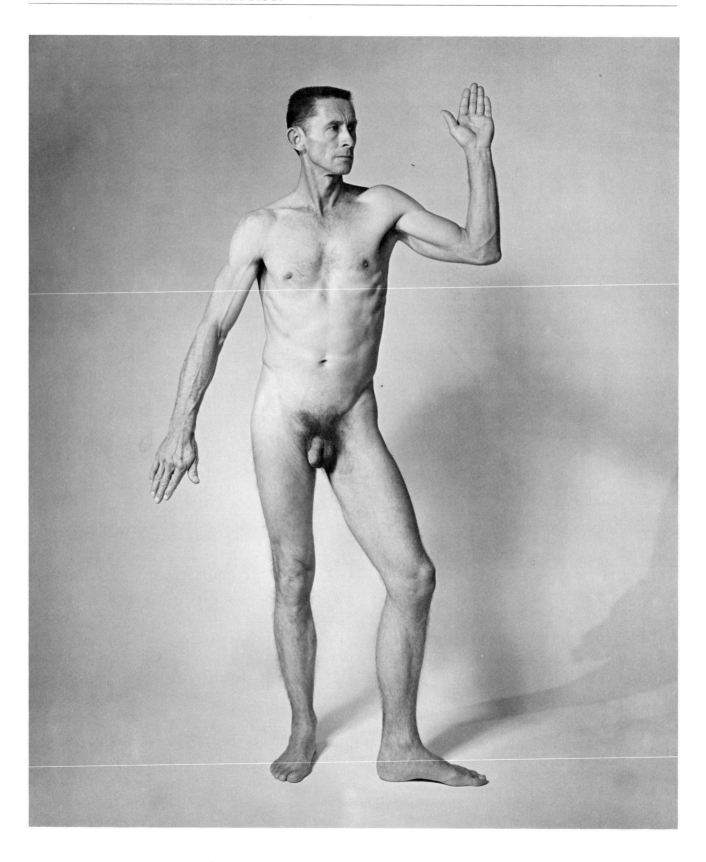

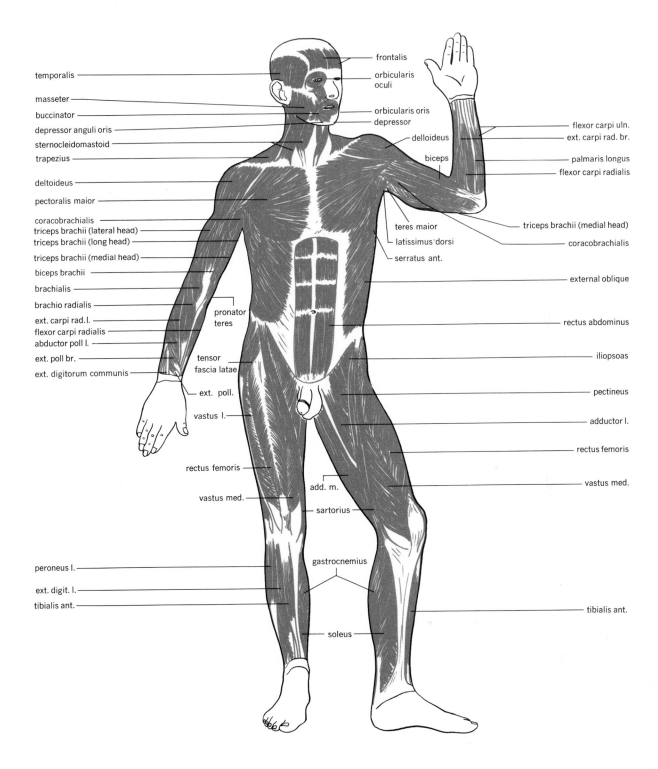

temporalis

masseter
buccinator
depressor anguli oris
sternocleidomastoid
trapezius

deltoideus

pectoralis maior

coracobrachialis
triceps brachii (lateral head)
triceps brachii (long head)

triceps brachii (medial head)

biceps brachii

brachialis

brachio radialis

ext. carpi rad.l.
flexor carpi radialis
abductor poll l.

ext. poll br.

ext. digitorum communis

pronator
teres

tensor
fascia latae

ext. poll.

vastus l.

rectus femoris

vastus med.

peroneus l.

ext. digit. l.

tibialis ant.

frontalis
orbicularis
oculi

orbicularis oris
depressor

delloideus

biceps

teres maior
latissimus dorsi
serratus ant.

add. m.

sartorius

gastrocnemius

soleus

flexor carpi uln.
ext. carpi rad. br.

palmaris longus
flexor carpi radialis

triceps brachii (medial head)

coracobrachialis

external oblique

rectus abdominus

iliopsoas

pectineus

adductor l.

rectus femoris

vastus med.

tibialis ant.

FIGURE 9–11A. General view of the muscular system.

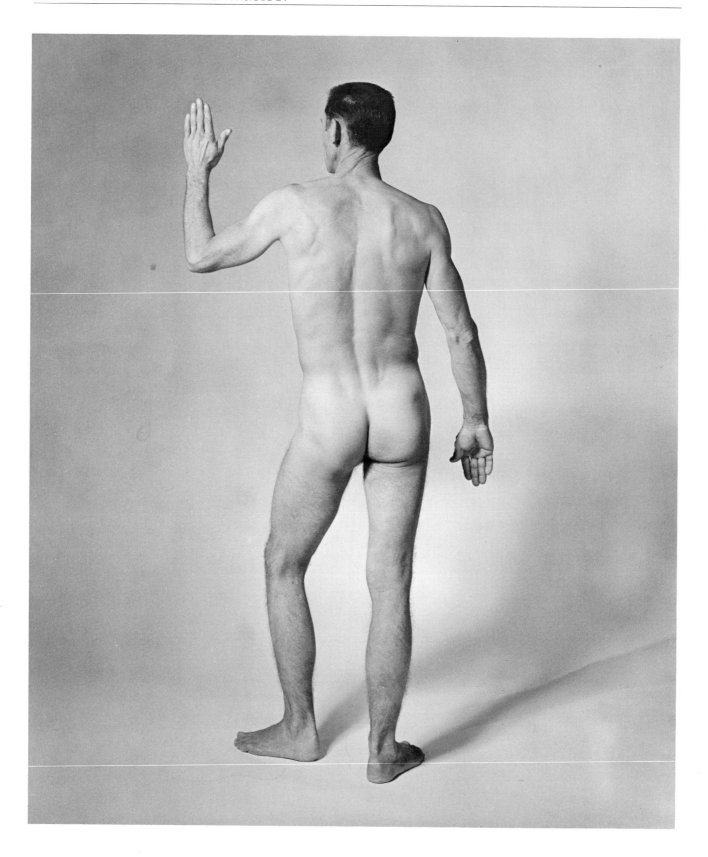

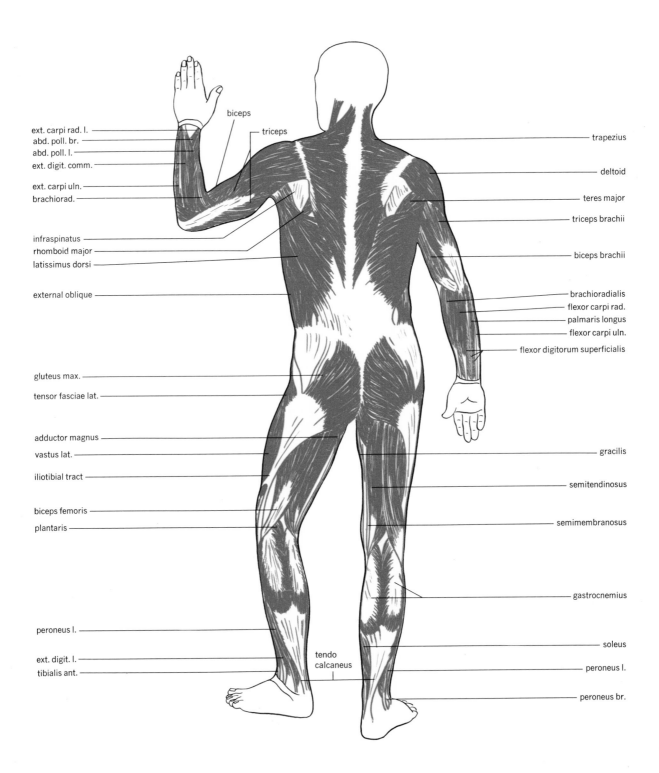

ext. carpi rad. l.
abd. poll. br.
abd. poll. l.
ext. digit. comm.

ext. carpi uln.
brachiorad.

infraspinatus
rhomboid major
latissimus dorsi

external oblique

gluteus max.

tensor fasciae lat.

adductor magnus

vastus lat.

iliotibial tract

biceps femoris
plantaris

peroneus l.

ext. digit. l.
tibialis ant.

biceps

triceps

tendo
calcaneus

trapezius

deltoid

teres major

triceps brachii

biceps brachii

brachioradialis
flexor carpi rad.
palmaris longus
flexor carpi uln.
flexor digitorum superficialis

gracilis

semitendinosus

semimembranosus

gastrocnemius

soleus

peroneus l.

peroneus br.

FIGURE 9–11B. General view of the muscular system.

MUSCLES OF THE HEAD AND NECK

Facial muscles (Fig. 9–12)

The facial muscles are the muscles of facial expression. Most take their origin from the facial bones or fascia, and insert into and move the skin to create the variety of expressions the face may present. The facial muscles include:

Orbicularis oculi	Triangularis
Orbicularis oris	Depressor labii inferioris
Levator labii superioris	Mentalis
Zygomaticus	Buccinator
Risorius	Corrugator

Orbicularis oculi arises from the medial wall of the orbit and traces a circular course around the orbit. The whole muscle winks the eye or closes it, and is involved in the narrowing of the visual opening between the eyelids as when we are exposed to strong light.

Orbicularis oris encircles the mouth. Its contraction causes the lips to "pucker."

Levator labii superioris includes several heads or parts which arise from the zygomatic bone and lower rim of the orbit, and pass downward to insert into the upper lip. Contraction causes elevation of the upper lip.

Zygomaticus arises from the zygomatic bone and passes to the corners of the mouth. Its contraction draws the corners of the mouth up and backward as in smiling and laughing.

Risorius passes almost horizontally from the ramus of the mandible to the corner of the mouth. Its contraction draws the mouth laterally.

Triangularis takes its origin from the oblique line of the mandible and passes to the lower corner of the mouth. Its contraction depresses the lower lip.

Depressor labii inferioris arises from the mental process and inserts on the skin of the lower lip. Its contraction depresses the lower lip.

FIGURE 9–12. Facial muscles.

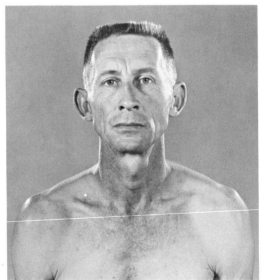

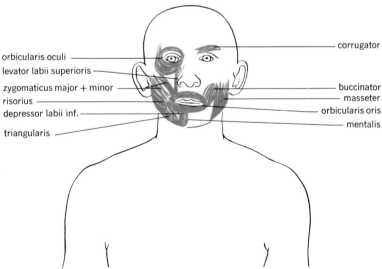

orbicularis oculi
levator labii superioris
zygomaticus major + minor
risorius
depressor labii inf.
triangularis

corrugator
buccinator
masseter
orbicularis oris
mentalis

Mentalis arises from the anterior part of the mandible and passes to the skin of the chin. Its contraction elevates and protrudes the lower lip and wrinkles the skin of the chin.

Buccinator forms the lateral walls of the oral cavity, and lies beneath the above mentioned muscles. It compresses the cheek, therefore aiding chewing, and forces air from the mouth as in playing a wind or brass musical instrument.

Corrugator passes from the medial side of the orbit to the skin over the middle of the orbit. It produces the vertical "frown lines."

Cranial muscles (scalp muscles) (Fig. 9–13)

These muscles include:
    Epicranius, divisible into:
        Frontalis
        Occipitalis
    The auricular muscles

Epicranius includes the *frontalis,* overlying the frontal bone, and the *occipitalis,* overlying the occipital bone. Between the two muscular portions is a broad, flat tendon covering the top and sides of the skull, the *galea aponeurotica.* The galea is in turn overlain by the scalp. Contraction of the frontalis draws the scalp forward, raises the eyebrows, and produces the transverse wrinkles in the forehead. Contraction of the occipitalis draws the scalp backward.

The auricular muscles are rudimentary in man. Three separate muscles are included, attaching to the pinna of the ear. They cause the ear to be moved slightly in an anterior, superior, or posterior direction.

Muscles of mastication (Figs. 9–13 and 9–14)

Four muscles are most important in chewing and biting. They are:
    Temporalis   Medial (internal) pterygoid
    Masseter   Lateral (external) pterygoid

FIGURE 9–13. Cranial muscles and superficial muscles of mastication.

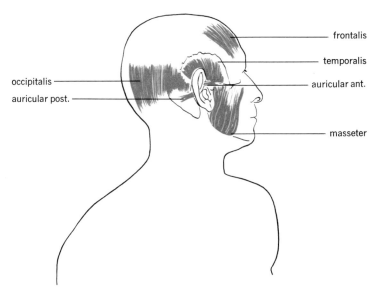

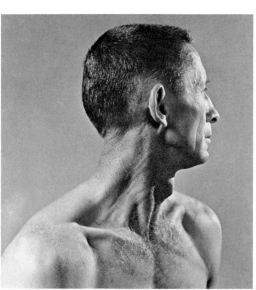

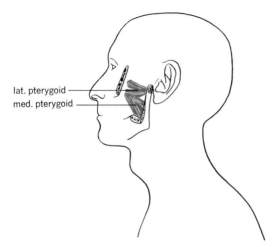

lat. pterygoid
med. pterygoid

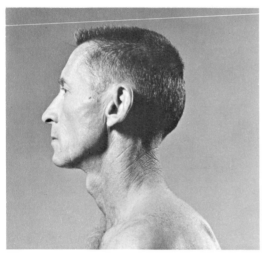

FIGURE 9–14. Deep muscles of mastication.

Temporalis occupies the temporal fossa on the sides of the skull and passes beneath the zygomatic arch to the coronoid process of the mandible. It elevates or closes the jaw.

Masseter arises from the zygomatic arch and inserts on the ramus of the mandible. It also elevates or closes the jaw.

Medial pterygoid arises from the pterygoid process and inserts on the medial surface of the ramus. Contraction elevates the jaw and pulls the mandible to the opposite side to give lateral grinding movements.

Lateral pterygoid is oriented about 90 degrees to the other muscles of mastication, and passes from the pterygoid process to the neck of the condyloid process of the mandible. Its contraction protrudes the mandible and opens the mouth.

Muscles associated with the hyoid bone

These muscles may be divided into several groups according to their location relative to the hyoid bone. The *suprahyoid* muscles (Fig. 9–15) lie in the floor of the mouth above the hyoid bone and include:

Digastric (2 bellies)
Stylohyoid
Mylohyoid

Digastric. The posterior belly of the muscle passes from the mastoid process to the hyoid bone; the anterior belly passes from hyoid anteriorly to the mandible.

Stylohyoid. As the name suggests, this muscle passes from styloid process to hyoid bone.

Mylohyoid forms the floor of the oral cavity, passing between the two limbs of the mandibular body.

These muscles elevate the hyoid bone, or pull it forward or backward. They are important in swallowing.

The *infrahyoid muscles* (Fig. 9–16) lie below the hyoid bone, mostly in the neck. They include:

Sternohyoid
Sternothyroid
Thyrohyoid
Omohyoid (two bellies)

The muscles are named for origin and insertion.

These muscles depress the hyoid bone, or in the case of the omohyoid, draw the hyoid backwards.

Neck muscles (Fig. 9–17)

The *sternocleidomastoid,* originating from sternum and clavicle, and passing to the temporal mastoid process is perhaps the most obvious muscle of the neck. Operating singly, it turns the head to the opposite side, while together, the muscles flex the head on the chest. Overlying the entire anterior part of the neck is the *platysma.* Its contraction draws the lower lip downwards. The deeper neck muscles are responsible for lateral deviation of the neck and for extending the neck. Three groups of neck muscles are recognized:

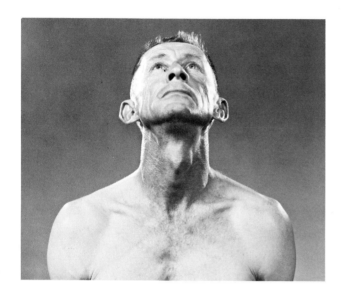

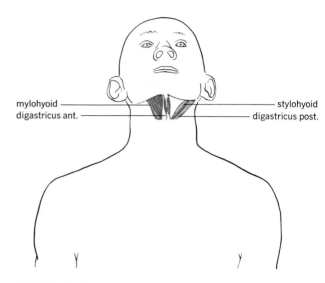

FIGURE 9–15. Suprahyoid muscles.

mylohyoid — stylohyoid
digastricus ant. — digastricus post.

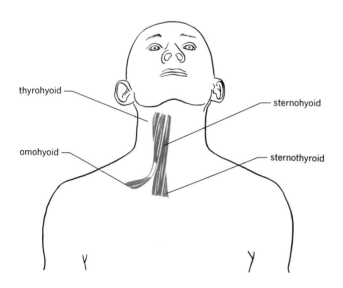

FIGURE 9–16. Infrahyoid muscles.

thyrohyoid — sternohyoid
omohyoid — sternothyroid

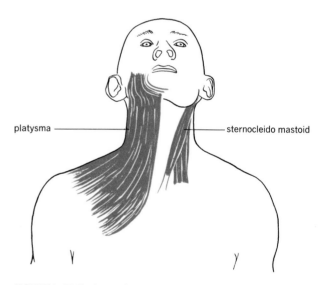

FIGURE 9–17. Neck muscles.

platysma — sternocleido mastoid

1. The *anterior group* (Fig. 9–18) includes
    Longus colli
    Longus capitus
    Rectus capitus
    These muscles flex the neck
2. The *lateral group* (Fig. 9–19) includes
    Scalenes
      Anterior
      Middle
      Posterior
    These muscles bend the vertebral column to one side.

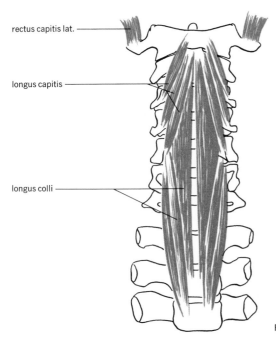

rectus capitis lat.

longus capitis

longus colli

FIGURE 9–18. Anterior neck muscles.

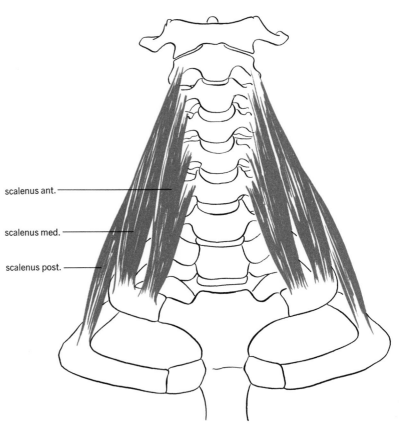

scalenus ant.

scalenus med.

scalenus post.

FIGURE 9–19. Lateral neck muscles.

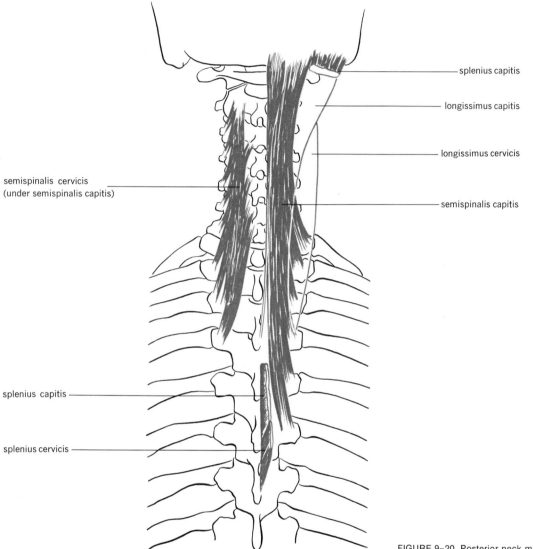

splenius capitis

longissimus capitis

longissimus cervicis

semispinalis capitis

semispinalis cervicis
(under semispinalis capitis)

splenius capitis

splenius cervicis

FIGURE 9–20. Posterior neck muscles.

3. The *posterior group* (Fig. 9–20) includes
    Semispinalis
    Longissimus
    Splenius
  These muscles extend the head and neck.

    The scalenes lie in the neck, and run between the cervical vertebrae and the first two ribs. Their action is to aid in rib elevation.

    A cross section of the neck is shown in Fig. 9–21, to illustrate the manner in which these muscles are layered.

    The trapezius overlies the posterior group of neck muscles. Its insertion, however, is upon the scapula and it will be considered with the muscles operating the scapula.

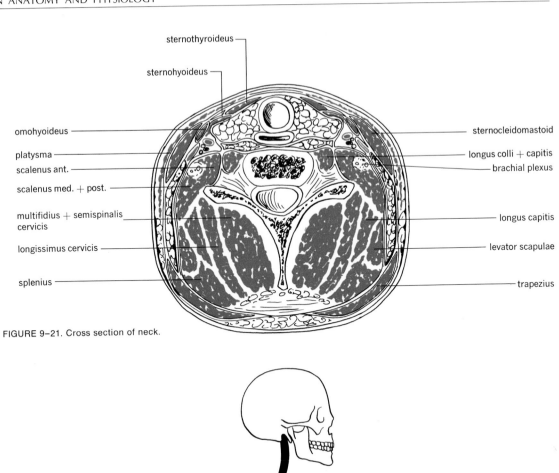

FIGURE 9-21. Cross section of neck.

Labels (left side): sternothyroideus, sternohyoideus, omohyoideus, platysma, scalenus ant., scalenus med. + post., multifidius + semispinalis cervicis, longissimus cervicis, splenius

Labels (right side): sternocleidomastoid, longus colli + capitis, brachial plexus, longus capitis, levator scapulae, trapezius

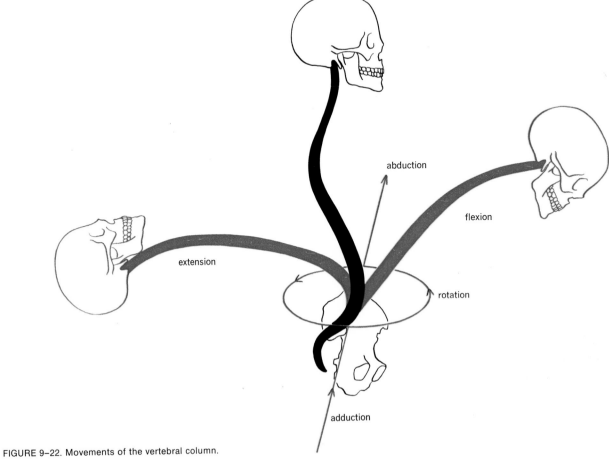

FIGURE 9-22. Movements of the vertebral column.

abduction, flexion, extension, rotation, adduction

MUSCLES OPERATING THE VERTEBRAL COLUMN

Movements possible (Fig. 9–22)

The column may be *flexed* or bent forwards, *extended* or bent back-wards, *abducted* or *adducted* in a side-to-side movement, and *rotated*. Some of the muscles responsible for these motions have other func-tions, such as forming the abdominal wall. Their names will appear again in another Section.

Flexors of the spine (Fig. 9–23)

Three muscles are most important in flexing the vertebral column, exclusive of the neck.

Rectus abdominus

Psoas }
Iliacus } Iliopsoas

Rectus abdominus. The rectus abdominus extends from the pubic symphysis to the xiphoid process and cartilage of ribs 5 to 7. Its con-traction flexes the column as in doing a "sit-up."

Psoas. The psoas originates from the bodies of the lumbar verte-brae and passes to the femur. If the thigh is fixed, the psoas aids in flexing the lumbar area of the column.

Iliacus. This muscle is really a muscle operating the hip. However, if the thigh is fixed, the iliacus may aid flexion of the column (weakly) by pulling upon the iliac bones.

Extensors of the spine (Fig. 9–24)

The muscles extending the vertebral column are all located on the posterior aspect of the thorax, spinous and transverse processes of the vertebrae, and posterior sacrum and ilium. The muscles are numerous and their arrangement is complicated. Perhaps it is best to visualize them as forming a series of ladder-like slips extending the length of the column, with three main groupings evident. All the muscles together are designated the *erector spinae (sacrospinalis).* The three groupings are:

Iliocostalis (laterally placed)
Longissimus (intermediate in placement)
Spinalis (medially placed)

Contraction of the muscles as a whole straightens or extends the spine. The iliocostal groups are lateral enough in placement to bend the spine laterally if only one side contracts.

Abductors and adductors of the spine (Fig. 9–25)

Sideward movement of the column is achieved by:
Iliocostalis
Quadratus lumborum

Iliocostalis. As indicated above, iliocostalis can bend the column.

Quadratus lumborum. This muscle extends between the ilium and the transverse processes of thoracic vertebrae 12 to 14. Its contraction

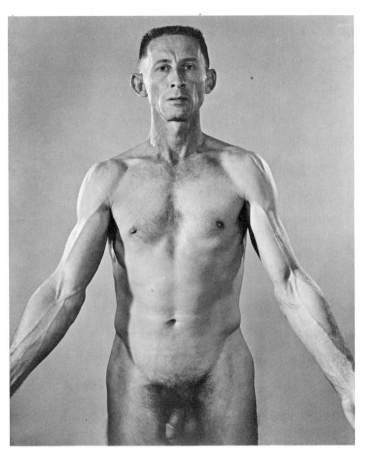

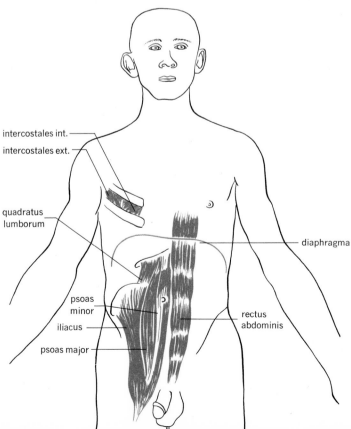

intercostales int.

intercostales ext.

quadratus
lumborum

diaphragma

psoas
minor

rectus
abdominis

iliacus

psoas major

FIGURE 9–23. Flexors of vertebral column and muscles
of resp.

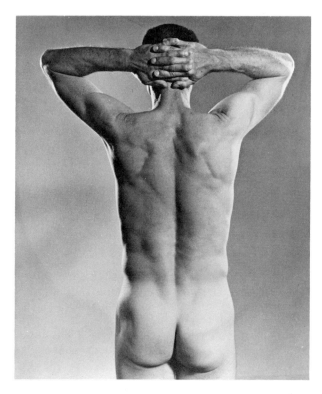

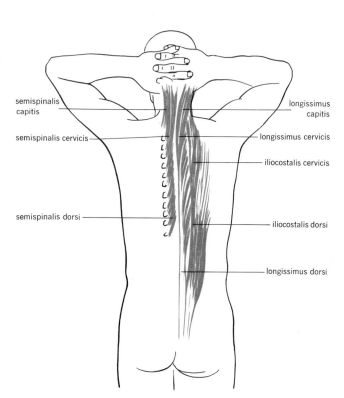

semispinalis capitis

longissimus capitis

semispinalis cervicis

longissimus cervicis

iliocostalis cervicis

semispinalis dorsi

iliocostalis dorsi

longissimus dorsi

FIGURE 9–24. Extensors of the vertebral column.

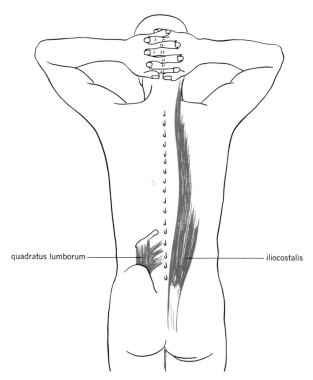

quadratus lumborum

iliocostalis

FIGURE 9–25. Abductors and adductors of the spine.

bends the lumbar area to the side and causes the remainder of the column to follow. The direction of pull of the muscle also causes it to exert some opposite rotation on the column.

Rotation of the spine

The muscles of the abdomen achieve the remaining rotatory movement. Three muscles are involved:
    External oblique
    Internal oblique
    Transverse abdominal

External oblique. The muscle takes its origin from the lower eight ribs of the abdomen and inserts in the anterior midline of the abdomen. Its direction of pull is thus from front to back. A moment's reflection will reveal that the contraction of one external oblique muscle will rotate the column to the same side and will bring the opposite shoulder forward.

Internal oblique. This muscle has its fibers oriented at about 90 degrees to those of the external, and is the second layer of muscle in the abdomen. It arises from ilium and anterior abdomen and passes around the body to attach to the posterior surface of the last three to four ribs. Direction of pull this time is from back to front. Singly, the muscle rotates the column to the opposite side and brings the same shoulder forward.

Transverse abdominal. The fibers of this muscle run nearly circularly around the abdomen and form the third and deepest layer of the abdomen. The muscle arises posteriorly and passes to insert anteriorly. Singly, it rotates the column to the same side. In addition, all three of these muscles are important in compressing the abdomen.

MUSCLES OF THE THORAX

The muscles of the thorax are those which insert on the rib cage, and which are involved in breathing.

Movements possible (Fig. 9–26)

The ribs may be *elevated* and *depressed,* and the *vertical dimension* of the thorax may be *increased.*

Muscles involved

Muscles involved are (Fig. 9–23):
    External intercostals
    Internal intercostals
    Diaphragm
    Serratus posterior

External intercostals. Eleven pairs of external intercostal muscles lie superficially between the ribs. Their origins are from the lower border of the rib above and they insert on the upper border of the rib below. If the first rib is fixed in position, contraction elevates the ribs in a

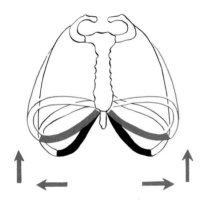

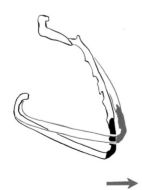

FIGURE 9–26. Movements of the ribs and thorax.

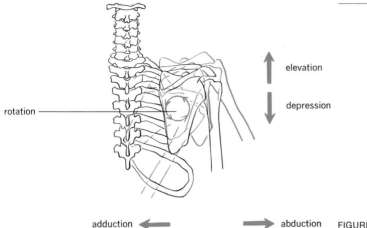

elevation

depression

rotation

adduction ← → abduction   FIGURE 9-27. Movements of the scapula.

"chain-like" fashion increasing both the lateral and anteroposterior dimensions of the thorax.

Internal intercostals. Also eleven pairs in number, the internal intercostals lie deep to the externals and their fibers are at 90 degrees to those of the externals. Origin of the internal intercostals is from the superior border of the rib below, and insertion is upon the lower border of the rib above (just opposite to the externals). Contraction thus depresses the ribs.

Diaphragm. The diaphragm is a dome-shaped muscle forming the floor of the thoracic cavity. Origin is from a circle at the level of the 10th rib, including ribs, xiphoid process and lumbar vertebrae. Insertion is upon a centrally placed tendon (central tendon). The muscle fibers are arranged more or less radially around the tendon, so that contraction tends to flatten the dome and increases the vertical dimension of the thorax.

Serratus posterior. This is a thin sheet of muscle on the posterior aspect of the thorax, extending from the spine to the posterior surface of the ribs. By contracting, the muscles tend to elevate the ribs.

MUSCLES OPERATING THE PECTORAL GIRDLE

Movements possible

The pectoral girdle is composed of scapulae and clavicles. Of the two bones, the clavicles are able to move only slightly, inasmuch as they have bony attachments on both ends. The clavicles may be *elevated* as the shoulders are elevated, and may be *depressed.* On the other hand, the scapulae have a bony attachment only at the clavicle, and exhibit a wide range of motion (Fig. 9-27).

Muscles involved

The muscles which operate the girdle are as follows:
    Posterior group (Fig. 9-28)
        Trapezius
        Rhomboid
        Levator scapulae

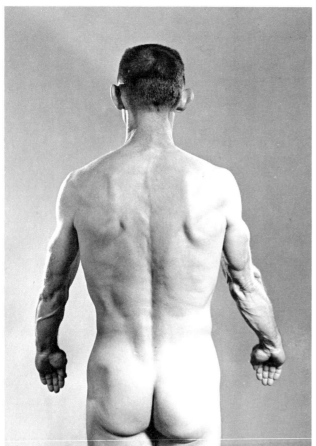

FIGURE 9-28. Posterior muscles operating the scapula.

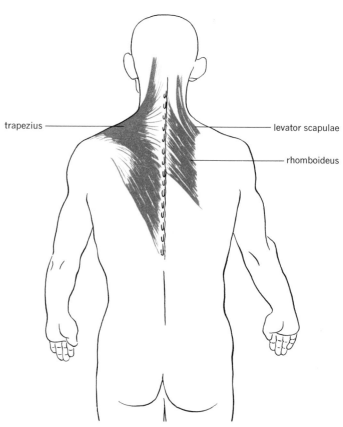

trapezius — levator scapulae — rhomboideus

Anterior group (Fig. 9–29)
Pectoralis minor
Serratus anterior
Subclavius

Trapezius. The trapezius is the large superficial muscle of the upper back. It is named for its trapezoidal shape. The origin is extensive, extending from the occipital bone to the 12th thoracic vertebra. It passes to insert upon the clavicle, and acromion and spine of the scapula. Its fibers are disposed in three directions, labeled on Fig. 9–28. The upper group (part I) elevates the clavicle and "shrug" the shoulder. The middle group (part II) is probably the strongest and adducts the scapulae, as in "bracing" at attention. The lower group (part III) depresses the scapula.

Rhomboid. The rhomboids lie deep to part III of the trapezius, taking origin from the 7th cervical to 5th thoracic vertebral spines. The insertion is the vertebral border of the scapula. The fibers run at an angle and contraction thus elevates *and* adducts the scapulae.

Levator scapulae. This muscle lies deep to part I of the trapezius. It extends from the transverse processes of the first four cervical vertebrae to insert on the superior angle of the scapula. The muscle runs nearly vertically up the neck, and thus its contraction elevates the scapula.

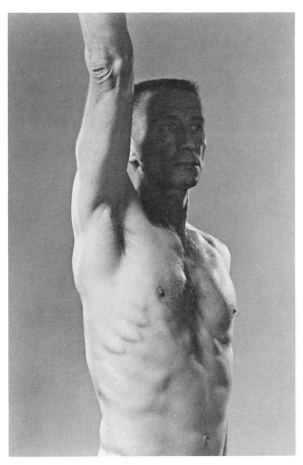

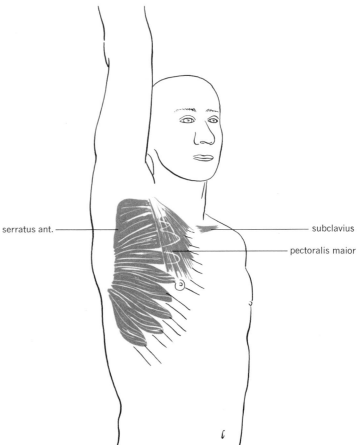

FIGURE 9–29. Anterior muscles operating the scapula.

Pectoralis minor. Pectoralis minor is a small muscle lying deep to the large pectoralis major on the chest. The pectoralis minor extends from the 3rd, 4th, and 5th ribs to the coracoid process of the scapula. Its action is to draw the scapula forwards and down. In most cases, these movements are largely prevented by other muscles, and the pectoralis minor becomes an elevator of the ribs, as in forced inspiration.

Serratus anterior. Serratus anterior arises from the lateral aspect of the upper nine ribs and sweeps around the thorax beneath the scapula to insert upon the vertebral border of the scapula. The direction of pull is thus one which abducts the scapulae. The muscle is strongly utilized in pushing motions.

Subclavius. This is the only muscle directly operating the clavicle. It originates from the 1st rib and inserts on the clavicle. It depresses and fixes the clavicle in position. It acts mostly as a synergistic muscle.

MUSCLES OPERATING THE SHOULDER JOINT

Movements possible

The shoulder is one of the freely movable joints in the body (probably *the* most freely movable joint). It exhibits the following movements (Fig. 9–30). The movements are chiefly those involving the humerus.

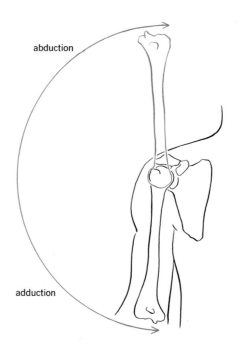

FIGURE 9–30. Movements of the shoulder joint.

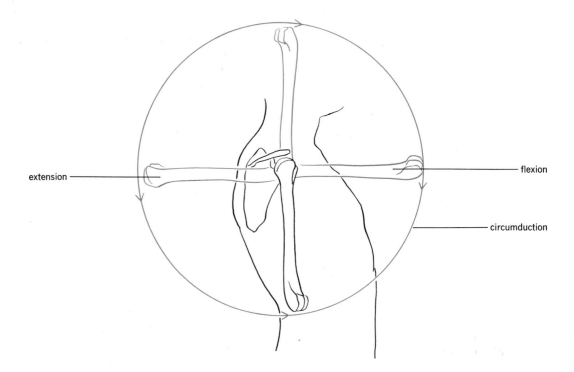

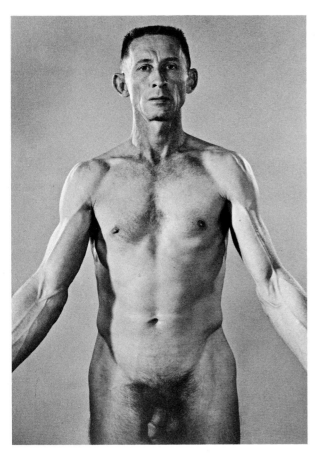

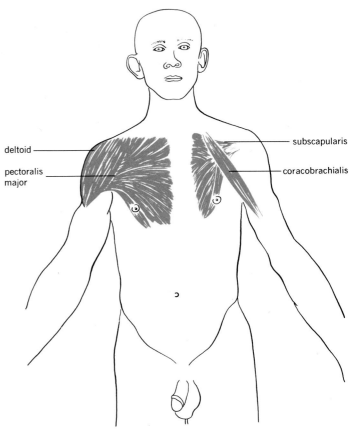

FIGURE 9–31A. Anterior muscles operating the shoulder joint.

Muscles involved

The muscles operating the shoulder joint (Fig. 9–31A and B) may be grouped by their primary actions as follows:

Flexors
    Pectoralis major
    Coracobrachialis
Extensors
    Latissimus dorsi
    Teres major
Abductors
    Supraspinatus
    Deltoid
Medial Rotators
    Subscapularis
    (Latissumus dorsi)
    (Pectoralis major)
Lateral rotators
    Infraspinatus
    Teres minor

Pectoralis major. Pectoralis major is the large, superficial, fan-shaped muscle of the chest. It originates from the clavicle, sternum and

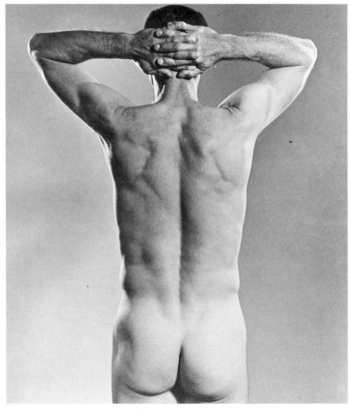

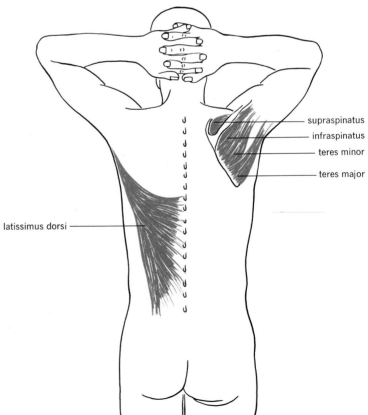

supraspinatus

infraspinatus

teres minor

teres major

latissimus dorsi

FIGURE 9–31B. Posterior muscles operating the shoulder joint.

costal cartilages of ribs 2 to 6 and inserts on the greater tubercle of the humerus. With the arms at the sides, contraction brings the humerus forward (flexion), the main action. However, the muscle also rotates the humerus medially. If the arm is horizontal, the pectoralis will bring the arm forwards and down. Working with the latissimus, pectoralis major will forcibly adduct the humerus.

Coracobrachialis. This muscle runs from the coracoid process of the scapula to the humerus. It is a weak flexor and adductor of the humerus.

Latissimus dorsi. Latissimus dorsi originates on a line beginning at the 6th thoracic spinous process, and continuing down the back to the sacrum. The fibers sweep to a tendon which inserts into the bicipital groove of the humerus. Latissmus extends, and rotates the arm medially, and, with pectoralis major, adducts the humerus.

Teres major. Teres major arises from the inferior angle of the scapula and inserts just below latissimus dorsi. It also extends, medially rotates, and adducts the humerus.

Supraspinatus. This muscle arises from the supraspinous fossa of the scapula, passes over the top of the shoulder joint, and inserts on the top of the greater tuberosity. The muscle abducts the humerus.

Deltoid. Deltoid overlies the shoulder joint. It has three sections, designated the anterior, middle, and posterior groups of fibers. The anterior fibers arise from the clavicle, the middle group from the acromion process, and the posterior group from the scapular spine. The muscle, acting as a whole, abducts the humerus. The anterior deltoid flexes the humerus, and the posterior deltoid extends it.

Subscapularis. Subscapularis fills the subscapular fossa and passes across the front of the shoulder joint to insert on the lesser tubercle of the humerus. The action of the muscle is to medially rotate the humerus.

Infraspinatus. Infraspinatus occupies the infraspinous fossa, and is inserted into the base of the greater tubercle. Action of the muscle is to laterally rotate the humerus.

Teres minor. Teres minor arises from the inferior angle of the scapula and inserts just below infraspinatus. It also laterally rotates the humerus.

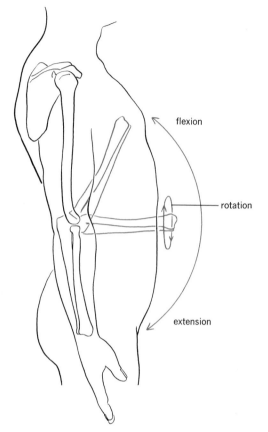

FIGURE 9–32. Movements of the forearm.

MUSCLES OPERATING THE FOREARM

Movements possible

The muscles operating the forearm originate for the most part on the scapula or humerus and insert on the radius or ulna. There are two joints involved in the elbow area: (1) the elbow joint proper, between the humerus and ulna, and (2) the joint between the humerus and radius. The two joints allow flexion, extension, and both medial and lateral rotation (pronation and supination respectively) (Fig. 9–32).

Muscles involved

The flexors (Fig. 9–33) lie on the anterior aspect of the humerus and include:
    Biceps brachii
    Brachialis
    Brachioradialis

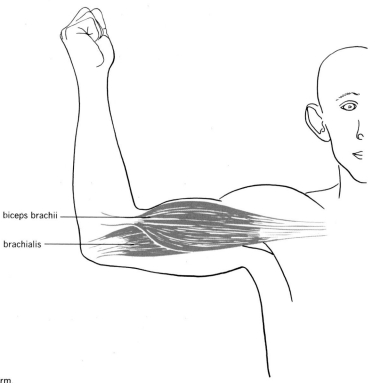

biceps brachii

brachialis

FIGURE 9–33. Flexors of the forearm.

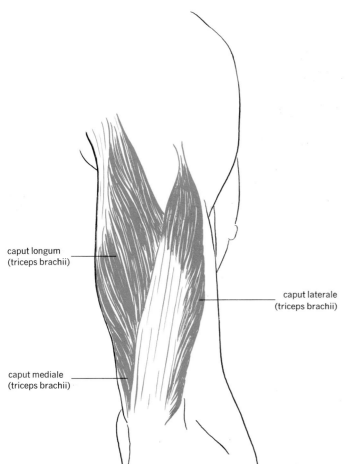

caput longum
(triceps brachii)

caput laterale
(triceps brachii)

caput mediale
(triceps brachii)

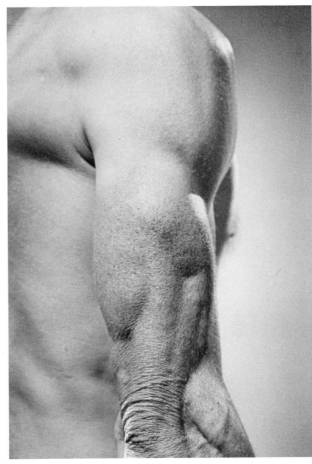

FIGURE 9–34. Extensors of the forearm.

The extensor (Fig. 9–34) lies on the lateral, posterior, and medial aspect of the humerus. It is the triceps brachii.

The rotators (Fig. 9–35) include the biceps brachii and

Supinator

Pronator teres

Pronator quadratus

Biceps brachii. As the name suggests, the biceps is a muscle having two heads. The *short head,* medially placed, arises from the coracoid process, while the *long head,* laterally placed, arises from the su-praglenoid tubercle of the scapula. A single tendon inserts on the radial tuberosity. Biceps flexes the forearm, and strongly rotates it laterally (supination) as in using a screwdriver.

Brachialis. Brachialis arises from the entire lower half of the anterior humerus and inserts on the coronoid process of the ulna. It lies beneath and is wider than the biceps and thus appears on either side of the biceps when viewed from the front. Brachialis is a pure flexor of the forearm.

Brachioradialis. Brachioradialis appears to arise from the brachialis and inserts on the radius. It flexes the forearm and returns it to a mid-point position from either full supination or full pronation.

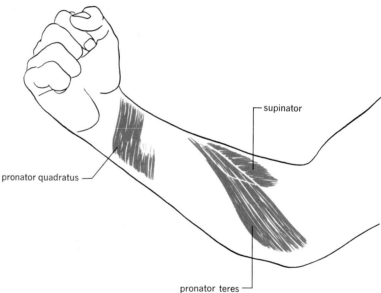

FIGURE 9-35. Rotators of the forearm.

Triceps brachii. Triceps is a three-headed muscle. The lateral head originates from the posterolateral surface of the distal humerus, the long head from the infraglenoid tubercle, and the medial head from the lower half of the posteromedial surface of the humerus. A common tendon inserts on the olecranon process of the ulna. All three heads extend the forearm.

Supinator runs from the lateral epicondyle of the humerus to the radius and laterally rotates (supinates) the forearm.

Pronator runs between the medial epicondyle and radius and pronates the forearm.

Pronator quadratus extends between the distal ends of radius and ulna and pronates the forearm.

MUSCLES OPERATING THE WRIST AND FINGERS

Movements possible (Fig. 9–36)

The *wrist* may be *flexed* and *extended, abducted* and *adducted.* The *fingers* are *extended* and *flexed* by muscles lying within the forearm Abduction and adduction of the fingers occurs by muscles within the hand itself.

Muscles involved

The muscles to be discussed next are many and varied, with difficult names. To understand their placement and actions, these muscles may be considered as forming two main groups. The muscles in the *an-*

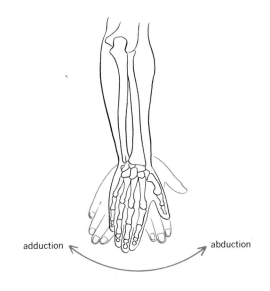

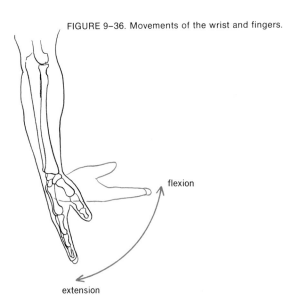

FIGURE 9–36. Movements of the wrist and fingers.

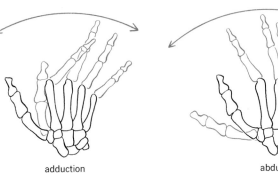

posterior view of the hand

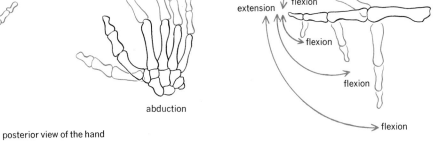

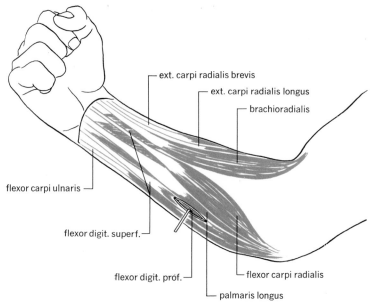

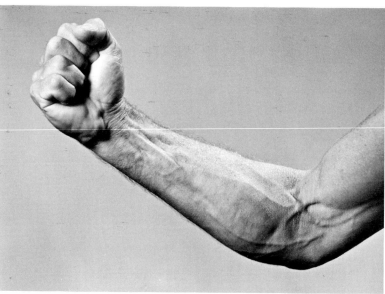

FIGURE 9–37. Muscles of the anterior forearm.

terior group originate primarily from the medial epicondyle of the humerus, lie on the anterior or palm side of the forearm, and are flexors of wrist and fingers. They are disposed in three layers relative to the radius and ulna (Fig. 9–37). The muscles, listed from lateral to medial, include:

    Anterior group
      Superficial layer
           Flexor carpi radialis ⎫
           Palmaris longus   ⎬ Flexors of wrist
           Flexor carpi ulnaris ⎭
    Intermediate layer:
      Flexor digitorum superficialis (sublimis)—flex fingers

Deep layer:

Flexor digitorum profundus—flex fingers

These muscles will not be described as to individual origins and insertions. Their actions however require additional consideration.

Flexor carpi radialis is located on the anterior lateral "corner" of the wrist. Its action, exerted singly, would be to both flex and abduct the wrist. *Palmaris longus* is the only muscle having only wrist flexion as its action. *Flexor carpi ulnaris,* operating singly, would flex and adduct the wrist. We thus see that the wrist may be moved forward (flexed) or this action may be combined with sideward motion. The tendons of the flexor digitorum superficialis arise from four separate heads and terminate at the distal end of the second phalanx of each finger. The tendons of *flexor digitorum profundus* go to the last phalanx of each finger. Both muscles are required therefore to "make a fist."

The *posterior group* (Fig. 9–38) of forearm muscles is disposed in a single layer. The muscles originate from the lateral epicondyle of the humerus and pass over the posterior aspect of the forearm to insert on the wrist or fingers, and extend those parts. The muscles, listed from lateral to medial, include:

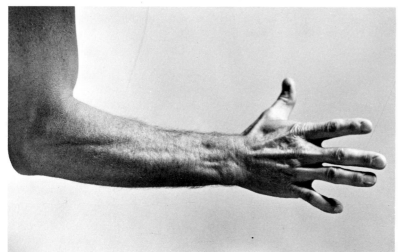

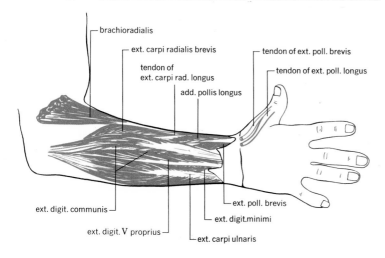

FIGURE 9–38. Posterior forearm muscles.

Extensor carpi radialis longus ⎫
Extensor carpi radialis brevis ⎭ extend wrist
Extensor digitorum communis—extend first three fingers
Extensor digiti minimi—extend little finger
Extensor carpi ulnaris—extend wrist

Again, it may be noted that the wrist extensors are located at the corners of the wrist. The wrist may therefore be caused to circumduct according to the interplay of both flexors and extensors.

THE THUMB

The thumb has its own supply of muscles separate from those operating the fingers. The muscles are in four groups, allowing for *flexion, extension, abduction* and *adduction* of the thumb, and circumduction by a sequence of contractions. The muscles are shown in Fig. 9–39.

MISCELLANEOUS MUSCLES OF THE HAND

The *intrinsic muscles* of the hand include the *lumbricales* and *interossei,* shown in Figure 9–39. These muscles enable abduction and adduction of the fingers to occur.

FIGURE 9–39. Muscle of the hand.

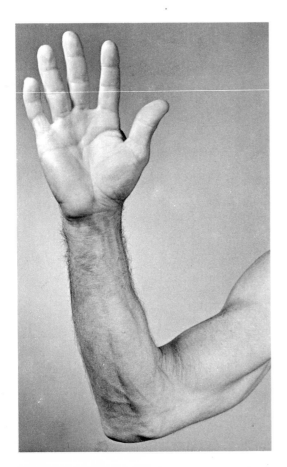

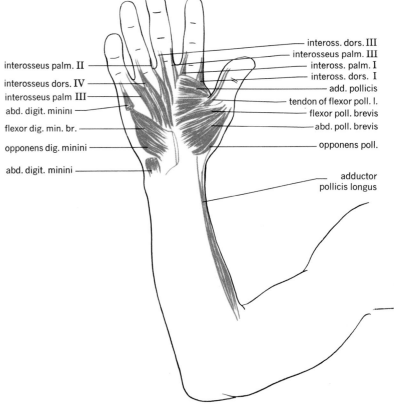

interosseus palm. II
interosseus dors. IV
interosseus palm III
abd. digit. minini
flexor dig. min. br.
opponens dig. minini
abd. digit. minini

inteross. dors. III
interosseus palm. III
inteross. palm. I
inteross. dors. I
add. pollicis
tendon of flexor poll. l.
flexor poll. brevis
abd. poll. brevis
opponens poll.

adductor pollicis longus

MUSCLES OPERATING THE HIP AND KNEE JOINT

Movements possible

The muscles operating the thigh and leg include a number of "two-joint" muscles, that is, muscles which cross two joints and may operate both simultaneously. Accordingly, it is somewhat difficult to consider the muscles as operating only a single joint. The movements of the hip and knee are shown in Fig. 9–40.

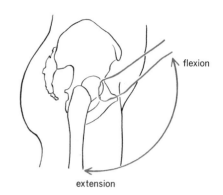

Muscles involved

A convenient functional classification of the muscles involved is given as follows:

Group 1. *Flexors of the hip and/or extensors of the knee.*

Iliacus ⎫
Psoas ⎬ Iliopsoas
Rectus femoris ⎫
Vastus lateralis ⎪
Vastus intermedius ⎬ Anterior thigh muscles
Vastus medialis ⎪
Sartorius ⎭

Group 2. *Extensors of the hip and/or flexors of the knee*

Gluteus maximus
Biceps femoris ⎫
Semitendinosus ⎬ "Hamstrings" ⎫ Posterior thigh
Semimembranosus ⎭ ⎬ muscles

Group 3. *Adductors of the hip*

Adductor magnus ⎫
Adductor longus ⎬ Medial thigh muscles
Adductor brevis ⎪
Gracilis ⎭

Group 4. *Abductors of the hip*

Gluteus medius ⎫
Gluteus minimus ⎬ Posterior hip
Tensor fascia lata ⎭

Group 5. *Lateral rotators of the hip*

Piriformis ⎫
Gemellus ⎪
Obturator ⎬ Posterior pelvis
Quadratus ⎭

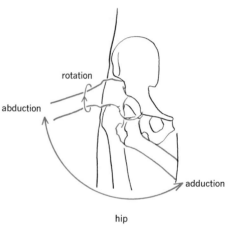

It may be noted that no separate group of muscles exists to medially rotate the thigh. This action is exhibited by muscles belonging to the other groups, primarily group 1.

Group 1. Muscles flexing the hip and/or extending the knee (Fig. 9–41).
Iliacus and Psoas. These two muscles have a common insertion; some investigators describe the muscle as iliopsoas, with two heads. The iliacus occupies the iliac fossa, while psoas arises from the lateral aspect of the last five

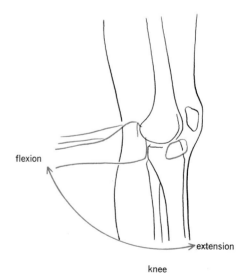

FIGURE 9–40. Movements of the hip and knee.

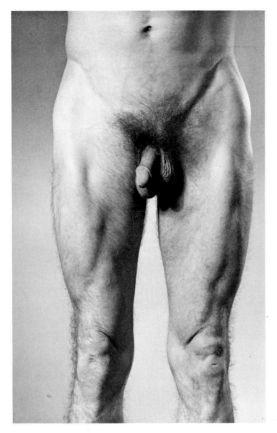

thoracic and all lumbar vertebrae. Insertion is on the lesser trochanter of the femur. The muscles flex the hip and rotate it medially.

Rectus femoris originates from the anterior inferior iliac spine and inserts on the patella. It lies superficially on the anterior aspect of the thigh. The muscle flexes the hip and extends the knee.

Vastus lateralis originates on the lateral margin of the linea aspera and occupies most of the lateral aspect of the thigh. It inserts on the patella, and extends the knee.

Vastus intermedius originates on the anterior aspect of the femur and inserts on the patella. It extends the knee. The muscle lies deep to the rectus femoris.

Vastus medialis originates on the medial margin of the linea aspera and inserts on the patella. It also extends the knee. The muscle forms a prominent bulge on the medial side of the lower part of the femur.

Sartorius is the exception to the general rule that two-joint muscles have opposite actions on the joints they cross. Sartorius originates on the anterior superior iliac spine, traverses the thigh from lateral to medial, and inserts upon the medial head of the tibia. It flexes both hip and knee. Sometimes called the "tailor's muscle," it is instrumental in crossing the knees in the manner of ancient tailors who used to sew upon their knees.

Group 2. Muscles extending the hip and/or flexing the knee (Fig. 9–42)

Gluteus maximus. The muscle of the buttock, gluteus maximus has an extensive origin extending from the posterior iliac crest onto the sacrum. It inserts upon the gluteal tuberosity of the femur. It extends and rotates the thigh laterally.

The next three muscles, the "hamstrings," are listed in order from lateral to medial.

Biceps femoris. Biceps femoris originates from the ischial tuberosity and inserts on the lateral tibial head. It extends the hip and flexes the knee.

Semitendinosus. This muscle also originates on the ischial tuberosity, and inserts on the medial tibial head. It, too, extends the hip and flexes the knee.

Semimembranosus. Origin is from ischial tuberosity and insertion is on medial tibial head. It, too, extends the hip and flexes the knee.

The hamstrings are important muscles of walking.

Group 3. Adductors of the thigh (Fig. 9–43). The adductors all originate on the pubis and/or ischium, and insert on the medial aspect of the femur or tibia.

Adductor magnus. This is the largest, having an extensive insertion along the entire medial border of the femur. It is largely hidden by the adductor longus and adductor brevis.

Adductor longus. This muscle lies superficially on the others, about half way down the thigh.

Adductor brevis. Brevis is superior and deep to the longus.

Gracilis. Gracilis is a strap-like muscle originating on the

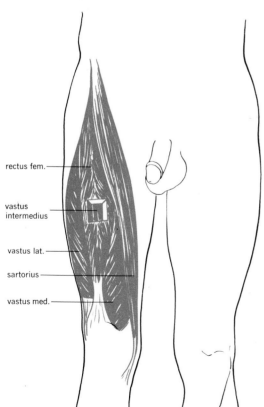

rectus fem.

vastus intermedius

vastus lat.

sartorius

vastus med.

FIGURE 9–41. Anterior thigh muscles.

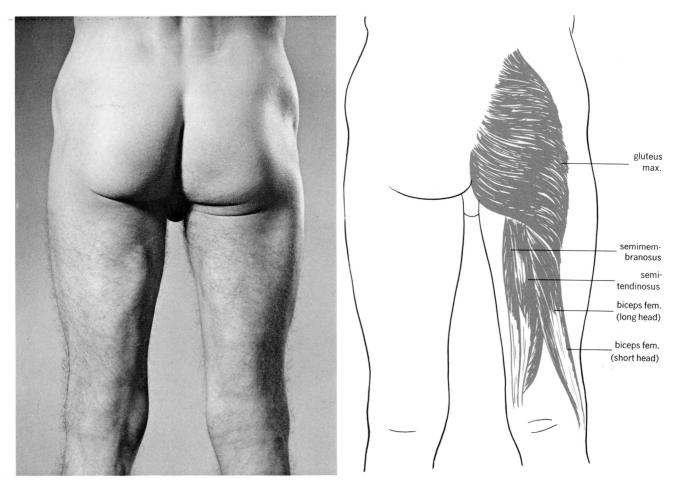

gluteus max.

semimem-
branosus

semi-
tendinosus

biceps fem.
(long head)

biceps fem.
(short head)

FIGURE 9–42. Posterior thigh muscles.

pubis and inserting on the medial tibial head. It is directly on the medial aspect of the thigh.

Pectineus lies superior to the longus and adducts the thigh.

Group 4. Abductors of the thigh (Fig. 9–44).

Gluteus medius. This muscle lies on the posterior surface of the ilium, partially covered by the gluteus maximus. It inserts on the greater trochanter, and it abducts and rotates the thigh medially.

Gluteus minimus. Gluteus minimus lies beneath the medius and has a similar insertion and identical action.

Tensor fascia lata. This muscle originates from the iliac crest and inserts in the upper end of the fascia lata. The fascia then passes to the lateral tibial head. The muscle abducts, slightly flexes, and rotates the thigh medially.

Group 5. Lateral rotators of the thigh (Fig. 9–44).

The piriformis, gemellus, obturator and quadratus form a superior to inferior group of muscles on the posterior aspect of the pelvis. All may be considered to have origins on the posterior sacrum, insertions on the greater trochanter, and the common action of lateral rotation of the thigh.

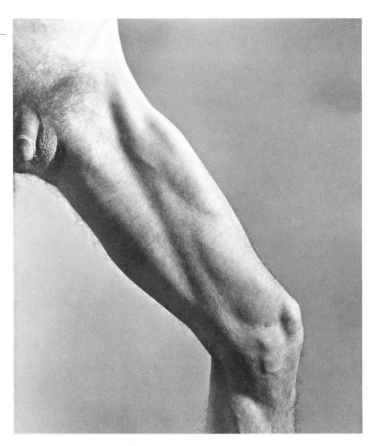

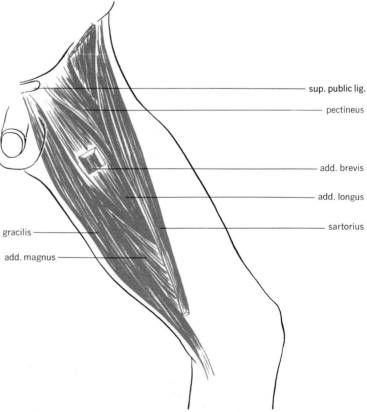

sup. public lig.

pectineus

add. brevis

add. longus

sartorius

gracilis

add. magnus

FIGU5E 9–43. Medial thigh muscles.

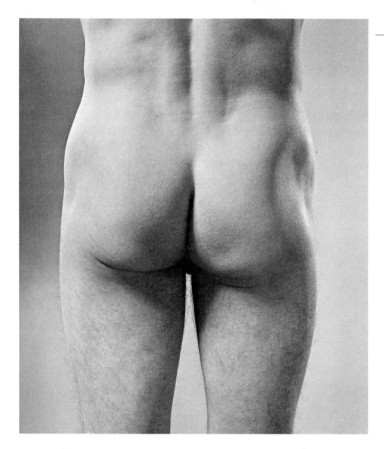

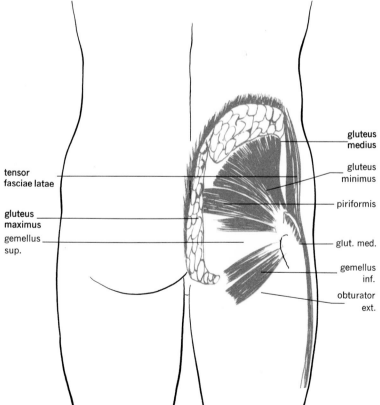

tensor
fasciae latae

gluteus
maximus

gemellus
sup.

gluteus
medius

gluteus
minimus

piriformis

glut. med.

gemellus
inf.

obturator
ext.

FIGURE 9–44. Posterior sacral muscles.

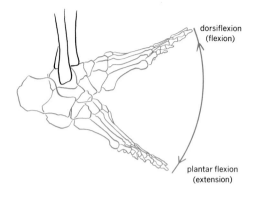

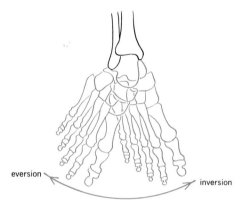

FIGURE 9–45. Movements of the ankle and foot.

MUSCLES OPERATING THE ANKLE AND FOOT

Movements possible (Fig. 9–45).

The ankle may be *flexed (dorsiflexion)* and *extended (plantar flexion)*. *Inversion* and *eversion* of the foot occurs as the tarsals move upon one another. The *toes* are *flexed* and *extended* by muscles lying on the leg; abduction and adduction of the toes occurs by intrinsic foot muscles. The muscles operating the ankle and foot are largely on the legs and send their tendons to operate the respective portion of ankle or foot.

Muscles involved

The muscles may be grouped into three masses as follows:

Anterior crural muscles (Fig. 9–46)

Peroneus tertius
Tibialis anterior
Extensor digitorum longus
Extensor hallucis longus

These muscles occupy the anterolateral aspect of the leg. They dorsiflex the ankle, invert the foot, and extend the toes.

Posterior crural muscles (Fig. 9–47)
Superficial
Gastrocnemius
Soleus
Plantaris
Deep
Tibialis posterior
Flexor digitorum longus
Flexor hallucis longus

These muscles occupy the posterior aspect of the leg. They plantar flex the ankle and flex the toes.

Lateral crural muscles (Fig. 9–46)

Peroneus longus
Peroneus brevis

These occupy the lateral aspect of the leg. They plantar flex and evert the foot.

Anterior crural muscles.
Tibialis anterior originates from the lateral surface of the tibia; the tendon of this muscle crosses the front of the ankle and inserts on the first cuneiform and metatarsal. It dorsiflexes and inverts the foot.

Extensor digitorum longus arises from tibia and fibula, and is partially beneath the tibialis anterior. It is a single muscle, giving rise to four tendons which pass over the top of the foot to insert upon the phalanges of the toes. When the muscle contracts, all four toes are extended together. There is no independent control of toe movement.

Extensor hallucis longus lies medial and deep to the tibialis anterior. Its tendon inserts on the top of the great toe and extends it.

Peroneus tertius lies on the dorsum of the foot. Its tendon inserts into the 5th metatarsal and the muscle dorsiflexes the foot.

Posterior crural muscles (superficial).

Gastrocnemius. Originating from the posterior aspect of the femur condyles, the two-headed gastrocnemius forms the great muscle of the calf. It inserts by way of the tendocalcaneus (tendon of Achilles) into the heel. The muscle is the most powerful plantar flexor of the ankle. Running, jumping, and walking cannot be performed if the tendon of this muscle is severed.

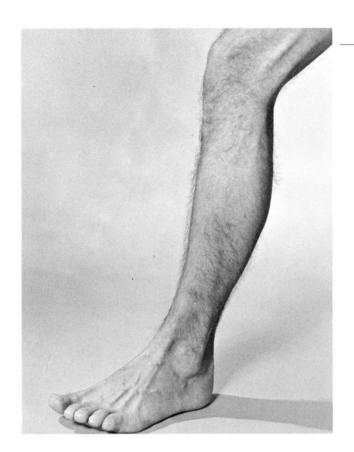

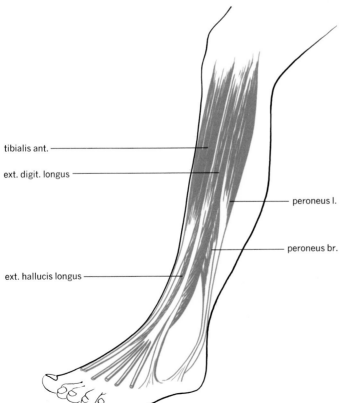

tibialis ant.

ext. digit. longus

peroneus l.

peroneus br.

ext. hallucis longus

FIGURE 9–46. Anterior crural and lateral crural muscles.

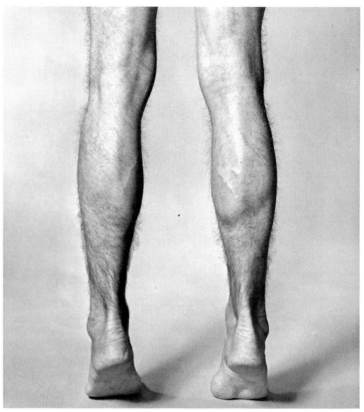

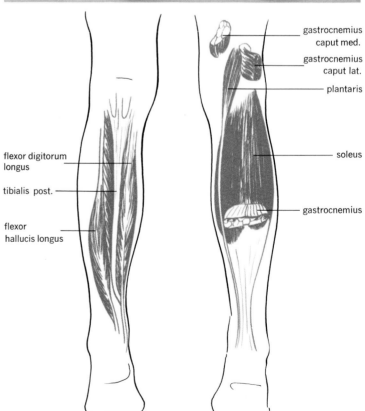

gastrocnemius
caput med.

gastrocnemius
caput lat.

plantaris

soleus

gastrocnemius

flexor digitorum
longus

tibialis post.

flexor
hallucis longus

FIGURE 9–47. Posterior crural muscles.

Soleus. Soleus originates from the upper posterior portion of the tibia. It has a common insertion and action with grastrocnemius.

Plantaris. Plantaris originates from the lateral condyle of the femur and inserts into the heel. It is a weak plantar flexor.

Posterior crural muscles (deep).

Tibialis posterior lies on the posterior side of the tibia, where it takes its origin. The tendon passes beneath the medial malleolus and inserts upon the under surface of navicular, cuneiform, and cuboid. The muscle plantar flexes the ankle.

Flexor digitorum longus arises from the upper posterior portion of the tibia. Its tendon passes beneath the medial malleolus, splits into four parts to insert on the under surface of the phalanges of the toes. It flexes all four toes at once.

Flexor hallucis longus arises from the posterior surface of the fibula. Its tendon passes beneath the medial malleolus and inserts upon the under surface of the great toe. It flexes the great toe.

Lateral crural muscles.

Peroneus longus and brevis. These muscles originate from the lateral surface of the fibula. Their tendons pass beneath the lateral malleolus to insert on the metatarsals 1 and 5. They plantar flex and evert the foot.

INTRINSIC FOOT MUSCLES

The foot, like the hand, is possessed of muscles originating and inserting within the foot. These muscles are labeled on Figs. 9–48 and 9–49. The muscles create a stable platform for support and locomotion.

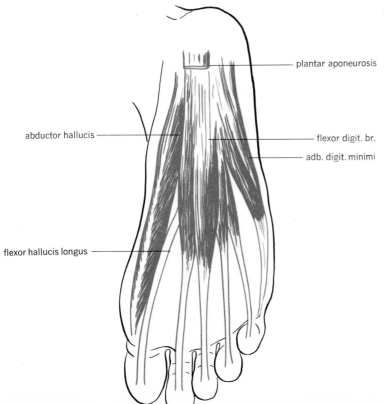

FIGURE 9–48. Superficial muscles of the sole of the foot.

plantar aponeurosis

abductor hallucis

flexor digit. br.

adb. digit. minimi

flexor hallucis longus

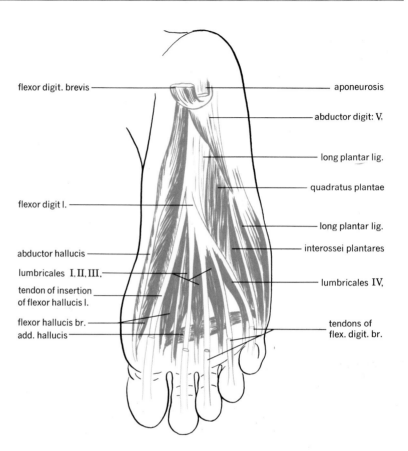

flexor digit. brevis — — aponeurosis

— abductor digit: V.

— long plantar lig.

— quadratus plantae

flexor digit I. — — long plantar lig.

— interossei plantares

abductor hallucis — — lumbricales IV.

lumbricales I, II, III,

tendon of insertion
of flexor hallucis I.

flexor hallucis br. — tendons of
add. hallucis — flex. digit. br.

FIGURE 9–49. Deep muscles of the sole of the foot.

TABLE 9–2

Muscles of the Body Arranged Alphabetically
(Includes those Mentioned or Diagrammed in the Text)

Muscle	Page Text	Page Figure	General Location	Origin	Insertion	Action
Abductor digiti minimi (foot)		48	Sole of foot	Calcaneus	First phalanx 5th toe	Abducts 5th toe
Abductor hallucis		48, 49	Sole of foot	Calcaneus	First phalanx great toe	Abducts 1st toe
Abductor pollicis brevis	258	39	Base of thumb on palm	Scaphoid multangular	First phalanx thumb	Abducts thumb
Adductor brevis	260	43	Medial thigh	Inferior pubic ramus	Upper third linea aspera	Adducts thigh (some flexion and medial rotation)
Adductor hallucis		49	Sole of foot	Metatarsals	First phalanx great toe	Adduction great toe

TABLE 9–2 *con't.*

Muscles of the Body Arranged Alphabetically
(Includes those Mentioned or Diagrammed in the Text)

Muscle	Page Text	Page Figure	General Location	Origin	Insertion	Action
✓ Adductor longus	260	43	Medial thigh	Crest and symphysis of pubis	Middle third linea aspera	Adducts thigh (some flexion and medial rotation)
✓ Adductor magnus	260	43	Medial thigh	Ramus of pubis ischium	Lower third linea aspera	Adducts, flexes, medially rotates thigh
Adductor pollicis	258	39	Between thumb and index finger palm of hand	Capitate, multangulars, metacarpals 2 to 4	First phalanx thumb	Adducts thumb
Auricularis Anterior Inferior Superior	235	13	Anterior to pinna Posterior to pinna Superior to pinna	Fascia of scalp Fascia of scalp Fascia of scalp	Pinna Pinna Pinna	Moves pinna forward Elevates pinna Moves pinna backward
✓ Biceps brachii	253	33	Anterior upper arm	Long head; supra glenoid tubercle Short head; coracoid process	Radial tuberosity	Flexes and supinates forearm
✓ Biceps femoris	260	42	Posterior thigh	Long head; ischial tuberosity Short head; linea aspera	Lateral head fibula	Extends thigh, flexes knee
✓ Brachialis	253	33	Anterior upper arm	Lower half anterior humerus	Coronoid process (ulna)	Flexes forearm
✓ Brachioradialis	253	37	Lateral forearm	Lateral supracondylar ridge (humerus)	Styloid process (radius)	Flexes forearm (semi-pronates semisupinates)
✓ Buccinator	235	12	Cheek of face	Alveolar processes of maxillae and mandible	Orbicularis oris fibers	Compresses cheek
✓ Coracobrachialis	251	31	Medial upper arm	Coracoid process	Midhumerus	Flexes and adducts humerus
Corrugator	235	12	Forehead	Inner end superciliary ridge	Skin above orbit	Produces frown

TABLE 9–2 *con't.*

Muscles of the Body Arranged Alphabetically
(Includes those Mentioned or Diagrammed in the Text)

Muscle	Page Text	Figure	General Location	Origin	Insertion	Action
√ Deltoid	251	31	Top of shoulder	Clavicle, acromion and spine of scapula	Deltoid tuberosity	Abducts, flexes extends humerus
Depressor labii inferioris	234	12	Inferior to lower lip	Mental process	Skin of lower lip	Depresses lower lip
√ Diaphragm	245	23	Between thoracic and abdominal cavities	Xiphoid process, last 6 ribs, lumbar vertebrae	Central tendon	Increases vertical dimension of thorax
Digastricus	236	15	Runs from mastoid process to hyoid	Posterior belly: mastoid notch	Greater horn, hyoid	Elevates hyoid, opens mouth
				Anterior belly: mandible	Body of hyoid	Draws hyoid forward
Epicranius Frontalis	235	13	Forehead	Muscles above orbit	Galea	Raises eyebrows
Occipitalis	235	13	Base of skull	Superior nuchal line and mastoid temporal	Aponeurotica	Pulls scalp backwards
Extensor carpi radialis brevis	258	38	Posterior forearm	Lateral epicondyle humerus	Top of 3rd metacarpal	Extends and abducts wrist
Extensor carpi radialis longus	258	38	Posterior forearm	Lateral epicondyle humerus	Top of 2nd metacarpal	Extends and abducts wrist
Extensor carpi ulnaris	258	38	Posterior forearm	Lateral epicondyle humerus	Base of 5th metacarpal	Extends and adducts wrist
Extensor digiti minimi	258	38	Posterior forearm	Lateral epicondyle, humerus	Top of 1st phalanx, 4th finger	Extends 4th finger
Extensor digitorum communis	258	38	Posterior forearm	Lateral epicondyle, humerus	2nd and 3rd phalanges of fingers	Extends fingers
Extensor digitorum longus	264	46	Anterolateral leg	Lateral tibial condyle, distal fibula	2nd and 3rd phalanges, four outer toes	Extends four outer toes
Extensor hallucis longus	264	46	Anterior fibula	Anterior surface of fibula	Top of great toe	Extends great toe

TABLE 9–2 *con't.*

Muscles of the Body Arranged Alphabetically
(Includes those Mentioned or Diagrammed in the Text)

Muscle	Page Text	Page Figure	General Location	Origin	Insertion	Action
Extensor pollicis brevis			Posterior radius	Dorsal surface of radius	1st phalanx of thumb	Extends thumb
Extensor pollicis longus			Posterior ulna	Dorsal surface of ulna	2nd phalanx of thumb	Extends thumb
✓ External intercostals	244	23	Between ribs— superficial	Inferior border of rib	Superior border of rib below	Elevates ribs
✓ External oblique	244	11	Abdomen— superficial	Anterior inferior surface, lower 8 ribs	Linea alba, pubis, iliac crest	Flexion, rotation of spine
Flexor carpi radialis	257	37	Anterior forearm superficial	Medial epi- condyle, humerus	Base 2nd metacarpal	Flexes and abducts wrist
Flexor carpi ulnaris	257	37	Anterior forearm superficial	Medial epicondyle of humerus, distal ulna	Base 5th metacarpal pisiform	Flexes and adducts wrist
Flexor digitorium longus	267	47	Posterior tibia	Posterior surface tibial shaft	Distal phalanges 4 outer toes	Flexes toes
Flexor digitorum profundus	257	37	Anterior forearm deep	Proximal three- fourths ulna	Base of distal phalanges	Flexes fingers
Flexor digitorum superficials	257	37	Anterior forearm middle	Medial epi- condyle, humerus; coronoid process, ulna; radial shaft	Second phalanges of fingers	Flexes fingers
Flexor hallucis longus	267	47	Posterior fibula	Lower two-thirds fibula	Undersurface of phalanges of great toe	Flexes great toe
Flexor pollicis brevis	258	39	Base of thumb on palm	Multangulars	Base of 1st phalanx of thumb	Flexes and adducts thumb
Flexor pollicis longus			Base of thumb on palm	Anterior surface, radius; coronoid process, ulna	Base of termi- nal phalanx of thumb	Flexes thumb
✓ Gastrocnemius	264	47	Calf of leg, superficial	Posterior surface, condyles of femur	Calcaneus	Plantar flexion flexes knee

TABLE 9-2 *con't.*

Muscles of the Body Arranged Alphabetically
(Includes those Mentioned or Diagrammed in the Text)

Muscle	Page Text	Figure	General Location	Origin	Insertion	Action
Gemellus	259	44	Posterior pelvis	Inferior: Ischial tuberosity Superior: Ischial spine	Greater trochanter	Rotates thigh laterally
Gluteus maximus	260	42	Buttocks, superficial	Gluteal line of ilium	Gluteal tuberosity, femur	Extends and laterally rotates thigh
Gluteus medius	261	44	Buttock, intermediate	Outer surface of ilium	Greater trochanter	Abducts and medially rotates thigh
Gluteus minimus	261	44	Buttock, deep	Outer surface of ilium	Greater trochanter	Abducts and medially rotates thigh
Gracilis	260	43	Medial thigh	Symphysis pubis	Medial head tibia	Adducts thigh
Iliacus	241 259	23	Iliac fossa	Iliac fossa and crest, sacrum	Lesser trochanter	Flexes and medially rotates thigh
Iliocostalis	241	24, 25	Midback	Iliac crest, ribs	Ribs and cervical transverse processes	Extends spine
Infraspinatus	251	31	Scapula below spine	Infraspinous fossa	Greater tubercle, humerus	Lateral rotator of humerus
Internal intercostals	245	23	Between ribs, deep	Inner surface of rib	Superior surface of rib below	Depresses ribs
Internal oblique	244		Abdomen, middle layer	Iliac crest, fascia of back	Lower 3 ribs, linea alba, xiphoid process	Flexes spine, compresses abdomen
Interossei Foot	268	49	Between metatarsals	Adjacent sides of metatarsals	Bases 1st phalanges	Abducts toes flexes "knuckles"
Hand	258	39	Between metacarpals	Adjacent sides of metacarpals	Bases 1st phalanges	Abducts fingers flexes "knuckles"
Latissimus dorsi	251	31	Lower back, superficial	Spinous processes T6-L5, iliac crest last 3 ribs	Bicipital groove, humerus	Extension, adduction, medial rotation humerus
Levator labii superioris	234	12	Above upper lip	Lower rim of orbit	Skin of upper lip	Elevates upper lip

TABLE 9-2 *con't.*

Muscles of the Body Arranged Alphabetically
(Includes those Mentioned or Diagrammed in the Text)

Muscle	Text	Figure	General Location	Origin	Insertion	Action
Levator scapulae	246	21, 28	Posterior neck	Transverse processes C1-4	Superior angle, scapula	Elevates scapula
Longissimus	239 241	20, 21 24	Midback	Transverse processes C5-T5	Mastoid process, skull	Extends spine, bends it to side
Longus capitus	237	18, 21	Lateral neck	Transverse processes C3-6	Occipital bone	Flexes neck
Longus colli	237	18, 21	Anterior neck, deep	Transverse processes and bodies C3-T7	Atlas, cervical bodies	Flexes and rotates neck
Lumbricales Foot	268	49	Plantar surface metatarsals	From flexor digitorum longus	Tendons of extensorium longus	Flexes toes
Hand	258	39	Anterior surface metacarpals	Tendons of flexor digitorum profundus	Tendons of extensor digitorum communis	Flexes knuckles
✓ Masseter	236	12, 13	Side of mandibular ramus	Zygomatic process of maxilla, zygomatic arch	Ramus, angle and coronoid process of mandible	Elevates mandible
Mentalis	235	12	Chin	Mental symphysis	Skin of chin and lower lip	Depresses lip, wrinkles chin
Mylohyoid	236	15	Floor of mouth	Mylohyoid lines	Hyoid body	Elevates hyoid and tongue
Obturator	259	44	Posterior pelvis	Externus: Rim of obturator foramen Internus: Rim of obturator foramen, pubis, ischium	Greater trochanter	Laterally rotates thigh
Omohyoid	236	16, 21	Shoulder to hyoid	Coracoid process	Body of hyoid	Depresses hyoid
✓ Orbicularis oculi	234	12	Around orbit	Medial surface of orbit	Skin of eyelid	Closes eye
✓ Orbicularis oris	234	12	Around mouth	Skin of lips, fibers of other facial muscles	Corners of mouth	Closes and puckers lips
Palmaris longus	257	37	Anterior forearm	Medial epicondyle humerus	Palm of hand	Flexes wrist

TABLE 9–2 *con't.*

Muscles of the Body Arranged Alphabetically
(Includes those Mentioned or Diagrammed in the Text)

Muscle	Page Text	Page Figure	General Location	Origin	Insertion	Action
√ Pectoralis major	249	31	Chest, superficial	Clavicle, sternum, costal cartilages true ribs	Bicipital groove	Flexes, adducts, medially rotates humerus
√ Pectoralis minor	247	29	Chest, deep	Ribs 3 to 5	Coracoid process	Depresses shoulder
Peroneus brevis	267	46	Lateral leg	Lower two-thirds fibula	Base 5th metatarsal	Plantar flexes and everts foot
Peroneus longus	267	46	Lateral leg	Upper two-thirds fibula	Base 1st metatarsal, 1st cuneiform	Plantar flexes and everts foot
Peroneus tertius	264		Anterior ankle	Anterior fibula	Dorsal surface 5th metatarsal	Dorsiflexion of foot
Piriformis	259	44	Posterior pelvis	Anterior sacrum	Greater trochanter	Lateral rotation thigh
Plantaris	267	47	Posterior lower thigh	Linea aspera	Calcaneus	Plantar flexion
Platysma	236	17, 21	Anterior neck	Fascia of deltoid and pectoralis major	Skin of lower face	Depresses lower lip
Pronator teres	254	35	Anterior upper forearm	Medial epicondyle, humerus; and coronoid of ulna	Midradius	Pronation
Psoas	241 259	23	Posterior wall of pelvic cavity	Transverse processes lumbar vertebrae	Lesser trochanter	Flexes thigh
Pterygoid	236	14				
Lateralis			Medial to ramus of mandible	Great wing of sphenoid, lateral pterygoid lamnia	Condyloid process	Protrudes and opens jaw
Medialis			Medial to lateralis	Lateral pterygoid lamnia, palatine, maxilla	Ramus and angle of mandible	Closes jaw
√ Quadratus	259		Posterior pelvis	Ischial tuberosity	Femur	Laterally rotates thigh
Quadratus lumborum	241	25	Between last rib and iliac crest	Iliac crest, lower lumbar vertebrae	Rib 12 and upper lumbar vertebrae	Flexes lumbar spine

TABLE 9–2 *con't.*

Muscles of the Body Arranged Alphabetically
(Includes those Mentioned or Diagrammed in the Text)

Muscle	Page Text	Page Figure	General Location	Origin	Insertion	Action
Quadratus plantae		49	Sole of foot in front of heel	Calcaneus	Tendons of flexor digitorum longus	Flexes 4 outer toes
√Rectus abdominus	241	23	Midabdomen	Pubic crest	Cartilages of ribs 5 to 7	Flexes spine
Rectus capitus	237	18	Anterior neck, deep	Atlas	Base of occipital bone	Bends of neck laterally
√Rectus femoris	260	41	Anterior thigh, superficial	Anterior inferior iliac spine, acetabulum	Patella	Flexes hips, extends knee
√Rhomboid	246	28	Back, deep	Spinous processes T1-5	Lower third vertebral border of scapula	Adducts and elevates scapula
Risorius	234	12	Lateral to mouth	Buccinator fascia	Skin of corners of mouth	Pulls mouth laterally
√Sartorius	260	41, 43	Anterior thigh	Anterior superior iliac spine	Medial head, tibia	Flexes hip and knee
Scalenes	239	19, 21	Lateral neck	Transverse processes cervical vertebrae	1st and 2nd ribs	Elevate ribs and rotate neck
√Semimembranosus	260	42	Posterior thigh	Ischial tuberosity	Medial condyle of tibia	Extends hip, flexes knee
Semispinalis	239 241	20, 21 24	Midback	Transverse processes lower cervical and thoracic vertebrae	Spinous processes cervical vertebrae and occipital	Extends and rotates spine
√Semitendinosus	260	42	Posterior thigh	Ischial tuberosity	Medial tibial shaft	Extends hip, flexes knee
√Serratus anterior	247	29	Lateral thorax	Upper 8 to 9 ribs	Vertebral border, scapula	Abducts scapula
Serratus posterior	245		Back, deep	Spinous processes thoracic and lumbar vertebrae	Ribs	Increases lateral dimensions of thorax

TABLE 9–2 con't.

Muscles of the Body Arranged Alphabetically (Includes those Mentioned or Diagrammed in the Text)

Muscle	Page Text	Page Figure	General Location	Origin	Insertion	Action
Soleus	267	47	Calf, deep	Head of fibula, upper tibia	Calcaneus	Plantar-flexion
Spinalis	see semispinalis		Posterior neck	Same as semi-spinalis	Same as semi-spinalis	Extends spine
Splenius	239	20, 21	Posterior neck	Spinous processes lower cervical and upper thoracic vertebrae	Nuchal lines and cervical transverse processes	Extension of neck, rotation
√ Sternocleidomastoid	236	17, 21	Lateral neck	Sternum and clavicle	Mastoid temporal	Rotates head, flexes neck
Sternohyoid	236	16, 21	Anterior neck	Sternum	Body of hyoid	Depresses hyoid
Sternothyroid	236	16, 21	Anterior neck	Sternum	Thryoid carti-lage of larynx	Depresses larynx
Stylohyoid	236	15	Runs from styloid process of skull to hyoid bone	Styloid process	Body of hyoid	Elevates and retracts hyoid
Subclavius	247	29	Beneath clavicle	Middle of 1st rib	Clavicle	Depresses clavicle
√ Subscapularis	251	31	Scapula, anterior	Subscapular fossa	Lesser tubercle, humerus	Medially rotates humerus
√ Supinator	254	35	Anterior upper forearm	Lateral epicondyle humerus	Radial shaft	Supination
√ Supraspinatus	251	31	Scapula, above spine	Supraspinous fossa	Greater tubercle, humerus	Abducts humerus
√ Temporalis	236	13	Lateral skull	Temporal fossa	Coronoid and ramus, mandible	Closes jaws
√ Tensor fascia lata	261	44	Lateral hip	Anterior superior iliac spine	Fascia lata	Abducts and flexes thigh
√ Teres major	251	31	Inferior angle of scapula to humerus	Inferior angle of scapula	Lesser tubercle, humerus	Adduction, extension, medial rotation humerus
Teres minor	251	31	Above teres major	Axillary border, scapula	Greater tu-bercle, humerus	Lateral rotation of humerus

TABLE 9–2 *con't.*

Muscles of the Body Arranged Alphabetically
(Includes those Mentioned or Diagrammed in the Text)

Muscle	Text	Figure	General Location	Origin	Insertion	Action
Thyrohyoid	236	16	Larynx to hyoid	Thyroid cartilage of larynx	Greater horn of hyoid	Elevates larynx
Tibialis anterior	264	46	Anterolateral leg	Upper two-thirds tibia	First cuneiform and 1st meta-tarsal	Dorsiflex and invert foot
Tibialis posterior	267	47	Posterior leg	Shaft of tibia and fibula	Navicular, cal-caneus, all cuneiforms	Plantar flexion and inversion of foot
Transverse abdominal	244		Abdomen, deep	Iliac crest, lumbar fascia, last 6 ribs	Xiphoid process linea alba, pubis	Compresses abdomen
Trapezius	239 246	21 28	Upper back, superficial	Superior nuchal line, ligamentum nuchae, spines C7-T12	Clavicle, spine and acromion of scapula	Elevates, adducts depresses scapula
Triangularis	234	12	Below corners of mouth	Body of mandible	Skin of lower lip	Depresses lower lip
Triceps	254	34	Posterior humerus	Lateral head humerus Long head scapula Medial head humerus	Olecranon process	Extends elbow
Vastus intermedius	260	41	Anterior thigh, deep	Anterior and lateral femur shaft	Patella	Extends knee
Vastus lateralis	260	41	Lateral thigh	Upper half, lateral linea aspera	Patella	Extends knee
Vastus medialis	260	41	Medial thigh	Upper half, medial linea aspera	Patella	Extends knee
Zygomaticus	234	12	Above corners of mouth	Zygomatic bone	Skin of corners of mouth	Draws mouth up and back

CHAPTER 10

Blood

Blood will tell, but often it tells too much. MARQUIS

THE CIRCULATORY SYSTEM

Introduction

Single-celled or very simple organisms living in a fluid environment utilize that environment as a source of nutrients and as the route of excretion of cellular wastes. As numbers of cells in the organism increases, it becomes inevitable that deeper lying cells are farther and farther removed from contact with the environment. Supplying such cells with nutrients and removing their wastes becomes an increasingly difficult task. In primitive animals, such as the jellyfish, these problems are partially solved by providing channels among the cells which com-

281

municate with the environment. The fluid is caused to ebb and flow through the channels as the jellyfish pulses. A primitive, open circulatory system has been provided. As demands for nutrients increase, an open, ebb and flow type of system becomes insufficient to supply adequate levels of nutrients. The fluid must be circulated more rapidly and must be directed in its flow. By closing the vessels or channels from direct contact with the environment, and by providing a pump to circulate the contained fluid, the necessary criteria may be met.

The three components that have evolved then are: a *fluid* to circulate, a *pump* to circulate the fluid, and *vessels* to contain the fluid. The human circulatory system is a *closed system* and has two subdivisions: (1) the *blood vascular system,* including the blood, heart, and blood vessels, and (2) the *lymph vascular system,* including the lymph, lymph vessels, and lymph organs.

THE BLOOD

General description of properties and functions

Blood comprises about 8 percent of the total body weight of an adult individual, and amounts to between 5 and 6 liters in volume in an average-sized man. Its main functions center about its ability to dissolve or suspend substances and therefore *transport* such materials through the body. The functions of the blood include:

1. *Transport*
    (a) Of *nutrients* acquired from digestive organs, such as amino acids, fats, and sugars.
    (b) Of *oxygen* acquired in the lungs.
    (c) Of *wastes,* acquired from the cells and carried to appropriate organs of excretion, such as carbon dioxide, urea, and uric acid.
    (d) Of *hormones,* acquired from specialized endocrine glands, to all body cells.
    (e) Of *enzymes,* some necessary to specialized reactions occurring within the blood, and others necessary to body cells in general.
2. *Regulation of the water content of the cells and tissues.* The blood stream, or more properly the plasma, is the source of all the fluids in other body fluid compartments. According to its osmotic and hydrostatic pressures, water flow into or out of the plasma will occur.
3. *Buffering.* The content in the blood of materials capable of taking up H^+ is important in regulation of body pH.
4. *Regulation of body temperature.* The water content of the blood enables it to absorb heat and carry the heat to areas of elimination.
5. *Prevention of blood loss* through injured vessels. By its ability to *coagulate* or clot, the blood changes from a liquid to a gelated condition and may prevent blood loss through "damming" a leak in a vessel.
6. *Protection.* Through chemical substances (antibodies) and specialized cellular elements found in the blood stream, protection against toxins or microorganisms is afforded.

Components of the blood

Circulating blood, and blood appearing at the body surface seems to be a homogeneous red fluid. Centrifugation will separate the blood into two phases, a straw-colored liquid, the *plasma,* and a mass of cells or cell-like bodies, the *formed elements.* The volume of packed red cells which results from centrifugation is designated as a packed cell volume (PCV), or as the *hematocrit.* If allowed to clot, a third phase, the *serum,* will be formed. Serum is plasma with some clotting elements (chiefly fibrinogen) removed. Plasma composes, on the average, 55 to 57 percent of the total blood volume; the formed elements, 43 to 45 percent.

PLASMA

Composition

Plasma composition closely resembles that of the living substance of cells themselves.

1. *Water,* 91 to 92 percent. As indicated in Chapter 2, water is a good *solvent* or suspending medium, *absorbs much heat* with only slight alterations in its own temperature, and *moves easily* under pressure. These properties are essential to the transport of substances and regulation of temperature.
2. *Inorganic substances,* 0.9 percent. The inorganic materials are summarized in Table 10–1.
3. *Organic substances.*
   A. Plasma proteins, 7 to 9 percent. Three major protein fractions may be separated by precipitation or electrophoresis.* These fractions are designated as *albumins, globulins,* and *fibrinogen.* As a group, these proteins *confer viscosity* to the blood, create oncotic pressure, *give suspension stability* to the blood, serve as a *reserve of amino acids,* and are involved in the *clotting* process.

Albumins

Albumins comprise 55 to 64 percent of all plasma proteins, and have a concentration range of 3.3 to 4.1 g/100 ml blood. They are the smallest of the plasma proteins, with molecular weights of 69,000 to 70,000. They contribute most of the oncotic pressure of the plasma.

Globulins

Globulins comprise about 15 percent of the plasma proteins and have concentrations of 2.23 to 2.39 g/100 ml. They may be separated electrophoretically into several subgroups:

1. *Alpha globulins.* The smallest of the globulins, alpha globulins

*Since proteins carry an electrical charge at normal body pH, they will migrate under the influence of an electrical current. According to size and degree of charge, they will migrate at different rates. A separation may thus be achieved in the plasma proteins.

TABLE 10–1

Concentration of Cations and Anions in Plasma

	Ion	Plasma (meq/l)
Cations	Na$^+$	142.0
	K$^+$	4.0
	Ca^{++}	5.0
	Mg^{++}	2.0
	Total	153.0
Anions	Cl$^-$	102.0
	HCO$_3^-$	26.0
	PO$_4^-$	2.0
	Other	6.0
	Protein	17.0
	Total	153.0
Total mOsm/l		306.0

have molecular weights of 150,000 to 160,000. They serve the general functions of the proteins.

2. *Beta globulins.* Molecular weights vary from 160,000 to 200,000. These also serve general functions, but contain small amounts of antibody activity.

3. *Gamma globulins.* In this fraction is found most of the antibody activity of the plasma. These proteins are characterized also as immunoglobulins (Ig) because of their functions concerned with protection against foreign substances. Gamma globulins may in turn be subdivided into the following categories:

   (a) *IgG,* molecular weight 150,000; 75 percent of gamma globulins. They contain most of the natural and acquired antibodies, leading to the production of immunity to certain diseases.

   (b) *IgA,* molecular weight about 150,000; 21 percent of gamma globulins. IgA is a general lytic globulin produced by many externally secreting glands.

   (c) *IgD,* molecular weight about 160,000; 0.2 percent of gamma globulins. Function unknown.

   (d) *IgM (macroglobulin),* molecular weight about 900,000. IgM is the first globulin to appear after antigenic stimulation.

   (e) *IgE,* molecular weight about 150,000. This globulin appears after long-term exposure to antigens.

Fibrinogen

Fibrinogen is a soluble plasma protein having a molecular weight of about 200,000 and a concentration range of 0.34 to 0.43 g/100 ml. Fibrinogen is converted to insoluble fibrin which forms the basis of a blood clot.

B. Organic materials other than plasma proteins. Enzymes, hormones, and other plasma protein fractions, as well as nitrogenous wastes, are found in variable amounts in the plasma. Glucose, prothrombin, creatinine, urea, amino acids, lipoprotein lipase, and lipids are examples of such materials.

Sources of plasma components

Water, inorganic substances and amino acids are derived chiefly by *absorption* from the digestive tract. The protein components are contributed by the *liver,* and by *destruction of blood cells.* Wastes (urea, creatinine) are contributed by all body cells as they metabolize nutrients. Table 10–2 summarizes the plasma components.

THE FORMED ELEMENTS

Erythrocytes *(red blood cells)*

Mature erythrocytes in peripheral blood are described as biconcave, non-nucleated discs, averaging 8.5 microns* in diameter. On a dried smear, they are about 7.75 microns in diameter. Normal erythrocytes have a remarkably constant size, and may thus be used as measures to determine approximate sizes of other formed elements. Though lacking a nucleus, mature erythrocytes are alive in the sense that they

*1 micron (μ) = $1/1000$ mm, or approximately $1/25,000$ inch.

TABLE 10–2

Summary of Plasma Components

Component	Examples	Sources	Characteristics	Functions
Water	– – –	Absorption from gut (90%) Metabolism (10%)	Liquid portion of blood, 91 to 92% of the plasma	Transport, heat absorption
Electrolytes	$Na^+, K^+,$ $Ca^{++},$ $Mg^{++},$ Cl^-, HCO_3^-	Absorption from gut	Inorganic solids of plasma	Establish osmotic pressure, aid in maintenance of irritability, buffering
Plasma proteins		Liver, certain body cells	7 to 9% of plasma solids	Give viscosity, reserve of amino acids, clotting, antibodies
	Albumins	Liver	Molecular weight 69,000 to 70,000; 55 to 64% of proteins	General functions
	Globulins	Plasma cells, lymphocytes	Molecular weight 150,000 to 900,000; 15% of proteins	
	Alpha		Molecular weight 150,000 to 160,000	General functions
	Beta		Molecular weight 160,000 to 200,000	General functions. Some antibody activity
	Gamma G A D M E		Molecular weight 150,000 Molecular weight 150,000 Molecular weight 160,000 Molecular weight 900,000 Molecular weight 150,000	Natural and acquired antibodies General lytic globulin Function unknown First to appear Appear with chronic antigen exposure
	Fibrinogen	Liver	Molecular weight 200,000	Clotting, fibrin formation
Other organic substances	Enzymes	Cells	Proteins of high molecular weight	Catalyze chemical reactions
	Hormones	Endocrine glands	Proteins, polypeptides, amines, steroids	Control body function
	Nitrogenous wastes, urea, creatinine	Cellular metabolism	– – –	None; waste products

metabolize glucose, consume ATP and oxygen, and release carbon dioxide, although at very low levels. This metabolism is indicative of the operation of active transport mechanisms in the membrane, and of limited enzymatic reactions within the cell itself. The framework or *stroma* of the unit is composed of an albumin-like protein and lipids (chiefly cholesterol, lecithin, and cephalin). The structure is spongelike and confers a great deal of plasticity to the element. The ability to change shape is necessary, inasmuch as some of the tiny blood vessels through which the elements must pass are smaller than the erythrocytes themselves.

Normal adult peripheral blood contains about 4.5 to 5.5 million erythrocytes per cubic millimeter. Females have a lower count than males.

Life history

In the embryo and fetus, erythrocytes are produced in the yolk sac, spleen, liver, and red bone marrow. At or shortly before birth, all areas except the red bone marrow cease to function as sites of production. The cells originate from a "stem cell" known as the hemocytoblast, and pass through a series of stages leading to the formation of mature elements (Fig. 10–1). In the latter stages of development, the nucleus condenses and is lost, and the cell accumulates a respiratory pigment, *hemoglobin.* Mature elements are shed into the blood stream, at rates estimated at 2,000,000 per second, where they remain functional for 90 to 120 days. Aged cells are removed at rates equivalent to their produc-

FIGURE 10–1. Formation of blood cells.

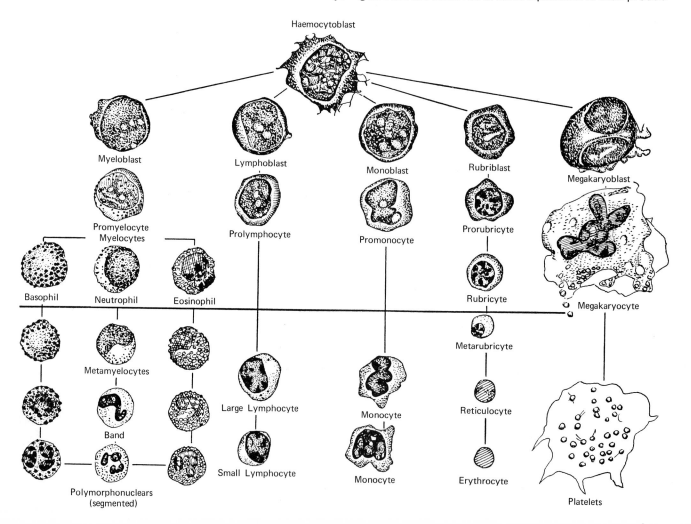

Haemocytoblast

Myeloblast

Lymphoblast

Monoblast

Rubriblast

Megakaryoblast

Promyelocyte
Myelocytes

Prolymphocyte

Promonocyte

Prorubricyte

Basophil   Neutrophil   Eosinophil

Metamyelocytes

Band

Large Lymphocyte

Monocyte

Rubricyte

Megakaryocyte

Metarubricyte

Reticulocyte

Polymorphonuclears
(segmented)

Small Lymphocyte

Monocyte

Erythrocyte

Platelets

tion, chiefly by phagocytosis in the liver, spleen, and bone marrow. These phagocytes not only remove worn-out cells from the circulation, but in large measure conserve, for further body use, the chemical components of the erythrocyte (lipids, protein, the iron of the hemoglobin).

Hemoglobin

Normal erythrocytes contain, to the extent of about 33 percent of their weight, a colored substance known as hemoglobin. Hemoglobin is one of a number of substances capable of combining reversibly with oxygen. Substances exhibiting these properties are known as respiratory pigments. The erythrocyte's primary function, that of oxygen transport through the body, is due to its content of hemoglobin. The cell also transports carbon dioxide, and is involved in the buffering mechanisms of the blood stream.

■ Hemoglobin consists of a pigment, heme, combined with a protein, globin. Heme is synthesized (Fig. 10–2) in the developing erythrocyte from glycine, an amino acid, and succinic acid. Through a series of enzymatically controlled steps, a precursor pyrolle ring is formed. Four such rings are combined to form a polypyrolle or porphyrin ring (uroporphyrogen III). Once the porphyrin is formed, another series of enzymes converts it to a specific porphyrin known as uroporphyrin III. This step is a critical one, for two enzymes are involved. Lack of an

$$\rightarrow HO_2C(CH_2)_2CO_2H$$

FIGURE 10–2. Hemoglobin synthesis. (con't. on next page)

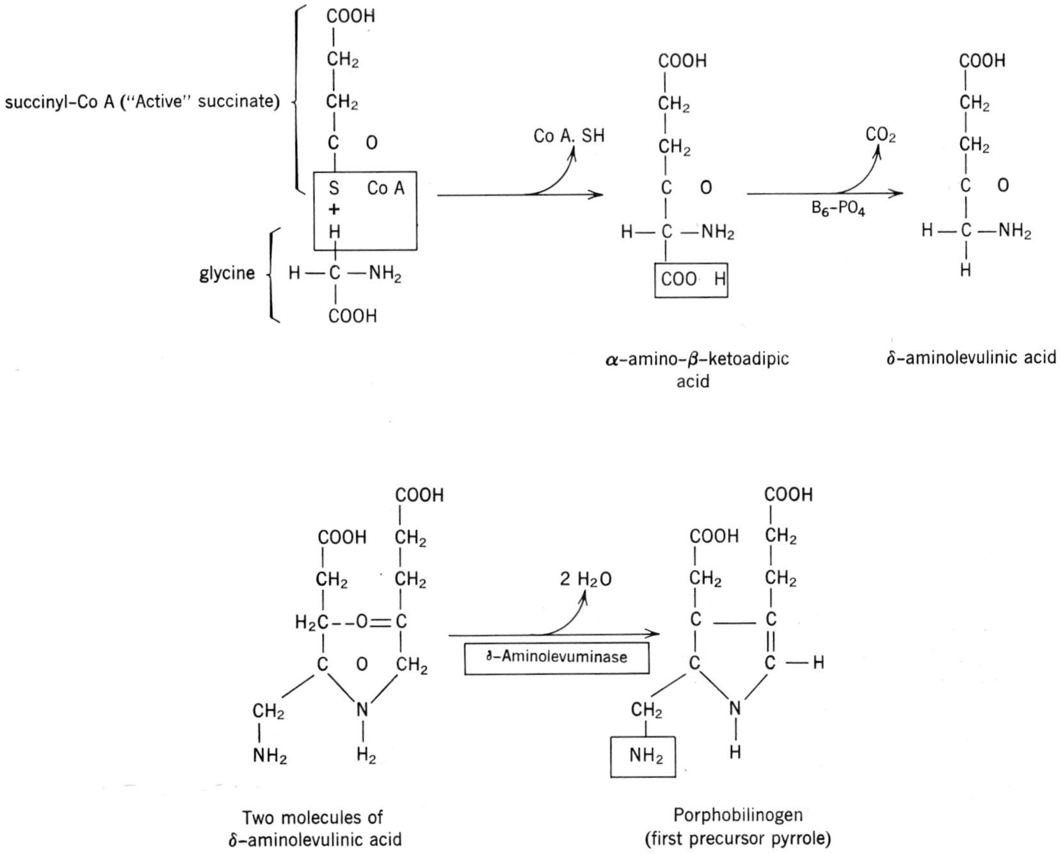

succinyl-Co A ("Active" succinate)
glycine
α-amino-β-ketoadipic acid
δ-aminolevulinic acid

Two molecules of δ-aminolevulinic acid
Porphobilinogen (first precursor pyrrole)

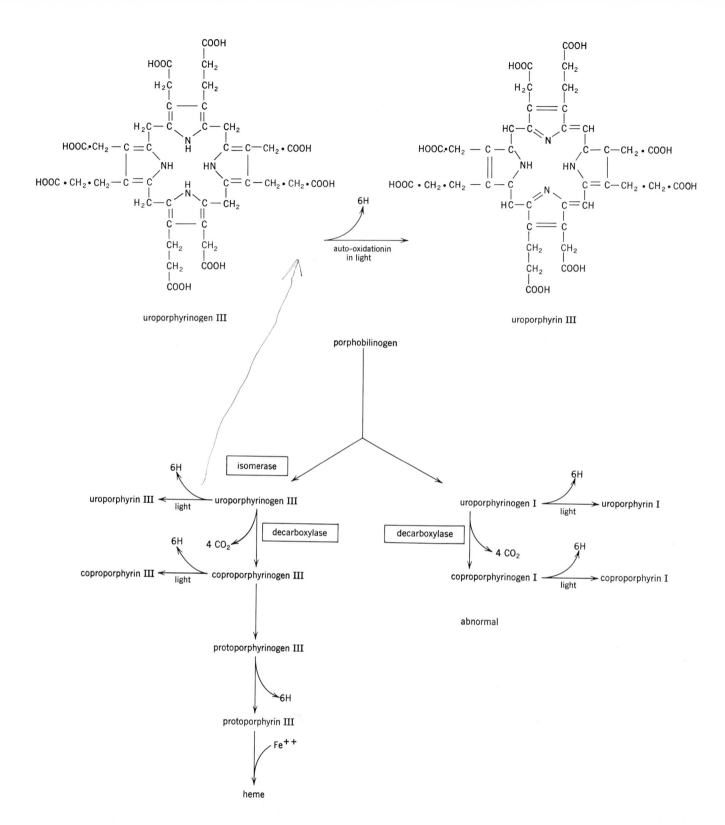

FIGURE 10–2. Hemoglobin synthesis. (con't. from previous page)

enzyme, or failure of one of these enzymes to act, will result in the production of an abnormal hemoglobin. Several chemical groups are next added to the uroporphyrin III, converting it to protoporphyrin III. An atom of iron is added next, and heme is formed. Heme is combined with a protein, globin, and hemoglobin is the result. Finally, four hemoglobin molecules are joined in a "packet" having a molecular weight of about 68,000, large enough to insure retention within the cell. A normal amount of hemoglobin is 14.5 to 16.0 grams per 100 ml of blood. The presence of the pigment raises the oxygen-carrying capacity of the blood by 60 times. Whereas 100 ml of water can carry only $\frac{1}{3}$ ml of oxygen, 100 ml of blood can carry 20 ml of oxygen. Other materials necessary for normal erythrocyte and hemoglobin formation include folic acid, copper, and vitamin B_{12}. Such materials are essential for normal maturation of cells and for normal sequences of hemoglobin formation. ■

Abnormalities in erythrocyte formation include those processes which result in abnormal shape of cells *(poikilocytosis)*, or abnormal size of cells *(anisocytosis)*. *Anemia* is a deficiency of hemoglobin in the blood, arising either from loss of cells, or insufficient hemoglobin accumulation within individual cells. Table 10–3 summarizes some of the types, causes and characteristics of common anemias.

TABLE 10–3

	Type of anemia	Causes	Characteristics	Symptoms	Treatment
INCREASED LOSS	Hemorrhagic				
	Acute	Trauma Stomach ulcers Bleeding from wounds	Cells normal	Shock	Transfusion
	Chronic (iron deficiency)	Stomach ulcers Excessive menses	Cells small (microcytes), deficient in hemoglobin content	None or Fatigue	Iron administration
	Hemolytic	Defective cells Destruction by parasites, toxins, antibodies	Young cells (reticulocytes) very prominent Serum haptoglobin reduced. Morphology of cells may be abnormal when defective	None or Fatigue	Dependent on cause
DECREASED FORMATION	Deficiency states				
	Folic acid	Nutritional deficiency	Cells large (macrocytes), with normal hemoglobin content	None or Fatigue	Folic acid
	Vitamin B_{12}	Lack of intrinsic factor in stomach (pernicious anemia)	Cells large with normal hemoglobin content		Administration of vitamin B_{12}
	Hypoplastic-aplastic	Radiation Chemicals Medications	Bone marrow hypoplastic	None or Fatigue	Transfusion Androgens Cortisone remove cause

The leucocytes or white blood cells

Leucocytes differ from erythrocytes in two important respects. First, they all have *nuclei*; second, they *lack hemoglobin.* Normal numbers range between 5000 and 9000 cells/mm³ of blood. Leucocytes are formed in red bone marrow and in the lymphatic tissues of the body, and remain in the blood stream for variable lengths of time. Life span is estimated to range from only a few hours for some cells to as much as 200 days for others. The cells then fragment in the blood stream or migrate through the walls of the digestive tract and are eliminated via the feces.

All leucocytes possess, to some degree, three basic properties:

1. *Ameboid movement.* This property confers independent motion to the cells, and they are able to move through capillary walls, connective tissue, and epithelial coverings in the body.
2. *Phagocytosis.* The cells are capable of engulfing particulate matter and, in most cases, digesting it. Chemical molecules may also be engulfed and neutralized.
3. *Chemotaxis.* During disease processes or inflammation, leucocytes are attracted to the abnormal area by some chemical produced at the site. A substance designated *leucotaxin,* a polypeptide, has been claimed to be the stimulus promoting leucocyte migration to a diseased area. *Nucleic acids,* arising from cell nuclei during destructive processes, also appear to act as effective stimuli.

Origin, morphology, and functions of individual leucocytes

Five varieties of leucocytes may be recognized on a Wright's stained peripheral blood smear. They include three cell types originating from the bone marrow or myeloid tissue (neutrophil, eosinophil, basophil) and two originating from lymphoid tissue (lymphocyte, monocyte). All cells have a single nucleus, but those of the myeloid elements have lobes or segments. The nucleus consists of several large masses connected by very thin (sometimes nearly invisible) strands of nuclear substance. The nuclei of the lymphoid elements may be indented, but are never lobed. In addition, myeloid elements always have *specific granules* in their cytoplasm. These are granules of constant occurrence, size, shape, and staining reaction in a given cell type. Both categories of cells may contain azure granules, granules which do not meet the criteria set forth above in reference to specific granules. The morphology of the cells is presented in Fig. 10–3, their development in Fig. 10–1.

The neutrophil *(heterophil).*   The neutrophil is a leucocyte of 12 to 14 microns in diameter, having a *lobed nucleus* and *fine specific granules* staining both blue and pink. The synonym heterophil refers to the staining of granules by both colors, that is, they exhibit no preference for stain. Neutrophils comprise 60 to 70 percent of all leucocytes. They are very active and have great powers of phagocytosis. They are the first cells to arrive at an area of tissue damage, and may remain to aid in healing of the tissues after the disease is controlled.

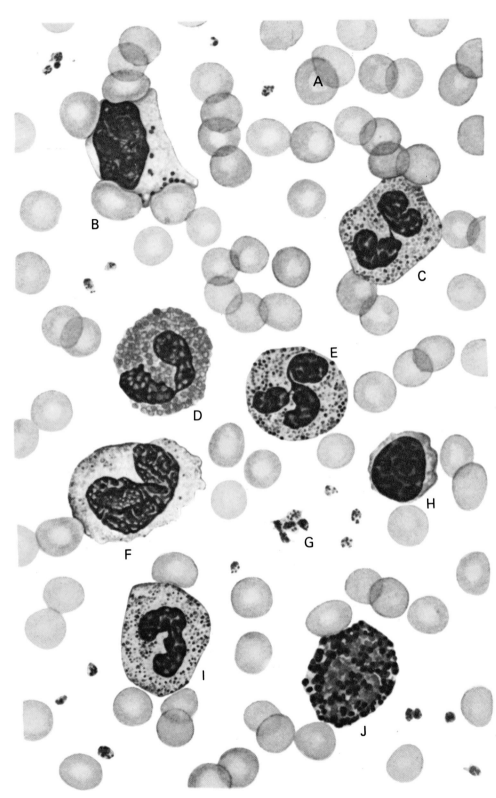

FIGURE 10–3. Morphology of blood elements.

Legend key.
CELL TYPES FOUND IN SMEARS OF PERIPHERAL BLOOD FROM NORMAL INDIVIDUALS. The arrangement is arbitrary and the number of leukocytes in relation to erythrocytes and thrombocytes is greater than would occur in an actual microscopic field.
A Erythrocytes
B Large lymphocyte with azurophilic granules and deeply indented by adjacent erythrocytes
C Neutrophilic segmented
D Eosinophil
E Neutrophilic segmented
F Monocyte with blue gray cytoplasm, coarse linear chromatin and blunt pseudopods
G Thrombocytes
H Lymphocyte
I Neutrophilic band
J Basophil

The eosinophil. Eosinophils also have a diameter of about 12 microns. They have a *lobed nucleus* and *large, shiny, red or yellow-orange staining specific granules* in their cytoplasm. They comprise 2 to 4 percent of all leucocytes. They are poor movers and phagocytes. Their number increases in allergic and autoimmune conditions, and in certain parasitic infections such as schistosomiasis, trichinosis, and strongyloidiasis. Analysis of the specific granules shows the presence of many enzymes, including peroxidases, oxidases, trypsin, and alkaline and acid phosphatases. Such enzymes may render the cell capable of detoxifying foreign proteins and other substances. They may therefore afford important protection against chemical intrusion.

The basophil. Basophils have a diameter of 9 to 12 microns. They have an *obscure, S-shaped nucleus,* and *large, dull, purple staining specific granules* in the cytoplasm. They comprise 0.15 percent or less of all leucocytes. These cells show little ameboid or phagocytic powers. Analysis of basophil specific granules reveals the presence of heparin (an anticoagulant), histamine (a vasodilator), and serotonin (a vasoconstrictor). The importance of these chemicals to the blood is minimal. Tissue basophils (*mast cells*) may originate from blood basophils which have migrated into the tissues. In the tissues, the chemicals mentioned above may assume more important functions related to viscosity of the ground substance and blood supply to the tissue.

The lymphocyte. The lymphocyte occurs in two sizes, small and large. The small lymphocyte has a diameter of about 9 microns, the large one about 14 microns. Both cells have a *nucleus* which *is round* or very slightly indented, and a clear, blue or blue-green staining cytoplasm. In small lymphocytes, the nucleus nearly fills the entire cell, while the large lymphocyte shows much more cytoplasm. Lymphocytes comprise 20 to 25 percent of all leucocytes. They are quite active and possess good phagocytic abilities. They are usually the second cell at a site of tissue injury. They contain much globulin in their cytoplasm which may be contributed to combating chemicals produced in disease processes.

The monocyte. The monocyte is the largest leucocyte, with a 20 to 25 micron diameter. The *nucleus is kidney or horse-shoe shaped,* and the cytoplasm stains a gray color. Azure granules are common. Active in movement and phagocytosis, monocytes comprise 3 to 8 percent of the leucocyte population.

Thrombocytes or platelets

Thrombocytes originate in the bone marrow from a cell known as a *megakaryocyte* (Fig. 10–1). They are actually a portion of granular cytoplasm of the cell of origin. Thrombocytes are 2 to 4 microns in diameter, and consist of a dark staining granule, the *chromomere (colored body)* and a surrounding light staining *hyalomere (clear body).* They are not nucleated, and carry one of several materials essential to the coagulation of blood. Thrombocytes are present to the extent of 250,000 to 350,000 units/mm^3 of blood. Table 10–4 summarizes information relative to the formed elements.

Physiological increases in leucocyte numbers to greater than 10,000 cells/mm^3 constitute a *leucocytosis.* A decrease in numbers is designated *leucopenia. Leukemia* is a pathological change in leukocyte formation resulting in either an excessive number of cells, immaturity of cells, or both. Such conditions may be designated more specifically

TABLE 10–4

Summary of Formed Elements

Element	Normal numbers	Origin (area of)	Diameter (microns)	Morphology	Function(s)
Erythrocytes	4.5 to 5.5 million/mm³	Myeloid (marrow)	8.5 (fresh) 7.5 (dry smear)	Biconcave, non-nucleated disc; flexible	Transports O_2 and CO_2 by presence of hemoglobin, buffering
Leucocytes	6000 to 9000/mm³		9 to 25		
Neutrophil	60 to 70% of total number	Myeloid	12 to 14	Lobed nucleus; fine heterophilic specific granules	Phagocytosis of particles, wound healing
Eosinophil	2 to 4% of total number	Myeloid	12	Lobed nucleus; large, shiny red or yellow specific granules	Detoxification of foreign proteins?
Basophil	<0.15% of total number	Myeloid	9	Obscure nucleus; large, dull, purple specific granules	Control viscosity of connective tissue ground substance?
Lymphocyte Small	20 to 25% of total number	Lymphoid	9	Nearly round nucleus filling cell, cytoplasm clear staining	Phagocytosis of particles, globulin production
Large			12 to 14	Nucleus nearly round, more cytoplasm	
Monocyte	3 to 8% of total number	Lymphoid	20 to 25	Nucleus kidney or horse-shoe shaped, cytoplasm looks dirty	Phagocytosis, globulin production
Thrombocytes	250,000 to 350,000/mm³	Myeloid	2 to 4	Chromomere and hyalomere	Clotting

by indicating the cell involved; for example, lymphocytic leukemia, neutrophilic leucocytosis or lymphopenia, neutropenia.

ANTIGEN-ANTIBODY REACTIONS

The serum proteins of blood are involved in a number of antigen-antibody reactions. To be able to understand the role of the blood in these reactions a general discussion of the topic is introduced at this time.

An *antigen* is a substance that can stimulate the body to produce a chemical designated as an antibody, which reacts with that particular antigen. Antigens usually enter the body by a route other than the digestive tract *(parenterally),* such as the respiratory system, through the skin or through a wound. Antigens differ in their ability to stimulate antibody formation, that is, they differ in *antigenicity.* Some antigens (haptens) must combine with a protein or other substance to exhibit antigenicity. Antigens usually have molecular weights of 10,000 or more. They usually have a large surface area with many "reactant sites" upon the surface. Branched molecules are usually more antigenic than straight chains. Particular chemical groups on the antigen molecule apparently aid in determining the specificity of the antigen, that is, the ability to engender the production of a specific antibody. Chemically, antigens may consist of protein, polysaccharide, nucleic acid; rarely lipids, although the latter may act as haptens. They may exhibit antigenicity by themselves, or may have to combine with one another to become effective antigens.

An *antibody* is a modified blood globulin, either present because of inheritance or formed in response to an antigen. It also exhibits specificity in the sense that it usually reacts only with the antigen which occasioned its production. The organism's body is able to discriminate between antigens which are a normal part of the body and those which are foreign. Immune antibodies are ordinarily formed only against foreign antigens.

The site of antibody production has been a subject for debate for some years. Phagocytic cells of the reticulo-endothelial system (that is, the fixed, phagocytic cells of the body), lymphocytes, and plasma cells have been investigated as the site of antibody production. At the present time, *plasma cells (plasmocytes)* seem most strongly implicated as the primary source of antibody. The plasma cells are regarded as a derivative of lymphocytes. The source of the antibody producing cells is the lymphatic tissue of the body. The role of the thymus in the development of immunity is to secrete chemical substances which insure development or maturation of lymphocytes into plasma cells. Animals whose thymus is removed at birth fail to develop normal numbers of plasma cells and hence are deficient in immunologic response.

The mechanism of antibody formation also has been a topic of debate for some years. Two general theories have been advanced. One designated as the *template* or *instructional theory* postulates that the antigen must actually contact the antibody producing site (direct template) or induce the cell to produce a copy of the antigen which may be used as a stimulant to antibody production (indirect template). According to this theory, the antigen and antibody have reverse structural images which fit together like a lock and key. The second, or *selective hypothysis,* postulates that the *ability* to produce antibody exists in the organism before the antigen arrives and, when it does, the antigen merely stimulates a capacity already in existence. The clonal selection theory, included in the selective hypothesis, suggests that antibody producing cells are derived from embryonic mesenchyme cells which are undergoing constant mutation. Such cells migrate to a body area, settle down and multiply. The cell or group of cells so formed constitutes a *clone.* Ten thousand or more clones may be formed during the life of the animal. As the body develops further, it produces antigens which react with competent clones. The clone produces an antibody, the two react, and the clone is destroyed in the reaction. For this reason, the organism lacks the ability to form antibodies to its own substances. Unchallenged clones remain to respond to foreign antigens.

The *antigen-antibody reaction* consists of a union between the two followed by some physical change. This change may not be visible, as occurs in the development of immunity, or it may produce a visible reaction. The visible change may be precipitation, agglutination, or lysis. These involve, respectively, the formation of a chemical precipitate, the formation of an alternate antigen-antibody chain between antibody and red cells, and destruction of the cell.

TRANSPLANTATION

Modern man's average life span is less than 75 years. Organs wear out, become diseased or otherwise become incapable of maintaining normal function. Replacement of defective organs with healthy organs from other individuals, transplantation, is an exciting new field of medicine. At present, only limited survival is possible with organ transplants, due to the *rejection phenomenon.*

The rejection phenomenon is the result of an antigen-antibody reaction, with the transplanted organ serving as the antigen to call forth the production of antibodies by the recipient's body. To minimize or reduce the production of antibodies and prolong the operation and life of the transplanted organ, several procedures may be employed, either singly or in various combinations:

1. X-ray or gamma radiation kills the cells responsible for antibody production. Radiation is, however, indiscriminate in terms of cell destruction and may damage bone marrow and lymphoid organs and lower resistance to disease to a point where ordinary stressors may prove fatal to the recipient.
2. Tissue matching, utilizing white cells in a type of cross match, determines how close the antigens of the donor match those of the prospective recipient. The closer the match, the less treatment will be required to combat the rejection.
3. Drug therapy, using cortisone-like materials, is utilized to reduce inflammatory reactions arising from the transplant.
4. Antilymphocyte serum, prepared by injecting human lymphocytes into horses, is used to react with human lymphocytes in the recipient. If the cells are destroyed, they cannot become plasma cells and thus produce antibodies.
5. Lymph, collected from the lymph ducts, may be treated to remove white cells from the fluid, and the fluid returned to the recipient. The lymphocytes are thus not available as potential antibody producers.
6. Development of tolerance to foreign organ antigens through the injection of small doses of these antigens over long periods of time may reduce the necessity for any external therapy and is one of the most promising avenues of investigation currently in vogue.

It is said that within the next ten years, the rejection phenomenon will be conquered, and we may have only to visit an organ bank and order a replacement for an ailing part.

THE BLOOD GROUPS

On the surface of the erythrocytes are found genetically determined chemical substances which act as antigens. Because they are substances unique to the bloodstream, these materials are designated as *isoantigens, isoagglutinogens,* or simply *agglutinogens.* Two main categories of substances are recognized; one group is designated as the ABO system, the other as the Rh system.

The ABO blood group

Two basic isoantigens are involved, designated A and B. With two antigens, four possible combinations may exist. The cell may have one, the other, both, or neither antigen on its surface. The particular antigen present determines the blood type.

Antigen Present	Blood Type
A	A
B	B
AB	AB
Neither A nor B	O

Hereditarily speaking, A and B inheritance is dominant to O, and each type is the result of two allelic genes. Therefore, the following genotypes could exist.

Blood Type	Genotype
A	AA, AO
B	BB, BO
AB	AB
O	OO

The plasma contains genetically determined antibodies *isoantibodies, isoagglutinins,* or *agglutinins* which correspond to the antigens described above. They are designated by the terms a, Anti-A, or alpha, and b, Anti-B, or beta. In the blood of any one given individual, the antibody present is always the reciprocal of the antigen. Recall the clonal hypothesis, which states that the cells capable of forming corresponding antibodies to the blood antigens are destroyed, embryonically, avoiding a reaction as new antigens are produced. The setup of both antigen and antibodies in the various blood types would be as follows.

Blood Type	Dominant Antigen	Antibody
A	A	b
B	B	a
AB	AB	none	univ. recip.
O	None	ab	" donor

The importance of the blood groups becomes apparent when one considers the necessity of transfusing blood between one individual and another. It must be remembered that the mixing of red cells having a given antigen with plasma containing a corresponding antibody must result in an antigen-antibody reaction. The bloods must "match" or correspond as far as the antigens are concerned if such a reaction is to be avoided. The blood may be typed and cross matched to determine the compatibility or "likeness" of blood given during transfusion. Typing for the ABO system involves placing pure agglutinin (a or b, available commercially) on a microscope slide, mixing with it the unknown blood, and watching for a visible antigen-antibody reaction. It is important to remember that when typing the blood one looks for a reaction, while in the actual transfusion one avoids it. When inspecting the slide of blood and antibody, one should ask himself the following question: What *had* to be present on the erythrocytes of the unknown blood in order to give an antigen-antibody reaction with a known antibody? The following chart summarizes the possible combinations which may result (plus indicates reaction; minus indicates no reaction):

Antibody		Type and antigen present is therefore:
a	b	
+	−	A
−	+	B
+	+	AB
−	−	O

The Rh system

This system of antigens and antibodies was discovered in the Rhesus monkey, hence the designation as *Rh* system. This system is composed of three allelic genes for each type. The dominant genes are designated conventionally as CDE, and the recessives as cde. Eight different genotypes are possible, and are grouped into two categories designed as Rh positive and Rh negative. In general, if the individual has at least two dominant genes, he will be Rh positive; if at least two recessive genes, Rh negative. For example:

$$
\left.\begin{array}{l} \text{CDE} \\ \text{cDE} \\ \text{CdE} \\ \text{CDe} \end{array}\right\} \text{Rh}^+
$$

$$
\left.\begin{array}{l} \text{cdE} \\ \text{Cde} \\ \text{cDe} \\ \text{cde} \end{array}\right\} \text{Rh}^-
$$

An individual who is Rh positive has antigen on the erythrocyte and has no corresponding antibody in his plasma. An individual who is Rh negative has no antigen on the cell, and likewise has no antibody in the plasma. However, the negative person has never had his clones challenged and still possesses the ability to form antibody if positive cells ever enter his body. There are two general sets of circumstances under which positive cells may enter. (1) The first is transfusion. Blood is typed for Rh factor in the same manner as for ABO factors, that is, by placing antibody on a slide, adding blood to it and watching for a reaction. A reaction means Rh positive; no reaction, Rh negative. If the person typing the blood makes an interpretive error, positive cells could be transfused into a negative person. (2) The other situation involves a possible reaction between an Rh positive child *in utero* and an Rh negative mother. The inheritance of Rh factor depends upon two genes, with Rh positive dominant to Rh negative. The following examples will illustrate the conditions under which a reaction between child and mother could occur.

Genotype of			Interpretation
Father	Mother	Child (1 gene from each parent)	
RhRh	RhRh	RhRh	Child and mother are Rh$^+$. No reaction possible.
Rhrh	Rhrh	RhRh or Rhrh or rhrh	Child is either Rh$^+$ or Rh$^-$ relative to a positive mother. No reaction possible.
Rhrh or RhRh	rhrh	Rhrh or rhrh	Child is either Rh$^+$ or Rh$^-$, mother is Rh$^-$. Possibility of reaction exists.

Normally, the blood supplies of the mother and child are separated by membranes in the placenta. If the membranes become permeable to the child's red cells, positive cells could enter the blood of the negative mother and cause her body to produce antibodies to the child's cells. The antibodies then make their way back into the child where a reaction may occur if the antibody concentration rises above a critical value. The disease, *erythroblastosis fetalis,* results from the destruction of red cells in the child. The reverse situation, that is, a positive mother and negative child, results in no damage to the mother since the child's body cannot produce enough antibody to overcome the dilution effect produced by the mother's blood stream. These days it is possible to transfuse blood *in utero* to the child if erythroblastosis is suspected. Blood is injected into the abdominal cavity of the child from which it is absorbed. Also, complete exchange transfusions may be carried out after birth if the child is born with erythroblastosis.

Research has shown that Rh antibody concentrations rise in the bloodstream of the Rh negative mother only *after* the birth of her Rh positive child. This result is interpreted to mean that it is at placental separation (afterbirth) that fetal blood may enter the maternal circulation, presumably through wounds created by placental separation. To desensitize the mother, that is, to remove these antigens, massive doses of anti D gamma globulin may be given immediately after childbirth. The antibody reacts with any antigens present in the maternal circulation to form a complex which is excreted through the kidney. The mother's blood is thus cleared of antigens and the chance that a second baby will be born with erythroblastosis is reduced.

Other blood groups have been found, usually in family groups or restricted populations, and are named, in many cases, for the family. The following list is not all-inclusive, but will serve to indicate the extent of present knowledge.

Kell	PS
Duffy	hR
Diego	Lutheran
Sutter	Lewis
MN	Kidd

Cross matching is a procedure whereby cell and plasma of both donor and recipient are mixed and a reaction watched for. This procedure determines compatibility for all factors, even those not typed for. The combination of donor's cells mixed with recipient's plasma constitutes the "major side" of the cross match. Compatibility (no reaction) must be assured on this combination; because each red cell may have 1000 antigens on it, and a transfusion adds millions of red cells, the possibility of a reaction is very great here. The combination of donor's plasma and recipient's cells constitutes the "minor side." Compatibility should be assured here also. However, if small quantities of blood are being transferred, incompatibility can be tolerated on the minor side. All types but AB have antibodies capable of reacting with other types. If small quantities of plasma are involved, these antibodies are usually diluted to the point where a reaction possibility is diminished.

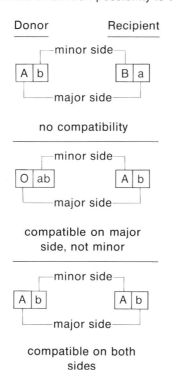

Transfusion fluids

Blood is obviously the ideal transfusion fluid, inasmuch as it supplies all needed blood components. It cannot, however, be stored for more than 2 to 3 weeks, because the cellular elements begin to disintegrate. Other substances may be employed. Plasma, (blood with formed elements removed), is used to aid in maintenance of blood volume. It contains all the organic materials of the blood, such as protein, and thus aids in maintaining blood volume by drawing water osmotically into the capillaries from the tissue. Plasma may be stored indefinitely. Solutions of colloids, such as albumin, may be transfused to maintain blood volume. These are artificial solutions and also draw water by osmosis into the blood vessels. They expand the volume of the plasma. Solutions of crystalloids such as NaCl may be employed to temporarily increase the blood volume. Their effect is only temporary because the particles are too small to remain in the capillaries and are filtered rapidly into the tissues.

HEMOSTASIS

Those reactions which occur in the body to prevent or aid in preventing blood loss through a wound constitute hemostasis. Two categories of reactions are involved. First, *vascular reactions* usually result in a vasoconstriction or narrowing of caliber of vessels to the injured area. This reduces the blood flow to the area. Second, the blood is capable of changing from its normal liquid condition to a gelated state in the process of *coagulation.* The clot bridges the rupture in the blood vessel and also aids in preventing blood loss to the tissues. A large number of chemicals have been implicated in the clotting process. New ones seem to appear almost daily. Factors definitely implicated in the process are listed below. Others which may be listed in some texts may be involved in the process as accelerators of the chemical reactions, or have a role as yet not definitely fixed in the scheme.

International Committee Designation	Synonyms	Origin	Location in Blood
Factor I	Fibrinogen	Liver	A plasma protein
Factor II	Prothrombin	Liver	A plasma protein
Factor III	Thromboplastin	By series of reactions in blood; also found, as such, in cells	Not present as such Produced in scheme
Factor IV	Calcium	Food and drink, from bones	As Ca^{++} in plasma
Factor V	Labile factor	Liver	Plasma protein
Factor VI	Proaccelerin	Liver	Plasma protein
Factor VII	SPCA (serum prothrombin conversion accelerator)	Liver	Plasma
Factor VIII	AHF (Antihemophilic factor)	Liver	Plasma

International Committee Designation	Synonyms	Origin	Location in Blood
Factor IX	AHG (Antihemophilic globulin) PTC (plasma thromboplastin component)	Liver	Plasma
Factor X	Stuart-Prower factor; develops full factor III power	Liver	Plasma
Factor XI	PTA (plasma thromboplastin antecedent)	Liver	Plasma
Factor XII	Hageman factor; contact factor; initiates reaction	?	Plasma
Factor XIII	Fibrin stabilizing factor; renders fibrin insoluble in urea		Plasma
Platelet Factor	Cephalin	Marrow	Platelets

Three phases in the scheme of clotting are recognized. A rupture of a blood vessel creates a roughened surface to which platelets adhere, partially plugging the break. Clotting then proceeds as follows.

Phase I—Development of thromboplastic activity

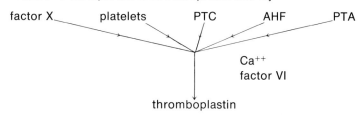

Phase II—Generation of thrombin

$$\text{thromboplastin} + \text{prothrombin} \xrightarrow[\text{V, SPCA}]{\text{X, Ca}^{++}} \text{thrombin}$$

Phase III—Generation of fibrin

$$\text{thrombin} + \text{fibrinogen} \xrightarrow{\text{Ca}^{++}} \text{fibrin}$$

The clotting of blood may therefore be viewed as a series of auto-catalytic reactions culminating in the formation of an insoluble protein network, fibrin. A "white clot" is first formed of fibrin, which subsequently entangles platelets, and white and red cells. The normal color of a clot is thus due to the trapped elements. The platelets have the additional function of causing the clot to shrink (syneresis), and an even tighter bond across an injured area is formed. An internal clot is eventually destroyed by the activity of neutrophils and by an enzyme known as plasmin.

Anticoagulants

Any procedure or chemical which renders the blood incapable of co-agulation is an anticoagulant. One may state, that if one of the necessary materials in the process is removed, the process will not go to completion. Of the various materials present, Ca^{++} is probably the easiest to remove. Mixing withdrawn blood with sodium oxalate or citrate, or ammonium oxalate or citrate, results in an exchange of Na^+ or NH_4^+ for Ca^{++}. The blood has been decalcified. Heparin is another anticoagulent which may be injected into the body and which acts to decrease thromboplastin generation. It also is antithrombo-plastic and interferes with the reactions of Phase II. Dicoumarol, a product first isolated from spoiled sweet clover, interferes with liver synthesis of prothrombin and Factors V and VII.

Disorders of clotting

The *hemophilias* are genetically determined conditions which result in failure of the blood to clot. *Hemophilia A,* or classical hemophilia, is mediated by a sex-linked recessive gene and results in failure to synthesize AHF. *Hemophilia B,* or Christmas disease, is also transmitted by a sex-linked recessive gene and results in failure of PTC synthesis. *Hemophilia C* is transmitted by an autosomal dominant gene and results in failure of PTA synthesis. *Afibrinogenemia,* or, more correctly, deficiency of fibrinogen in the blood, may be either congenital or acquired following severe liver damage.

Blood may spontaneously clot within the vessels to give an *intravascular clot* or *thrombus.* Slow blood flow, and the presence of roughened surfaces on a blood vessel appear to predispose to intravascular clotting. A thrombus may be quite serious if thrombosis of a vital vessel occurs; for example, thrombosis of coronary or cerebral vessels. An *embolus* is a floating clot. Embolism is the condition resulting from such a clot. The danger here is that the clot may lodge in a vital vessel. Clots floating from the lower appendages appear particularly dangerous because they have a good chance of passing through the right side of the heart and lodging in the lungs to create pulmonary embolism.

QUESTIONS

1. What requirements led to the development of a closed, high-pressure circulatory system?
2. Why should plasma so closely resemble cytoplasm in composition? Compare and explain.
3. What devices does the blood possess for protection against invasion by microorganisms or foreign chemicals?
4. Contrast the development of white and red cells.
5. What criteria must "respiratory pigments" meet in order to function efficiently? Describe hemoglobin.
6. In what ways are antigen-antibody reactions in general similar to and different from the reactions between the blood antigens and antibodies?
7. What is a "cross match," and what is its value?
8. If the surface of the body is penetrated or cut, what reactions will occur in order to prevent loss of blood? Describe how each occurs and the end results.

CHAPTER 11

The Heart

The human heart is like Indian rubber: a little swells it, but a great deal will not burst it.
A. BRONTE

THE HEART

Introduction

The transport function of the blood cannot be carried out unless the blood is caused to circulate through the body. Sufficient pressure to circulate the fluid implies the presence of a pump to create pressure, and a closed system of tubes to permit a greater pressure to be developed. The *heart*, a four-chambered, double muscular pump, creates the pressure necessary to circulate the blood in sufficient volume to meet the cells' requirements for nutrients and waste removal.

Size, position, and relationships

The human heart approximates the size of the clenched fist of the owner. At birth it measures about 2 cm (¾ inch) in depth, width, and length, and weighs about 20 grams. The adult heart measures 6 cm (2½ inches) in depth, 9 cm (3½ inches) in width, 12 cm (4¾ inches

305

in length, and weighs about 300 grams. The total increment in weight is 15 times; in length, about 6 times; with the greatest growth periods occurring between the ages of 8 and 12 years and between 18 and 25 years. The heart lies obliquely in the central cavity of the thorax *(mediastinum),* at the level of the 5th to 8th thoracic vertebrae. About 65 percent of its mass lies to the left of the midsternal line, 35 percent to the right. The broad *base* is directed towards the right clavicle while the narrowed *apex* lies at the level of the 5th intercostal space about 8 cm to the left of the midsternal line. The *axis* of the heart, established by a line which passes through the center of the base to the apex, is directed from upper right to lower left. The organ is enclosed by a double-walled *pericardial sac,* which attaches to the sternum and diaphragm and so aids in maintaining the normal position of the heart.

Gross Anatomy (Figs. 11–1, 11–2, and 11–3)

The mammalian heart is a four-chambered organ consisting of two upper chambers, or *atria,* and two lower chambers, or *ventricles.* Externally, the boundaries between the chambers are marked by the *coronary (atrioventricular) sulcus* between atria and ventricles and the *anterior* and *posterior longitudinal sulci* between right and left ventricles. Internally, the atria are separated by the *interatrial septum,* and the ventricles by the *interventricular septum.* The right atrium receives blood by way of the venae cavae after it has traversed the body generally. The right atrium is thin walled and is adapted as a volume rather than a pressure chamber. About half way down its interior is located a fold of atrial tissue (valve of the inferior vena cava) serving to direct blood from the inferior vena cava towards the interatrial septum. In the fetus, the valve is thought to direct incoming blood towards the fetal opening in the septum and thus bypass the right ventricle. It is of no importance in the adult. Beneath the medial margin of this valve is found the opening of the *coronary sinus,* the latter vessel draining about 70 percent of the blood circulated to the heart itself. This coronary sinus orifice is also provided with a fold of atrial tissue *(Thebesian valve)* which tends to close as the atrium contracts and prevent backflow into the sinus. In the adult heart the interatrial septum bears an oval depression, the *fossa ovalis,* which indicates the position of the foramen ovale of the fetus. The foramen permits shunting of blood from the right atrium to the left atrium in the fetus, tending to bypass the nonfunctional lung. The internal wall of the atrium is ridged by numerous bands of muscle, the *musculi pectinati.* The exit from the right atrium is provided by the *right atrioventricular orifice.* The orifice opens into the *right ventricle* and is guarded by the *right atrioventricular* or *tricuspid valve.* The valve consists of three unequal flaps of tissue attached to a fibrous ring surrounding the orifice. The free edges of the flaps project into the right ventricle. The *right ventricle* is a triangularly shaped chamber occupying most of the anterior surface of the heart. Entrance to the chamber is through the aforementioned atrioventricular orifice, while the exit is through the narrowed *conus arterious* to the pulmonary trunk. The inner surface of the right ventricle is ridged with muscle, the *trabeculae carnae.* From some of these ridges project *papillary muscles,* from which run small strands of collagenous tissue known as the *chordae tendinae.* The chordae attach to the free edges of the tricuspid valve cusps and prevent the valve from reversing as pressure rises in

All arteries in red, veins in blue.

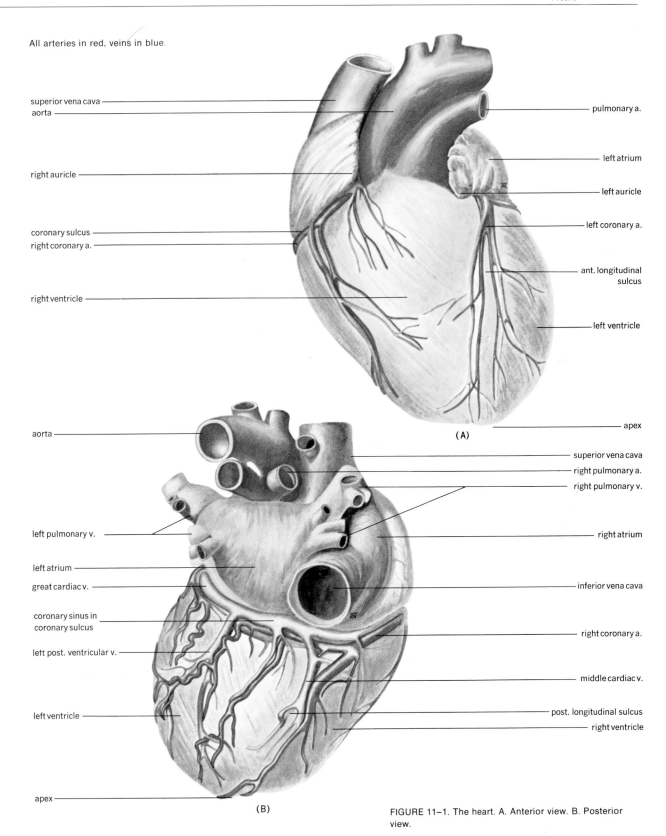

superior vena cava
aorta
right auricle
coronary sulcus
right coronary a.
right ventricle

pulmonary a.
left atrium
left auricle
left coronary a.
ant. longitudinal sulcus
left ventricle
apex

(A)

aorta
left pulmonary v.
left atrium
great cardiac v.
coronary sinus in coronary sulcus
lett post. ventricular v.
left ventricle
apex

superior vena cava
right pulmonary a.
right pulmonary v.
right atrium
inferior vena cava
right coronary a.
middle cardiac v.
post. longitudinal sulcus
right ventricle

(B)

FIGURE 11-1. The heart. A. Anterior view. B. Posterior view.

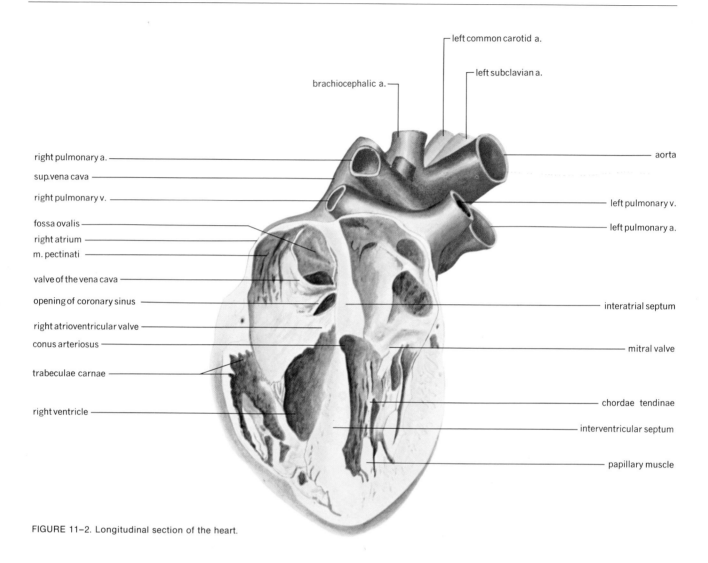

right pulmonary a.

sup. vena cava

right pulmonary v.

fossa ovalis

right atrium

m. pectinati

valve of the vena cava

opening of coronary sinus

right atrioventricular valve

conus arteriosus

trabeculae carnae

right ventricle

brachiocephalic a.

left common carotid a.

left subclavian a.

aorta

left pulmonary v.

left pulmonary a.

interatrial septum

mitral valve

chordae tendinae

interventricular septum

papillary muscle

FIGURE 11–2. Longitudinal section of the heart.

the right ventricle and closes the valve. A *semilunar valve* guards the opening into the pulmonary trunk. This valve consists of three pocket-like cusps which are pressed against the vessel wall as blood flows out of the ventricle. When the ventricle relaxes, the blood in the artery attempts to return to the ventricle and, in so doing, fills the cusps and closes the valve. The right ventricle and pulmonary trunk thus constitute the pump and main channel of the *pulmonary circulation,* a system designed to convey oxygen-poor, carbon dioxide-rich blood from the tissues to the lungs for re-oxygenation and carbon dioxide elimination.

After passing through the lungs, the blood is returned through four *pulmonary veins* to the *left atrium* of the heart. No valves are present in the pulmonary veins. The left atrium is smaller and thicker-walled than the right atrium. Musculi pectinati are also present in the left atrium, although smaller than in the right atrium, and the fossa ovalis is visible from the left atrium as well as the right. The exit from the left atrium is provided by the *left atrioventricular orifice,* again guarded by a valve, the *left atrioventricular (bicuspid, mitral) valve.* The structure of this valve is similar to that on the right side but contains only two flaps or cusps. The *left ventricle* has a wall about three times as thick as the right ventricle. The wall is provided with *trabeculae carnae, papillary muscles* and *chordae tendinae* to the valve. The exit from the left ventricle is provided by the *aortic orifice* leading to the *aorta.* This opening is guarded by the *aortic valve* constructed in the same fashion as the valve of the pulmonary trunk. The left ventricle and aorta form the pump and main channel of the *systemic circulation* supplying blood to the entire body. The area served is greater than that served by the right side of the heart, hence, a greater pressure is required to move the blood through the systemic portion. For this reason, the left ventricle has a thicker-walled chamber.

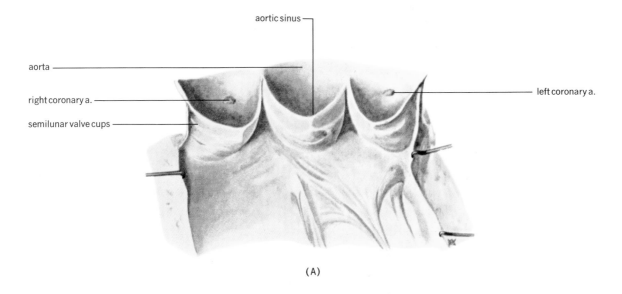

aortic sinus

aorta

right coronary a.

semilunar valve cups

left coronary a.

(A)

FIGURE 11–3. A. Structure of the semilunar valves. B. Origin of the coronary arteries.

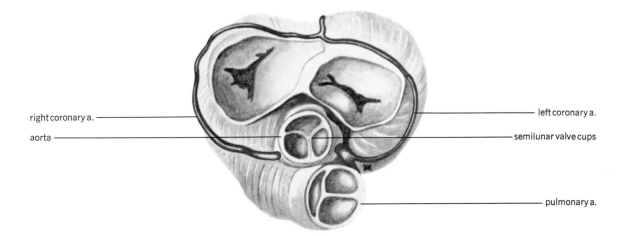

right coronary a.

aorta

left coronary a.

semilunar valve cups

pulmonary a.

(B)

TISSUES COMPRISING THE HEART

Categories of tissues

Several types of tissues are found in the heart. The inner surfaces of the chambers and valves are lined by *endothelium*. This tissue provides a smooth surface over which the blood flows. Beneath the endothelium is a *subendothelial layer* of collagenous connective tissue. Next come layers of *elastic fibers* and *smooth muscle*. Collectively, the layers of tissue just listed form the *endocardium* or inner layer of the heart wall. A *subendocardial layer* contains the nodal tissue of the organ. The central *myocardium* constitutes most of the width of the heart wall and is composed of *cardiac muscle*. The outer *epicardium (visceral pericardium)* consists of a subepicardial connective tissue layer and a covering of *mesothelium*. It is reflected at the bases of the aorta and pulmonary artery to form the thicker parietal pericardium. The potential space or *pericardial cavity* which lies between the visceral and the parietal layers of the pericardium is lubricated by a serous secretion, allowing nearly frictionless movement as the heart contracts. A *fibrous skeleton* provides the basic framework of the heart and affords attachment to the muscle bundles and support for the valves.

Cardiac Muscle

The fibers of cardiac muscle (Fig. 11–4 A to D) are intermediate in both morphology and functional properties between skeletal and smooth muscle.

The muscle is *striated* as is skeletal muscle, but it is *involuntary*, as is smooth muscle. Contraction is assured by built-in forces and is not controlled willfully. Although individual muscle fibers can be recognized in the tissue, the mass behaves as though it was a *syncytium*. That is, stimulation of one part is transmitted without hindrance

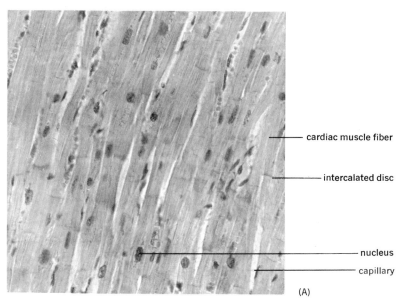

— cardiac muscle fiber

— intercalated disc

— nucleus
— capillary

FIGURE 11–4A. Cardiac muscle. Longitudinal section.

(A)

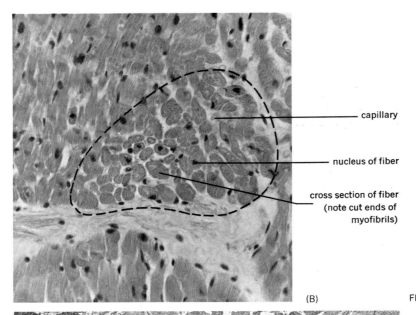

capillary

nucleus of fiber

cross section of fiber
(note cut ends of
myofibrils)

(B)   FIGURE 11–4B. Cardiac muscle. Cross section.

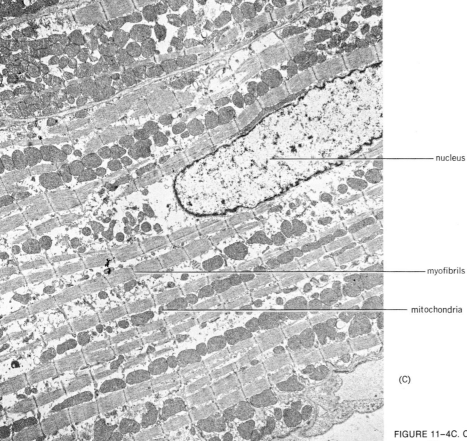

nucleus

myofibrils

mitochondria

(C)

FIGURE 11–4C. Cardiac muscle cell from laboratory
mouse. (Photograph supplied by Norton B. Gilula,
Department of Physiology-Anatomy, University of
California, Berkeley. X 4,000.

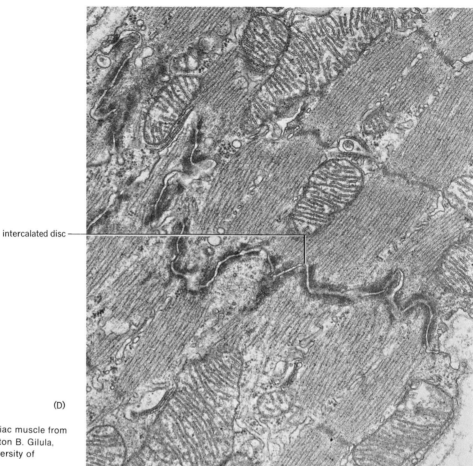

intercalated disc —

(D)

FIGURE 11–4D. Intercalated disc of cardiac muscle from lab mouse. (Photograph supplied by Norton B. Gilula, Department of Physiology-Anatomy, University of California, Berkeley.) × 4,000.

through adjacent fibers, so that the entire mass contracts nearly simultaneously. Cardiac muscle *contracts rapidly, continuously,* and *rhythmically,* and *fatigues slowly* if adequately nourished. The fibers follow the *all-or-none law,* which states that "according to the conditions at the time, a contraction is a maximal effort or does not occur at all." This does not mean that pumping action cannot be altered. Such alteration is necessary because the body requires greater or lesser amounts of circulation, depending upon activity level. These alterations are achieved by changes in the chemical environment of the organ. To a point, increased tension on the muscle fibers results in a stronger contraction, a phenomenon embodied in the so-called *law of the heart.* This factor also is involved in alteration of pumping action. The cardiac fibers possess a *long refractory period,* a time during which they are not capable of responding to another stimulus. This period extends part way into the relaxation phase of the heart's action, and therefore *no tetanus* or sustained contraction may result.

The nodal tissue

Developing embryologically from cardiac tissue, the nodal tissue has lost its contractile power and has developed to a high degree the properties of irritability and conductivity. It is the nodal tissue which generates the electrical impulses necessary to inititate the contraction of the muscle. The tissue consists of discrete nodes and bundles of tissue with a morphology different from that of cardiac muscle (Fig. 11–5). The *sinoatrial (SA) node* lies imbedded in the cardiac muscle, approximately at the junction of the superior vena cava and right atrium. It is about 6 by 18 mm (1/4 by 3/4 inches) in size. In the area of the interatrial septum near the coronary sinus is the *atrioventricular (AV) node.* The node gives rise to an attached *AV bundle* which passes downward for a short distance in the interventricular septum, and then branches to form a *right and left bundle branch.* The bundle branch is passed into the ventricular cavities and supplies *Purkinje fibers* to the cardiac muscle itself.

All parts of the nodal system are capable of originating impulses by spontaneous depolarization. The *pacemaker*, or mass which sets the rate of the heart beat, is that region with the highest rate of deploriza-

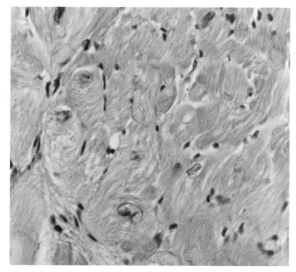

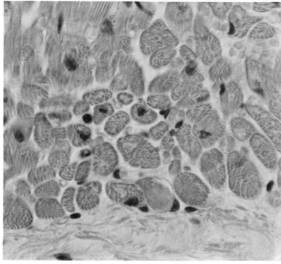

FIGURE 11–5. Nodal tissue.

tion. In the mammalian heart the SA node normally acts as the pacemaker. The student should note that no direct connection by way of nodal tissue exists between the SA and AV nodes. The major pathway available for the impulse to travel is through the cardiac muscle. As it passes through the muscle of the atria, at about the rate of 1 meter (about 3 feet) per second, it causes contraction of atrial muscle fibers. Upon reaching the AV node, the impulse causes deplorization of that mass and further spread of the depolarization wave. However, it should be noted that the distance from the SA node to the AV node is less than the distance from the SA node to the furthest reaches of the left atrium. If the AV node were to pass the impulse immediately to the rest of the system and the ventricles, the ventricles might contract before the atria completed their action. Accordingly, the impulse is delayed as it passes through the AV node. The conduction velocities here are very slow, about $1/10$ meter per second. Once into the AV bundle, the impulse travels about 2 meters per second and arrives at all ventricular muscle fibers nearly simultaneously. Obviously, a near-simultaneous contraction of the ventricles is essential to the development of a pressure high enough to adequately circulate the blood. The spread of the electrical disturbance through the ventricular muscle occurs at about $4/10$ meter per second. A time lapse of about 80 milliseconds ($8/100$ second) is required for the entire ventricular muscle mass to become excited in the human heart.

The net result of all this activity is the assumption of a basic rate of beat set by the SA node. An orderly, rhythmical progression of contraction, involving first the atria, then the ventricles, is created, the chambers being excited by impulses reaching them through the nodal tissue and spreading through the syncytium itself.

THE CARDIAC CYCLE

Timing

If we begin at any one point in the series of events of a single heart beat and return to the same point, a single cardiac cycle has been completed. The logical starting point is with the impulse created by the depolarization of the SA node. At a normal resting rate of about 70 beats per minute, one cardiac cycle takes about $8/10$ second. The contraction and relaxation phases of atrial and ventricular activity occupy times shown in the following diagram:

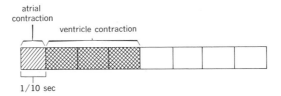

The atria are contracted only $1/10$ second and are not active for $7/10$ second; the ventricles are contracted $3/10$ second and are not active for $5/10$ second. The heart as a whole is therefore actively contracting during only half of the total cycle. The relaxation and rest periods assure adequate filling of the ventricular chambers and allow flow through the coronary vessels to occur.

Events during the cardiac cycle

Three primary events take place during the cycle. (1) Recordable electrical disturbances are produced as the wave of deplorization initiated by the SA node sweeps over the heart. (2) Contraction of the chambers results in pressure and volume changes within the heart and the great vessels leaving the heart. (3) Sounds are created.

Electrical disturbances; electrocardiogram (ECG). Generation and transmission of a depolarization wave over the heart results in changes in voltage potential which may be recorded with appropriate instruments. A modern electrocardiograph produces a written record of these voltage fluctuations as an *electrocardiogram* (Fig. 11-6). On such a record the various voltage fluctuations are recorded as deviations in a vertical plane. The paper upon which the record is inscribed moves through the machine at a constant rate of speed (1 cm per second), so that the time relationship of the various events may be determined. The following major components are seen during each cardiac cycle: The "*P-wave*" represents atrial excitation. The "*QRS complex*" represents ventricular excitation, while the "*T-wave*" is indicative of the repolarization of the ventricle. Several intervals and segments are also labeled in Fig. 11-6.

The value of the ECG lies in its use in the diagnosis of improper heart depolarization or timing, and as a tool to aid in following recovery from various types of heart disease. Analysis of the ECG has proceeded to the point where it is possible to demonstrate definitively different ECG patterns in many different types of heart disease. Several abnormal ECGs. shown in Fig. 11-7. Compare them with the normal one and see if you can detect how the patterns are altered.

Pressure and volume changes (Fig. 11-8). Contraction of the heart chambers produces changes in three pressures, within the heart and the arteries leaving the ventricles. The pressure within the right atrium *(intraatrial pressure),* rises to about 7 mm Hg and then declines to a low value of 3 mm Hg. The left atrium, contracting a bit more strongly, achieves a peak pressure of about 10 mm Hg. Atrial contraction provides only a small push to blood movement from atrium to ventricle, the main impetus being provided by the lowered pressure of the relaxing ventricle. After completion of atrial contraction, ventricular contraction will ensue, and the pressure within the ventricles *(intraventricular pressure)* begins to rise. Initially less than intraatrial pressure to permit blood flow from atrium to ventricle, the intraventricular pressure rises rapidly. At the point where it exceeds the intraatrial pressure, the valves between atrium and ventricle are closed. The aortic and pulmonary valves are already closed from a previous cycle. All four valves being closed renders the ventricles closed chambers. Pressure is rising, but no blood is moving out of the ventricles; the ventricles are in a period of *isometric contraction.* This period continues until the ventricle generates enough pressure to overcome the pressure remaining in the pulmonary artery and aorta *(intraarterial pressure)* and eject blood from the ventricle.

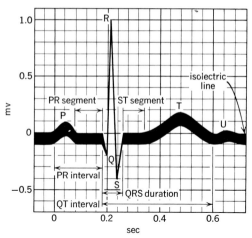

FIGURE 11-6. The normal electrocardiogram.

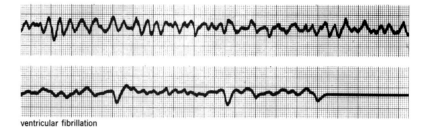

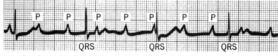

ventricular fibrillation

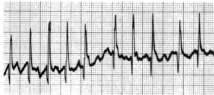

complete heart block (atrial rate, 107; ventricular rate, 43)

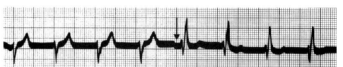

atrial fibrillation

FIGURE 11–7. Some abnormal electrocardiograms in various conditions.

intermittent right bundle branch block

The right ventricle must create a pressure of at least 18 mm Hg to open the pulmonary valve; the left ventricle must generate some 80 mm Hg to open the aortic valve. Pressure continues to rise, with *ejection* of blood from the ventricles increasing. Peak pressures of 30 mm Hg and 120 mm Hg are reached by the right and left ventricles. Ventricular relaxation ensues, with falling intraventricular pressures. As the ventricle relaxes, its pressure will eventually fall to less than that within the great arteries. At this point, the semilunar valves are caused to close by the attempted return of blood to the ventricle. The ventricles are again closed chambers and are in the period of *isometric relaxation.* Intraventricular pressure ultimately declines to less than intraatrial pressure, the valves between atria and ventricles are opened, and blood again fills the ventricles. During the ventricular ejection the volume of those chambers are reduced from about 100 ml capacity, to about 40 ml; this indicates that, at rest, the ventricle ejects about 60 percent of its contents. It should thus become clear that the blood always moves from the higher to the lower area of pressure and that the operation of the valves is entirely passive. The valves simply respond by moving one way or the other, depending upon which side is subjected to the greater pressure. **Heart sounds.** The placing of a sensitive microphone on the chest as the heart works, reveals the presence of four heart sounds. A clue as to what causes these sounds may be gained by noting what event occurs during the same period of time as the sound is created. The *first heart*

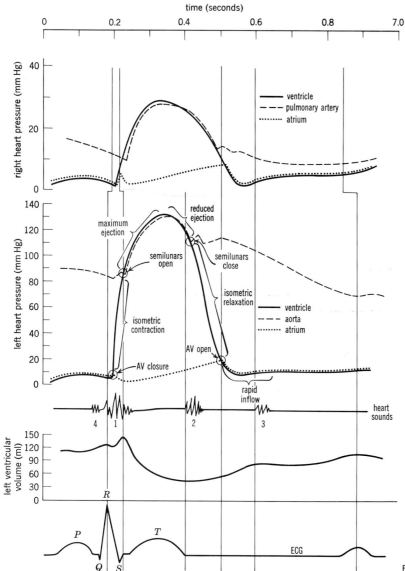

FIGURE 11–8. Composite diagram of heart activity.

sound occurs at the time of AV valve closure, and is attributed to the tensing of the valve leaflets as they are closed by the rising ventricular pressure, and to the vibration of the chordae tendinae. The sound is low pitched, of a definite duration, and is phonetically approximated by the word "lubb." The *second sound* occurs at semilunar valve closure, and is attributed to tensing of the valve leaflets after closing, and subsequent vibration created in the arterial wall and blood column. The sound is higher pitched and of shorter duration than the first sound and is approximated by the word "dup." The *third heart sound* occurs during or slightly after atrial contraction and is attributed to tensing of the ventricular wall due to the rush of blood from atrium to ventricle. A *fourth heart sound* may be appreciated during atrial contraction and is attributed to the muscular contraction of those chambers. Ordinarily only the first two sounds are heard well with the stethoscope. The third sound may be heard under certain conditions in normal people, but is heard more frequently and clearly in disease states. The fourth sound is rarely heard except in disease.

NOURISHING THE HEART

Coronary circulation (Fig. 11–1)

The myocardium requires continued supplies of nutrients and oxygen if its activity is to continue normally. The coronary circulation provides the necessary substances. A pair of *coronary arteries* arise just behind the cusps of the aortic valve. The *right coronary artery* follows the right half of the atrioventricular sulcus onto the posterior surface of the heart. A marginal branch supplies blood to the anterior portions of right atrium and ventricle, and a posterior descending branch supplies the posterior portions of both right and left ventricles. The *left coronary artery* follows the left half of the atrioventricular sulcus forward to the anterior aspect of the heart where it forms a large *anterior interventricular branch (anterior descending)*. A posteriorly directed branch *(posterior circumflex artery)* follows the coronary sulcus to the posterior surface. It gives off a posterior interventricular branch and then anastomoses with the right coronary artery. Smaller branches are given off to the atria from the encircling vessels, and interventricular branches supply the ventricles. Blood next passes through a system of smaller arteries and capillaries and is collected in a system of *coronary veins* located alongside of and named in the same fashion as the arteries. A *great cardiac vein* occupies the anterior coronary sulcus and receives blood from the anterior aspect of the heart. The *middle cardiac vein* follows the posterior sulcus and receives blood from the posterior aspect of the heart. These vessels terminate in opposite ends of the *coronary sinus* on the posterior aspect of the heart. The sinus, as previously described, empties into the right atrium. Anastomoses between branches of the arteries permit a slow plugging of a vessel to be compensated for. A sudden blockage may, however, precipitate a heart attack by depriving the tissues beyond the blockage of vital nutrients. It has been determined statistically that approximately $\frac{1}{3}$ of the population has a heart where right and left coronary arteries are balanced, in other words, where the right coronary arteries supply the right side of the heart and the left coronary arteries the left side of the heart. Approximately 48 percent of the population have what is termed a right coronary predominance, where the right coronary arteries supply more than just the right side of the heart. Some 18 percent of the population have a left coronary predominance.

About 70 percent of the blood nourishing the heart is carried by the vessels described above. The remaining 30 percent passes through the coronary arteries to terminate in arterial sinusoids, located among the cardiac fibers. From the sinusoids, the blood empties into the cavities of the ventricles, bypassing the coronary veins and sinus.

Flow through the system

During the period of ventricular contraction, the smaller branches of the coronary arteries and the capillaries and sinusoids between the myocardial cells themselves are squeezed shut. There is therefore, during the contraction phase of heart activity, little flow through the coronary vessels. It is only when the myocardium begins to relax, or when it is not actively contracting, that flow through the vessels can replenish the myocardial cells with the necessary nutrients. Mentioned earlier in this Section was the fact that the heart was relaxing or at rest during approximately one half of the cardiac cycle. Again, we may draw your attention to the fact that it is during this period of relaxation and rest that coronary flow, as well as filling of the ventricles, takes place.

Metabolism of the heart

Carbohydrates account for about 35 percent of the energy supply of the human heart. Materials utilized include glucose, pyruvic acid and lactic acid. About 60 percent of the energy supply is obtained from the metabolism of fatty acids. The remaining 5 percent of the energy requirement is met through the metabolism of amino acids and ketone bodies. Oxygen consumption at rest is about 9 ml of oxygen per 100 grams of heart tissue per minute, and increases with increasing activity levels. Noting the pressure–volume work output of the heart as compared to the oxygen requirement, efficiency of heart action may be calculated at about 25 percent. Other materials necessary for normal heart action include sodium, potassium, calcium, and magnesium ions. Lowering of the sodium levels to about 100 mg/L results in failure of myocardial cells to depolarize and they therefore do not contract. Increase up to 200 mg/L has little effect. Increase or decrease in potassium ion results in difficult depolarization and interference with repolarization, respectively. Only very high calcium levels interfere with normal activity and cause fibrillations or tachycardia (increased heart rate). Low magnesium levels lengthen a beat, while high concentrations cause cardiac arrest. The pH tolerance of the heart is quite wide, beats being maintained between 5 and 10 pH. In summary, the heart can utilize a wide variety of materials as energy sources and can tolerate rather great alterations in the environment before abnormal function occurs. This obviously contributes to the continued action of this vital organ.

THE ROLE OF THE HEART IN CREATION AND MAINTENANCE OF BLOOD PRESSURE

Introduction

Five factors are responsible for the origin and maintenance of the blood pressure: (1) The *pumping action* of the heart. By injecting into the arterial system a variable quantity of blood per minute, a basic pressure is created. The term *cardiac output* refers to the volume of blood pumped per minute by one ventricle (2). The *peripheral resistance.* Outflow of blood to the tissues from the larger arteries is accomplished through small, muscle-surrounded vessels. Change in ease of outflow (peripheral resistance) depends upon the diameter of these vessels. (3) The *elasticity* of the large arteries near the heart. These vessels expand as the ventricle ejects blood into them and the elastic tissue of their walls stores energy. When the ventricles relax, elastic recoil acts as a secondary pump to move the blood onwards, maintaining a continuous though pulsatile flow. (4) The *viscosity* of the blood. A more viscous or thicker fluid requires a higher pressure to move it. Thus, as viscosity increases so does the pressure. (5) The *volume* of blood in the system. Any increase of fluid in a system which already has a certain capacity will raise the pressure. Of these five factors only the first two are subject to any degree of rapid physiological control and therefore become the most important in terms of immediate alteration of blood pressure.

The cardiac output

The volume of blood ejected from a ventricle per minute is the product of two other factors. First, the *rate of beat (stroke rate)*; second, the *volume ejected per beat (stroke volume)*. Thus:

Cardiac output (CO) in ml/min = stroke rate (SR)
in beats/min × stroke volume (SV) in ml/beat.
(Beats will cancel, leaving ml/min)

The cardiac output therefore may be altered by changing either the stroke rate, stroke volume, or both. All other factors remaining the same, the blood pressure will therefore vary directly with the amount of blood entering the arterial system per unit of time. It may be noted that cardiac output cannot increase *ad infinitum*. At fast heart rates (greater than 150), little filling occurs between beats, and output may decrease.

Control of stroke rate. Three primary factors determine changes in stroke rate. While the SA node is capable of spontaneous depolarization at a basic rate of 70 to 80 times per minute, this rate may be altered by changes in temperature, chemical factors, and nervous influences.

Temperature effects. It is not known specifically what reactions occur in the SA node which lead to its spontaneous activity. One may logically suppose that this activity involves enzyme-dependent chemical reactions. Such reactions are influenced by temperature in a direct fashion. Heat, either applied directly to the SA node by artificial means or brought about by a rise in the temperature of the blood entering the right atrium, increases the rate of discharge of the SA node. The reverse, of course, is also true. The primary cause of an increase of temperature to the node would be the assumption of a higher level of muscular activity which consumes more fuel, produces more heat, and therefore raises the temperature of the blood returning to the right atrium. The SA node will double its rate of activity for a 10°C rise in temperature ($Q_{10} = 2$).

Chemical effects. A wide variety of chemical substances may alter heart rate. *Acetylcholine* tends to reduce heart rate, presumably by slowing repolarization of the myocardium. *Epinephrine (adrenalin)* elevates heart rate by accelerating repolarization. *Thyroxin* increases overall body metabolism, an effect in which the heart also shares. These two methods of rate control are very crude and not susceptible to a fine degree of control.

Nervous effects provide the basis of fine and continuous monitoring to control rate. Arising from the brain stem and passing to the SA and AV nodes of the heart are nerve fibers of the *vagus* (cranial nerve X). If these nerves are severed, the rate of heart beat will increase. This fact indicates that the normal effect of these fibers is to inhibit the rate. Also, the accelerating response indicates that the effect is a continuous or tonic one. Arising from the upper segments of the spinal cord are other nerve fibers *(cervical sympathetics)* which pass to the heart, and which if cut result in a fall of heart rate. One may conclude, therefore, that these fibers tend to elevate the heart rate. The vagal fibers to the heart are termed *cardio-inhibitory,* while the sympathetic fibers are *cardio-acceleratory.* These fibers produce acetylcholine and norepinephrine, respectively, at their terminations. The nerve fibers may be traced to nerve cell bodies within the central nervous system. These groups of cells constitute *cardio-inhibitory* and *cardio-acceleratory centers.* These centers do not on their own discharge spontaneously over their nerves, but are to a large degree themselves controlled by impulses arriving at them from other body areas. A *reflex control* is thus set up, involving a *receptor* somewhere in the periphery of the body which responds to some alteration in the internal environment. An *afferent nerve* carries impulses from the receptor to the *center.* An *efferent nerve* proceeds from the center to the *heart* itself. Receptors for reflexes which exert primary control on heart rate are found in the walls of the carotid sinus and in the aortic arch (Fig. 11–9). The receptors here are sensitive to stretch imposed upon them by an increased blood pressure in the respective area. Receptor impulses pass to the centers, where activity of the cardio-inhibitory center is increased and that of the cardio-acceleratory center is diminished. The effect is thus to slow the heart rate. A peripheral vasodilation (opening) of muscular vessels also occurs and the blood pressure will fall. These nerves constitute a mechanism of utmost importance in controlling the arterial blood pressure and circulation to the brain.

■   Other possible sites of pressure-sensitive receptors have been studied in the arteries of the neck, thorax and viscera, lungs, right and left atrium, left ventricle, and coronary arteries. It has been shown that increasing the volume of fluid in the right atrium accelerates the heart rate, an effect which is abolished by cutting the vagus nerve. Elevation of pulmonary pressure reduces the heart rate, an effect which is also abolished by cutting the vagal nerves. Increased pressure in the left ventricle depresses heart rate and this too, is abolished by cutting the vagi. The coronary reflex (Bezold-Jarish reflex) results in slowing of heart rate when the pressure rises in the left coronary artery. Again, vagal section prevents the change. These reflexes indicate that control of heart rate revolves primarily around cardiac inhibition and to a very large degree is dependent upon intact vagal nerves to carry out their effect. ■

Control of stroke volume.   Nerves appear to have little effect upon stroke volume. Normal alterations in this component of the cardiac output depend upon *chemical* and *mechanical* factors. These two work together, and no attempt will be made to describe them separately. Stroke volume is determined by the *degree of filling* of the ventricle during relaxation and rest, and by the *strength* of the contraction expelling blood from the ventricles. A normal ventricular volume is about 100 ml. Resting stroke volume is about 70 ml. Obviously, increasing capacity or more complete ejection (to a maximum of ventricular capacity) will boost output. Increased filling may be accomplished by a rise of blood CO_2 levels which bring about a greater ventricular relaxation. Increase in hydrogen ion also has the same effect. The increased massaging action of the skeletal muscles during exercise increases blood return to the right heart, which increases ventricular filling. Increased filling in turn produces an increased tension or stretch on the ventricular muscle. This results in a stronger contraction (see Law of the Heart). Epinephrine stimulates the force of contraction, also increasing ejection by a more complete emptying. The role of nerves in control of stroke volume is limited to the operation of the carotid and aortic body reflexes. The action of these reflexes will be described in detail in the chapter on respiration, since these receptor mechanisms have as their primary effect control of respiration rather than stroke volume.

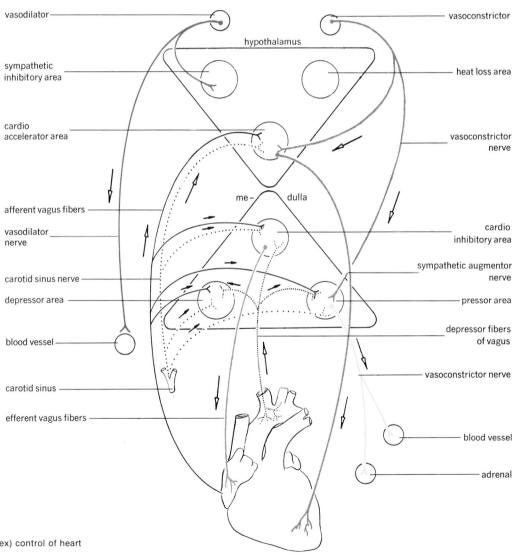

FIGURE 11-9. The nervous (reflex) control of heart action.

QUESTIONS

1. What is the significance of the statement that ''the heart is a double pump, pumping into two separate systems of vessels?''
2. What properties of the nodal tissue–cardiac muscle ''team'' especially fit the heart to act as a rhythmical, adjustable, pressure-creating pump?
3. Describe in detail the interrelationships between ECG, pressure changes, volume changes, and heart sounds. Pay special attention to the causes of valve action and heart sounds.
4. By what routes, and using what materials, does the heart sustain its action?
5. What is the role of the cardiac output in the creation and maintenance of the blood pressure? What enables alterations in cardiac output? Explain the latter fully.

CHAPTER 12

Blood Vessels. Regulation of Blood Pressure.

A human being is an ingenious assembly of portable plumbing.

MORLEY

THE BLOOD VESSELS

QUESTIONS

THE BLOOD VESSELS

Introduction

The blood vessels act as the conduits through which the blood is pumped by the heart. The vessels fall naturally into three general categories:

1. *Arteries* are vessels which convey blood away from the heart and towards the tissues. According to size and structure, large, medium and small arteries (arterioles) are recognized. A transition vessel between arteries and capillaries is formed by the meta-arteriole. Because of the content of smooth muscle in their walls, medium and smaller arteries play an important role in the regulation of blood pressure and blood flow.
2. The *capillaries* permeate the body organs and tissues and act as the mechanism for exchange of materials between blood and cells.
3. *Veins* convey blood from the tissues and towards the heart. They act as volume conduits, rather than pressure vessels.

The general plan of these vessels in the body is presented in Fig. 12–1.

Smooth muscle

We encounter smooth muscle for the first time as a component of an

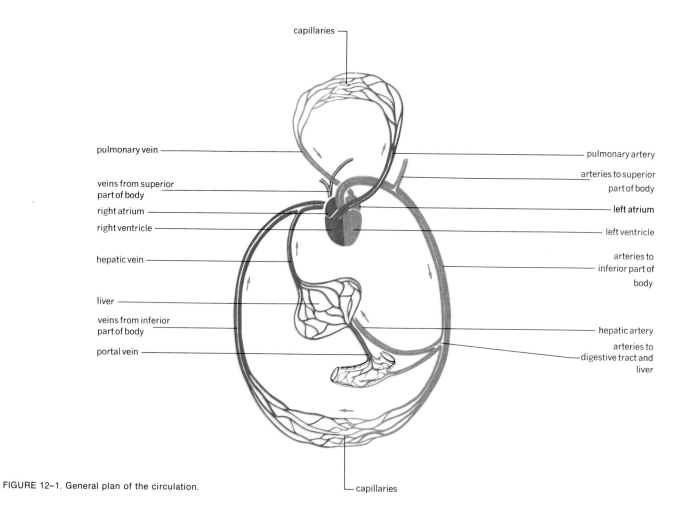

capillaries

pulmonary vein

veins from superior
part of body

right atrium

right ventricle

hepatic vein

liver

veins from inferior
part of body

portal vein

pulmonary artery

arteries to superior
part of body

left atrium

left ventricle

arteries to
inferior part of
body

hepatic artery

arteries to
digestive tract and
liver

capillaries

FIGURE 12–1. General plan of the circulation.

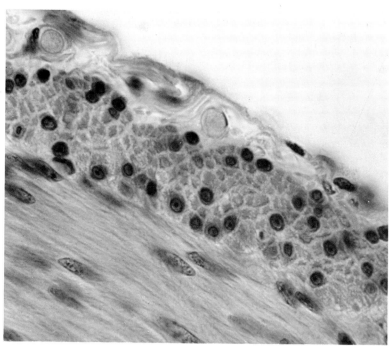

FIGURE 12–2A. Smooth muscle, cross and longitudinal
sections.

organ in the blood vessels. Accordingly, a description of its morphology and properties is presented in this chapter.

Smooth muscle (Fig. 12–2A and B) occurs in the form of spindle-shaped cells measuring 40 to 100 microns in length and 3 to 8 microns wide. Each cell has a single, centrally placed nucleus and the cytoplasm (sarcoplasm) appears to be very finely granulated with a few, small, unstriated myofibrils within. In organs, the smooth muscle cells are bound together into sheets by reticular fibers. The sheets are disposed largely in circular fashion around the blood vessels, creating the potential for increasing and decreasing the diameter of the vessel. Physiologically, the muscle is *involuntary,* that is, it does not depend upon outside nerve supply to initiate activity. It is particularly sensitive to chemical changes in its environment, and normally receives innervation via the autonomic nervous system for control of activity. The muscle contracts and relaxes at very slow rates; a full cycle may take from 3 to as long as 20 seconds.

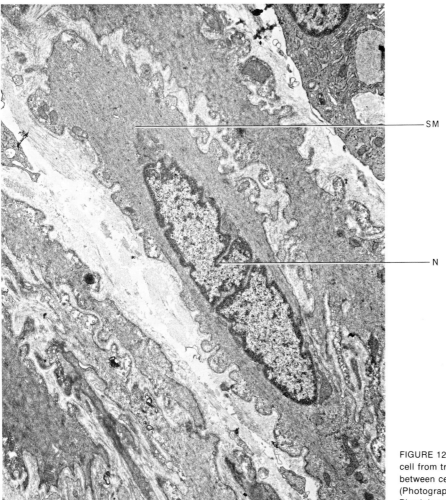

FIGURE 12–2B. Electron micrograph of smooth muscle cell from trachea of lab mouse (connective tissue between cells). SM, smooth muscle cell; N, nucleus. (Photograph supplied by Norton B. Gilula, Department of Physiology-Anatomy, University of California, Berkeley.) X 8,000.

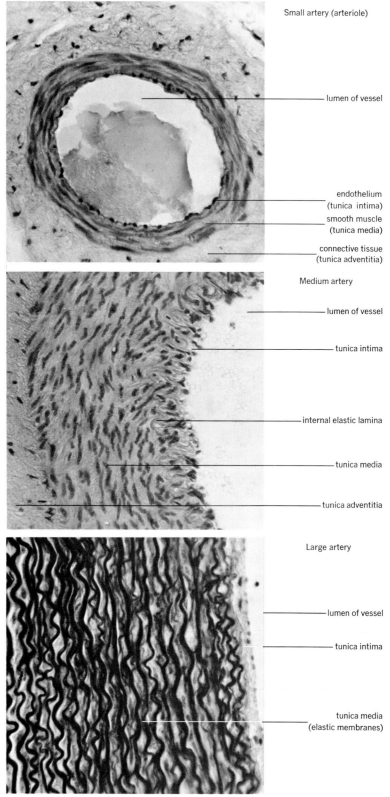

Small artery (arteriole)

lumen of vessel

endothelium
(tunica intima)
smooth muscle
(tunica media)
connective tissue
(tunica adventitia)

Medium artery

lumen of vessel

tunica intima

internal elastic lamina

tunica media

tunica adventitia

Large artery

lumen of vessel

tunica intima

tunica media
(elastic membranes)

FIGURE 12–3. Artery histology.

The histology of arteries (Fig. 12–3)

Arteries vary in size and structure. The aorta measures 25 mm (1 inch) in diameter; small arteries, 0.5 mm in diameter. Three coats of tissue are present in all but the smallest arteries. The *tunica intima* or innermost coat is composed of a lining endothelial layer and an underlying connective tissue layer. The *tunica media* is usually the thickest layer of tissue, and forms the middle layer of the arterial wall. In larger arteries, the media is separated from the intima by a well-defined *internal elastic lamina* or membrane. In the large arteries (aorta, pulmonary artery, carotid arteries) the predominant tissue of the arterial media is elastic connective tissue. As the ventricle contracts and ejects blood into these vessels, they expand to contain the extra blood volume and this stores potential energy in the elastic tissue. As the ventricle relaxes, elastic recoil forces the blood onward through the vessels, insuring a continuous though pulsatile blood flow. The *tunica adventitia* forms the outermost coat of the arteries. It is usually separated from the media by an external elastic membrane and consists of loose connective tissue.

As the vessels branch and become smaller and more numerous, there is a gradual replacement of elastic tissue in the media by smooth muscle oriented circularly. The named arteries of the appendages are classed as medium-sized or muscular arteries. The smaller muscular arteries are known as arterioles. They are still more numerous and retain a muscular media and three recognizable tissue layers. The arterioles are the vessels most involved in controlling blood flow and pressure, and determine the amount and direction of blood runoff from the larger vessels. Arterioles give way to metaarterioles or precapillary sphincters, having one layer of smooth muscle or only a scattered muscle cell or two around their circumference. These vessels do not have a three-layered structure, but do control the entry of blood into into the capillary beds.

Systemic arteries

Aorta and its branches.   The aorta, exiting from the left ventricle, is the vessel supplying blood to the entire systemic circulation. The superiorly directed portion is designated as the *ascending aorta.* The vessel curves laterally to the left as the *arch of the aorta,* then passes inferiorly as the *descending aorta* to the level of the fourth lumbar vertebra, where it terminates by dividing into the common iliac arteries. The descending portion is known as the *thoracic aorta* as it passes through the chest, and as the *abdominal aorta* as it passes through the abdomen. A number of branches arise from these sections of the aorta, and pass to the thoracic and abdominal viscera. The branches are named in Table 12–1 and are diagrammed in Fig. 12–4.

Arteries of the upper appendage (Fig. 12–5).   The *subclavian arteries* form the primary vessels supplying blood to the upper appendages. Four main branches arise from it: the *vertebral, internal mammary, thyrocervical* and *costocervical* arteries. These vessels supply blood to the brain, mammary glands, and muscles of the chest and neck, respectively. As the vessel passes out of the thorax into the axilla (armpit) it is known as the *axillary* artery. Three important branches, the *long thoracic, ventral thoracic,* and *subscapular,* supply muscles of chest and scapula. Passing onto the upper arm, the vessel becomes the *brachial artery* and gives off branches to the muscles on the humerus.

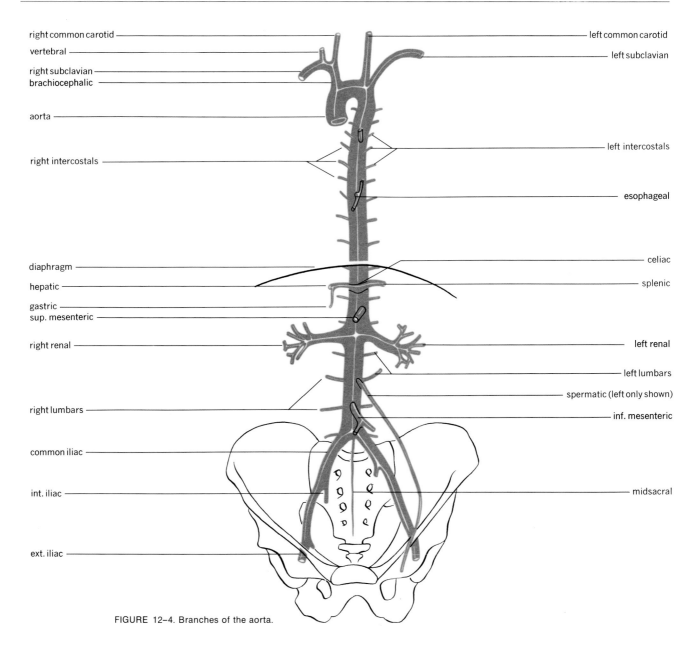

right common carotid
vertebral
right subclavian
brachiocephalic
aorta
right intercostals
diaphragm
hepatic
gastric
sup. mesenteric
right renal
right lumbars
common iliac
int. iliac
ext. iliac

left common carotid
left subclavian
left intercostals
esophageal
celiac
splenic
left renal
left lumbars
spermatic (left only shown)
inf. mesenteric
midsacral

FIGURE 12-4. Branches of the aorta.

Just below the elbow joint, the vessel branches to form *radial* and *ulnar* arteries, following the respective bones of the forearm. Branches are given off to the muscles of the forearm from these two arteries. At the wrist, radial and ulnar arteries are connected by the *superficial and deep volar arches,* from which arise the *digital arteries* to the fingers and thumb.

Arteries of the neck and head. The *common carotid arteries* course up the anterolateral sides of the neck to level of the thyroid cartilage. Approximately at the level of the upper border of the cartilage, the artery branches to form an *external* and *internal carotid* artery. The external carotid supplies branches, providing blood to the structures externally on the cranium, face, and neck (Fig. 12–6). The *internal*

TABLE 12–1

Branches of the Aorta and Organs Supplied

Arising from	Branch	Area supplied by branch
Ascending aorta	Coronary	Heart
Arch of aorta	Brachiocephalic; gives rise to:	Right side of head and neck; right arm
	Right common carotid	Right side head and neck
	Right subclavian	Right arm
	Left common carotid	Left side head and neck
	Left subclavian	Left arm
Descending aorta:		
Thoracic portion	Intercostals	Intercostal muscles, chest muscles, pleurae
	Superior phrenics	Posterior and superior surfaces of the diaphragm
	Bronchials	Bronchi of lungs
	Esophageals	Esophagus
	Inferior phrenics	Inferior surface of diaphragm
Abdominal portion	Celiac; gives rise to:	
	Hepatic	Liver
	Left gastric	Stomach and esophagus
	Splenic	Spleen, part of pancreas and stomach
	Superior mesenteric	Small intestine, cecum, ascending and part of transverse colon
	Suprarenals	Adrenal glands
	Renals	Kidneys
	Spermatics (male)	Testes
	or ovarians (female)	Ovaries
	Inferior mesenteric	Part of transverse colon, descending and sigmoid colon, most of rectum
	Common iliacs, which give rise to:	Terminal branches of aorta
	External iliacs	Lower limbs
	Internal iliacs	Uterus, prostate gland, buttock muscles

carotid artery, together with the vertebral, forms the blood supply of the brain and spinal cord (Fig. 12–7). They center at the circle of Willis on the underside of the brain.

Arteries of the lower appendage (Fig. 12–8) The *external iliac arteries,* derived from the common iliac arteries, pass across the hip joint and become the *femoral* arteries. Branches pass to the skin and muscles of the thigh, lower abdomen, and external genitalia.

Just above the knee, the vessel becomes the *popliteal* artery, passes across the posterior aspect of the knee joint, and branches to form an *anterior tibial, posterior tibial,* and *peroneal artery.* These supply branches to skin and muscles in the leg. The anterior tibial artery becomes the *dorsalis pedis* artery on the upper surface of the foot; the posterior tibial becomes the *plantar artery* on the sole of the foot. The *plantar arch* connects these arteries, and gives rise to *digital arteries* to the toes.

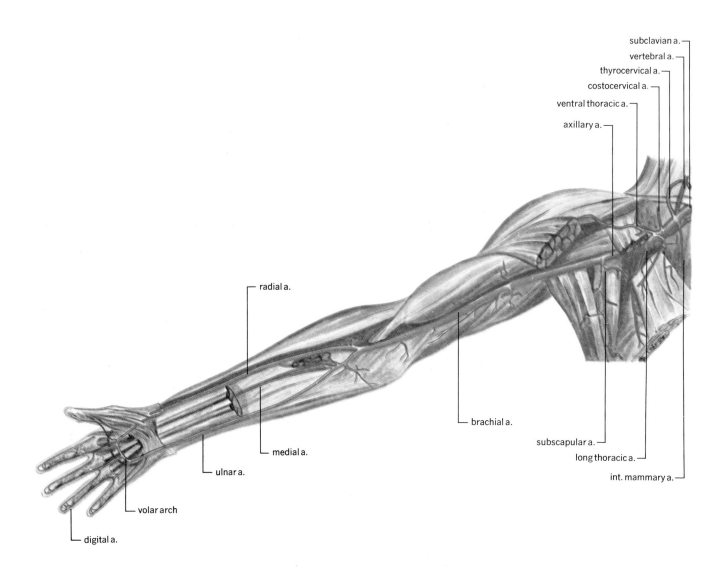

subclavian a.
vertebral a.
thyrocervical a.
costocervical a.
ventral thoracic a.
axillary a.

radial a.

brachial a.

medial a.

ulnar a.

subscapular a.
long thoracic a.
int. mammary a.

volar arch

digital a.

FIGURE 12–5. Arteries of the upper appendage.

Fetal circulation

A child in utero obtains its necessary nutrients and oxygen through the placenta, and does not utilize its lungs or digestive system as a source for such material. Accordingly, the heart and blood vessels are constructed so as to essentially bypass a nonfunctional lung and digestive tract. Blood leaves the placenta through a single *umbilical vein,* which enters the body of the fetus through the umbilical cord and umbilicus. It gives off several branches through the liver. At the porta of the liver it divides into two main channels, one of which joins the portal vein, the other, the ductus venosus, passing to the inferior vena cava.

Blood entering the right atrium is diverted by the valve of the vena cava towards an opening in the interatrial septum known as the *foramen ovale*. Most of the right atrial blood thus passes to the left atrium, to the left ventricle and out through the aorta. Inevitably, some blood will pass through the right atrioventricular valve into the right ventricle. Blood entering this chamber of the heart passes through the pulmonary trunk, a small amount actually passing to the lungs to nour-

334

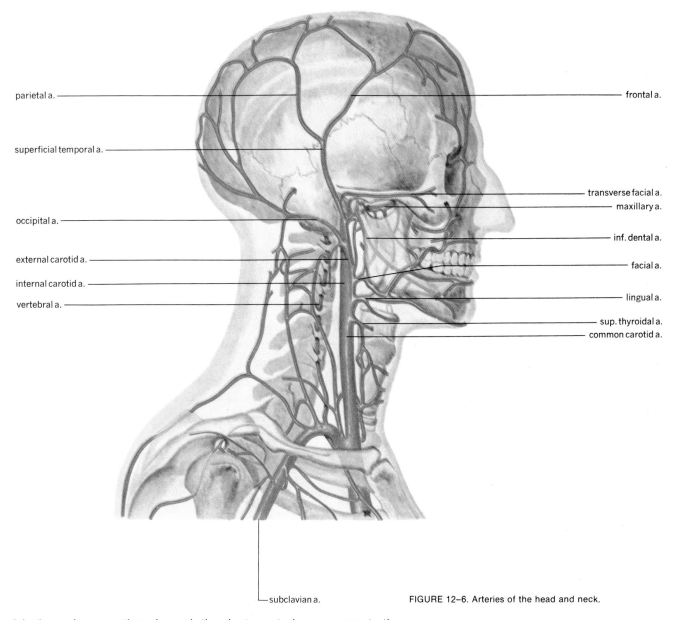

parietal a.

superficial temporal a.

occipital a.

external carotid a.

internal carotid a.

vertebral a.

frontal a.

transverse facial a.

maxillary a.

inf. dental a.

facial a.

lingual a.

sup. thyroidal a.

common carotid a.

subclavian a.

FIGURE 12–6. Arteries of the head and neck.

ish them. A connecting channel, the _ductus arteriosus,_ connects the pulmonary trunk to the aorta and so most of the blood passing through the pulmonary trunk is directed into the aorta. The blood passes through the aorta to the internal iliac arteries where a pair of _umbilical arteries_ pass to the placenta through the umbilical cord. Waste products are now passed to the mother's blood.

At birth, with the severing of the umbilical cord, the pressure relationships within the heart are altered so that the flaps of the foramen ovale are closed. They are held closed by these pressure differences and will eventually grow shut permanently. The higher oxygen content of the blood assured by the baby's drawing his own breath causes contraction of the muscle in the walls of the ductus arteriosus. Umbilical arteries and veins shut down, degenerate and are eventually replaced by collagenous tissue. The umbilical vein between the body wall and liver becomes the _ligamentum teres_ (round ligament); the ductus venous becomes the _ligamentum venosum._ The proximal portions of the um-

335

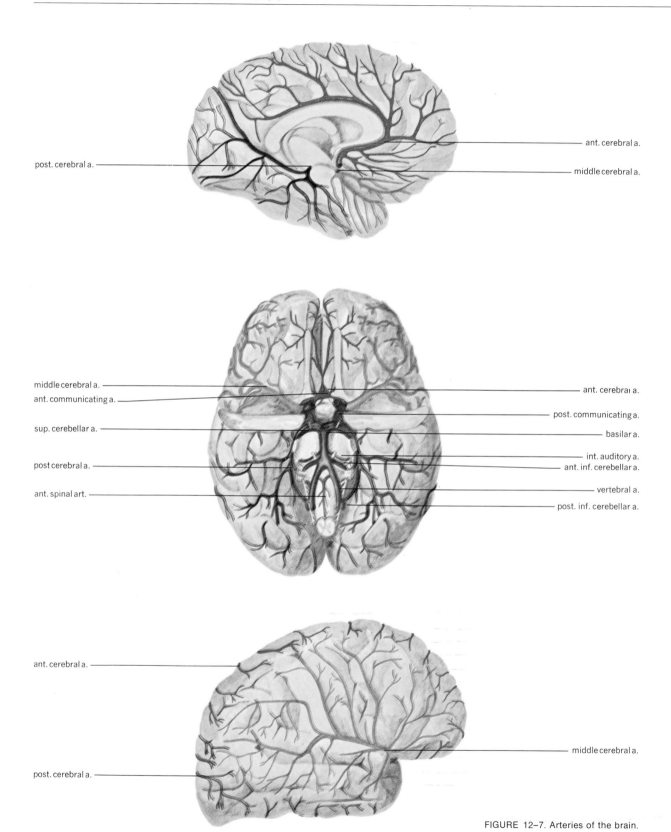

post. cerebral a.

ant. cerebral a.

middle cerebral a.

middle cerebral a.
ant. communicating a.

ant. cerebral a.

post. communicating a.

sup. cerebellar a.

basilar a.

int. auditory a.
ant. inf. cerebellar a.

post cerebral a.

ant. spinal art.

vertebral a.

post. inf. cerebellar a.

ant. cerebral a.

middle cerebral a.

post. cerebral a.

FIGURE 12-7. Arteries of the brain.

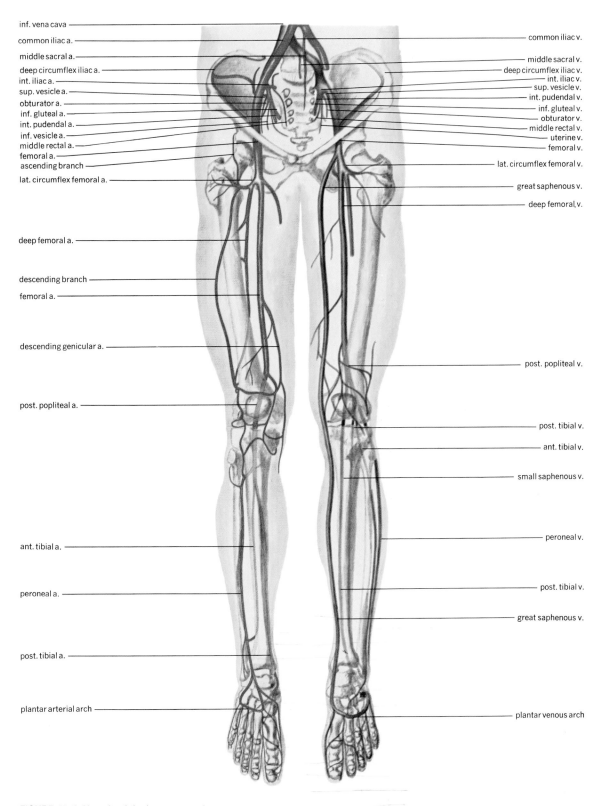

inf. vena cava
common iliac a.
middle sacral a.
deep circumflex iliac a.
int. iliac a.
sup. vesicle a.
obturator a.
inf. gluteal a.
int. pudendal a.
inf. vesicle a.
middle rectal a.
femoral a.
ascending branch
lat. circumflex femoral a.

deep femoral a.

descending branch
femoral a.

descending genicular a.

post. popliteal a.

ant. tibial a.

peroneal a.

post. tibial a.

plantar arterial arch

common iliac v.
middle sacral v.
deep circumflex iliac v.
int. iliac v.
sup. vesicle v.
int. pudendal v.
inf. gluteal v.
obturator v.
middle rectal v.
uterine v.
femoral v.
lat. circumflex femoral v.

great saphenous v.
deep femoral, v.

post. popliteal v.

post. tibial v.
ant. tibial v.
small saphenous v.

peroneal v.

post. tibial v.

great saphenous v.

plantar venous arch

FIGURE 12–8. Vessels of the lower appendage.

bilical arteries remain as the functioning internal iliac arteries; the distal portions extending between the urinary bladder and umbilicus under the peritoneum become the *lateral umbilical ligaments* of the bladder. In the adult, the ductus arteriosus remains as the *ligamentum arteriosum.* Figure 12–9 indicates the plan of the fetal circulation. Other differences in fetal circulation include the possession by the fetus of a hemoglobin which is chemically different than that found in the adult. This is designated hemoglobin F and has as its main advantage the fact that it will load oxygen at approximately half the tension present in the normal adult circulation. The fetus is thus adapted to exist and develop in much lower oxygen levels than is the adult. At birth and shortly thereafter, hemoglobin F tends to be replaced by the normal adult hemoglobin, hemoglobin A.

FIGURE 12–9. The fetal circulation.

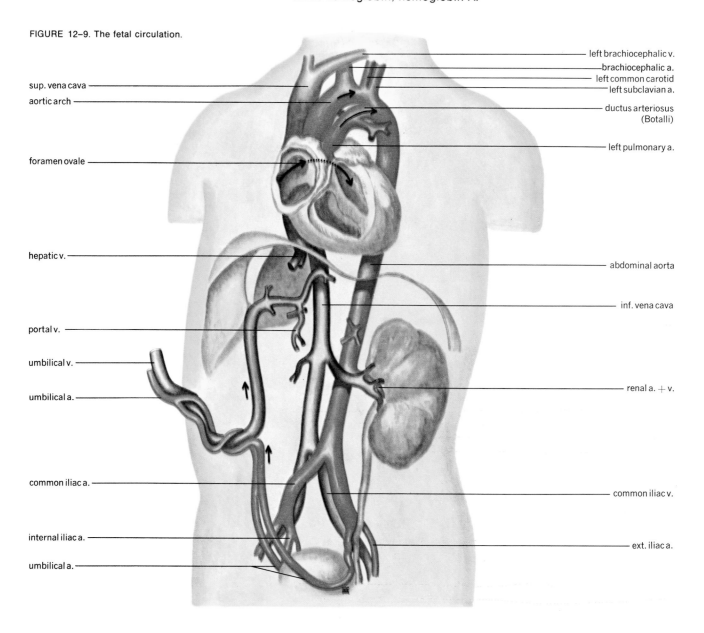

Role of arteries in pressure regulation and flow

Pressure, velocity, and flow of blood through the blood vessels are governed by a number of laws which apply to the flow of any fluid through any system of tubes. Pressure varies directly with the cardiac output and force of heart contraction, and inversely with the length and total cross-sectional area of the tubes through which the fluid passes. In short, a higher pressure is gained if more blood enters the system per unit of time, but falls as more and smaller vessels are encountered. Velocity is directly proportional to pressure and inversely proportional to total cross-sectional area. Flow varies as the fourth power of the radius of the tube.

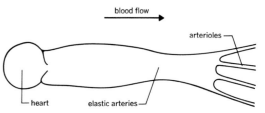

FIGURE 12–10. Arterioles as controllers of arterial outflow.

The involvement of the arteries in control of blood pressure and flow depends primarily upon the caliber of the smaller vessels. Reference to Fig. 12–10 indicates that these vessels act to control the escape of blood from the large and medium arteries into which the heart is pumping blood. Assuming a constant cardiac output, the height of pressure achieved in the proximal portion of the system will depend upon the ease with which blood escapes through the arterioles (peripheral resistance). Also, the pressure beyond the arterioles (in the capillaries) will depend upon the diameter of the arterioles. If more arterioles narrow their caliber while others open, a third thing will happen; that is, blood will be diverted towards the path of least resistance. The systolic pressure, therefore, or pressure which may be measured in the large and medium arteries during ventricular contraction, is largely determined by peripheral resistance, as is the depth to which this pressure will fall (diastolic pressure) during ventricular relaxation.

Factors determining arterial diameter

Local function dictates blood flow to specific body areas. High levels of activity require greater flow than low levels of activity. Flow, as we have seen, is dependent largely upon caliber of arterioles leading to a given mass of tissue.

Size of vessels may be altered passively, as in the large arteries and capillaries, by the pressure of the blood within them. Active change of size depends upon the presence of smooth muscle in the artery wall. The circular disposition of this muscle means that contraction reduces, and relaxation increases, the caliber of the vessels. Serotonin tends to constrict vessels, as does epinephrine. Acetyl choline tends to dilate vessels. By far the finest degree of control over vessel diameter is provided by the vasomotor nerves. Two types of fibers are described as reaching the muscle in arteriolar walls. *Vasoconstrictor nerves* stimulate contraction and narrow caliber, raising blood pressure generally. *Vasodilator nerves* inhibit contraction, enlarge the vessels, and generally cause a drop in blood pressure.

Vasoconstrictor nerves (Fig. 11–9) are derived from the sympathetic nervous system. They arise from the first thoracic to third lumbar segments of the spinal cord. Upon stimulation, these nerves release norepinephrine at their endings. This chemical generally causes constriction of vessels in the viscera and appendages while inhibiting contraction of the muscle in the walls of vessels to heart and skeletal muscle. Since more vessels constrict than dilate, the overall effect is usually one of a raised blood pressure. Dilator fibers arise chiefly from the parasympathetic nervous system and release acetyl choline at their endings. In general, this chemical has the exact opposite effect on the blood vessels and blood pressure as does norepinephrine.

Vasomotor fibers arise from areas of the central nervous system constituting *vasomotor centers* (Fig. 11–9). The centers in turn receive reflex fibers from the periphery and higher centers of the brain. The chief sources of peripheral receptors for this reflex control are derived from the aorta, carotid sinuses, and right atrium. These are areas, it should be remembered, which also control heart rate. In general, it may be stated that the effect of these fibers on the vasomotor center will be one which augments and compliments the effect on the cardiac centers to achieve the desired effect on blood pressure. For example, a stimulation of carotid or aortic pressoreceptors signals that pressure is high and needs to be reduced. Slowing the heart rate, and thus lowering the cardiac output, could achieve some reduction of pressure. If, at the same time, blood vessels could be dilated generally over the body, a more rapid adjustment of the pressure could be achieved. This is exactly what happens. In short, heart activity and blood vessel caliber are coupled to achieve the desired end result. Oxygen deficiency and carbon dioxide excess stimulate chemoreceptors in the carotid sinus and aorta known as the carotid and aortic bodies. These receptors send impulses over afferent nerves to the cardiac and vasomotor centers. The result is a stimulation of heart action, constriction of vessels, and the blood pressure is therefore raised. More rapid acquisition of oxygen and elimination of carbon dioxide is thus assured.

Velocity of blood flow in the arterial portion of the system is, as indicated earlier, dependent upon surface area and pressure. Flow is slowest where there is the greatest surface, or in other words, where there is the greatest friction presented to the flow of blood. This is in the capillary beds. Flow is slow in the area where exchange occurs so exchange can take place more efficiently.

The histology of capillaries (Fig. 12–11)

Capillaries are composed of an endothelial tube supported by a few reticular fibers. They measure about 7 to 9 microns in diameter and are the most numerous of the body's blood vessels. There are, quite literally, miles of capillaries in the body, and they present a very large surface area to the flow of blood.

Functions of capillaries

Because of their extreme thinness, capillaries serve as the vessels through which exchange of materials between cells and blood occurs. The large surface area of the capillaries insures a slow flow of blood through vessels, permitting time for exchanges to occur.

The histology of the veins (Fig. 12–12)

Venules or small veins drain the capillary beds. These vessels typically have two layers of tissue in their walls, endothelium and a surrounding layer of collagenous connective tissue. *Medium-sized veins* acquire a thin media containing scattered smooth muscle cells, and a prominent adventitia. *Large veins* are almost all adventitia. Veins have little pressure to withstand and are easily collapsed. Veins of the appendages

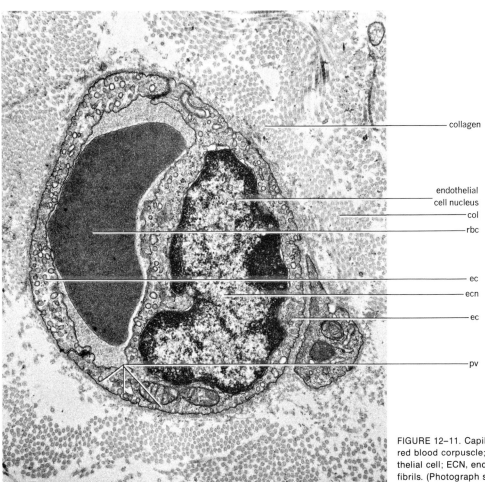

collagen

endothelial
cell nucleus
col
rbc

ec
ecn
ec

pv

FIGURE 12–11. Capillary from trachea of lab mouse. RBC, red blood corpuscle; PV, pinocytic vesicles; EC, endothelial cell; ECN, endothelial cell nucleus; Col, collagen fibrils. (Photograph supplied by Norton B. Gilula, Department of Physiology-Anatomy, University of California, Berkeley.) X 40,000.

and viscera are provided with valves to insure blood flow towards the heart. Medium and large vessels of both types possess a system of blood vessels which nourish the tissues in their walls. These constitute the vasa vasorum; literally, "blood vessels to blood vessels."

The systemic veins (Figs. 12–8, 12–13, and 12–14)

In general, veins accompany the arteries. In the appendages, there are two sets of veins: *deep veins* following along the arteries, and having the same names; *superficial veins* located just beneath the skin. Veins returning blood from the head and neck (Fig. 12–15), join with the veins from the upper appendage to form right and left *brachiocephalic (innominate)* veins. These two vessels join to form the *superior vena cava* draining into the right atrium. The superior vena cava also receives the *azygos vein,* composed in turn of hemiazygos, intercostal, esophageal, pericardial, and bronchial veins. This system of vessels provides a route which can, if necessary, drain the whole blood flow from body areas below the heart. The *inferior vena cava* forms the primary vessel draining blood from the abdominopelvic cavity and lower appendages. At its

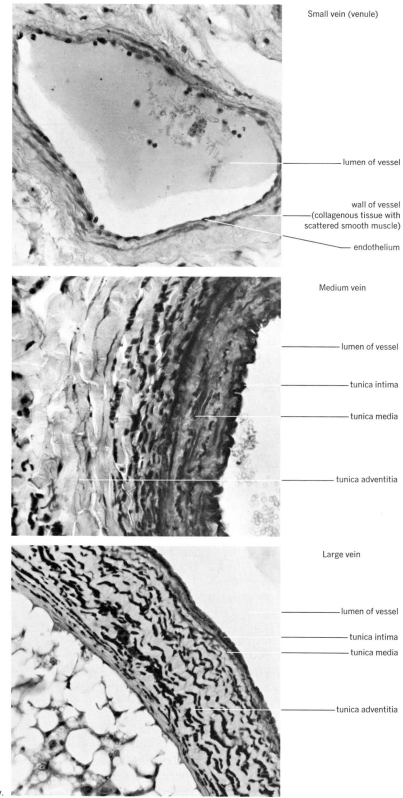

Small vein (venule)

— lumen of vessel

wall of vessel
(collagenous tissue with
scattered smooth muscle)

— endothelium

Medium vein

— lumen of vessel

— tunica intima

— tunica media

— tunica adventitia

Large vein

— lumen of vessel

— tunica intima
— tunica media

— tunica adventitia

FIGURE 12–12. Vein histology.

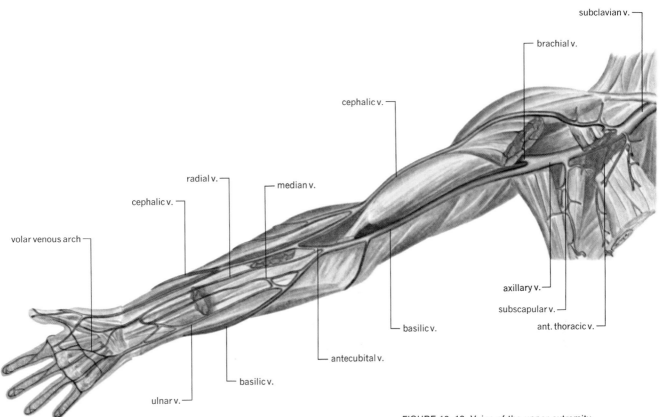

FIGURE 12–13. Veins of the upper extremity.

lower end, it is formed from common iliac and midsacral veins. As it ascends through the abdomen, the inferior vena cava receives spermatic or ovarian, lumbar, renal, suprarenal, and hepatic veins (Figs. 12–14 and 12–15). The vena cava empties into the right atrium. Notably lacking, as contributors to the inferior vena cava, are the veins corresponding to gastric, splenic, superior, and inferior mesenteric arteries. These vessels form the *portal system* (Fig. 12–14). This arrangement allows the liver first choice of the nutrients contained in portal blood, derived from the organs of digestion.

Venous blood flow

Blood pressure is largely lost in traversing the arterioles and capillaries. There is no pump provided beyond the capillaries. What then causes continuation of blood flow through the veins? Normal venous pressure is 14 mm Hg or less, the lesser values being found close to the heart. Some residual pressure *(vis a tergo)* remains upon the blood after it has passed through the capillaries. *Vis a tergo* represents all that is left of the pressure imparted to the blood by left ventricular contraction. *Gravity,* for vessels located above the heart, aids blood flow. The *massaging action* of skeletal muscle contraction, coupled with valves in the veins, assures flow towards the heart. *Respiratory movements* (inspiration) which lower the pressure in the thorax, enlarge the venae cavae and draw blood from outside the thorax into these vessels.

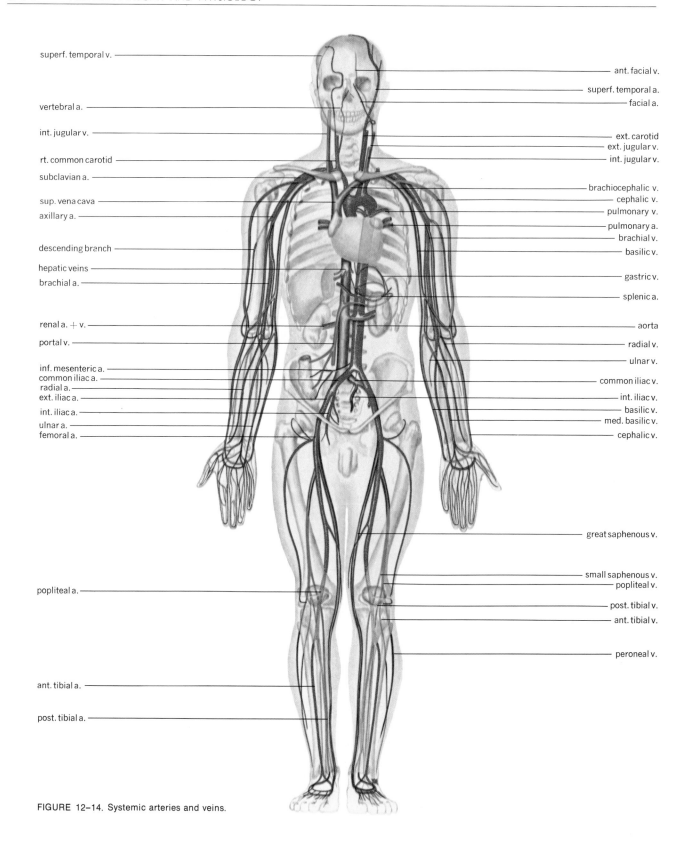

FIGURE 12-14. Systemic arteries and veins.

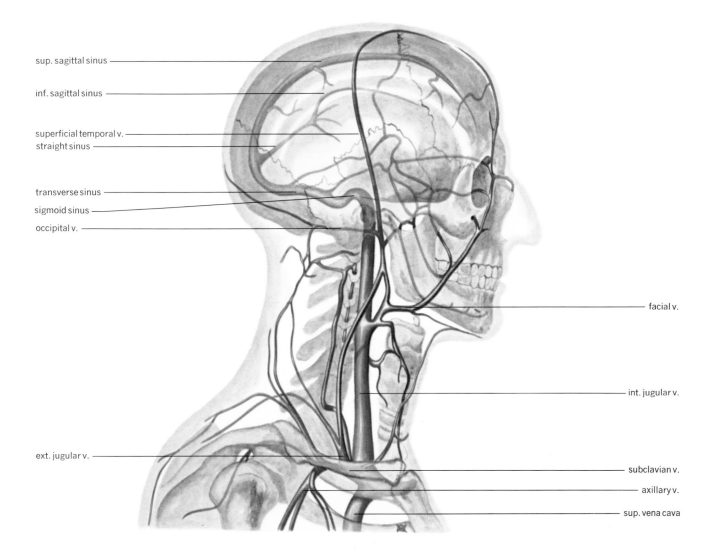

sup. sagittal sinus

inf. sagittal sinus

superficial temporal v.
straight sinus

transverse sinus
sigmoid sinus
occipital v.

facial v.

int. jugular v.

ext. jugular v.

subclavian v.

axillary v.

sup. vena cava

The aspirating action of ventricular relaxation draws blood from the atria which then transmits the lowered pressure to blood in the venae cavae. Figure 12–16 summarizes the relationships between pressure, velocity, surface area, and blood flow in the various vessels of the body.

FIGURE 12–15. Veins of the head and neck.

Circulation in special regions. Brain.

■ The striking thing about the cerebral circulation is its constancy under a wide variety of physiological conditions. Dropping normal arterial pressure to one-half normal is without effect upon cerebral circulation. Increase of intracranial pressure, a force normally bringing about collapse of cerebral vessels, does not interfere with flow until relatively high values (450 mm water) are reached. Nervous influences, although provided to the cerebral vessels, seem to be without effect in man.

Chemical influences appear to be most important in regulation of cerebral flow. Carbon dioxide is the most potent cerebral vasodilator

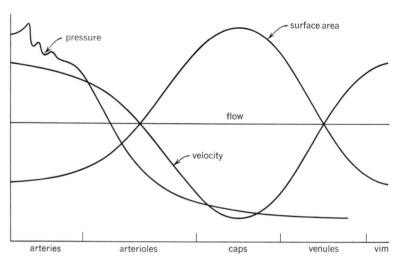

FIGURE 12–16. Relationship of flow, pressure, and velocity to surface area.

known. The rise of carbon dioxide content in cerebral blood to 5 to 7 percent occasions a 75 percent increase in cerebral blood flow. This effect is not related to associated changes in pH. Experiment has shown that pH, if controlled independently of carbon dioxide level, has exactly the opposite effect upon cerebral blood flow as does carbon dioxide. It is therefore, logical to assume that carbon dioxide plays the most important role in regulation of cerebral blood flow. Oxygen also plays a role in control of cerebral blood flow. Decrease of oxygen to one half normal value will occasion a 35 percent increase in cerebral vascular flow. In the normal body, oxygen content rarely decreases to this level, therefore oxygen deficiency does not assume an important role in the control of normal cerebral blood flow. Anxiety and sleep have little effect upon cerebral blood flow. Flow through the coronary circulation has already been described in connection with the heart. Other important areas, including the flow through the lung and the portal system, will be considered in appropriate chapters. ■

QUESTIONS

1. Compare the structure and physiological properties of skeletal, cardiac, and smooth muscle. What special properties fit each for the tasks it has to carry out in the body?
2. What is the relationship between pressure within a blood vessel and the type and thickness of tissues in its wall?
3. Compare arterial and venous systems in terms of total surface area, number of pathways to and from the tissues or organs, and pressure within the systems.
4. How are arteriolar diameter and cardiac output integrated to achieve alterations in blood pressure?
5. Compare the factors causing blood flow through the arteries and veins.
6. It has been said that the whole circulatory system exists to serve the capillaries. Discuss this statement.

CHAPTER 13

Lymph Vascular System

THE LYMPH VASCULAR SYSTEM

Introduction

The lymph vascular system is composed of a circulating fluid, lymph, a system of vessels to collect and transport the lymph, and a variety of lymph organs including the lymph nodes, spleen, thymus, and tonsils. The functions served by the system are related to maintenance of the water and protein balance of the tissues, and to protection against invasion by foreign chemicals and microorganisms.

Lymph

The fluid circulated through the system, lymph or tissue fluid, is derived from interstitial fluid. It is known as lymph once it has entered the lymph vessels.

Characteristics and composition. Lymph has a specific gravity of 1.015 to 1.023. It is low in protein concentration, lacks erythrocytes and thrombocytes, and contains little oxygen. Its content of inorganic substances resembles that of plasma. Content of organic molecules other than protein varies with the location of the lymph sample analyzed. Content of glucose and lipids is highest in lymph drawn from the lymphatics of the digestive tract. Lymph contains a variable number of leucocytes, the number increasing as the lymph drains into the blood vascular system.

349

Formation and functions. Lymph is formed from plasma by a process of filtration. This process accounts for the low protein content and lack of cells in the fluid. Since the filtration pressure in the capillaries exceeds the osmotic return back to the capillaries by about 7 mm Hg (32 to 25 mm Hg), there is a tendency for the tissue spaces to accumulate excess fluid. Removal of this fluid and its return to the vascular system through the lymphatic system tends to maintain normal tissue fluid concentrations. The small amounts of protein which filter through the capillaries must also be returned to the plasma. Additionally, microorganisms, dead cells, or foreign chemicals may be carried with the fluid out of the tissues and are removed or detoxified in the lymph nodes.

Lymph vessels (Fig. 13—1)

The system of lymphatic vessels begins within the tissues as blind-ended tubes known as lymph *capillaries.* (Lacteals in the intestine). These have an endothelial lining similar to that of blood capillaries. The vessels are most numerous in skin and digestive and respiratory tracts. Tissue fluid enters the lymph capillaries by diffusion, aided by a small indirect pressure gradient conferred by the blood pressure. Little resistance to entry of materials is offered by the lymph capillary walls; they are more permeable than blood capillaries. The tiny capillaries form larger vessels similar to the venules and veins of the blood vascular system. The walls are thinner than those of blood veins and valves are much more numerous. The larger lymph veins ultimately converge to form two large *lymph ducts.* The *thoracic duct* lies along the vertebral column extending from the second lumbar vertebra to its termination in the left subclavian vein. Its lower end is enlarged to form the *cisterna chyli.* The thoracic duct collects lymph from the left side of the head and thorax, the left arm, and the entire body below the ribs. The *right lymphatic duct* drains lymph from the remaining portions of the body and empties into the right subclavian vein.

Factors causing flow of lymph

The primary factor assuring lymph flow through the vessels is a pressure gradient which exists from tissues to subclavian vein. In the tissues, hydrostatic pressure approximates 25 mm Hg; at the subclavian vein, pressure may be zero or negative. Additionally, contractions of skeletal muscles act to massage the lymph through the vessels, backflow being prevented by valves. Respiratory movements also aspirate lymph into the upper portions of the ducts.

Lymphoid organs

Introduction. The lymphoid organs are areas of production of lymphoid leucocytes and phagocytosis of particulate matter in the lymph or blood. The lymph nodes are placed in the course of the lymph veins, and therefore act to filter the lymph as it passes through the nodes. The spleen is placed in the course of blood vessels, and serves functions of both a lymphatic and blood vascular nature. Thymus and tonsils are not placed along the course of any vessel, and serve only to produce lymphoid leucocytes and chemicals essential to the development of full immunological competency. Nodules are simply aggregates of lymphoid cells with no organized structure.

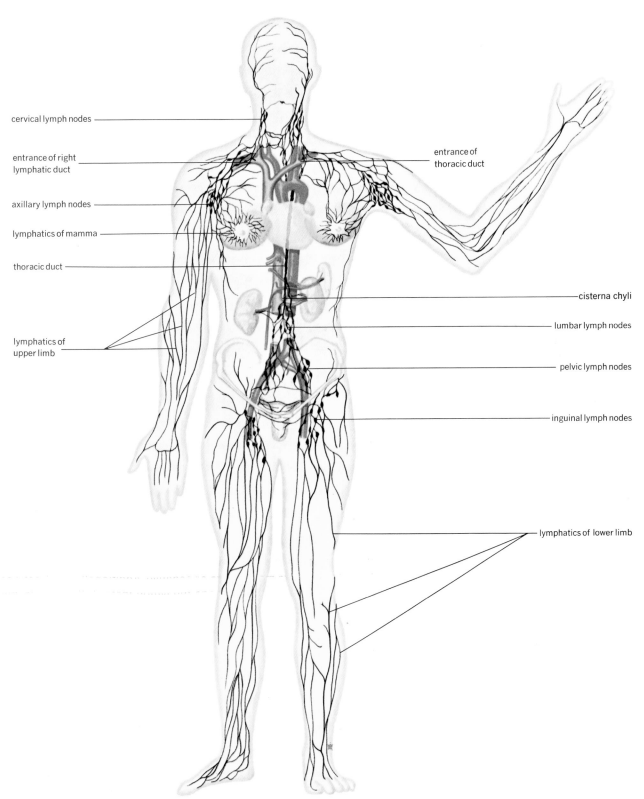

cervical lymph nodes

entrance of right
lymphatic duct

axillary lymph nodes

lymphatics of mamma

thoracic duct

lymphatics of
upper limb

entrance of
thoracic duct

cisterna chyli

lumbar lymph nodes

pelvic lymph nodes

inguinal lymph nodes

lymphatics of lower limb

FIGURE 13–1. The lymphatic vessels and nodes.

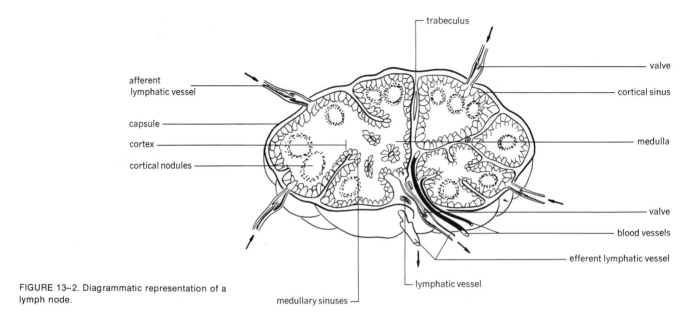

FIGURE 13–2. Diagrammatic representation of a lymph node.

Lymph nodes. (Figs. 13–2 and 13–3).   Lymph nodes vary in size from 1 to 2 mm to as much as 25 mm (1 inch) in length. Being placed along the course of the lymph veins, they are supplied with both afferent (incoming) and efferent (outgoing) lymphatics. Nodes are encapsulated, possessing a collagenous tissue capsule surrounding the organ. Beneath the capsule is a subcapsular sinus into which the lymph passes from the afferent vessels. Connective tissue septae (trabeculae) extend from the capsule into the node. The outer cortex of the node consists of densely packed lymphoid cells which may show areas of active lymphocyte proliferation (germinal centers). An inner medulla consists of cordlike masses of lymphoid cells. The cells of the node include lymphocytes, monocytes, plasmocytes, and fixed phagocytic

FIGURE 13–3. Lymph node.

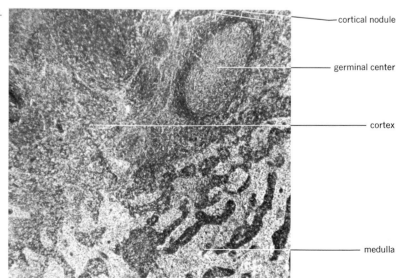

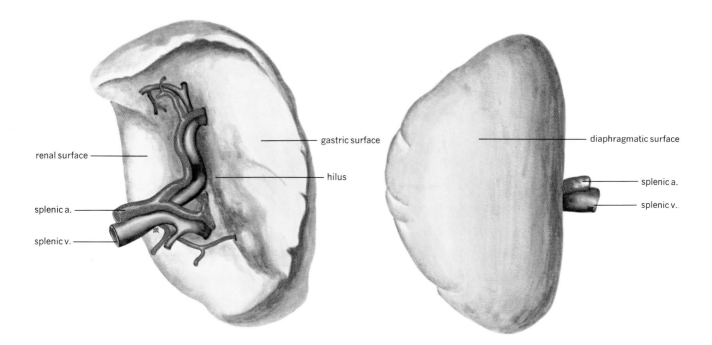

renal surface

gastric surface

hilus

splenic a.

splenic v.

diaphragmatic surface

splenic a.

splenic v.

cells. The node produces lymphoid leucocytes, immune globulins and
cleanses the lymph by removing particulate matter. The node also traps
cells and may become inflamed as infection occurs, or may be a site
where cancer cells may lodge.

Nodes are concentrated in several body regions (Fig. 13–1). Of these
regions, the axillary, inguinal, cervical, lumbar and sacral groups are
the largest.

Spleen (Figs. 13–4 and 13–5).   The spleen is an ovoid organ about
12 cm (4¾ inches) in length, located to the left and slightly behind the
stomach. An indentation, the *hilus,* marks the entry and exit of splenic
blood vessels and efferent lymphatics. An elastic capsule containing
smooth muscle surrounds the organ and gives rise to trabeculae which

FIGURE 13–4. The spleen.

FIGURE 13–5. Spleen.

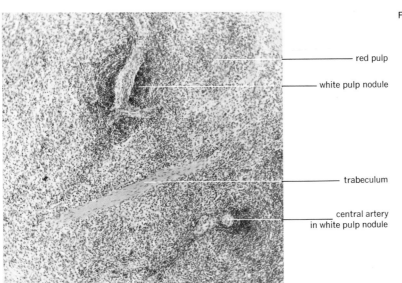

red pulp

white pulp nodule

trabeculum

central artery
in white pulp nodule

353

penetrate into the organ. Internally, splenic pulp of two varieties composes the organ:

1. *White pulp* surrounds the central arteries of the organ and is composed of masses of lymphocytes.
2. *Red pulp* occupies the remaining portion of the organ and consists largely of blood which is in transit through the organ or is being stored within the spleen.

Functionally, the spleen serves as a site of production of lymphoid leucocytes, as an area for phagocytosis of aged erthrocytes, as a blood reservoir, and as a site of fetal blood formation.

Thymus (Fig. 13–6 and 13–7).   The thymus is a two-lobed structure located behind the sternum. A collagenous capsule surrounds the organ and extends internally as septae which subdivide the organ into smaller lobules. A cortex of densely packed lymphoid cells surrounds a loosely packed medulla. Within the medulla are *thymic corpuscles,* function unknown. The thymus produces lymphoid cells and chemicals which insure development of immunological competency in plasmocytes. The organ increases in size until puberty, then undergoes an involution and replacement by fat and connective tissue.

Tonsils.   Three sets of tonsils (Fig. 13–8) are found in the body. *Palatine tonsils* are located on either side of the oral cavity at the level of the soft palate. *Pharyngeal tonsils* (adenoids) are found in the upper part of the throat (nasopharynx), and the *lingual tonsils* are located in the base of the tongue. All have a similar structure (Fig. 13–9). A partial capsule of collagenous tissue lies beneath the organ. Numerous fissure-like *crypts* extend into the organ from the surface. There is no organization into cortex and medulla, although germinal centers are usually present. The tonsils have only efferent lymphatics into which newly produced

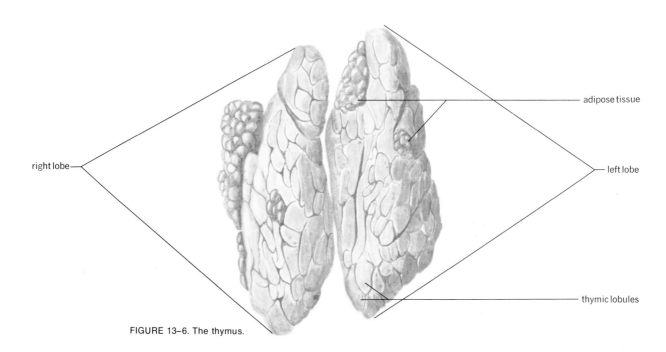

right lobe

adipose tissue

left lobe

thymic lobules

FIGURE 13–6. The thymus.

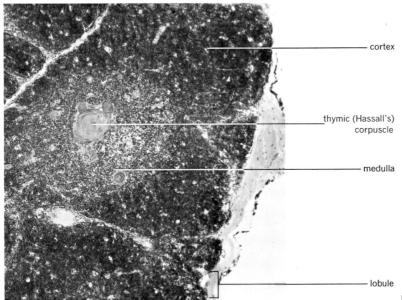

cortex

thymic (Hassall's)
corpuscle

medulla

lobule

FIGURE 13–7. Thymus.

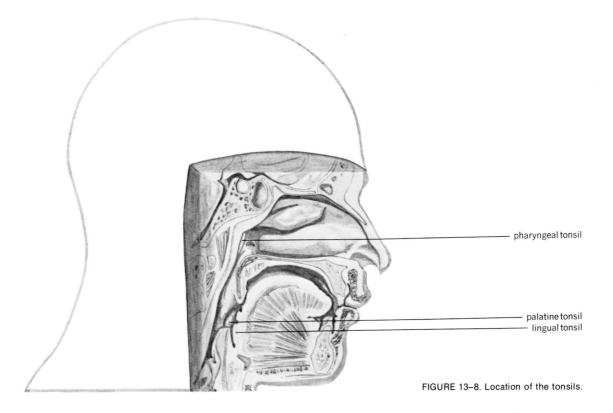

pharyngeal tonsil

palatine tonsil
lingual tonsil

FIGURE 13–8. Location of the tonsils.

lymphocytes are placed. The only function ascribed to the tonsils is that of lymphocyte manufacture.

Nodules.   In the walls of digestive and respiratory systems are found unencapsulated masses of lymphocytes known as lymph nodules. The nodules of the ileum receive the special name of Peyer's patches. The

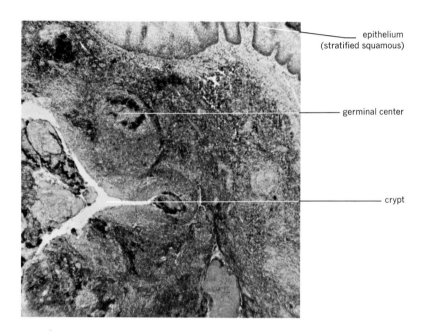

epithelium
(stratified squamous)

germinal center

crypt

FIGURE 13-9. Palatine tonsil.

TABLE 13-1

A Comparative Summary of Lymph Organs

Organ	Capsule present	Tissue type in capsule	Subcapsular sinus	Crypts	Cortex and medulla	Germinal centers	Pulp	Lymphatics		Corpuscles in medulla	Functions
								Afferent	Efferent		
Lymph node	Yes	Collagenous c.t.	Yes	No	Yes	Yes	No	Yes	Yes	No	Filter; lymphatic cell production
Spleen	Yes	Elastic c.t. and smooth muscle	No	No	No	Yes	Yes; white and red	No	Yes	No	Blood storage, RBC destruction, cell production
Thymus	Yes	Collagenous c.t.	No	No	Yes	No	No	No	Yes	Yes; "thymic corpuscles"	Cell production; immunity
Tonsils	Yes; partial	Collagenous c.t.	No	Yes	No	Yes	No	No	Yes	No	Cell production
Nodules	No	– – –	No	No	No	Yes	No	No	No	No	Cell production

nodules offer an additional source of lymphoid cells and create a line of defense against organisms which may enter the body through the walls of these systems.

Table 13–1 presents a comparative summary of the lymphoid organs.

QUESTIONS

1. Compare blood and lymph vascular systems as to the nature of the fluids circulating in each, the forces moving the fluids, and the functions served by each.
2. In what ways are the various lymphoid organs similar? Different?

CHAPTER 14

Respiratory System

Each person is born to one possession which outvalues all his others — his last breath.   MARK TWAIN

RESPIRATORY SYSTEM

Introduction

Metabolic reactions within cells require continual supplies of oxygen, and produce carbon dioxide as one of the products of that metabolism. It is the primary function of a respiratory system to provide a means of acquisition of oxygen and elimination of carbon dioxide. The system actually is exposed to the external environment and must be provided with a means to protect against changes or toxins in the environment. The elimination of carbon dioxide also provides a means of regulation of the pH of the internal environment. A respiratory system thus provides:

1. A maximum surface area for diffusion of the gases oxygen and carbon dioxide.
2. A means of constantly renewing the gases in contact with that surface (ventilation).

361

3. Means to protect that important surface membrane from harmful environmental factors such as microorganisms, airborne toxic particles, adverse temperatures, and drying.
4. A first line of defense against sudden shifts in pH of the blood and body fluids.

ANATOMY OF THE SYSTEM

Organs (Fig. 14–1)

The organs of the respiratory system may be divided into:

1. A *conducting division,* whose walls are too thick to permit exchange of gases between the air in the tube and the blood stream. The nostrils (nares), nasal cavities, pharynx, larynx, trachea, bronchi, bronchioles and terminal bronchioles are included in this division.
2. A *respiratory division,* whose walls are thin enough to permit exchange of gases between tube and blood capillaries surrounding

FIGURE 14–1. Organs of the respiratory system.

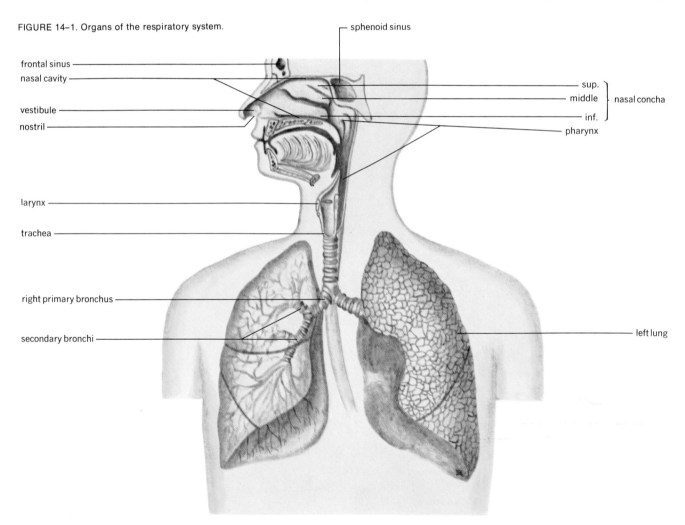

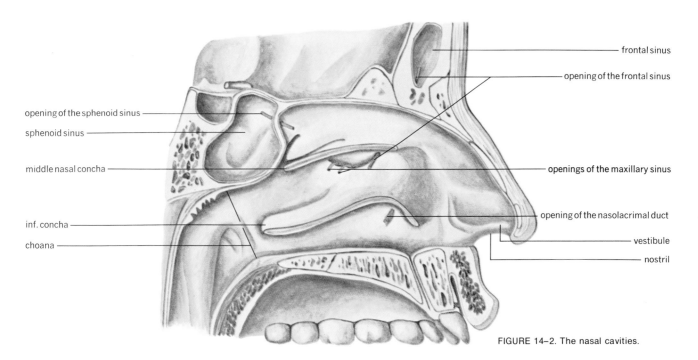

opening of the sphenoid sinus

sphenoid sinus

middle nasal concha

inf. concha

choana

frontal sinus

opening of the frontal sinus

openings of the maxillary sinus

opening of the nasolacrimal duct

vestibule

nostril

FIGURE 14–2. The nasal cavities.

them. The respiratory bronchioles, alveolar ducts, atria, and alveolar sacs comprise this division.

Essential to the exchange of air within both divisions are the muscles
of respiration (diaphragm and intercostal muscles), the ribs and
sternum.

The conducting division

Nasal cavities (Fig. 14–2).   The nasal cavities open to the exterior
through the nares or nostrils. The *vestibule* is a dilated area bearing
large hairs, lying immediately posterior to the nares. The nasal cavity
proper is delimited anteriorly by the nares; posteriorly by the *choanae*
or internal nares opening into the throat; laterally by the ethmoid,
maxillae, and inferior conchae; medially by the vomer and ethmoid perpendicular plate; and the floor is formed by the hard palate. The ethmoid conchae and inferior conchae project into the cavity proper and
create a turbulent air flow through the cavity, a device essential to
cleansing of incoming air. The paranasal *sinuses,* (frontal, sphenoid,
maxillary, ethmoid) open into the cavity (Fig. 14–2). In the apex of the
nasal cavity is the *olfactory area* wherein are found the receptors for the
sense of smell. The vestibule is lined by a stratified squamous epithelium. Except for the olfactory region, the remaining portion of the
nasal cavities and sinuses is lined by a mucous membrane closely applied to underlying bone or cartilage. The epithelium of the nasal
mucous membrane is pseudostratified ciliated, containing goblet cells.
A prominent basement membrane separates the epithelium from the
underlying lamina propria. The lamina contains many thin-walled veins,
and seromucous glands, giving a spongy nature to the membrane. The
blood vessels allow radiation of heat to warm incoming air. The viscous
mucus traps particles and the cilia move them toward the throat. The

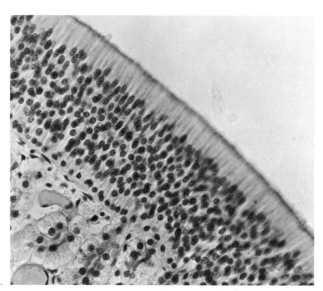

FIGURE 14-3. Olfactory epithelium.

olfactory region has a thicker epithelium devoid of cilia and goblet cells. (Fig. 14-3). Glands of Bowman are found in the lamina of the olfactory region, and they secrete a watery fluid which moistens the olfactory surface. The mucous membranes rest upon the bone or cartilage of the cavity wall.

Pharynx (Fig. 14-1). The nasal cavities communicate with the pharynx through the choanae. The pharynx is an organ common to both respiratory and digestive systems. It extends from behind the nasal cavities to the level of the larynx and is divided into three portions: nasal pharynx, oral pharynx, and laryngeal pharynx.

The *nasal pharynx* extends from the choanae to the inferior border of the soft palate and is respiratory in function. It receives in its lateral walls the openings of the *auditory (Eustachian) tubes* from the middle ear. In its posterior wall is found the *pharyngeal tonsil* (adenoid). The mucous membrane of the nasal pharynx consists of a pseudostratified ciliated epithelium with goblet cells and a typical lamina. The superior pharyngeal constrictor muscle forms the third layer of its wall.

The *oral pharynx* extends from the soft palate to the level of the hyoid bone. It is both respiratory and digestive in function. Its mucous membrane consists of a stratified squamous epithelium and a typical lamina. The middle constrictor of the pharynx forms the third layer of this region.

The *laryngeal pharynx* extends from the level of the hyoid bone to the beginning of the esophagus. It, too, is lined with stratified squamous epithelium and possesses a lamina. The inferior constrictor forms the third layer of the wall.

Larynx. The larynx is the anterior opening from the inferior portion of the laryngeal pharynx. Three single major cartilages establish the basic shape of the organ (Fig. 14-4 A-B). The largest, or *thyroid cartilage,* consists of two plates or laminae joined in the anterior midline. It forms the "Adam's apple." The *cricoid cartilage* lies inferior to the thyroid cartilage and is ring shaped, wider posteriorly. Thyroid and cricoid cartilages are composed of hyaline cartilage. The *epiglottis* is a large, leaf-

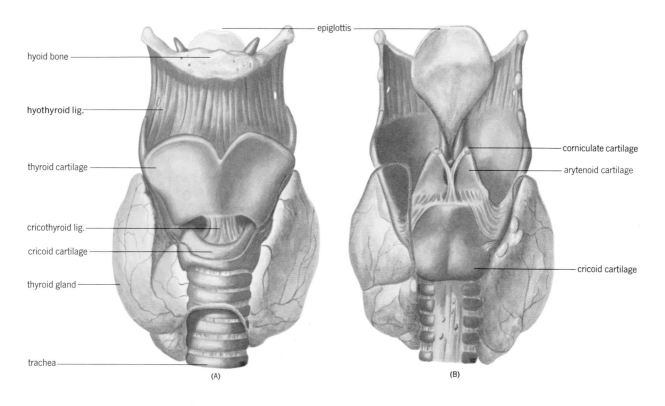

hyoid bone

hyothyroid lig.

thyroid cartilage

cricothyroid lig.

cricoid cartilage

thyroid gland

trachea

epiglottis

corniculate cartilage

arytenoid cartilage

cricoid cartilage

(A)

(B)

FIGURE 14–4. A. Larynx, anterior view. B. Larynx, posterior view. C. The vocal folds.

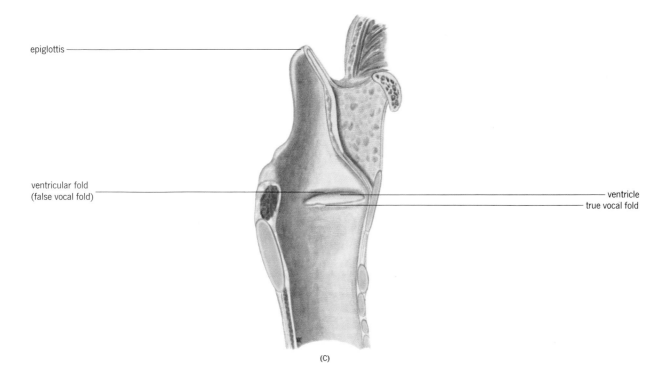

epiglottis

ventricular fold
(false vocal fold)

ventricle

true vocal fold

(C)

shaped mass of elastic cartilage attached to the superior internal surface of the thyroid cartilage.

Three pairs of accessory cartilages support the vocal folds (Fig. 14–4 B-C). The *arytenoid cartilages* attach to the superior aspect of the cricoid cartilage in the posterior larynx. The *corniculate cartilages* attach to the apices of the arytenoids, and the *cuneiform cartilages* are elongated cartilages lying anterior to the arytenoids.

The *hyothyroid ligament* connects the hyoid bone to the thyroid cartilage; the *cricothyroid ligament* connects thyroid to cricoid cartilage. Posteriorly, the larynx is closed by muscle. The vocal folds are formed by foldings of the mucous membrane. The *true vocal folds* also have an internal support of collagenous fibers. The epithelium of the larynx is pseudostratified, except over the vocal folds where stratified squamous is found. The lamina contains many seromucous glands. The *glottis* is the slit-like opening between the vocal folds. *Extrinsic muscles* of the larynx connect the organ to surrounding structures and aid in elevating and depressing it during swallowing. *Intrinsic muscles* connect the various cartilages, control the tension on the vocal cords, and open and close the glottis.

Trachea and bronchi.   The trachea extends about 11 cm into the thorax and terminates by dividing into the right and left bronchi. The trachea is 18 to 25 mm (¾ to 1 inch) in diameter. Its wall (Fig. 14–5) consists of a pseudostratified ciliated epithelium containing goblet cells, and a lamina with glands. The third layer is composed of C- or Y-shaped rings of hyaline cartilage, incomplete posteriorly. The rings support the trachea, keeping it open during breathing. The trachealis muscle connects the open ends of the tracheal rings. The bronchi are similar to the trachea in structure and are about 12 mm (½ inch) in diameter. The right bronchus is about one half as long as the left, and leaves the trachea at a lesser angle. Further branching of the bronchi results in a tree-like arrangement of more numerous and smaller tubes. These tubes are the secondary bronchi, bronchioles, and terminal bronchioles. Some 13 generations of dichotomous (1 tube gives rise to 2) branchings occur within the lung itself. Histologically, several major trends are seen as the branching occurs:

1. Rings of cartilage are replaced by plates of cartilage. The plates become smaller and finally disappear as the terminal bronchiole is reached.
2. As cartilage decreases, the proportion of smooth muscle increases, so that in the terminal bronchiole muscle forms the greatest thickness of the wall.
3. The epithelium changes from pseudostratified ciliated with goblet cells, to pseudostratified ciliated without goblet cells, to simple columnar ciliated as the terminal bronchiole is reached.
4. The diameters of the tubes decrease from about 12 mm in the bronchi to 0.7 mm in the terminal bronchioles. Surface area, because of branching, increases from 2.3 cm² to about 115 cm². Sections of bronchus, bronchiole, and terminal bronchiole are shown in Figs. 14–6 and 14–7.

The respiratory division (Fig. 14–8).

This division is composed of about 9 generations of branchings. The chief trends in this division are a thinning of the epithelium to simple

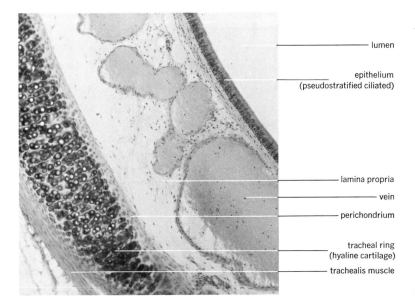

— lumen

— epithelium
(pseudostratified ciliated)

— lamina propria
— vein
— perichondrium

— tracheal ring
(hyaline cartilage)
— trachealis muscle

FIGURE 14–5. Trachea.

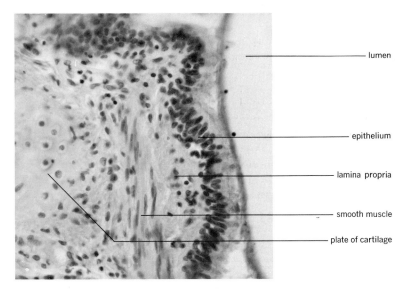

— lumen

— epithelium

— lamina propria

— smooth muscle
— plate of cartilage

FIGURE 14–6. Bronchiole.

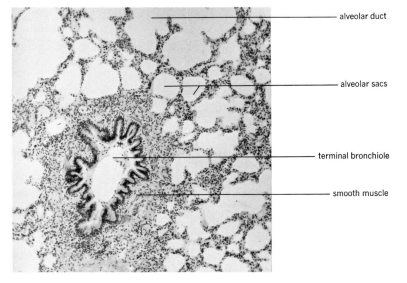

— alveolar duct

— alveolar sacs

— terminal bronchiole

— smooth muscle

FIGURE 14–7. Lung.

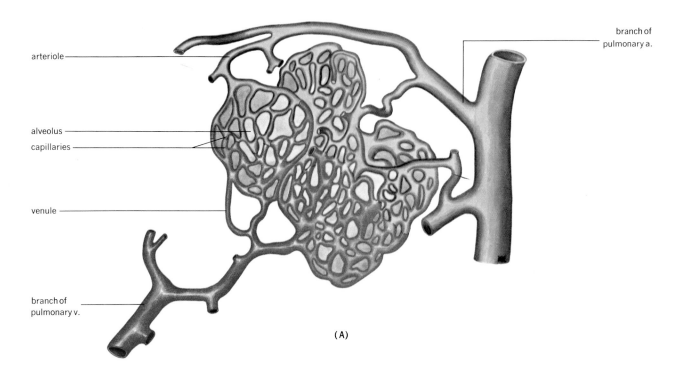

arteriole

alveolus

capillaries

venule

branch of
pulmonary v.

branch of
pulmonary a.

(A)

FIGURE 14–8. A. Blood supply to the alveoli. B. Lobules of the lung. 1–4, Secondary
lobule. 2–4, Primary lobule.

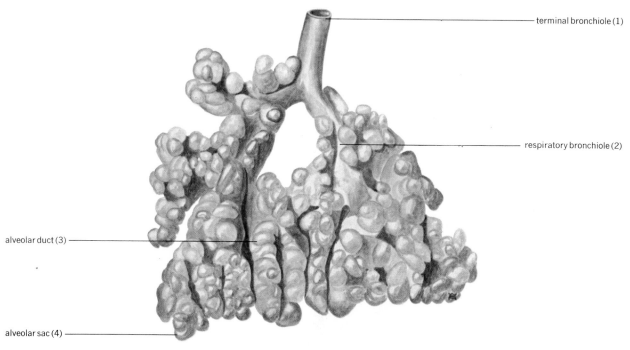

terminal bronchiole (1)

respiratory bronchiole (2)

alveolar duct (3)

alveolar sac (4)

(B)

squamous, and the appearance of alveoli in the sides of the tubes. Much elastic connective tissue is found in the walls, replacing the muscle of the conducting division. Increasing vascularity is also apparent in the respiratory division since gas exchange occurs between this division and the branches of the pulmonary arteries. Diameter of the tubes decreases little in the respiratory division, a fact which maintains air flow. Surface area, however, increases tremendously to achieve the surface necessary for efficient gas diffusion. Table 14–1 summarizes some of the basic facts concerning size and surface area of both divisions.

TABLE 14–1

Some Characteristics of the Respiratory System

Organ	Generation of branching	Number of organs	Diameter (mm)	Total cross-sectional area (cm²)
Trachea	0	1	18 to 25	2.5
Bronchus	1	2	12	2.3
Lobe bronchi	2	4 to 5	8	2.1
Small bronchi	5 to 10	1024	1.3	13.4
Terminal bronchioles	14 to 15	32,768	0.7	113.0
Respiratory bronchioles	16 to 18	262,000	0.5	534
Alveolar ducts	19 to 22	4.2 million	0.4	5880
Alveoli	24	300 million	0.2	(50-70m²)

The lungs.   All parts of the system beyond the bronchi are contained within the lungs (Fig. 14–9). The paired lungs lie within the two lateral *pleural cavities* of the thorax. A serous membrane, the *visceral pleura,* covers the lung surface, and a similar membrane, the *parietal pleura,* lines the thoracic cavity. These membranes are continuous with one another and are reflected upon each other at the *root* of the lung. With the lung in place and inflated, the pleural cavity is reduced to a fluid-lined potential space between the two membranes. The fluid allows easy slippage between lung and chest wall as breathing occurs, and, by the adhesive effect it creates, aids in keeping the lung expanded. The parietal pleurae are named specifically according to the surface of the thorax they line. Costal pleura lines the inner surface of the ribs and intercostal muscles; diaphragmatic pleura covers the superior aspect of the diaphragm; and the cervical pleura rises into the neck. The right lung averages 625 grams in weight, the left 562 grams. In the male, the lungs form about 1/37 of the body weight, in the female 1/43. The lungs are conical in shape with a narrow superiorly directed *apex,* and a broad inferiorly directed *base.* The *costal surfaces* are rounded to match the curvature of the ribs. The medial surface is indented to form a *hilum,* the point of entry and exit of bronchi, pulmonary vessels, and nerves. The medial surface of the left lung bears a concavity, the *cardiac impression,* which conforms to the shape of the heart. The right lung is

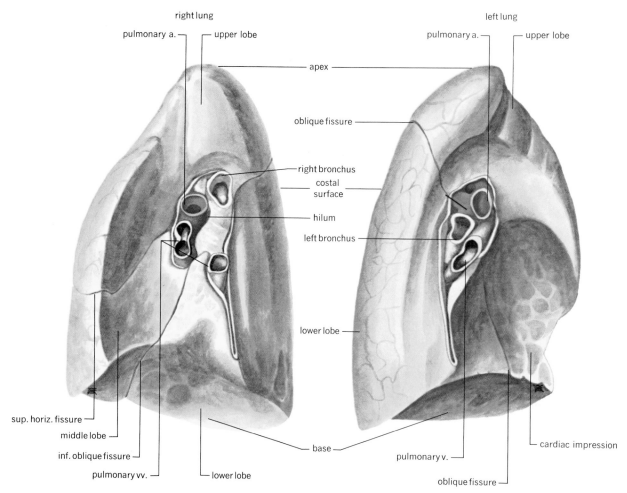

right lung

left lung

pulmonary a. — upper lobe

pulmonary a. — upper lobe

apex

oblique fissure

right bronchus
costal surface

hilum

left bronchus

lower lobe

sup. horiz. fissure
middle lobe
inf. oblique fissure
pulmonary vv.
lower lobe
base
pulmonary v.
oblique fissure
cardiac impression

FIGURE 14–9. The lungs.

FIGURE 14–10. The bronchopulmonary segments.

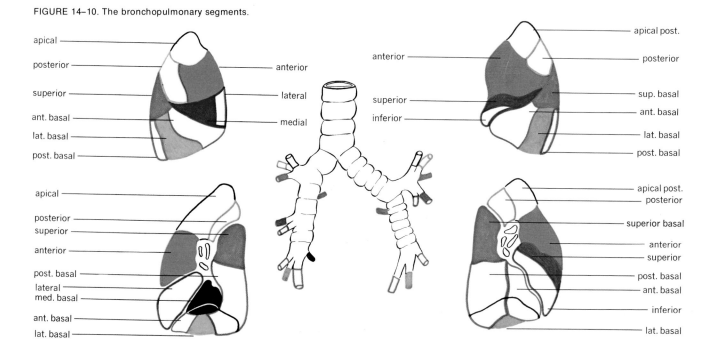

apical
posterior
superior
ant. basal
lat. basal
post. basal

anterior
lateral
medial

anterior

superior
inferior

apical post.
posterior
sup. basal
ant. basal
lat. basal
post. basal

apical
posterior
superior
anterior
post. basal
lateral
med. basal
ant. basal
lat. basal

apical post.
posterior
superior basal
anterior
superior
post. basal
ant. basal
inferior
lat. basal

divided into three lobes (upper, middle, and lower) by a superior horizontal fissure and an inferior oblique fissure. The left lung is divided into an upper and lower lobe by an oblique fissure only. Further subdivision of the lung substance into *bronchopulmonary segments* is provided by connective tissue and the branching of the bronchi (Fig. 14–10). The lung may also be subdivided into lobules (Fig. 14–8). A *secondary lobule* is an anatomical unit created by the fibrous septae which separate a group of terminal bronchioles and all their branches from other groups. A *primary lobule* is a functional unit consisting of one respiratory bronchiole and its divisions.

Blood supply

The conducting division receives nourishing blood by way of the *bronchial arteries,* which arise from the upper aorta and/or the upper intercostal arteries. The bronchial arteries follow the bronchi and bronchioles as far as the terminal bronchioles. The *pulmonary arteries* bring blood to the lungs for oxygenation and nourish the tissues of the respiratory division. The pulmonary arteries do not have well-developed muscular arterioles as are found in the systemic circulation. Therefore much less pressure is required to move the blood through the pulmonary circulation. The capillaries arise directly from the pulmonary arteries in the respiratory zone and form rich networks in the alveolar walls, where oxygenation of the blood occurs. Union of the *capillaries* from both the bronchial and pulmonary arteries forms the *pulmonary veins* which carry the oxygenated blood to the left atrium.

Nerves

Branches of the *vagus, phrenic,* and *intercostal* nerves reach the lungs and respiratory muscles. Their distribution and function will be considered in connection with the physiology of the system.

PHYSIOLOGY

The various stages or events occurring as gases are acquired and carried to the cells may be described by an ancient but nevertheless useful series of divisions:

1. Pulmonary ventilation. This term describes the largely mechanical series of events which draws air into (inspiration) and forces air out of (expiration) the lungs.
2. External respiration. This describes the exchange of gases between blood and lung.
3. Internal respiration. This is exchange of gases between blood and tissue.
4. Transport. Carriage of gases in the blood.

We shall also consider the control of respiration and the factors responsible for changing basic respiratory rhythm.

Pulmonary ventilation

Mechanics of breathing. The pleural cavities which house the lungs are closed cavities, that is, they do not communicate with the exterior.

Their limits are imposed by the rather rigid bony structures of the ribs and sternum, except at the inferior aspect, where they are closed by the diaphragm. The lungs normally contact the inner surfaces of the pleural cavities and diaphragm and are caused to adhere to these surfaces by a thin film of fluid between the visceral and parietal pleurae. This adhesion is one factor tending to maintain the lung in an expanded state, and also causes the lung to follow any alteration in the size of the pleural cavities. Measurement of the pressure within the cavities (intrathoracic pressure) shows it to be less than atmospheric pressure, to the extent of 3 to 5 mm Hg. This factor also aids in insuring expansion, inasmuch as the pressure within the lung is always greater than that around the lungs. Inspiration of air is achieved according to the following principle: if the volume of a closed cavity is increased, the pressure within it will fall.

Increase in volume of the thorax is brought about by two means:

1. Contraction of the external intercostals causes an elevation of the ribs. The ribs, because of their manner of attachment to the vertebral column and their shape, swing up and out as they are elevated. This increases the dorsoventral and lateral dimensions of the thorax.
2. Contraction of the diaphragm pulls its central region downwards, increasing the vertical dimensions of the thorax.

These movements increase the volume of the thorax by about 500 ml during normal breathing, and further decrease the pressure within by 3 to 5 mm Hg. The next problem is to transmit this increased size and decreased pressure to the lung itself. The lung is expanded during inspiration primarily because of its adhesion to the pleural cavity walls by the film and fluid already described. In other words, as the thorax expands the lung follows. This expansion creates a temporary less than atmospheric pressure in the lungs. Since the lungs communicate with the exterior, a less than atmospheric pressure within them will cause atmospheric air to rush into the lungs to fill the low pressure area. The flow is at first rapid, and then declines and stops as the pressure becomes equalized. As the lungs are expanded, the elastic tissues within them are stretched and potential energy is stored. Inspiration may be said to be "active" in the sense that it requires muscular contraction to bring it about. It must be emphasized, however, that the muscular contractions do not involve the lung, but the muscles of respiration. The lungs respond passively to the volume and pressure changes created by muscles of respiration.

Expiration is brought about as the muscles of respiration relax and the thorax returns to its original size. The potential energy stored in the elastic elements is converted to kinetic energy as the lung tissue recoils during decrease in thoracic volume and lung volume. The recoil creates a temporary pressure in the lungs higher than atmospheric pressure, which forces air out of the lungs. Expiration is normally passive, not involving muscular activity. It may become active through contraction of abdominal musculature which forces the viscera against the diaphragm and elevates it. Important to this discussion of intake and output of air in the lungs is to recognize that the lungs normally possess both stretchability (or compliance) and elasticity. The former term refers to the resistance to inflation posed by the lungs; the latter term represents the force of recoil of the elastic elements. Obviously, normal ventilation of the lungs with air may be changed if either of these factors is altered.

The alveoli present an air–liquid interface, because the epithelium is covered by tissue fluid. At such interfaces, fluids exert a surface tension, tending to pull the alveoli into a minimal surface for a given volume. This tension creates a tendency for alveolar collapse. Also, a fluid–air interface should present a more or less constant surface tension, whether the alveoli are expanded maximally or not. Measurements of surface tension within the lung indicate that it increases on inflation and decreases on deflation. This phenomenon is explainable only by the supposition that something other than water is present at the alveolar surface. Investigation of the problem disclosed the presence of a lipoprotein coating on the alveolar surface. It is named *surface active agent,* or *surfactant* for short. It aids in keeping alveoli from collapsing further when deflated, and aids in deflation when alveoli are expanded.

Factors affecting ventilation. The respiratory system is superbly organized to achieve maximal ventilation with minimal energy expenditure. The main factors determining the work of ventilation are:

1. Resistance to flow of air through the conducting division.
2. Compliance (stretchability) of the lung tissues.
3. The surface tension of the alveolar walls.

Airway resistance. According to the laws governing the flow of fluids through small rigid tubes, resistance to flow varies inversely with the fourth power of the radius of the tube (for example, if the radius of the tube is doubled, the resistance decreases by the fourth power, or 16). This rule applies (approximately) to the flow of air through the respiratory passageways, even though the tubes are not rigid and we are dealing with gases and not fluids. Reference to Table 14–1 indicates that as the conducting pathway branches into smaller and smaller tubes, the total cross-sectional area of the tubes increases. On the basis of an increased surface area, we would expect an increase in airway resistance, and therefore work, since there is a greater frictional surface presented to the air flow. However, little change in resistance actually occurs, because the collectively larger tubes (with greater area) act as a pathway of least resistance, and resistance to flow is raised little, if at all. Normally this resistance is very low; hence, the effort of breathing rarely comes to our notice. Any condition which reduces the cross-sectional area will increase resistance to airflow and therefore increase the effort of breathing. Some conditions having this effect are:

1. Constriction of bronchiolar smooth muscle, as in bronchial asthma.
2. Mucosal edema, congestion or inflammation, as in bronchitis.
3. Plugging of the tube by mucus or other material, as in bronchitis.
4. Collapse or twisting of the tubes, as in emphysema.

Bronchiolar smooth muscle is responsive to the autonomic nervous system and various chemical stimuli. The sympathetic nervous system and epinephrine relax bronchiolar muscle and dilate the bronchioles. The parasympathetic nervous system and acetylcholine contract bronchiolar muscle and constrict the bronchioles. Histamine is also a powerful bronchoconstrictor These effects also influence resistance to flow.   *antihistamine*

Compliance. Normally, the force required to stretch the pulmonary tissue is low (about 2.5 cm water for an average 500 ml breath). This can be changed by conditions which alter the amount or type of connective tissue in the lungs. Pulmonary emphysema, a condition in which there is loss of elastic fibers in the lung, results in a greater compliance (the lung is more easily stretched), but difficulty in expiring air (due to loss of elastic recoil). Diffuse increase of collagenous fibers in the lung results in a decrease of compliance, as the tough, nonstretchable fibers resist expansion.

Surface tension. As previously described, the force required to overcome surface tension in inflating the alveoli is determined by the fluid material covering the alveolar surface. The presence of the lipoprotein surfactant material at the surface, rather than water, reduces the alveolar surface tension and reduces the amount of effort required to breathe. Reduction of blood supply, in particular, reduces the production of surfactant.

Volumes of air exchanged. Normal breathing at rest causes an exchange of an amount of air which varies with size, age, and sex of the individual. In the average individual, approximately 500 ml of air is moved with each respiration. This volume is referred to as the *tidal volume* (TV). Of this volume, about 150 ml remains in the conducting division and is not available for exchange. This air is known as *dead air* and the tubes it occupies constitutes the *dead space.* If, at the end of a normal inspiration, one forces inspiration by a voluntary effort, an additional quantity of air, averaging 3000 ml, may be drawn into the lungs. This air constitutes the *inspiratory reserve volume* (IRV). If, at the end of a normal expiration, a forcible expiration is made, an additional quantity of air, averaging 1100 ml, may be expelled from the lungs. This is the *expiratory reserve volume* (ERV). A normal respiration, therefore, utilizes the central portion of the total exchangeable air and leaves both an inspiratory and expiratory reserve for added activity. Air still remains in the lungs even after the most forcible expiration. This air is termed *residual volume (RV)* and amounts to about 1200 ml. Collapse of the lung will drive about half of this air from the lung; what still remains is termed *minimal volume* (MV).

Various respiratory capacities may be calculated by adding these volumes in various combinations. *Inspiratory capacity* is the sum of tidal volume and inspiratory reserve volume. It represents the total inspiratory ability of the lungs. *Functional residual capacity* is the sum of expiratory reserve volume and residual volume. This is an important capacity physiologically because it represents the air from which the blood is oxygenated during expiration. *Vital capacity* is the sum of inspiratory reserve, tidal, and expiratory reserve volumes and represents the total exchangeable air of the lung. *Total lung capacity* is the sum of all volumes. The interrelationships between these volumes and capacities is presented in Fig. 14–11. Normal breathing rate in the adult is 12 to 16 complete excursions (inspiration and expiration) per minute. Assuming a normal tidal volume of 500 ml, multiplying rate times volume will give the *minute volume of respiration:*

MVR = 14 (average) × 500 (TV) or 7000 ml/min (MVR)

Of this volume, only 4900 ml reaches the alveoli; 150 ml × 14, or 2100 ml, will remain in the dead space.

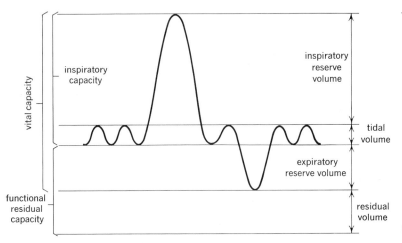

FIGURE 14–11. Respiratory volumes and capacities.

Composition of the respired air.   The term partial pressure is used to designate that portion of total atmospheric pressure contributed by a particular gas in a mixture of gases. Partial pressure is calculated by multiplying the total atmospheric pressure (usually corrected to standard temperature and pressure) by the percent of the gas in the mixture. The quantity is designated by a lower case p, followed by the formula of the gas; for example, pO_2, pCO_2, pN_2.

Analysis of the composition of inspired air, expired air, and alveolar air (the last air exhaled under a forced expiration) shows the compositions in Table 14–2. Studying the figures should indicate that some oxygen has disappeared from the inhaled air, while carbon dioxide has been added to it. Analysis of pO_2 and pCO_2 in the tissues shows average levels of 30 mm Hg, and 50 mm Hg, respectively. The movement of these gases proceeds, by the process of diffusion, from an area of higher partial pressure of the gas to the area of lower partial pressure of the gas. As seen from the preceding table, there exists a pressure gradient for oxygen from alveoli to blood to tissue, and a pressure gradient for carbon dioxide from tissue to blood to alveoli. Knowing the pO_2 and pCO_2 in lungs and tissues will enable prediction of approximate partial pressures of the two gases in arterial and venous blood.

TABLE 14–2

Gas	Inspired Air		Expired Air		Alveolar Air	
	%	Partial pressure (mm Hg)	%	Partial pressure (mm Hg)	%	Partial pressure (mm Hg)
Nitrogen	78.30	594.70	75.0	569	75.0	570
Oxygen	21.00	160.00	15.2	116	13.6	103
Carbon dioxide	0.05	0.30	3.6	28	5.2	40
Water vapor	0.65	(average) 5.00	6.2	47	6.2	47
Totals	100.00	760.00	100.0	760.	100.0	760.

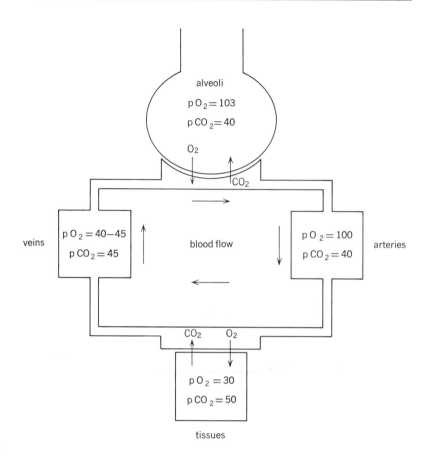

The above diagram may aid in the understanding of the problem.

Oxygen will diffuse into the blood from the alveoli until a theoretical maximum pO_2 of 103 mm Hg is reached (equal to pO_2 in alveoli). This value is not achieved for two reasons: (1) diffusion slows as equilibrium is approached, and (2) blood is moving rapidly through the capillaries and time is not provided for the attainment of complete equilibrium. Therefore the blood leaving the lungs has an actual pO_2 of 95 to 100 mm Hg, and is saturated to about 96 percent of its total oxygen-carrying capacity. Carbon dioxide, being twenty times more soluble than oxygen, diffuses much more rapidly into the alveoli. Complete equilibrium at pCO_2 of 40 mm Hg is easily achieved, and arterial blood leaves the lungs with this pCO_2. Arriving at the tissues, oxygen diffuses out of the blood and carbon dioxide diffuses into the blood until a pO_2 of 40 to 45 mm Hg, and a pCO_2 of about 45 mm Hg is achieved. The pO_2 falls to a lower value than might be expected on the basis of diffusion equilibrium due to the fact that the presence of carbon dioxide drives an additional quantity of oxygen from the blood (see below). The venous blood then is carried to the lungs where the process is repeated.

Transport of gases

Oxygen. Ninety-five percent of the oxygen transport in the blood is carried out by reversible combination of oxygen with hemoglobin (Hgb) according to the general equation:

$$Hgb + O_2 \rightleftharpoons HgbO_2$$
reduced Hgb   oxy Hgb

Loading of the hemoglobin with O_2 is not actually this simple. Hemoglobin molecules are arranged in packets of 4 in the erythrocyte. Loading thus takes place in stages:

$$Hb_4 + O_2 \longrightarrow Hb_4O_2$$
$$Hb_4O_2 + O_2 \longrightarrow Hb_4O_4$$
$$Hb_4O_4 + O_2 \longrightarrow Hb_4O_6$$
$$Hb_4O_6 + O_2 \longrightarrow Hb_4O_8$$

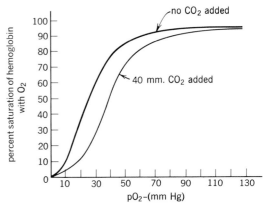

FIGURE 14–12. Oxygen dissociation curve.

The amount of hemoglobin available for O_2 transport also depends on whether its binding sites are occupied by other materials, chiefly CO_2. An oxygen dissociation curve (Fig. 14–12) shows the carrying capacity of hemoglobin, with, and without varying quantities of CO_2 present. The shifting of the curve to the right with increasing pCO_2 (the Bohr effect) is not necessarily an undesirable effect. Increased activity produces more CO_2 and requires more O_2. The presence of excess CO_2 actually drives more O_2 from the hemoglobin, insuring the tissues of an added O_2 supply. At the same time, it creates additional binding sites for CO_2 transport. The remaining 5 percent of O_2 carried in the blood is in dissolved form in the plasma.

Carbon dioxide. CO_2 is transported in three ways:

1. Sixty-four percent as bicarbonate, according to the following equation:

$$CO_2 + H_2O \longrightarrow H_2CO_3 \longrightarrow H^+ + HCO_3^-$$
carbonic   hydrogen   bicarbonate
acid   ion   ion

2. Up to 27 percent is carried on the hemoglobin as carbamino hemoglobin according to the following equation:

$$Hgb + CO_2 \rightarrow HgbCO_2$$

3. Nine percent is carried in simple solution in the plasma.

Bicarbonate formation occurs within the plasma as indicated, and within the erythrocyte in the process designated as the chloride shift:

$$CO_2 \longrightarrow + H_2O \longrightarrow H_2CO_3$$

H + HCO_3^- (increase $\longrightarrow$ HCO_3^-
                              creates   (loss of
                              diffusion   Negative
                              gradient)   ion)

(to maintain electrical balance) $\longleftarrow$ Cl$^-$

erythrocyte

The formation of bicarbonate also creates H^+, which must be buffered to avoid pH change. Systems available for buffering include:

1. Bicarbonate system buffers. These buffer fixed acids (HA) as follows:

$$HA + HCO_3^- \longrightarrow H_2CO_3 + A^-$$
$$\downarrow$$
$$H_2O + CO_2$$

   The carbon dioxide is then eliminated by breathing. Bicarbonate systems account for about 50 percent of the total buffering capacity (buffer base) of the blood.

2. Nonbicarbonate systems. These buffers include hemoglobin, protein, and phosphates, and account for the remaining 50 percent of the blood's buffering capacity. The reaction of these systems may be represented by the general equation

$$HA + Buffer^- \longrightarrow H\,Buffer + A^-$$

Carbonic acid may be buffered by nonbicarbonate systems as follows

$$H^+HCO_3^- + Buffer^- \longrightarrow H\,Buffer + HCO_3^-$$

The operation of this reaction results in no change in total buffer base, since the disappearance of 1 meq. of Buffer$^-$ is balanced by the production of 1 meq. of HCO_3^-. This is perhaps the only example in the body of a 100 percent efficiency.

Acid–base regulation by the lungs

The lung is the most important body organ involved in the regulation of acid–base balance. The production of carbon dioxide by body cells results in the addition of 13,000 to 20,000 meq. per day of H^+ to the body fluids. The reaction involved in the formation of H^+ from carbon dioxide is:

$$CO_2 + H_2O \longrightarrow H_2CO_3 \longrightarrow H^+ + HCO_3$$

This reaction is reversed in the lungs by the elimination of carbon dioxide, which shifts the chemical equilibria of the reaction towards the formation of carbonic acid:

$$H^+ + HCO_3^- \longrightarrow H_2CO_3 \longrightarrow H_2O + CO_2$$
$$\text{(eliminated)}$$

The rate of elimination of carbon dioxide is dependent upon the process of ventilation; and in turn the arterial pCO_2 is a function of carbon dioxide production and ventilation according to the equation

$$\text{arterial } pCO_2 = K\, \frac{CO_2\ \text{production}}{\text{alveolar ventilation}}$$

We thus have a system for self-regulation of arterial pCO_2 and therefore pH. The respiratory centers controlling respiration are stimulated by increase in pCO_2 and/or a decrease of pH, and elimination of carbon dioxide rises. Conversely, decreases of pCO_2 and/or rise of pH result in decreased stimulation of the centers and carbon dioxide elimination is decreased. The entire system is thus seen to be controlled within very narrow limits by a feedback mechanism which moniters the pCO_2 as determined by the balance between carbon dioxide production and elimination. Adjustments are made rapidly and accurately.

Control of respiration

Introduction.   The basic desire to breathe may be characterized as being due to involuntary spontaneous neuronal activity modified by chemical influences. A basic pattern is established which, within certain limits, may be voluntarily modified to permit talking, singing, hyperventilation, and breath holding. Reflex control of respiration is afforded by peripheral and central chemoreceptive sites which are triggered by changes in pH and gas levels in the blood and cerebrospinal fluids. These factors tend to adjust ventilation to levels designed to maintain a balance between gas utilization, production, and elimination.

The respiratory centers.   The neurones responsible for establishing the basic rhythm of breathing are located within the medulla and pons of the brain stem. There appear to be five functional groups of neurones in these areas, each contributing something to the establishment of a normal respiratory rhythm.

1. Paired *inspiratory centers,* located in the medulla, appear to have some degree of spontaneous activity, and initiate inspiration.
2. Paired *expiratory centers,* located dorsally to the inspiratory centers, initiate expiration.
3. Paired *pneumotoxic centers,* located in the anterior pons, contribute reinforcement to the expiratory center to achieve expiration.
4. Paired *nuclei of the vagus nerve* in the medulla provide additional reinforcement for expiration.
5. An unnamed and unidentified area in the region of the fourth ventricle of the brain apparently moniters the pH of the cerebrospinal fluid and stimulates respiration if pH falls.

The primary stimulus for respiration is provided by the presence of carbon dioxide and H^+ in the cells of the inspiratory center. Discharge of nerve impulses over phrenic and intercostal nerves to diaphragm and intercostal muscles causes inspiration. Inflation of the lungs causes stimulation of pulmonary receptors which send impulses over the vagus nerves to inhibit the activity of the inspiratory center (*Hering-Breuer reflex*). Additional input for expiration is provided by the pneumotaxic and expiratory centers. The result will be a periodic interruption of continuous inspiratory effort, causing expiration. It should be appreciated that the respirations themselves determine arterial pCO_2 and therefore H^+, and so a self-regulating system is created.

Additional control of respiration is provided by chemoreceptive cells in the *carotid* and *aortic bodies.* These moniter pCO_2 and pO_2 in the circulation to the brain and body generally. Increase of arterial pCO_2 by 0.5 percent, or a fall of pH reflexly increases rate and depth of breathing, eliminating the excess carbon dioxide.

Protective mechanisms of the system

Since the respiratory system transports gases from the atmosphere to the lungs, there may be inhaled a wide variety of potentially dangerous microorganisms, antigens, and particulate material. To aid in understanding the operation of the mechanisms of protection, recall that the greater part of the respiratory system is lined with ciliated epithelium. Also recall the presence of a mucus coating over the surface epithelium, provided by the externally secreting glands of the lamina propria and goblet cells. There are lymphatics, alveolar macrophages, and reflex mechanisms for dislodging large particles from the system.

The specific mechanisms involved are:

1. *The mucociliary escalator.* The presence of a mucus layer on the epithelium of the system allows trapping of particles as incoming air contacts the surface. Coordinated ciliary action moves the mucus layer posteriorly in the nasal cavities to the throat where the mass is usually swallowed. A similar mechanism operates from the terminal bronchioles upward through the conducting division. Continuous cleansing of the system above the terminal bronchiole is thus provided. In normal individuals, the rate of mucus production and removal by ciliary action is nicely balanced, and mucus rarely comes to our conscious attention.

2. *Alveolar macrophages.* Since both cilia and goblet cells are not found beyond the terminal bronchiole, a cleansing mechanism is provided for the respiratory division. Phagocytic cells (alveolar macrophages or dust cells) are found in the alveoli of this division. These cells engulf and destroy bacteria, particles of dust, foreign antigens, or other harmful substances. Their numbers are partially dependent upon the level of contamination of inhaled air, increasing as the load of pollutants increases. Their phagocytic ability may also vary, becoming less as the amounts of chemical contaminants increases. These cells thus provide an extremely important line of defense.

3. *Filtering.* The presence of the large hairs around the nostrils tends to restrict the entry of large objects into the system.

4. *Secretion of immune globulins.* The presence of immunoglobulin A (IgA) has been detected in the secretions of the laminal glands in all parts of the respiratory tree. The substance is a nonspecific antibody whose production is apparently triggered by a wide variety of antigenic challenges. The amount produced is directly proportional to the degree of antigenic challenge and forms an important defense mechanism against foreign chemicals.

5. *Lymphatics.* Numbers of lymphatics are somewhat greater in the respiratory and digestive system than elsewhere in the body. Recall that both systems contact the external environment. The combination of nodules in the mucous membranes and lymph vessels carrying fluid to lymph nodes aids removal of matter which enters the tissues themselves.

6. *Reflex protective mechanisms* include sneezing and coughing. Sneezing follows irritation of the nasal mucosa, and coughing follows irritation of the tracheobronchial mucosa. Both responses are dependent upon nervous pathways and result in a sharp inspiration followed by an explosive expiration. The force of the

expiration tends to blast the offending particles from the system.
7. Drying of the diffusing surface is prevented by the *humidifying* of the incoming air by the secretions of the mucosal glands.
8. *Warming* is provided by radiation of heat from the many blood vessels in the mucosa.

Respiratory abnormalities

Abnormalities of respiration may concern alterations in breathing patterns, diffusion of gases, failure of ventilation, or combinations of these. Commonly, it may be stated, any one given alteration will initiate a chain of events culminating in decreased blood oxygen saturation and increased blood carbon dioxide levels. Receiving much attention today are those disorders which are classed under the heading of obstructive pulmonary disease. Specifically, included are emphysema, bronchitis, asthma, and pneumonia.

Emphysema. This disorder is characterized by enlargement of the alveoli, atrophy of the septae between alveoli, loss of elastic tissue of the lung, and collapse and kinking of the terminal bronchioles with great increase in airway resistance. A decrease in diffusion surface results from loss of alveoli or their coalescence into larger units. Inspiration is usually made without difficulty, but expiration is made with extreme difficulty. The elastic recoil of the lungs is greatly diminished and the terminal bronchioles tend to collapse upon increase of pressure within the lung. Residual volume is thus increased, vital capacity is decreased 20 to 60 percent, and uneven aeration of alveoli occurs. Loss of pulmonary capillary beds may be demonstrated, further reducing ability to adequately oxygenate the blood and eliminate CO_2. Specific causative factors for the disease are lacking. A genetic tendency appears to exist but more important are vascular and degenerative changes initiated by inhalation of irritants. Strongly implicated as irritants are those contained within the smoke of tobacco. Tobacco smoke constricts pulmonary blood vessels, including bronchial and pulmonary artery branches. Deprivation of nutrients may lead to loss of elastic elements and to degeneration of smooth muscles. The circulatory changes lead to increased resistance in the lung, and a greater load is placed upon the right ventricle. Interestingly, normal compensatory mechanisms of the body increase red cell numbers, hemoglobin content, and therefore oxygen capacity. The respiratory centers become less sensitive to CO_2 and the individual shows little increase in respiratory minute volume with increasing CO_2 levels.

Bronchitis, strictly defined, refers to the inflammation of the larger tubes of the respiratory system. It, too, is due primarily to irritants or to bacterial or viral disease within the system. A chronic abcess is often seen in individuals who are exposed to polluted environments. Such individuals have a decreased capacity to clear their lungs of the irritants and commonly possess a chronic cough and diminished resistance to respiratory infection.

Asthma is the result of spasm of smooth muscle in the terminal bronchioles. A strong hereditary tendency exists for the disease. Bronchitis may be of reflex origin due to nervous reflexes originated by irritation of the mucosal surface of the bronchioles, or may be allergic as the result of sensitization to foreign proteins.

Pneumonia results from bacterial or viral infection of the lung, with production of exudate which fills the alveoli. Diffusion of gases is reduced, with the development of hypoxia and CO_2 retention. Respirations are rapid and shallow, leading to further reduction of gas ventilation and exchange. Recovery depends upon stopping the inflammatory process.

Other disorders. Atelectasis is collapse of alveoli. It may result from any condition which lowers pressure within the alveoli or increases pressure on the lung surface. Pleural oversecretion, pneumothorax (a channel through the chest wall equalizes pressure outside and inside the lung), or obstruction of terminal bronchioles may cause the condition.

Hypoxia results when there is inadequate supply of oxygen to the tissues. Some of the types and causes of hypoxias are summarized in Table 14–3.

TABLE 14-3

Hypoxias

Type	Causes	Comments

Hypoxic	1. Low O_2 in inspired air
	    (a) altitude
	    (b) rebreathing air
	 2. Decreased ventilation
	    (a) airway obstruction
	    (b) pneumothorax
	    (c) respiration paralysis
	 3. Abnormal lungs
	    (a) asthma
	    (b) emphysema
		All reduce arterial pO_2 and usually (2 and 3) cause a rise in CO_2
		 O_2 capacity of blood and blood flow essentially normal

Anemic	Lowered Hgb content, CO poisoning, abnormal Hgb with decreased O_2 capacity	All result in decreased pO_2 in blood; due either to less Hgb, or to occupation of Hgb by something forming a stronger bond than CO_2
Hypokinetic	Decreased rate of blood flow, as in cardiac failure, hemorrhage, or shock	Primarily a circulatory problem
Histotoxic	Poisoning of cells by drugs, leading to failure to utilize O_2	All other factors are normal; cyanide is a common histotoxic substance

QUESTIONS

1. What is the functional significance of the changes in epithelium type which occur as we pass from the nasal cavities to the alveoli?
2. Compare the blood supplies of conducting and respiratory divisions, and comment on the effect on the tissues that a decreased blood flow in each system might have.
3. Describe the typical pattern of tissue layers occurring in the wall of the respiratory system.

4. Compare lobes, bronchopulmonary segments, and primary and secondary lobules of the lungs in terms of anatomical extent and functional significance.

5. Commencing with the effect of carbon dioxide upon the inspiratory center, catalogue all events which occur to achieve inspiration and expiration.

6. Explain how acid–base regulation by the lung is self-controlling.

7. What factors govern air flow through the respiratory system? Which are changed, and how, in the disease emphysema?

8. Describe the protective mechanisms employed by the respiratory system. Against what type of threat is each designed to protect?

CHAPTER 15

The Digestive System

*Part of the secret of success in life is to eat what you like
and let the food fight it out inside.* MARK TWAIN

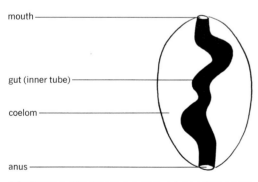

mouth

gut (inner tube)

coelom

anus

FIGURE 15-1. The digestive system, a tube within a tube.

388

THE DIGESTIVE SYSTEM

Introduction

The cells of the body require a wide variety of materials in order to carry out their normal functions. Amino acids, simple sugars, fatty acids and glycerol, salts, and water are essential to the continuation of cellular activity. *Ingestion* of the necessary nutrients is one of the several functions of the digestive system. Organic nutrients are not ingested in a form which makes them directly available for cellular use. Where the cells require amino acids, proteins are ingested; where simple sugars are required, polysaccharides are ingested. Some method must therefore be provided to reduce these complex materials to their constituent building blocks for cellular use. The process of *digestion* accomplishes the required reduction. Digestion is in part a mechanical subdivision, because the foods are chewed and mixed in the tract; but the major process is a chemical one, utilizing digestive enzymes to split the chemical bonds between the basic building units in a given foodstuff. The main portion of the digestive system, the alimentary tract, is a tube within a tube (Fig. 15-1), and materials which are within it are, in a sense, not inside the body. Only when a substance crosses the lining membrane of the tract can it truly be said to be inside the body. Hence, *absorption* of the digested food materials is another function of the system. Included in the food we eat is a certain amount of indigestible material, or material which is only partially digested in the tract. Such material, along with excess water and many bacteria, are *egested* or *eliminated* from the body through the feces.

Movement of the food mass through the tract is necessary in order that the various enzymes of the tract may act in orderly sequence to insure efficient digestion. *Motility* in the tract depends upon muscular activity, controlled by direct nervous stimulation, reflex mechanisms, and chemical factors. Without this motility, few if any of the functions of the system could be carried out effectively.

The organs of digestion (Fig. 15–2).

The organs of digestion include those forming the tube or *alimentary tract,* and *accessory structures* essential to some phase of the digestive process. The tract includes the mouth, pharynx, esophagus, stomach, small and large intestines, rectum, anal canal, and anus. The accessory structures are located outside the limits of the tract proper and empty into it through ducts. They are the salivary glands, liver, and pancreas.

The histological structure of the alimentary tract

The organs of the alimentary tract, particularly those from the esophagus to the anal canal, have the same basic arrangement of tissue layers in their walls. Each organ, however, has certain special structural features of its own which enable one to identify it histologically as well as grossly. These special features make possible the unique functions performed by each organ. Four basic layers of tissue are described (Fig. 15–3):

1. The *mucosa (tunica mucosa).* This layer consists of a lining *epithelium,* an underlying layer of connective tissue, the *lamina propria,* and a thin layer of nonstriated (smooth) muscle, the *muscularis mucosae.* The mucosa is glandular except in the front portion of the mouth and in the anal region.
2. The *submucosa (tunica submucosa).* This layer consists of vascular connective tissue. In the intestines, the submucosa is thrown into folds, the plicae circulares, which serve to increase the surface area of the gut. The submucosa contains glands in the esophagus and first part of the small intestine. An important nerve plexus, the *submucosal plexus (of Meissner)* is also located within this tissue layer.
3. The *muscularis (tunica muscularis* or *muscularis externa).* The muscular coat is typically composed of two layers of muscle. The inner layer is in the form of a tight spiral *(circular layer)* while the outer coat is a loose spiral *(longitudinal layer).* The muscle is nonstriated (smooth) except in the mouth, pharynx, upper esophagus, and anal canal. Between the two muscle layers is located the second major nerve plexus of the tract, the *myenteric plexus (of Auerbach).*
4. The *serosa* or *adventitia.* The outermost coat of the tract is the serosa on those organs which are suspended by mesenteries within a body cavity. Where the organ is not suspended by a mesentery, but is immediately surrounded by other organs or tissue,

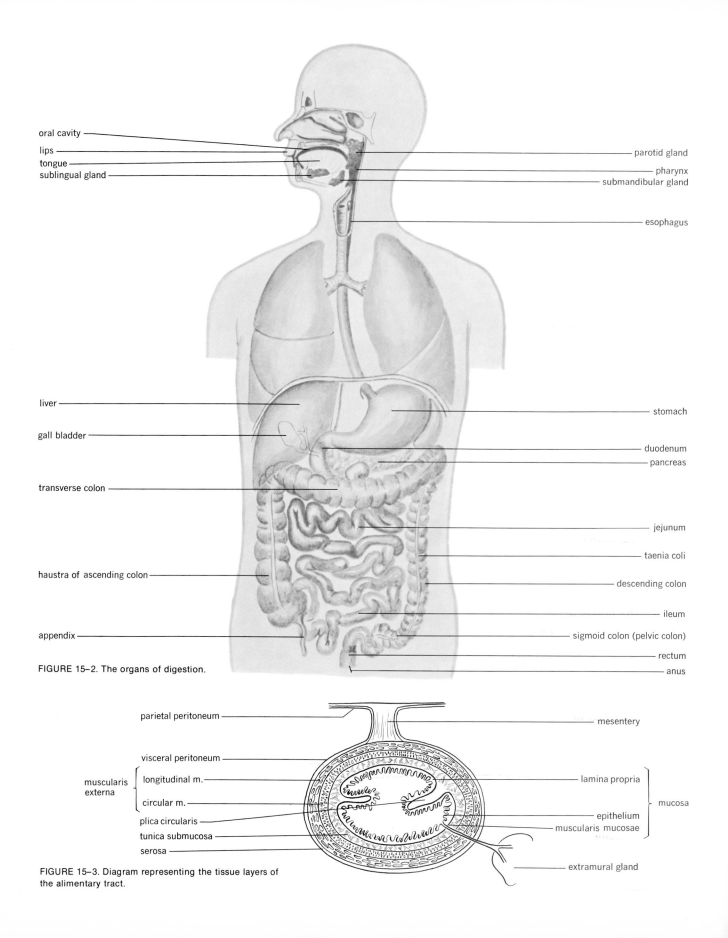

oral cavity

lips

tongue

sublingual gland

parotid gland

pharynx

submandibular gland

esophagus

liver

gall bladder

transverse colon

haustra of ascending colon

appendix

stomach

duodenum

pancreas

jejunum

taenia coli

descending colon

ileum

sigmoid colon (pelvic colon)

rectum

anus

FIGURE 15–2. The organs of digestion.

parietal peritoneum

mesentery

visceral peritoneum

muscularis externa { longitudinal m. circular m.

plica circularis

tunica submucosa

serosa

lamina propria

epithelium

muscularis mucosae

mucosa

extramural gland

FIGURE 15–3. Diagram representing the tissue layers of the alimentary tract.

there is no clear line of demarcation between the organ and the surrounding tissue. In this case, the outer coat is referred to as the adventitia.

Motility within the tract

The mouth and pharynx. Chewing (mastication) involves mechanical subdivision of foods by the action of the teeth. The degree of subdivision achieved is largely determined by the length of time the food is chewed. Chewing is begun voluntarily, and is continued, largely at the reflex level, as follows. Mechanical stimulation of the walls of the mouth causes a reflex inhibition of the mouth-closing musculature (masseter, temporalis, medial pterygoid) and stimulation of the digastricus. The mandible is depressed, and the mouth opens. The opening of the mouth reduces the inhibition of the closing musculature and they reflexly contract. The cycle is repeated in rhythmical fashion. Mastication not only reduces the size of the food particles, but also mixes the food with saliva for commencement of digestion and lubrication. Swallowing (deglutition) is initiated voluntarily by the collection of a formed mass of food (bolus) on the tongue, and the movement of that bolus by the tongue into the pharynx. The mandible is closed, the tongue is elevated to seal the mouth, and the soft palate is elevated to seal the nasal cavities. The tongue then moves backwards propelling the food into the pharynx by creating a positive pressure of 4 to 10 mm Hg. The larynx is elevated, the epiglottis is bent backwards over the glottis and the bolus moves into the lower pharynx. Once into the lower pharynx, further movement of the bolus is through reflex action. Stimulation of the pharyngeal wall results in the initiation of nervous impulses which pass to the medullary swallowing center over the glossopharyngeal (IX) and vagus (X) cranial nerves. The center discharges to the pharyngeal and esophageal musculature over the trigeminal, facial, vagus, accessory, and hypoglossal (V, VII, X, XI, and XII) cranial nerves. These control a complex series of actions which relax the upper esophageal sphincter, contract the lower pharyngeal musculature and move the bolus into the esophagus. Once in the esophagus and still under this control, a ring-like constriction is formed behind the bolus to move it forward.

Esophagus. In the esophagus, movement is completed by a *peristaltic wave* which sweeps from pharynx to stomach in about nine seconds. The musculature of the esophagus is serially innervated by the vagus nerve. Stretching of the esophagus by the bolus evokes discharges by the vagus which creates a strong contraction of the circular muscle layer. Stimulation of successive segments of the esophagus causes the wave to traverse the length of the esophagus, stripping the mucosa clean of food particles.

Stomach. The stomach tends to act as a unit in terms of its motility, since the muscle layers are essentially one continuous mass. When food enters the stomach under the impetus of the esophageal peristaltic wave, the stomach musculature undergoes a brief relaxation (*receptive relaxation*), followed by heightened activity. The volume of an empty stomach is about 50 ml, and the initial relaxation allows foods to enter. Further movements of the stomach are of two types: tonic contractions and peristalsis. *Tonic contractions* occur at the frequency of 15 to 20 per minute and are found in all parts of the stomach. They serve to mix and churn the contents of the stomach. *Peristaltic waves* originate near the upper part of the organ and sweep towards the small intestine at the rate of 1 to 2 per minute. Peristalsis moves small amounts of the stomach contents into the small intestine for further digestion. The basic motility of the stomach depends upon inherent activity within the submucous and myenteric plexuses, and is only modified by the action of extrinsic nerves. Sympathetic stimulation inhibits, and parasympathetic stimulation increases, gastric motility. During fasting, the vagus nerve, in response to a lowered blood sugar level, stimulates the muscle of the stomach to very strong contractions. Such contractions are termed "hunger contractions." Feeding is the normal response to these contractions. With food intake the blood sugar level is elevated and the contractions cease.

The rate of emptying the stomach is determined by the interaction of numerous factors. Under normal conditions, the stomach will empty itself in 3 to 5 hours.

1. Emptying time is increased in proportion to the *amount of fat* included in the meal. Fats entering the small intestine cause the liberation from the intestinal mucosa of enterogastrone. This is a hormonal material which enters the circulation and upon reaching the stomach, inhibits the contraction of the muscle.
2. *Products of protein digestion* initiate a nervous reflex which decreases motility. The reflex has been designated as the enterogastric reflex.
3. Cutting the *vagus nerve* decreases motility. The procedure decreases the sensitivity of the muscle to its own inherent electrical activity.
4. *Osmotic pressure* of the stomach contents, either greater or lesser than that of the blood, decreases motility until the contents are rendered isotonic by movement of water either into or out of the blood. This action suggests the presence of osmoreceptors in the stomach or duodenum which continually monitor the fluids leaving the stomach.
5. The *emotional state* of the individual, through modifying nervous responses, greatly influence gastric motility. Fear, rage, and anger of short duration decrease motility, while chronic stress increases it.

Small intestine. Circular and longitudinal muscle layers operate much more independently in this organ than they do in the stomach. In addition, the muscularis mucosae plays an important role in one of the several types of movement which occurs in the intestine.

1. *Peristalsis.* Peristaltic waves occur at a frequency of 17 to 18 per minute. They depend upon an intact myenteric (Auerbach's) plexus and occur primarily through activity of the circular muscle layer. The primary stimulus initiating peristalsis is a stretch of the intestine wall. The reflex contraction of the muscularis which results is designated the *myenteric reflex.* Examination of the fluids leaving the intestine after the myenteric reflex has been elicited shows a high content of a chemical substance, serotonin. This suggests serotonin may play a role in the reflex, but the details are largely unknown.
2. *Segmenting contractions.* Segmenting contractions are ring-like local contractions of the circular muscle layer occurring at the rate

of 12 to 16 per minute. The contractions appear to be due to inherent activity within the plexuses and muscle and serve to mix and churn the intestinal contents.

3. *Pendular movements.* Occurring primarily within the longitudinal muscle layer, pendular contractions shorten a length of the gut. They do not seem to have a particular frequency. They also serve to mix the intestinal contents.

4. *Villus contractions.* Shortening of the villi, as well as waving motions, are observed in the intestine. A material designated villikinin may be isolated from the blood leaving the intestine. Villikinin is produced when the upper part of the intestine is bathed by the digesting food mass. Villus movements stir the intestinal contents and aid the rate of absorption of materials by continually exposing the villus to "fresh" material. The muscularis mucosae is primarily responsible for villus movements.

Colon.   Movements of the colon are similar to those seen in the small intestine. *Tonic* and *segmenting* movements similar to those of the stomach and small intestine, are present. *Peristalsis* moves materials through the colon. Frequency of peristalsis is lower in the colon than elsewhere in the tract, varying between 3 and 12 per minute. A mass movement, or very strong peristaltic wave occurs 3 to 4 times a day and drives material into the pelvic colon.

Defecation. Material moving into the rectum from the pelvic colon distends the rectum and initiates the act of defecation. Distension results in a reflex contraction of the rectal musculature which tends to expel the rectal contents. Expulsion cannot occur until the external sphincter, composed of voluntary striated muscle, is relaxed. Defecation is thus a reflex act which can be voluntarily inhibited. Another reflex can also stimulate defecation. The *gastrocolic reflex* occurs when the stomach is distended with food and stimulates contraction of the rectal musculature. The uninhibited operation of the reflex is perhaps best seen in an infant, where feeding is almost invariably followed by defecation within 15 to 20 minutes.

A summary of motility in the alimentary tract is given in Table 15–1.

THE ANATOMY AND FUNCTIONS OF THE DIGESTIVE SYSTEM

The mouth and oral cavity

The mouth.   The mouth (Fig. 15–4) is bounded anteriorly by the lips, laterally by the cheeks, posteriorly by the soft palate and uvula, inferiorly by the tongue, and superiorly by the hard palate. The *vestibule* lies just inside of the lips and external to the teeth and gums. The *oral cavity* forms the major part of the mouth and lies centrally to the gums and teeth.

The lips and cheeks.   The lips are two highly mobile, vascular, and sensitive structures which surround the orifice of the mouth. They are covered externally by thin skin, and internally by a soft (uncornified) stratified squamous epithelium. This type of epithelium is the most resistant to mechanical abrasion. The cheeks are similar in structure. Together, these structures are important in moving food between the teeth during chewing, in transferring food to different parts of the oral

TABLE 15-1

A Summary of Motility in the Alimentary Tract

Area	Type of Motility	Frequency	Control Mechanism	Result
Mouth	Chewing	Variable	Initiated voluntarily proceeds reflexly	Subdivision, mixing with saliva
Pharynx	Swallowing	Maximum 20 per min	Initiated voluntarily reflexly controlled by swallowing center	Clears mouth of food
Esophagus	Peristalsis	Depends on frequency of swallowing	Initiated by swallowing	Transport through esophagus
Stomach	Receptive relaxation	Matches frequency of swallowing	Unknown	Allows filling of stomach
	Tonic contraction	15 to 20 per min	Inherent by plexuses	Mix and churn
	Peristalsis	1 to 2 per min	Inherent	Evacuation of stomach
	"Hunger contractions"	3 per min	Low blood sugar level	"Feeding"
Small intestine	Peristalsis	17 to 18 per min	Inherent	Transfer through intestine
	Segmenting	12 to 16 per min	Inherent	Mixing
	Pendular	Variable	Inherent	Mixing
	Villus movements shortening and waving	Variable	Villikinin	Facilitates absorption
Colon	Peristalsis	3 to 12 per min	Inherent	Transport
	Mass movement	3 to 4 per day	Stretch	Fills pelvic colon
	Tonic	3 to 12 per min	Inherent	Mixing
	Segmenting	3 to 12 per min	Inherent	Mixing
	Defecation	Variable 1 per day to 3 per week	Reflex triggered by rectal distension	Evacuation of rectum

cavity, and in the articulation of speech. The redness of the mouth linings is due to the translucent epithelium and the highly vascular nature of the underlying connective tissue. Absorption of materials can occur from the mouth into these blood vessels. Medicines are often held under the tongue, and the dissolved products are directly absorbed into the vessels of the mucous membrane.

The gums and teeth.   The gums (gingivae) are a continuation of the mucous membrane of the mouth over the alveolar margins of the mandible and maxillae. The gums attach to the teeth, aiding in holding them in their sockets, and are continuous with the periosteum (*periodontal membrane*) lining the socket of the tooth.

There are four varieties of teeth in the mouth. In each half of the upper and lower jaw there are, anteriorly, two *incisors*. These teeth are chisel-shaped and exert a shearing or scissors-like action useful in biting. Next to the incisors is a single *canine* or eyetooth. Canines are conical, are prolonged in some animal forms into "fangs," and are used for tearing or shredding food. Next are two *premolars (bicuspids),*

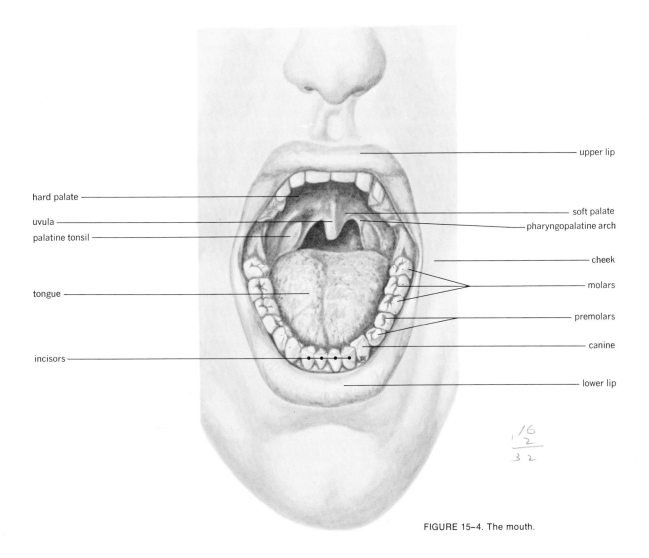

hard palate

uvula

palatine tonsil

tongue

incisors

upper lip

soft palate

pharyngopalatine arch

cheek

molars

premolars

canine

lower lip

$\dfrac{16}{2}$
$3\ 2$

FIGURE 15–4. The mouth.

typically having two roots and two grinding surfaces or cusps on their tops. Three *molars (tricuspids)* follow. Premolars and molars are teeth specialized for grinding foods, and are responsible for causing the finest mechanical subdivision of the food. The numbers of teeth indicated lead to the establishment of the dental formula for the human of 2 + 1 + 2 + 3, or a total of 8 teeth, in each quarter of the jaw. The first five of the eight teeth (deciduous teeth) are replaced by permanent teeth. The individual thus normally acquires a total of 52 teeth during his lifetime.

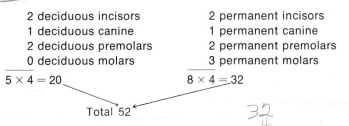

2 deciduous incisors	2 permanent incisors
1 deciduous canine	1 permanent canine
2 deciduous premolars	2 permanent premolars
0 deciduous molars	3 permanent molars
$5 \times 4 = 20$	$8 \times 4 = 32$

Total 52

$\dfrac{32}{4}$
28

Time of Eruption of Teeth

Deciduous teeth	Months
Lower central incisors	6–8
Upper central incisors	9–12
Upper lateral incisors	12–14
Lower lateral incisors	14–15
First molars	15–16
Canines	20–24
Second molars	30–32

Permanent teeth	Years
First molars	6
Central incisors	7
Lateral incisors	8
First premolars	9–10
Second premolars	10
Canines	11
Second molars	12
Third molars	17–18

A section through a typical tooth (Fig. 15–5A and B) shows its parts and the materials of which it is composed. The tooth consists of two major portions, the *crown* and *root,* connected by a slightly constricted region, the *neck (cervix).* The crown may be further subdivided into a *clinical crown,* the visible part, and the *anatomical crown,* which includes the clinical crown and a part normally covered by the gums. The root carries an apical foramen, which opens through a *root canal* into the *pulp cavity* of the tooth. The tooth is composed of the following structural materials:

1. *Enamel.* Enamel covers the anatomical crown. It consists of 95 to 97 percent inorganic material (chiefly calcium phosphate), and is the hardest substance in the body. It is however, rather brittle.
2. *Cementum.* Cementum covers the root of the tooth, and is similar to bone in composition.
3. *Dentin.* Dentin forms the greater mass of the tooth, and also is bonelike in composition.
4. *Pulp.* Pulp fills the tooth cavity and is a vascular connective tissue liberally supplied with nerves and lympatics. The pulp provides the means of nourishing the tooth during development and in adult life.

The tongue.   The tongue (Fig. 15–6A and B) is a muscular organ covered with connective tissue and a stratified, partially cornified, squamous epithelium. The musculature of the tongue may be divided into *extrinsic muscles,* which originate outside of the tongue and insert within it, and *intrinsic muscles* which both originate and insert within the organ. Extrinsic muscles are responsible for the gross movements of the tongue (in and out, side to side). Such movements are important in guiding food between the teeth for chewing and in swallowing. The extrinsic muscles include the hyoglossus, genioglossus, and styloglossus. The intrinsic musculature is responsible for the changes in shape of the tongue during speech and swallowing, and includes the transverse, vertical, and longitudinal lingual muscles. The tongue is at-

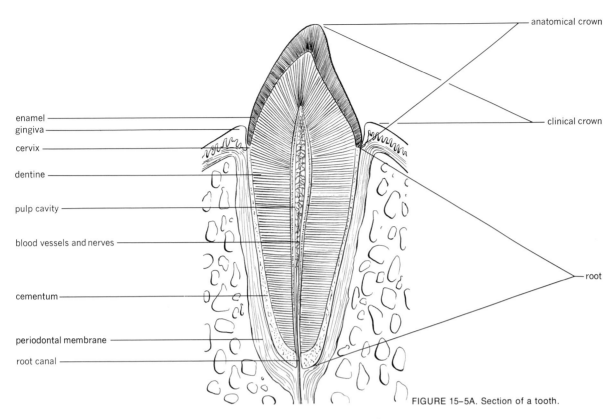

enamel

gingiva

cervix

dentine

pulp cavity

blood vessels and nerves

cementum

periodontal membrane

root canal

anatomical crown

clinical crown

root

FIGURE 15–5A. Section of a tooth.

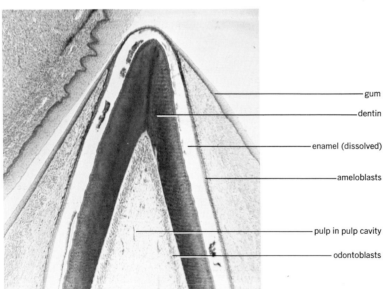

gum

dentin

enamel (dissolved)

ameloblasts

pulp in pulp cavity

odontoblasts

FIGURE 15–5B. Tooth, longitudinal section.

tached to the floor of the oral cavity by the *frenulum*, a membrane lying in the midline of the cavity floor.

The upper surface and sides of the tongue carry a variety of *papillae* which are folds of lamina propria covered with epithelium. *Filiform papillae* are conical projections distributed evenly over the anterior two-thirds of the tongue. *Fungiform papillae* are rounded elevations most common on the sides and top of the tongue. These two types of papillae roughen the surface of the tongue, increasing its efficiency

397

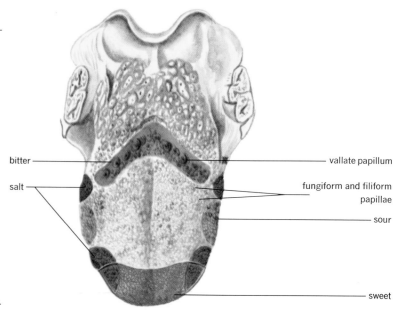

bitter — — vallate papillum

salt — — fungiform and filiform papillae

— sour

— sweet

FIGURE 15-6A. The tongue.

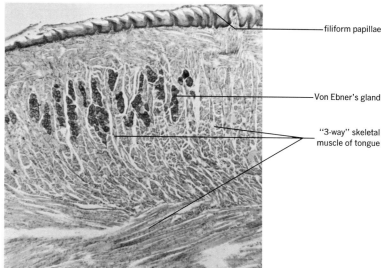

— filiform papillae

— Von Ebner's gland

— "3-way" skeletal muscle of tongue

FIGURE 15-6B. Tongue.

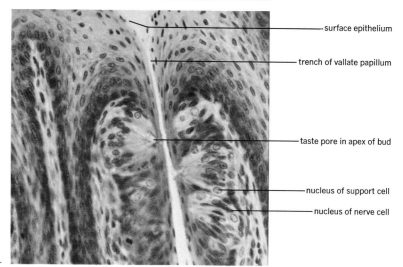

— surface epithelium

— trench of vallate papillum

— taste pore in apex of bud

— nucleus of support cell
— nucleus of nerve cell

FIGURE 15-7. Taste buds.

during manipulation of food in the mouth. *Circumvallate (vallate) papillae* are embedded in the tongue and are provided with a trench or moat around the papillum proper. Vallate papillae are found on the posterior surface of the tongue and are arranged in the form of a V. *Taste buds* (Fig. 15–7) are found in the fungiform and vallate papillae, and on the walls of the soft palate. Emptying into the trenches of the vallate papillae are the ducts of the *von Ebner's glands*. The secretory portions of the glands lie between the muscle bundles of the tongue. These glands produce a watery fluid which aids in moistening foods and in dissolving them so that they may be tasted.

The salivary glands. (Fig. 15–8A and B). Three pairs of salivary glands are located outside the mouth and empty their secretions by ducts into the mouth. Two varieties of cells may be found in the glands, *serous cells* and *mucous cells.* The former are small, very granular cells which secrete a watery fluid rich in salivary amylase. The latter are larger, pale-staining cells which produce a viscous mucus lacking in amylase. Table 15–2 summarizes the important facts concerning the salivary glands.

It can be seen that the secretion of each gland is different. The composite fluid of all three sets of salivary glands constitutes the saliva.

Saliva. Saliva is an *acidic* fluid (pH 6.35 to 6.85) produced in amounts up to 2500 ml per day. An average figure for daily volume would be 1000 to 1500 ml. The saliva has the following composition:

Water, 99.5 percent
Dissolved or suspended solids, 0.5 percent:
1. Salts. $NaCl$, KCl, $NaHCO_3$, $KHCO_3$, Na_3PO_4, K_3PO_4. These salts buffer the saliva.
2. Gases. O_2, N_2, CO_2
3. Organic substances
   (a) Urea and uric acid.
   (b) Proteins. Albumins and globulins.
   (c) Mucus. Lubrication.
   (d) Salivary amylase (ptyalin). Enzyme.
   (e) Lysozyme. Bacteriolytic.
   (f) Bradykinin. A vasodilating agent. May control blood flow through gland.
   (g) ABO blood group substances.

Saliva functions to *moisten* and *soften* ingested foods, to *lubricate* them for swallowing, to *cleanse* the mouth and teeth, to moisten the mucous membranes of the mouth, and acts as a route for the *excretion* of many materials.

The digestive function of saliva centers around its content of *salivary amylase (ptyalin)*. This enzyme attacks the chemical bonds between the simple sugar units in cooked starch (carbohydrates), and, by hydrolysis, splits the starch into smaller units known as dextrins. The amylase also has the power, given sufficient time, to break the dextrins into disaccharides. However, foods do not normally remain in the mouth long enough for the disaccharide stage to be reached by more than 3–5 percent of the dextrins.

$$\text{starch} \xrightarrow[\text{amylase}]{\text{salivary}} \text{dextrins (95–97\%)} \xrightarrow[\text{amylase}]{\text{salivary}} \text{disaccharide (3–5\%)}$$

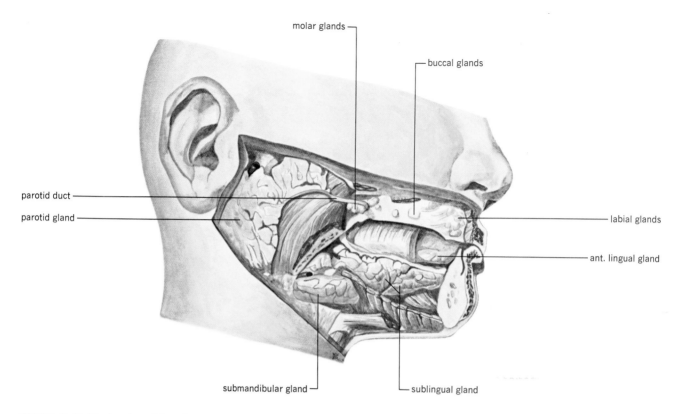

molar glands

buccal glands

parotid duct

parotid gland

labial glands

ant. lingual gland

submandibular gland

sublingual gland

FIGURE 15–8A. The location of the salivary and oral glands.

Control of salivary secretion. There is a continuous production of saliva necessary for the moistening and cleansing functions it serves. The greatest volume of saliva is produced when food is present in the mouth. The control mechanism for secretion (Fig. 15–9) is a nervous reflex originating in the taste buds and the walls of the mouth. The buds act as receptors, initiating nerve impulses when dissolved food enters. Mechanical stimulation of the walls of the mouth causes nerve impulses to be formed in the touch receptors located there. Afferent (incoming) impulses pass over the facial, glossopharyngeal and vagus nerves to salivary centers in the brain stem. Efferent (outgoing) im-

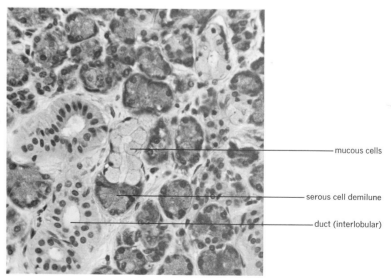

mucous cells

serous cell demilune

duct (interlobular)

FIGURE 15–8B. Submandibular salivary gland.

TABLE 15-2

The Salivary Glands

Name of gland	Location	Cellular composition	Name of duct	Entry of duct into mouth	Secretion contains
Parotid	Side of mandible in front of ear	All serous	Stensen's	Lateral to upper 2nd molar	Water, salts enzyme
Submandibular	Beneath the base of the tongue	Mostly serous, some mucous	Wharton's	Papillum lateral to frenulum	Water, salts enzyme, some mucus
Sublingual	Anterior to submandibular under tongue	Mostly mucous, some serous	Rivinus'	With duct of submandibular	Mostly mucus, a little water, salt and enzyme

pulses pass from the superior salivary nucles to the parotid gland, and from the inferior salivary nucleus to the submandibular and sublingual glands. The glands act as effectors and secrete saliva. The nature of the secretion, as well as its quantity, can be altered by this mechanism. For

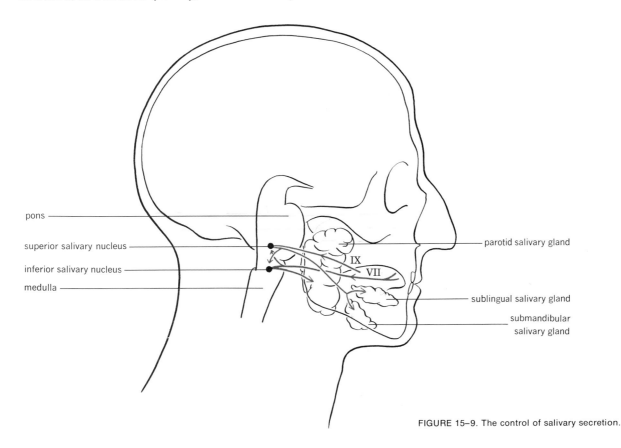

FIGURE 15-9. The control of salivary secretion.

instance, soft foods occasion the production of saliva having a smaller volume and lower water content than does a dry food. This mechanism represents the most effective means, but not the only one, by which salivary secretion may be augmented. The sight, sound, and smell of food in preparation, or the mere thought of food, may cause the "mouth to water." Therefore, there exist what are termed "higher cerebral influences" which may activate the brain stem salivary centers.

The pharynx

The pharynx (Fig. 15–10) is an organ common to both the respiratory and digestive systems. It lies behind the nasal cavities, oral cavity, and larynx, and is divided into the corresponding *nasal, oral,* and *laryngeal portions.* Three layers of tissue form its wall: a mucous membrane, a fibrous layer, and muscle. The muscle is skeletal in type. Two sets of tonsils are located within the pharynx and were considered in the chapter on the lymphatic system. The involvement of the pharynx in the di-

FIGURE 15–10. The muscles of the pharynx.

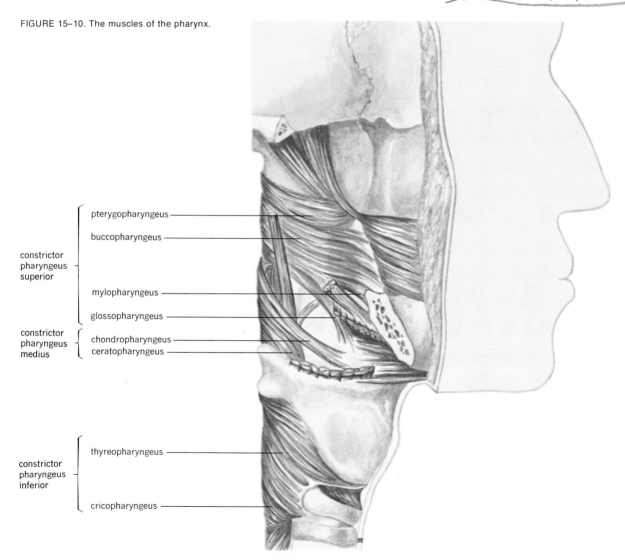

constrictor pharyngeus superior
pterygopharyngeus
buccopharyngeus

mylopharyngeus
glossopharyngeus

constrictor pharyngeus medius
chondropharyngeus
ceratopharyngeus

constrictor pharyngeus inferior
thyreopharyngeus

cricopharyngeus

gestive process is to act as a transmitting tube for food transfer from the oral cavity to the esophagus, during swallowing.

The esophagus

The esophagus is a muscular tube about 25 cm (10 inches) in length which connects the pharynx to the stomach. The tissue plan follows that stated earlier. The epithelium is stratified squamous. The muscularis is skeletal in the upper portion, mixed skeletal and smooth in the middle portion, and all smooth in the lower portion. The outermost layer is an adventitia except on the lower 1 to 2 cm, where it is replaced by a serosa. The esophagus conducts food to the stomach and has no digestive function. The remaining organs of the alimentary tract lie below the diaphragm in the abdominopelvic region.

Mesenteries and omenta

The abdominopelvic or peritoneal cavity, and some of the organs within it, are lined and covered by serous membranes known generally as the *peritoneum.* The *parietal peritoneum* lines the walls of the cavity proper, while the *visceral peritoneum* covers the organs as part of their serosa. Both membranes consist of a simple squamous mesothelium supported by a submesothalial connective tissue layer.

As the liver, stomach, intestine, and some of the reproductive organs develop, they grow from the tissues which lie external to the peritoneum into the cavity, pushing ahead of them the lining membrane (Fig. 15–11). The organs eventually sever their connections with their

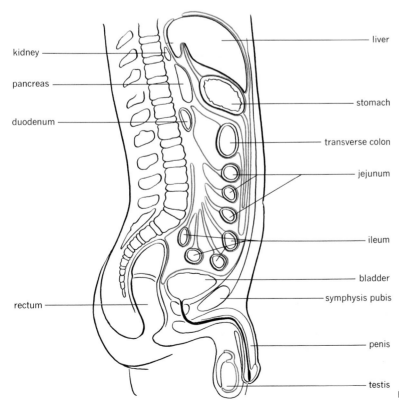

FIGURE 15–11. The mesenteries.

points of origin and become suspended within the cavity by a double-layered fold of the peritoneum. This double-layered suspending structure is a *mesentery.* The mesenteries transmit numerous blood vessels to the organs they suspend. Additionally, specific mesenteries are often named according to the organ they suspend, for example: the mesogastrium (stomach), mesocolon (large intestine), mesovarium (ovary).

Between the liver and stomach is a double-layered fold of tissue, the *lesser omentum.* Since it does not suspend one organ, it is not a true mesentery. Between the stomach and the first portion of the small intestine is a double-layered fold of tissue, the *greater omentum.* This membrane becomes greatly elongated, folds upon itself, and forms a four-layered curtain which overlies the intestines. The greater omentum commonly serves as a storage depot for fat. It also contains many phagocytic cells, which are important in keeping infections from becoming established within the cavity.

Some organs such as the pancreas, kidney, ureters, and urinary bladder remain essentially against the body wall and are not suspended within the cavity by mesenteries. The parietal peritoneum passes across their surface. Such organs are said to be *retroperitoneal,* that is, lying behind the peritoneum. These relationships are shown in Fig. 15–11.

Peritonitis is an inflammation of the peritoneum. It is a potentially dangerous condition, since the inflammation may spread throughout the peritoneal cavity and infect all its organs. The reason for easy spread of infection revolves around the fact that the membranes are continuous throughout the cavity.

The stomach

Anatomy. The stomach is a J-shaped organ lying under the diaphragm slightly to the left of the midline of the abdomen. The gross anatomy of the organ is illustrated in Fig. 15–12A and B. Microscopically, the stomach wall follows the basic plan given earlier. The epithelium is simple columnar, in which the outer portion of the cells contain mucin (mucus). The mucin provides a continuous covering protecting the interior of the organ from auto (self) digestion. Extending into the lamina propria from the foveolae or gastric pits are gastric glands. These are estimated to number 35 million and are lined with four cell types:

1. *Mucous neck cells.* These occur in the upper or neck portion of the gland and are cuboidal cells with flattened or crescent-shaped nuclei. They produce mucus and intrinsic factor.
2. *Zymogenic (chief) cells.* These cells are cuboidal granular units which produce pepsinogen, an inactive form of the enzyme pepsin. The cells also produce rennin and gastric lipase, though in much smaller quantities than pepsin.
3. *Parietal cells.* Parietal cells produce hydrochloric acid and are distinguished by being spherical units with centrally placed nuclei.
4. *Argentaffin cells.* Scattered among the other cells, argentaffin cells have basal acidophilic granules and apical nuclei. Their exact function is unknown, but they have been shown to contain serotonin.

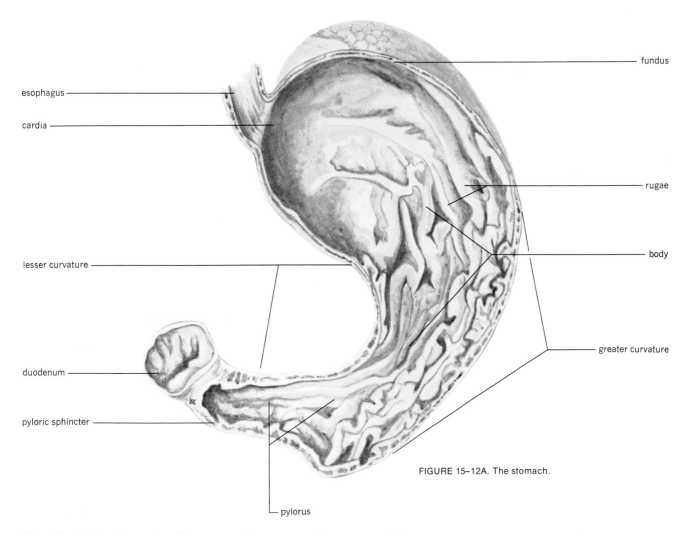

esophagus

cardia

lesser curvature

duodenum

pyloric sphincter

pylorus

fundus

rugae

body

greater curvature

FIGURE 15–12A. The stomach.

Digestion in the stomach. The stomach is an organ for storage of the food consumed during a meal. During the time that the foods are in the stomach, they are subjected to the action of the *gastric juice.* The juice is a watery solution of hydrochloric acid containing pepsin, mucin, and small quantities of rennin and lipase. In the infant stomach, which contains much less acid than that of the adult, the pH is about 5. Rennin and lipase operate optimally in this pH. The infant's diet consists chiefly of milk, and the two enzymes contribute greatly to the digestion of that milk. Rennin coagulates the milk protein, while lipase digests the butterfat in the milk. In the adult stomach, where the pH is about 1, the only important digestive action is that of pepsin. Recall that the zymogenic cells secrete pepsinogen which must be activated to pepsin. The sequence of events leading to the digestive action of pepsin may be diagrammed as follows:

within parietal cell   secreted   reabsorbed

$$CO_2 + H_2O + NaCl \longrightarrow \quad HCl \quad + \quad NaHCO_3$$

$$HCl + pepsinogen \longrightarrow \quad pepsin$$

$$protein \xrightarrow[HCl]{pepsin} proteoses \text{ and peptones (units of 4–12 amino acids)}$$

405

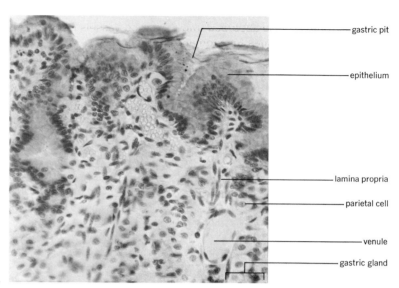

gastric pit

epithelium

lamina propria

parietal cell

venule

gastric gland

FIGURE 15–12B. Stomach.

The acid therefore provides a suitable environment for the proteolytic action of pepsin. It activates pepsinogen and, in addition, terminates the action of salivary amylase.

Control of gastric secretion. Secretion of gastric juice occurs even during fasting and is augmented when food is ingested. Secretion is controlled in three ways:

1. The *cephalic (psychic) phase.* Food in the mouth enters taste buds, afferent impulses pass to the brain stem and from the brain stem to the stomach via the vagus nerve. The stomach responds directly to the vagal stimulation by the secretion of 50 to 150 ml of gastric juice. Vagal stimulation also releases gastrin (see below).

2. The *gastric phase.* This phase is primarily under hormonal control and occurs when swallowed food contacts the gastric juice produced during the cephalic phase. Proteins in the food are broken down into proteoses and peptones by the action of pepsin. The proteoses and peptones act as *secretagogues (secretory stimulators)* and cause cells of the stomach mucosa to produce *gastrin.* Gastrin enters the blood vessels leaving the stomach and is distributed to the entire body including the stomach. Upon reaching the stomach, gastrin stimulates the production of 600 ml or more of gastric juice over a period of 3 to 4 hours.

3. The *intestinal phase.* Food placed directly into the small intestine in experimental animals has been shown to cause gastric secretion. Secretion has been attributed to the production of a hormonal material. The nature of the hormone is unknown. It is characterized as being "gastrin-like," and may be secretin. Some investigators contend that no hormone is involved and that the chemical is an absorbed product of digestion acting as a secretagogue.

Inhibition of gastric secretion occurs in several ways. The digesting food mass tends to neutralize the acids of the gastric juice, and the pH rises. Increase of pH results in a decrease of pepsin action and the gastric phase is essentially diminished. Fats, entering the small intestine from the stomach, cause the production of the hormone gastrone (enterogastrone). This is absorbed into the blood vessels of the intestine and ultimately reaches the stomach, where it inhibits gastric juice secretion and simultaneously decreases gastric motility.

Absorption from the stomach.   The stomach wall is rather impermeable to the passage of materials. The molecules resulting from digestion are still quite large. Therefore, absorption from the stomach is limited, and confined to the passage of water, salts, and small molecules such as alcohol.

Small intestine

Anatomy.   The small intestine extends from the pyloric valve of the stomach some 3.3 to 4 meters (10 to 12 feet*) to the ileocecal valve of the colon. It is divisible into three parts: the *duodenum,* the first 25 to 30 cm (10 to 12 inches); the *jejunum,* the next 1 to 1.5 meters (3 to 4 feet); and the *ileum,* the remaining 2 to 2.5 meters (6 to 7 feet). All parts have a typical and similar structure (Fig. 15–13A–C). The divisions between the parts can be appreciated only microscopically.

Circular folds of the submucosa *(plicae),* fingerlike projections of the mucosa *(villi),* and mucosal glands *(crypts of Lieberkühn)* are characteristic features of the entire intestine. Plicae and villi provide for increased surface area for absorption and secretion. These two devices are estimated to result in about 10 square meters of surface. The surfaces of the epithelial cells lining the intestine are provided with microvilli (Fig. 3–3). These are tiny cytoplasmic projections from the free surface of the cell, and they further increase the absorptive surface. The three sections of the intestine may be distinguished microscopically by:

1. The presence of submucosal mucous glands (Brunner's) in the duodenum.
2. The highest plicae and thinnest wall in the jejunum.
3. The presence of aggregated lymphoid nodules (Peyer's patches) in the submucosa of the ileum.

Digestion in the small intestine.   The small intestine receives not only the secretions of its own glands *(succus entericus)* but also the secretions of the pancreas and liver. From the standpoint of logical order in the digestive process, the secretions of the pancreas and liver will be considered here and a discussion of the anatomy of these organs deferred until later.

Pancreatic juice is an alkaline fluid (pH 7.1 to 8.2) which creates the proper environment for the action of all enzymes operating in the intestine, and it stops the action of pepsin. The alkaline reaction of the fluid is due to the bicarbonate content. The juice contains enzymes active on all three foodstuff groups; carbohydrates, fats and proteins:

A. Protein and nucleic acid enzymes

1. Trypsin, secreted as trypsinogen, and activated by enterokinase and trypsin itself (autocatalytic).
2. Chymotrypsin, secreted as chymotrypsinogen, and activated by trypsin.

* The small intestine is this long in the living human. Post-mortem measurements may be 7 meters (20 to 21 feet).

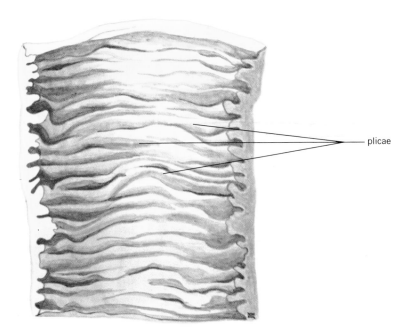

plicae

FIGURE 15–13A. General view of the wall of the small intestine.

FIGURE 15–13B. Blood and lymph vessels of the villus.

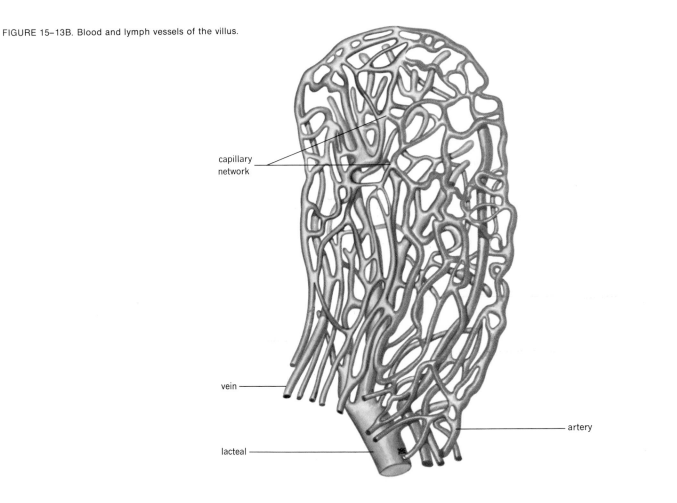

capillary
network

vein

artery

lacteal

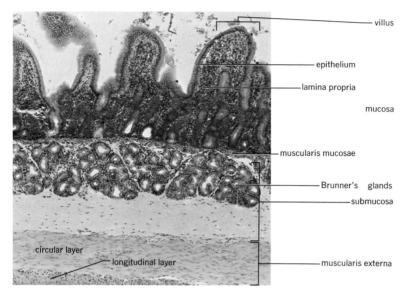

FIGURE 15–13C. Small intestine (duodenum).

3. Carboxypolypeptidase, secreted as procarboxypepti-
dase and also activated by trypsin.

4. Deoxyribonuclease, secreted in active form, which digests
the DNA present in the cells of the foods eaten.

The interrelationships between these enzymes and their effects may be
diagrammed as follows:

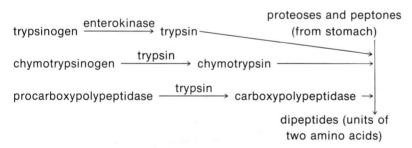

B. Pancreatic amylase (amylopsin) attacks the dextrins resulting
from salivary digestion and converts them to disaccharides:

$$\text{dextrins} \xrightarrow{\text{pancreatic amylase}} \text{disaccharides} \begin{cases} \text{maltose} \\ \text{lactose} \\ \text{sucrose} \end{cases}$$

C. Pancreatic lipase (steapsin), in one step, splits triglyceride into its
component fatty acids and glycerol.

The *intestinal juice* produced by the intestinal glands completes the
digestion of the proteins and carbohydrates:

A. Proteolytic enzymes
    1. Erepsin, a generic name for a group of enzymes, splits re-
maining dipeptides into amino acids.

B. Specific carbohydrases attack their respective disaccharides and
convert them to simple sugars.

$$\text{maltose} \xrightarrow{\text{maltase}} \text{2 glucose molecules}$$

$$\text{sucrose} \xrightarrow{\text{sucrase}} \text{glucose and fructose}$$

$$\text{lactose} \xrightarrow{\text{lactase}} \text{glucose and galactose}$$

Control of secretion of pancreas and small intestine. Release of pancreatic juice is primarily under the control of hormonal (chemical) mechanisms. The most effective stimulant to pancreatic secretion is the presence in the duodenum of HCl from the stomach. Other materials including water, meat juices, bread, fats, soaps, and alcohol stimulate pancreatic secretion. All these substances cause the intestinal mucosa to produce a material called *secretin.* This enters the blood stream and upon reaching the pancreas causes the secretion into the pancreatic duct of a buffered watery solution deficient in enzymes. Acid alone is the most effective stimulant to the production of another hormone, *pancreozymin.* This also enters the blood vessels and upon reaching the pancreas stimulates the secretion of enzymes into the pancreatic duct. Stimulation of the *vagus nerve* also causes enzyme secretion by the pancreas. Thus, two chemicals are necessary for total pancreatic secretion, and enzyme secretion alone is stimulated by nerves.

Secretion of intestinal juice may be elicited by a wide variety of mechanical, nervous and chemical stimuli to the wall of the intestine. A hormone designated as *enterocrinin* is produced when the intestine is stroked, touched, stretched, or flooded with liquid. The hormone enters the circulation and is distributed to the entire small intestine, where it stimulates succus entericus production. A nervous influence is also seen in the production of intestinal juice. Parasympathetic stimulation increases, and sympathetic stimulation decreases intestinal secretion.

The function of the liver in the digestive process. The liver produces the *bile,* which, though not an enzyme, aids pancreatic lipase in the digestion of fats. Bile coats the small fat droplets and prevents their coalescence into larger drops. This enhances the ability of the lipase to digest the drops. This action is termed the *emulsifying action* of the bile. Fatty acids produced by the action of the lipase are insoluble in the water of the digestive fluid, and in this state are difficult to absorb. The bile salts react chemically with the fatty acids and convert them to water-soluble compounds which are much easier to absorb. This action of the bile is termed the *hydrotropic action.* Although produced continually by the liver, the bile is stored in the gall bladder until required for digestion. Release from the gall bladder is caused by *cholecystokinin.* The latter is another intestinal hormone produced when fats are in the duodenum. It causes the contraction of the gall bladder musculature, emptying the stored bile into the small intestine. Table 15–3 summarizes the actions of the various chemicals produced by the organs of digestion.

Absorption from the small intestine. Simple sugars are rapidly absorbed from the intestine at constant rates, even against concentration gradients. This suggests that they are being actively transported. They are not, however, all absorbed at the same rate. Galactose passes most rapidly (77 grams per hour), followed by glucose (70 grams per hour), fructose (43 grams per hour), mannose (19 grams per hour), xylose (15 grams per hour), and arabinose (9 grams per hour). These sugars are

TABLE 15-3

Chemicals Produced by the Organs of Digestion

Material	Site of production	Stimulus to production	Substrate or organ affected	Effect or end product
Salivary amylase	Salivary glands	Nervous	Cooked starch	Dextrins 95–97%; disaccharides 3–5%
HCl	Stomach, parietal cell	Nervous and chemical	Pepsinogen	Pepsin
Pepsin	Stomach, zymogenic cell (as pepsinogen)	Nervous and chemical	Protein	Proteoses and peptones
Rennin	Stomach, zymogenic cell	Nervous and chemical	Milk protein	Curdles protein
Gastric lipase	Stomach, zymogenic cell	Nervous and chemical	Fats	Fatty acids and glycerol*
Enterokinase	Duodenal mucosa	Food entering duodenum	Trypsinogen	Trypsin
Trypsin	Pancreas (as trypsinogen)	Primarily chemical	Proteoses and peptones and inactive pancreatic proteoses	Dipeptides and active proteases
Chymotrypsin	Pancreas (as chymotrypsinogen)	Primarily chemical	Proteoses and peptones	Dipeptides
Carboxypolypeptidase	Pancreas (as procarboxypolypeptidase)	Primarily chemical	Proteoses and peptones	Dipeptides
Pancreatic amylase	Pancreas	Primarily chemical	Dextrins	Disaccharides
Pancreatic lipase	Pancreas	Primarily chemical	Fats	Fatty acids and glycerol*
Erepsin	Small intestine	Primarily chemical	Dipeptides	Amino acids*
Sucrase	Small intestine	Primarily chemical	Sucrose	Glucose and fructose*
Maltase	Small intestine	Primarily chemical	Maltose	2 glucose*
Lactase	Small intestine	Primarily chemical	Lactose	Glucose and galactose*
Bile	Liver	Chemical	Fats	Emulsification solubilizes fatty acids
Gastrin	Stomach mucosa	Secretagogues	Stomach	Secretion of gastric glands
Unnamed	Duodenal mucosa	Food in duodenum	Stomach	Secretion of gastric glands
Secretin	Duodenal mucosa	HCl in duodenum	Pancreas and liver	Secretion of pancreas stimulate bile production
Pancreozymin	Duodenal mucosa	HCl in duodenum	Pancreas	Secretion of pancreatic enzymes
Enterocrinin	Duodenal mucosa	Stretch or contact in duodenum	Small intestine	Secretion of intestinal glands
Gastrone	Duodenal mucosa	Acid in duodenum	Stomach	Inhibits gastric secretion
Cholecystokinin	Duodenal mucosa	Fats in duodenum	Gall bladder	Expulsion of stored bile
Mucus	All organs	Probably inherent programming in cell	None specific	Lubrication

*End products of chemical digestion.

actively absorbed into the blood vessels of the villus and are conducted to the liver before passing to the body generally. Disaccharides, probably because there are no carriers to transport them, are absorbed to only a limited extent.

Amino acids are also actively absorbed from the intestinal contents (chyme). To a small extent, dipeptides, proteoses, and other larger units are also absorbed, probably passively. Some undigested protein also apparently passes across the wall, but in extremely small amounts. As with the simple sugars, the amino acids are absorbed into the blood vessels of the villus and are conducted to the liver.

Glycerol, being a water-soluble compound, is easily absorbed passively, chiefly into the lacteals of the villi. Fatty acids as such are water insoluble and are absorbed with difficulty. However, they can and do react with bile acids to form soluble soaps, which are then passively absorbed into the lacteal. Once in the lacteals, glycerol and fatty acid react with one another to form triglycerides again. They are then coated with protein, which prevents the coalescence of the fat droplets formed, and, as tiny droplets (chylomicrons), they pass through the thoracic duct and empty into the subclavian veins. The fats thus bypass the liver on the first trip. After a meal rich in fats, the blood may actually be turned white from the heavy accumulation of fats. This is called lipemia. The chylomicrons are removed from the blood to storage areas or are made ready for metabolism by the clearing factor (lipoprotein lipase) of the plasma.

Salts are actively transported through the wall and water follows osmotically. Some pinocytosis may account for a small amount of water absorption. The trace substances also appear to pass actively, especially vitamins.

The large intestine (colon)

Anatomy.   The large intestine (Fig. 15–14A and B) is about 1½ meters (5 feet) long and is composed of cecum, colon, rectum, and anal canal. It is larger in diameter than the small intestine and is easily recognized by its *sacculations* or *haustra.* The *cecum* is a large blind pouch below the entrance of the ileum and has arising from it the *vermiform appendix,* The *colon* consists of an *ascending* portion passing upward along the right side of the abdominal cavity. Near the liver it bends, forming the *right colic (hepatic) flexure,* and passes across the abdominal cavity as the *transverse colon.* Near the spleen it again bends, forming the *left colic (splenic) flexure,* and passes downwards along the left side of the abdominal cavity as the *descending colon.* The *sigmoid colon* begins at the pelvic brim and is an S-shaped bend returning the colon to the midline postion. At the level of the 3rd sacral vertebra it becomes continuous with the *rectum.* The rectum (Fig. 15–15) is distinguished from the colon proper by the absence of haustra, and by the presence of the *rectal columns.* The latter are longitudinal folds of tissue in the rectal wall, containing hemorrhoidal arteries and veins. If the veins become enlarged, they may give rise to the condition known as hemorrhoids.

Microscopically, the large intestine shows no villi and a simple columnar epithelium with many goblet cells. The mucus produced by these cells lubricates the drying mass of fecal material as it passes through the colon. The lamina propria contains many long crypts of Lieberkuhn. The submucosa is typical, while the muscularis externa has a complete inner circular layer and a discontinuous outer longitudinal layer. The longitudinal layer is in the form of three strips of smooth muscle, the *taenia coli.* The taenia are shorter than the colon,

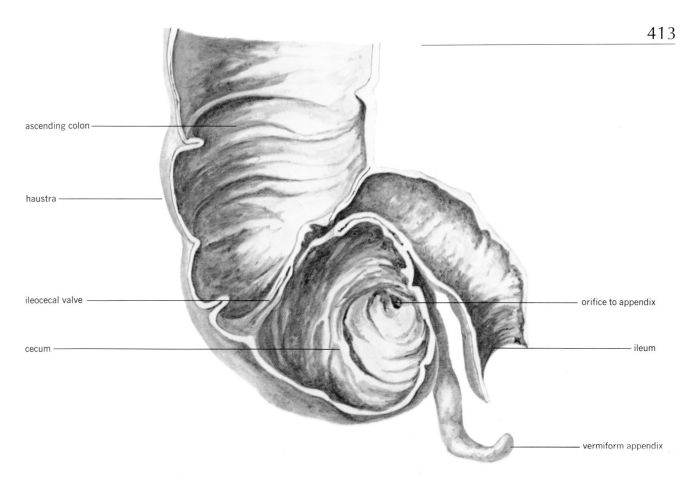

ascending colon

haustra

ileocecal valve

cecum

orifice to appendix

ileum

vermiform appendix

and to make the colon conform to the taenia length, it must be accordion pleated, hence the haustra.

Absorption in the colon. The chief substance absorbed in the colon is water. It has been estimated that in man, the colon absorbs 300 to 400 ml of water per day. The colon also maintains a considerable flora of microorganisms and these produce a variety of materials which are

FIGURE 15–14A. The ileocecal region.

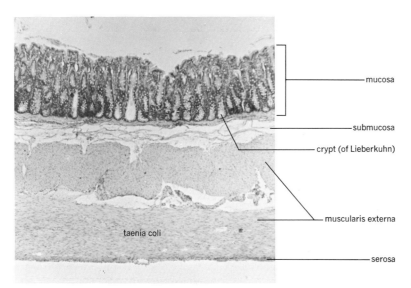

mucosa

submucosa

crypt (of Lieberkuhn)

muscularis externa

taenia coli

serosa

FIGURE 15–14B. Large intestine.

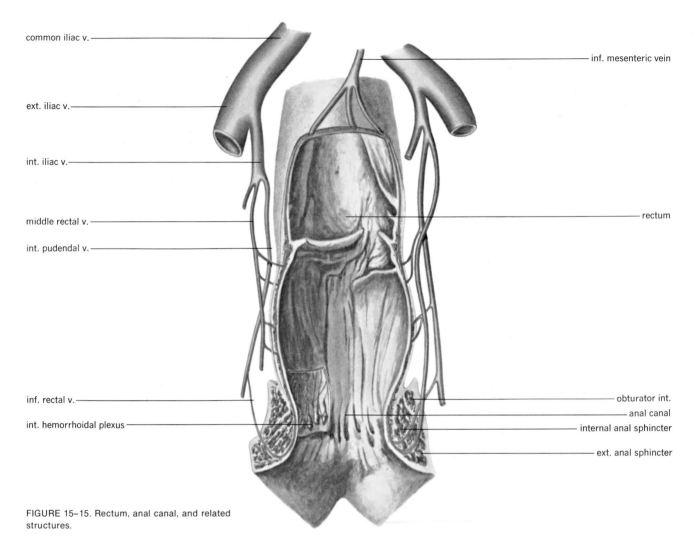

common iliac v.

ext. iliac v.

int. iliac v.

middle rectal v.

int. pudendal v.

inf. rectal v.

int. hemorrhoidal plexus

inf. mesenteric vein

rectum

obturator int.

anal canal

internal anal sphincter

ext. anal sphincter

FIGURE 15–15. Rectum, anal canal, and related structures.

nutritionally important. Vitamins K and B and amino acids are synthesized by these organisms and are absorbed by the host.

Feces formation. The material which is of no further use to the body is eliminated as feces. Feces consist of approximately 60 percent solid matter, including microorganisms, residues of digestive secretions, and undigested material. The remaining 40 percent is water.

The liver

Anatomy. The liver (Fig. 15–16) is the largest gland in the body and is located in the right upper quadrant of the abdomen, against the diaphragm. It consists of two main *lobes,* a large *right* one and a smaller *left* one. Two small lobes, the *caudate* and *quadrate,* are associated with the right lobe. Located between the right and left lobes is the *falciform ligament,* which attaches the liver to the anterior abdominal wall and diaphragm. The *coronary ligament* superiorly and posteriorly, and the right and left *triangular ligaments* laterally, also fix the liver to the diaphragm. The liver is covered with a *fibrous capsule,* the capsule

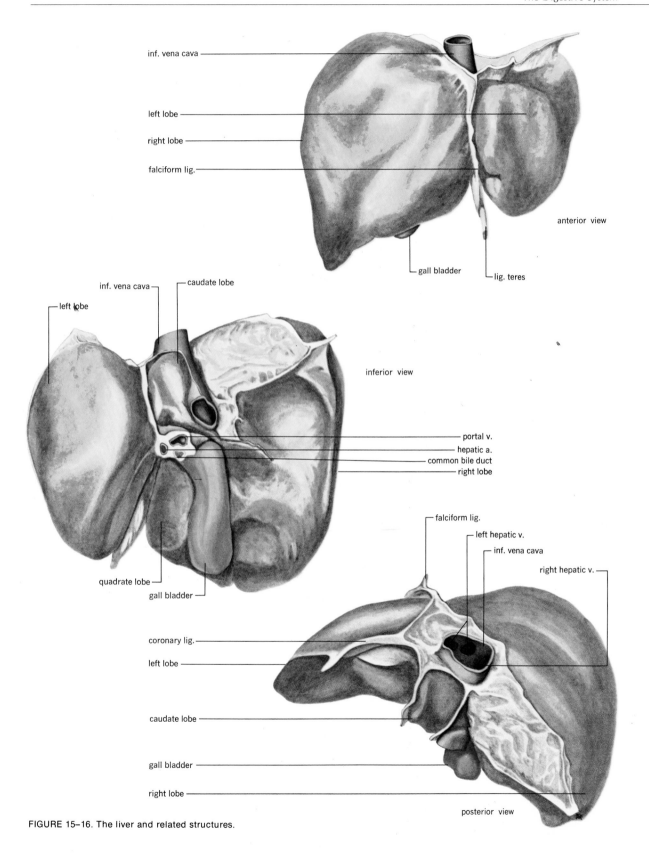

inf. vena cava

left lobe

right lobe

falciform lig.

anterior view

gall bladder

lig. teres

inf. vena cava

caudate lobe

left lobe

inferior view

portal v.

hepatic a.

common bile duct

right lobe

quadrate lobe

gall bladder

falciform lig.

left hepatic v.

inf. vena cava

right hepatic v.

coronary lig.

left lobe

caudate lobe

gall bladder

right lobe

posterior view

FIGURE 15–16. The liver and related structures.

of Glisson. Associated with the liver is the gall bladder and its ducts. Leaving the liver are the two *hepatic ducts* which fuse to form the *common hepatic duct*. Branching from this duct is the *cystic duct* to the gall bladder. From the junction of the common hepatic and cystic ducts, the *common bile duct* passes to the small intestine.

The unit of structure of the liver is the *lobule* (Fig. 15–17A, B, and C). These are five- to seven-sided cylindrical units, 1 to 2½ mm in diameter, which (in the human) are incompletely invested by connective tissue. The center of the lobule is formed by the *central vein,* from which radiate interconnecting *cords* or *plates of liver cells.* Between the cords are irregular blood spaces, or *sinusoids,* lined with epithelial and Kupffer cells. The latter are fixed phagocytes, which are important in the destruction of worn out erythrocytes. Blood reaches the liver through two afferent channels, the hepatic artery and the portal vein. The hepatic artery brings fresh (oxygenated) blood to the organ, while the portal vein, draining the intestines, brings blood rich in amino acids, sugars, vitamins, and other nutrients to the liver for processing and storage. They both empty into the sinusoid. The smaller interlobular branches of these vessels may be seen running together with the interlobular bile duct in the loose connective tissue surrounding each lobule. When these three structures are seen together in microscopic sections of the lobule, they are referred to as the *portal area.* The entire course of these vessels into the liver is known as the *portal canal.*

The bile system (Figs. 15–17A and 15–18) begins as tiny bile capillaries or canaliculi around each and every hepatic cell. These drain into the interlobular duct mentioned above, and ultimately into the small intestine, through the system of extrahepatic ducts already described.

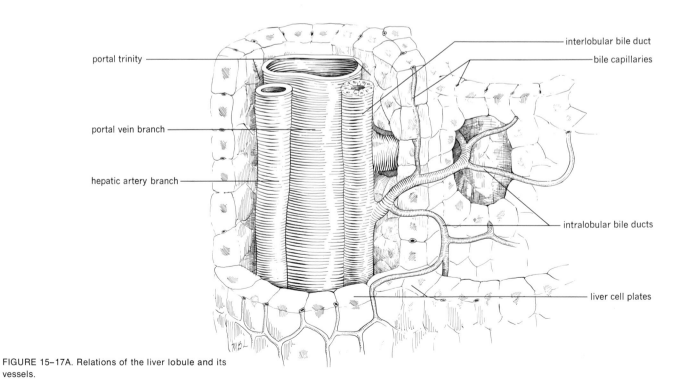

portal trinity

portal vein branch

hepatic artery branch

interlobular bile duct

bile capillaries

intralobular bile ducts

liver cell plates

FIGURE 15–17A. Relations of the liver lobule and its vessels.

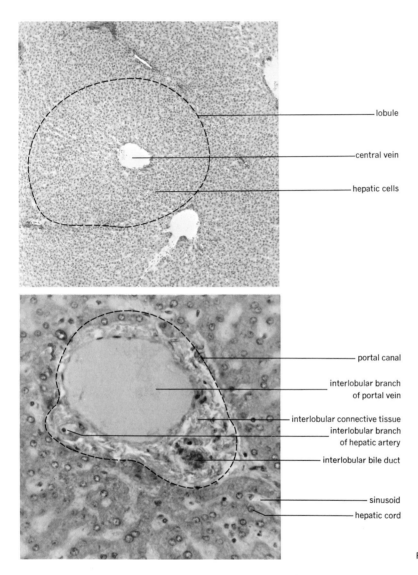

FIGURE 15–17 B and C. Liver lobules.

Functions of the liver. The liver carries out a greater variety of activities than any other body organ. These activities are listed below:

1. *Synthetic reactions*
   (a) Bile production
   (b) Amino acids and proteins, including serum albumins and globulins, prothrombin, fibrinogen, antibodies
   (c) Glycogen
   (d) Phospholipids
   (e) Fatty acids
2. *Metabolic reactions*
   (a) Catabolism of glucose, glycogen, fatty acids, glycerol, amino acids
   (b) Interconversion of proteins, carbohydrates, fats
   (c) Formation of ATP
3. *Formation* of blood cells (embryo)

4. *Destruction* of erythrocytes
5. *Storage*
   (a) Glycogen
   (b) Amino acids
   (c) Fats
   (d) Vitamins (A, B complex, D)
   (e) Iron and copper
   (f) Hemopoietin
6. *Detoxification;* conversion of toxic substances to harmless compounds

Bile is continually produced by the liver. As noted before, until it is needed, it is stored in the gall bladder. This bag-like organ can store up to 1 to 2 ml of bile per kilogram of body weight, and concentrates the bile by the removal of water. Gallstones may result if the solution within the bladder becomes supersaturated, and materials begin to precipitate.

The pancreas

The functions of the pancreas relative to the process of digestion have already been discussed.

Anatomy. The pancreas (Fig. 15–18) is a carrot-shaped gland located along the greater curvature of the stomach, between it and the duodenum. It possesses a *head, neck, body,* and tapering *tail.* It usually empties through a single duct, the *pancreatic duct (Wirsung),* into the small intestine in company with the common bile duct. An *accessory duct* (of *Santorini*) may in some cases, be found emptying separately into the small intestine above the main duct.

FIGURE 15–18. The bile ducts, gall bladder and pancreas.

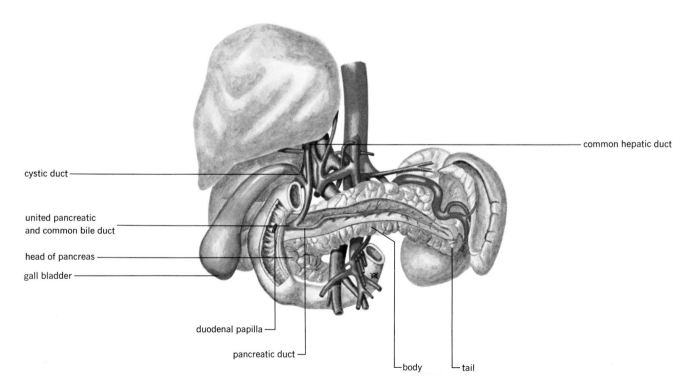

cystic duct

united pancreatic
and common bile duct

head of pancreas

gall bladder

duodenal papilla

pancreatic duct

common hepatic duct

body

tail

The pancreas consists of an *exocrine* portion, producing digestive enzymes, and an *endocrine* portion, producing hormones. The exocrine portion is supplied with ducts and consists of many small spherical masses of cells *(acini)* attached to the duct system much in the same fashion as grapes hanging in a bunch. Scattered amongst the acini are islets of endocrine tissue *(islets of Langerhans)* (Fig. 15–19), which are not ducted and which place their products directly into the blood stream. The islets consist of two main cell types called alpha and beta cells.

Functions of the pancreas other than those associated with digestion. The alpha cells produce a hormone known as *glucagon.* This material tends to cause the conversion of stored glycogen in the liver and muscle into glucose, and so raises the blood sugar level. The beta cells produce the more familiar hormone *insulin.* Insulin lowers the blood sugar level by causing a greater entry of glucose into the body cells, and by increasing the storage of glucose in the liver as glycogen. A rather delicate balance exists between the production and action of these two hormones, so that the blood sugar level normally remains within rather narrow limits.

Failure of the islets to produce sufficient insulin leads to a great increase in the blood sugar level and the development of *diabetes mellitus,* or sugar diabetes.

DISORDERS OF THE DIGESTIVE SYSTEM

Dental caries. Cavity formation in the teeth appears to be one of the chief penalties of civilization and the consumption of soft sweet foods. While no one cause can be advanced to explain cavity formation, the common denominator agreed upon is acid production by bacterial action on foods in the mouth. Excess acid appears to speed the demineralization process in the teeth, creating areas of weakness which are more susceptible to further change. Many host factors, such as the general nutritional state, the availability of calcium in the diet, dental hygiene practices, and heredity, influence the susceptibility of the teeth

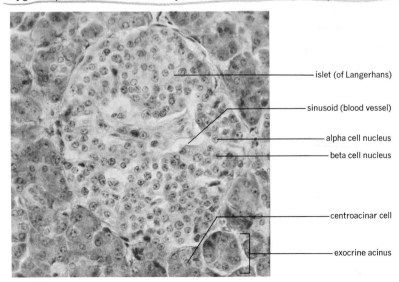

islet (of Langerhans)

sinusoid (blood vessel)

alpha cell nucleus

beta cell nucleus

centroacinar cell

exocrine acinus

FIGURE 15–19. Pancreas.

to acid damage. These are mainly involved in the formation and maintenance of good tooth materials. Fluoride appears to be of value when administered as part of the diet during the time that the teeth are forming. It promotes the formation of more perfect crystalline structures which are more resistant to dislocation and damage. Direct application of the fluorides to tooth surfaces, as in dentifrices, can apparently increase the resistance of the teeth to damage. The main benefit to be derived from any program of oral hygiene is to cleanse the teeth and mouth of adherent food particles, and so remove sites where bacteria and their products can accumulate.

Mumps. Mumps is a disease of viral origin affecting primarily the parotid salivary glands. The gland becomes swollen, painful, infiltrated with lymphocytes, and engorged with blood. The disease is transmitted by droplets of saliva containing the virus. The disease runs its course in about three weeks.

Gastritis. In inflammation of the stomach, the mucosal cell cementing substance is damaged by irritant foods, alcohol, aspirin, or many other substances. HCl and pepsin may then diffuse into the mucosa causing further damage. Pain and discomfort results, and further action of the stomach contents on the mucosa may create ulcerations in the stomach wall.

Peptic ulcer. The specific causes of local erosion of the stomach and duodenal wall are still a mystery, but there appear to be three common denominators present in most ulcer patients: (1) the presence of a stressful situation, which results in (2) overproduction of gastric juice, and (3) the secretion of excess pepsin and hydrochloric acid in the stomach when food is not present.

Inflammation of the small intestine. Duodenitis, jejunitis, and ileitis consist of inflammation of the corresponding segments of the small intestine, and are most commonly due to bacterial or parasitic infection.

Appendicitis. Being a blind-end portion of the alimentary tract, the appendix is very likely to accumulate bacteria and food materials. The organ can normally cleanse itself of these collections. When their exit is blocked, inflammation may result, with development of nausea, vomiting, and pain in the right lower quadrant. Surgical removal is usually indicated before rupture and peritonitis results.

Cirrhosis of the liver involves fibroid invasion of the organ with decrease of the number of functioning hepatic cells. Causes include malnutrition, overconsumption of alcohol, and various disease processes (hepatitis) which affect the liver and cause degeneration of cells.

Pancreatitis. This disease appears to have as its common denominator the release of pancreatic enzymes into the gland, as in obstruction of the duct, or direct injury to the gland. Onset is sudden, and pain is experienced in the upper abdomen.

Typhoid fever. Typhoid fever is due to a bacillus transmitted by contamination from infected feces. The transmitting agent may be a fly going from feces to food, or the unclean hands of an infected food handler. The organism enters through the mouth, multiplies in the tonsils, and invades the blood stream. From here, it lodges in the lymph nodes, lungs, bone marrow, spleen, and liver. Increase in the size of the nodes and particularly the Peyer's patches of the ileum is seen. With modern antibiotic therapy, the disease is usually controlled.

Diarrhea and constipation. These are usually the result of other abnormal processes in the body and should be considered symptoms rather than disease entities themselves. Diarrhea means the evacuation of a watery and unformed stool. It may be caused by bacterial infection, food sensitization of the tract, poisons (metals especially), vitamin lack (B group), circulatory disturbances, and emotional disturbances. *Constipation* refers to the passage of hard, dry feces. It should be emphasized that consistency of the mass is the critical criterion, and not the frequency of evacuation. Many people pass a stool of normal size and consistency only a few times a week. Constipation appears to be associated with slow passage of feces through the colon, allowing a greater than normal water absorption to occur. The condition is generally eased by an increase in the consumption of laxative foods and water. Laxatives themselves more often irritate than aid, and should be used sparingly.

Jaundice. Though most often associated with the liver, jaundice is basically defined as the presence of excess bile pigment in the blood stream. Enough is present to yellow the sclerae of the eyes. Blockage of the bile ducts (preventing the normal flow of bile into the small intestine), liver damage, and excess destruction of erythrocytes all may cause jaundice.

QUESTIONS

1. Now that you have studied the entire digestive system, explain the concept that the inside of the digestive organs is a part of the outside environment.
2. What is indicated by the phrase that in its lumen each digestive organ had its "own peculiar ecology?"
3. What are some of the "built-in" features that a digestive system must have in order to carry out digestion and absorption efficiently?
4. Name the types of teeth in the human mouth and describe the anatomy of a tooth.
5. What is the relationship between calcium and tooth development and decay?
6. Name the important functions of the tongue. Account for its versatility of movement.
7. Name the tunics in the walls of a typical hollow, tubular, organ of the digestive system, and their tissue components.
8. What is different about the muscle tissues in the esophagus relative to that in the small intestine?
9. Describe the glands of the stomach.
10. List some of the ways by which the absorptive surface of the small intestine is increased.
11. What are some of the movements involved in the small intestine during digestion and absorption? How is each controlled?
12. What is the function of bile in the digestive system?
13. Elaborate the functions of the small intestine.
14. Describe the functions of the large intestine.
15. Name and locate the lobes of the liver.
16. Describe a liver lobule and its relationships to bile and blood channels.
17. Explain the statement: "The liver has a double blood supply."
18. The pancreas is a double gland with a two-fold function. Explain this statement.
19. Trace, in proper order, the enzymes acting upon starch; list the products formed at each step. Do the same for protein.
20. What is a peptic ulcer?
21. Describe two viral diseases affecting the digestive system.

CHAPTER 16

Metabolism and Nutrition

METABOLISM AND NUTRITION

Introduction

Energy for continued cellular function is obtained from the degradation of chemical bonds in the foodstuffs we eat. Energy released by breaking these bonds may be utilized to run the body's operations, such as muscular contraction, secretion, maintenance of membrane transport systems, and many other functions. Or, it may be utilized for the synthesis of new substances by the cell. By artificial means, an ex-

425

perimenter can determine just how much energy a given amount of a foodstuff contains. The basic procedure employed is to ignite the substance and measure the amount of heat, in calories, given off by the combustion. Cells cannot and do not release the energy in a foodstuff molecule in one large "batch," for to do so would overwhelm the chemical reactions leading to physiological response. Cells degrade and synthesize substances in a stepwise fashion, releasing or adding energy a little at a time so as to channel as much of it as possible into specific physiological mechanisms. The study of the reactions which cells employ in the degradation and synthesis of materials forms that area of physiology known as intermediary metabolism.

Expression of energy content

The term *free energy* is used to describe the amount of usable energy that is released by the complete degradation of a foodstuff. It is expressed in terms of *calories* per mole of a food substance. A calorie, spelled with a small c, is a unit of heat energy defined as that amount of heat necessary to raise the temperature of 1 gram of H_2O one degree centigrade (for example, from 15°C to 16°C). The Calorie, spelled with a capital C, is 1000 times as large as a small calorie; it is the amount of heat necessary to raise the temperature of one kilogram of H_2O through the same temperature interval. It is also known as a kilocalorie. All references to energy release or energy requirements made in this chapter will be expressed in terms of kilocalories.

Coupling the energy in a foodstuff to a physiological mechanism

Energy in the chemical bonds of a sugar, fat, or amino acid molecule is not directly available to a cell's chemical processes. The energy must first be transferred to a compound which will then pass it to the specific reaction desired. A group of organic compounds formed from attaching phosphate residues to an organic substance serve as coupling agents. Energy released from the combustion of foods goes primarily towards the synthesis of such compounds. The energy required to form such phosphate bonds is relatively high (10 to 12 kcal) and upon hydrolysis, the bond releases 7.6 to 7.8 kcal. These compounds are known as *high energy phosphate compounds*. Among the specific high-energy compounds found in the body are *adenosine triphosphate* (ATP), *uridine diphosphate* (UDP), and *creatine phosphate* (CP). ATP (Fig. 16–1) is of universal occurrence in body cells and is regarded as the storage form of immediately available energy for cell reactions.

ATP is a *nucleotide* containing three phosphate radicals, the last two of which are attached to the rest of the molecule by high-energy bonds ($\sim$). The energy of the bond is transmitted to a particular cellular reaction by transferring the entire phosphate radical to another compound, in the process known as *phosphorylation*. A specific enzyme is required to achieve the transfer. As an example:

ATP	+	X	$\longrightarrow$	XPO$_4$	+	ADP
adenosine triphosphate		organic compound		phosphorylated compound		adenosine diphosphate

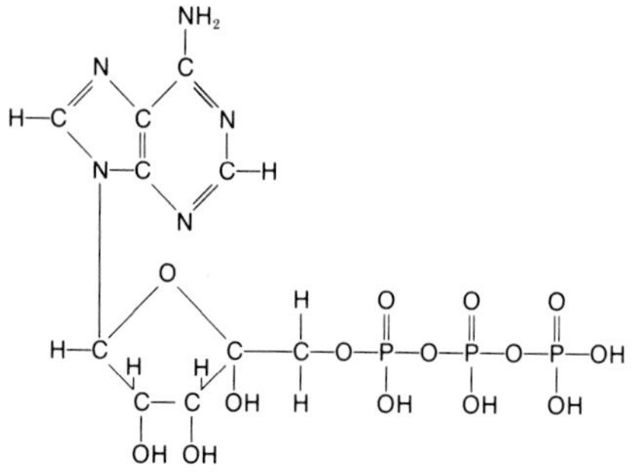

FIGURE 16–1. Adenosine triphosphate (ATP).

The transfer may result in an immediate observable physiological reaction, such as muscle contraction, or it may activate the recipient molecule, enabling it to proceed further in some series of chemical reactions.

In summary, synthesis of ATP represents the fate of some 90 percent of the energy inherent in foodstuff molecules. ATP in turn passes its contained energy to cellular reactions by transferring its phosphate group to another molecule.

CARBOHYDRATE METABOLISM

Transport into cells

It may be recalled that the processes of digestion produce a variety of monosaccharides, including glucose, fructose, and galactose. These are absorbed into the blood stream and distributed to body cells. To be utilized by cells, these sugars must pass through the cell membrane. Passage is by means of active transport systems and, once into the cell, the sugar is phosphorylated in the presence of ATP and specific enzymes:

$$\text{glucose} \xrightarrow[\text{glucokinase}]{\text{ATP} \quad \text{ADP}} \text{glucose-6-phosphate}$$

$$\text{fructose} \xrightarrow[\text{fructokinase}]{\text{ATP} \quad \text{ADP}} \text{fructose-6-phosphate}$$

$$\text{galactose} \xrightarrow[\text{galactokinase}]{\text{ATP} \quad \text{ADP}} \text{galactose-1-phosphate}$$

The phosphorylation of the sugar is irreversible except in liver and kidney tubule cells. In these areas, a phosphatase enzyme is present which splits the phosphate group from the sugar. This particular reaction frees the sugar to leave the cell and go elsewhere in the body. For example:

liver cell

Conversion of sugars

Glucose and fructose form the preferred sources of energy for cellular activity. There are no enzyme systems capable of metabolizing galactose efficiently. Galactose forms about one-third of the monosaccharides liberated by carbohydrate digestion in the alimentary tract. In order to use galactose in the body economy, it must be converted to a form which is metabolizable, that is, glucose. Liver cells possess a system of enzymes capable of achieving this conversion. An energy-rich phosphate compound, uridine diphosphoglucose (UDPG) is required in addition to the enzymes. The reactions occurring are as follows:

Reaction 1 activates the galactose, using ATP. In reaction 2, galactose replaces glucose on the uridine compound. In reaction 3, conversion of galactose to glucose occurs by rearrangement of atoms in the sugar molecule. In reaction 4, glucose is split from the uridine molecule with formation of another high-energy compound, UTP.

■   In the inherited metabolic disease galactosemia, there is accumulation of galactose in the blood due to an apparent inability to convert it to glucose. Deficiency of the enzyme catalyzing step 2 may be demonstrated. ■

Fructose may be metabolized by the same series of enzymes metabolizing glucose, and thus does not require conversion.

Formation of glycogen

Glycogen ("animal starch") represents a polysaccharide used as a storage form of glucose when the latter is present in the blood in greater amounts than the body can utilize. The formation of glycogen from glucose is designated as *glycogenesis.* Glucose-1-phosphate* is the starting material for glycogenesis. The latter material may be

*Hexose sugars contain 6 carbons, which are numbered from the aldehyde group, 1 to 6. A number appearing in the name of a compound designates the particular carbon to which the chemical group following the number is attached. For example:

formed from galactose, or by interconversion from glucose-6-phosphate, according to the reaction:

$$\text{glucose-6-phosphate} \underset{\text{phosphoglucomutase}}{\rightleftharpoons} \text{glucose-1-phosphate}$$

UDPG serves as the source of glucose to be polymerized into glycogen. It is formed, in reaction 1, by combining UTP and glucose-1-phosphate. Reaction 2 is the actual polymerization, and is influenced, in direct fashion, by the hormone insulin.

Breakdown of glycogen

Glucose may be recovered from glycogen by the process known as *glycogenolysis*, according to the following reactions:

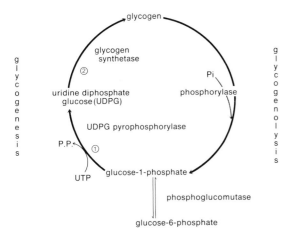

This reaction is directly influenced by the hormones glucagon and epinephrine. So far, we have established the following facts:

1. The main use to which carbohydrates and fats are put in the body is to release energy to synthesize ATP, the directly utilizable source of energy for cellular reactions.
2. Transport of sugars into cells requires ATP in order to activate the sugar.
3. Sugars other than glucose, specifically galactose, may be converted into glucose.
4. Glycogen may either be formed from glucose, or may be broken down into glucose. Both conversions are hormone-dependent.
5. Specific enzymes are required for each step in these reactions.

The interrelationships of the reactions presented so far are shown in Fig. 16–2. Enzymes are not included.

Glucose and fructose metabolism

Conversion of glucose and fructose to pyruvic acid is known as *glycolysis*. The process is anaerobic, that is, it can be carried out in the absence of oxygen. Glycolysis releases about 56 kcal of energy, and

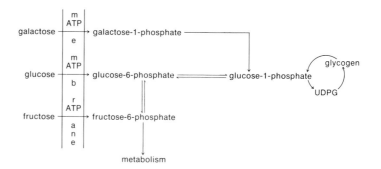

FIGURE 16–2. Interrelationships of monosaccharides and glycogen.

results in a net gain of two ATP molecules to the body. The reactions involved are shown on opposite page.

Reaction 1 utilizes one ATP molecule, and activates the sugar to enable it to continue its metabolism. Reaction 2 rearranges glucose into fructose. It is at this step that absorbed fructose may enter the scheme without being changed into glucose. Reaction 3 uses another ATP molecule, forming a 6-carbon compound with a phosphate on either end. Reaction 4 splits the 6-carbon fructose into two different 3-carbon compounds, each with a phosphate attached. The dihydroxy compound is then changed into the glyceraldehyde compound, and from this point onwards, two molecules are carried simultaneously. Reaction 5 phosphorylates each compound again, but uses inorganic phosphate (Pi) and not ATP. Reaction 6 releases enough energy to resynthesize two ATP molecules. Recall that to this point we have used two ATP molecules, so that the books are balanced. Reactions 7 and 8 again result in enough energy release to resynthesize two more ATP molecules. This represents a net gain. Reaction 9 occurs without enzymes and forms pyruvic acid.

Since the scheme releases 56 kcal and 2×7.7 kcal are trapped in the two ATP molecules formed, the efficiency of the operation in terms of energy trapping is $\dfrac{15.4}{56.}$ kcal $\times$ 100, or 27.5 percent.

Lactic acid formation

Further combustion of pyruvic acid to form, eventually, CO_2 and H_2O, requires the presence of oxygen. If the oxygen necessary for this combustion is not immediately available, pyruvic acid is changed to lactic acid. The lactic acid acts as a temporary storage form for the pyruvic, until O_2 becomes available, whereupon lactic is converted back to pyruvic and combusted. Lactic acid formation occurs as follows:

$$CH_3-\overset{\overset{\displaystyle O}{\|}}{C}-COOH \xrightarrow[\text{NADH}_2 \quad \text{NAD}]{+\ 2H} CH_3-\overset{\overset{\displaystyle OH}{|}}{\underset{\underset{\displaystyle H}{|}}{C}}-COOH$$

pyruvic acid   lactic acid

The hydrogens are contributed and picked up by hydrogen acceptors, molecules to be described later.

Glycolysis, Anaerobic Breakdown of Glucose (Embden-Meyerhof Pathway)

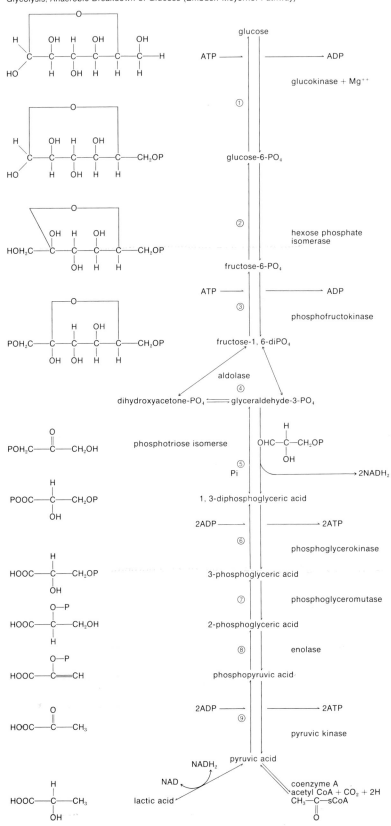

glucose

ATP ⟶   ⟶ ADP

glucokinase + Mg^{++}

①

glucose-6-PO$_4$

②

hexose phosphate isomerase

fructose-6-PO$_4$

ATP ⟶   ⟶ ADP

③

phosphofructokinase

fructose-1, 6-diPO$_4$

aldolase

④

dihydroxyacetone-PO$_4$ ⇌ glyceraldehyde-3-PO$_4$

phosphotriose isomerse

⑤

Pi

⟶ 2NADH$_2$

1, 3-diphosphoglyceric acid

2ADP ⟶   ⟶ 2ATP

⑥

phosphoglycerokinase

3-phosphoglyceric acid

⑦

phosphoglyceromutase

2-phosphoglyceric acid

⑧

enolase

phosphopyruvic acid

2ADP ⟶   ⟶ 2ATP

⑨

pyruvic kinase

pyruvic acid

NADH$_2$

NAD

lactic acid

coenzyme A
acetyl CoA + CO$_2$ + 2H
CH$_3$—C—sCoA
      ‖
      O

Further combustion of pyruvic acid

In order to enter the cycle which further combusts it, pyruvic acid must be changed to acetic acid. This is accomplished by removal of CO_2 (decarboxylation) from the pyruvic molecule. In the transformation, a pair of H^+ is released.

The acetic acid then reacts with another molecule, coenzyme A (CoA), which activates the acetic molecule.

$$\text{acetic acid} + \text{CoA} \longrightarrow \text{acetyl CoA}$$

Acetyl CoA then enters the Krebs cycle (citric acid cycle, tricarboxylic acid or TCA cycle). The main job that the Krebs cycle carries out is to produce 1 ATP molecule, 2 CO_2, and 8 H^+ per revolution:

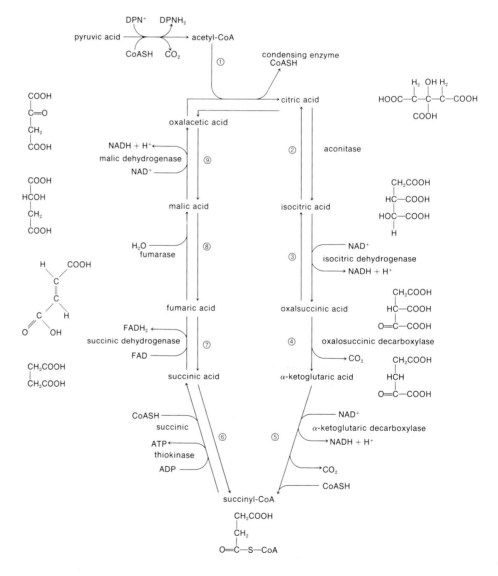

Reaction 1 couples the 2-carbon acetyl CoA to the 4-carbon oxaloacetic molecule to form a 6-carbon unit, citric acid. Reaction 2 rearranges the molecule. Reaction 3 removes 2 H. Reaction 4, by removing a CO_2, reduces the 6-carbon oxalosuccinic acid, to the 5-carbon acid, alpha-ketoglutaric. Reaction 5 removes an additional CO_2, as well as 2 H, and creates the 4-carbon compound, succinyl CoA. Studies involving isotope-labeled acetyl CoA have proven that the 2 CO_2 molecules released originate from the acetyl CoA. We can therefore state that it is the acetyl CoA (derived from pyruvic acid) which is undergoing degradation. Reaction 6 produces enough energy to synthesize a molecule of ATP from ADP. Reaction 7 produces 2 H. Reaction 8 is a rearrangement of the molecule. Reaction 9, in addition to releasing 2 H, regenerates oxaloacetic acid, which is then ready to accept another acetyl CoA molecule and repeat the cycle. It should be noted that the cycle is not completely reversible, owing to the fact that reactions 4 and 5 can proceed in only one direction. The cycle would be traversed twice in combusting the 2 pyruvics resulting from glycolysis, with a total production of 16 H, 2 ATP, and 4 CO_2 molecules.

Fate of hydrogens released in the cycles

Examination of the scheme of glycolysis, pyruvic → acetic transformations, and Krebs cycle discloses that a total of 24 hydrogens have been liberated (4 from glycolysis, 4 from pyruvic acid → acetic acid transformations, and 16 by two revolutions of Krebs). The hydrogens are picked up by molecules known as *hydrogen acceptors.* Three acceptors are known to exist:

1. diphosphopyridine nucleotide — DPN
             or
   nicotinamide adenine dinucleotide — NAD
2. triphosphopyridine nucleotide — TPN
             or
   nicotinamide adenine dinucleotide phosphate — NADP
3. flavine adenine dinucleotide — FAD

Hydrogens released by glycolysis and the pyruvic → acetic transformation are picked up by NAD, as are 6 of the 8 released by one revolution of Krebs cycle. FAD accepts hydrogens only from step 7 of the Krebs cycle. Remembering that to combust 2 pyruvic acid molecules requires two revolutions of the Krebs cycle, we can state that 20 of the 24 hydrogens produced are accepted by NAD, and 4 by FAD. The hydrogens are then carried to still another cycle known as oxidative phosphorylation, which converts the hydrogens to H_2O and releases energy for ATP formation.

Oxidative phosphorylation is shown at top of following page.

We may note that each compound on the left side of the scheme accepts hydrogens (or electrons), and each one on the right has hydrogen (or electrons) on it. Also, if hydrogens enter this scheme on NAD, they begin at the top and thus give rise to 3 ATP molecules for each pair of hydrogens. Remembering that 20 hydrogens arrive at the scheme on NAD, we can calculate that 30 ATP molecules (20 H = 10 pair; 3 ATP per pair or 3×10) will be produced. Four hydrogens come to this cycle on FAD. These give rise to 2 ATP per pair, a total of 4 ATP. A total of 34 ATP are thus produced by oxidative phosphorylation from the hydrogens released in other cycles.

Recalling that our starting material was glucose, and remembering

Fate of Hydrogens Released Via Krebs Cycle and Glycolysis

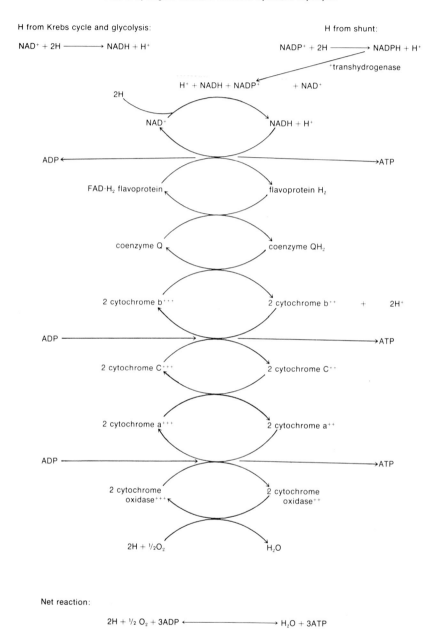

H from Krebs cycle and glycolysis:

$NAD^+ + 2H \longrightarrow NADH + H^+$

H from shunt:

$NADP^+ + 2H \longrightarrow NADPH + H^+$

$^+$transhydrogenase

$H^+ + NADH + NADP^+$ $+ NAD^+$

2H

NAD$^+$ NADH + H$^+$

ADP   ATP

FAD·H$_2$ flavoprotein   flavoprotein H$_2$

coenzyme Q   coenzyme QH$_2$

2 cytochrome b^{+++} 2 cytochrome b^{++} +   2H$^+$

ADP   ATP

2 cytochrome C^{+++} 2 cytochrome C^{++}

2 cytochrome a^{+++} 2 cytochrome a^{++}

ADP   ATP

2 cytochrome oxidase^{+++} 2 cytochrome oxidase^{++}

2H + $\frac{1}{2}$O$_2$ H$_2$O

Net reaction:

$2H + \frac{1}{2} O_2 + 3ADP \longleftarrow\longrightarrow H_2O + 3ATP$

that 2 ATP were produced by glycolysis, and 2 by the Krebs cycle in combusting 2 acetyl CoA molecules, we can conclude that the complete combustion of a molecule of glucose will create a net total of 38 ATP molecules to the body. These 38 molecules represent a storage of 292.6 kcal of energy (38×7.7 kcal per phosphate bond) in the body cells. Knowing that there is inherent in a glucose molecule 686 kcal of energy, the trapping represents an efficiency of 42.6 percent $\left(\frac{292.6}{686} \text{kcal} \times 100\right)$, a good figure. Recall again the statement that a main aim of combusting foodstuffs is to release energy for ATP synthesis. The combustion of a single glucose molecule creates a significant increase in the body's store of ATP.

Alternative pathways of glucose combustion

Two additional metabolic pathways exist for the degradation of glucose. The direct oxidative pathway (hexose monophosphate shunt or HMS) degrades glucose into 5-carbon sugars, glyceraldehyde, a 2-carbon fragment known as active glycolaldehyde, and utilizes NADP as the hydrogen acceptor. The 5-carbon sugars are utilized in nucleic acid synthesis. $NADPH_2$ serves as the hydrogen donor for fatty acid synthesis.

The reactions of the direct pathway are as follows:

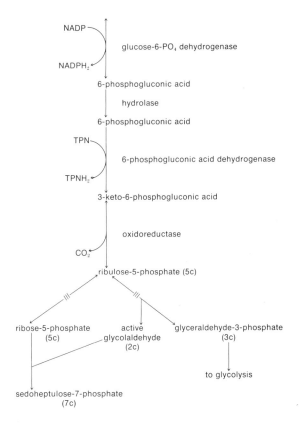

The uronic acid pathway creates UDPG for utilization in galactose → glucose conversion and glycogenesis, and creates glucuronic acid for utilization into connective tissue ground substance.

Of the two pathways, the shunt must be regarded as more important as a metabolic pathway for glucose degradation. It is an important route of glucose metabolism in the lactating mammary gland. It can handle up to 30 percent of the glucose metabolism in other cells.

FAT METABOLISM

Digestion of fats in the intestine results in the production of fatty acids and glycerol as the major end products. Absorption of these end products occurs primarily by physical processes, and the products are carried from the gut by way of the lymphatics (lacteals). As the fatty acids and glycerol pass through the intestinal mucosa, they are resynthesized into triglycerides (3 fatty acid molecules plus 1 glycerol molecule) and enter the lacteal as tiny droplets of fat known as chylomicrons. Coalescence of the chylomicrons is prevented by adsorption of lacteal protein upon the fat, creating a lipoprotein. The chylomicrons are then transferred to the venous circulation via the thoracic duct. After consumption of a meal rich in fat, the fat content of the blood may rise to where the blood becomes "milky." A lipemia has occurred. Over a period of 4 to 6 hours, the lipemia will disappear. Disappearance is due to absorption of chylomicrons by liver cells and subsequent metabolism, and by the action of a plasma enzyme, lipoprotein lipase (clearing factor), which hydrolyzes the fats again into free fatty acids and glycerol. The acids and glycerol may then enter storage depots in the body. The chief storage areas are the subcutaneous tissue *(panniculus adiposus),* the omentum, and in muscles.

Fate of glycerol

Glycerol, a 3-carbon compound, is easily converted to glyceraldehyde-3-phosphate, a compound found in the scheme of glycolysis, by the following reactions:

Fate of glycerol

glycerol $\rightleftharpoons$ dihydroxyacetone phosphate $\rightleftharpoons$ glyceraldehyde-3-phosphate
phosphatase   triose isomerase

The reactions are reversible, and since the compound glyceraldehyde-3-phosphate is common to the metabolism of two different foodstuff groups, it forms a method of linking the two schemes. In short, it is possible to convert lipid sources to carbohydrate and vice versa.
The ultimate fate of the glyceraldehyde will be either conversion to glucose, or metabolism via glycolysis.

Fate of fatty acids

Fatty acids undergo synthesis into neutral fats in depot areas, or undergo degradation via a process known as *beta oxidation.** In general terms, the process involves the degradation of a fatty acid by the progressive removal of 2-carbon segments in the form of acetyl CoA. Five steps occur:

* A fatty acid is an even-numbered (16 or 18) chain of carbon atoms. At one end is a carboxyl (COOH) group. Carbons in the chain are named alpha, beta, gamma, delta, etc., carbons, commencing with the first carbon beyond the carboxyl carbon. The changes occurring in beta oxidation take place on the beta carbon.

Beta Oxidation

1. $R-CH_2-CH_2-\overset{\overset{O}{\|}}{C}-OH + CoA-SH \xrightarrow[\text{Mg}^{++}]{\text{ATP + thiokinase}} R-(CH_2)_2-\overset{\overset{O}{\|}}{C}\sim SCoA + AMP + PP^*$

*pyrophosphate

2. $R-(CH_2)_2-\overset{\overset{O}{\|}}{C}\sim SCoA \underset{\text{FAD}\dashrightarrow\text{FADH}_2}{\overset{\substack{\text{acyl} \\ \text{dehydrogenase}}}{\rightleftharpoons}} R-CH=CH-\overset{\overset{O}{\|}}{C}\sim SCoA$

3. $R-CH=CH-\overset{\overset{O}{\|}}{C}\sim SCoA \overset{\text{crotonase}}{\rightleftharpoons} R-\overset{\overset{OH}{|}}{CH}-CH_2-\overset{\overset{O}{\|}}{C}\sim SCoA$

4. $R-\overset{\overset{OH}{|}}{CH}-CH_2-\overset{\overset{O}{\|}}{C}\sim SCoA \underset{\text{NAD}\dashrightarrow\text{NADH}_2}{\overset{\text{dehydrogenase}}{\rightleftharpoons}} R-\overset{\overset{O}{\|}}{C}-CH_2-\overset{\overset{O}{\|}}{C}\sim SCoA$

beta-keto fatty acid

5. $R-\overset{\overset{O}{\|}}{C}-CH_2-\overset{\overset{O}{\|}}{C}\sim SCoA + CoA-SH \xrightarrow{\text{thiolase}} \boxed{R-\overset{\overset{O}{\|}}{C}-SCoA} + CH_3-\overset{\overset{O}{\|}}{C}\sim SCoA$

acetyl CoA

Acetyl CoA is a compound found in the carbohydrate metabolism scheme at the beginning of the Krebs cycle. Again, a compound linking two foodstuff metabolism schemes is present. Accordingly, AcCoA formed from fats may enter Krebs for combustion, or be synthesized into glucose.

Synthesis of fats (triglycerides)

To synthesize a triglyceride, glycerol and fatty acids are required. Glycerol may be made from glyceraldehyde-3-phosphate. Fatty acids are synthesized by two routes:

1. *Beta reduction,* a reversal of the steps of beta oxidation. $NADPH_2$ from the shunt provides the source of hydrogen needed to create the fatty acid (step when FAD picked up H_2). This pathway is found in the mitochondria.
2. A cytoplasmic system, independent of the mitochondria, carboxylates (adds CO_2) to acetyl CoA to form a 3-carbon unit, malonyl CoA. Malonyl CoA is then added to a fatty acid chain (even number of carbons) and, during the addition, CO_2 is removed. A beta-keto fatty acid results. From here, reversal of steps 4 to 1 in beta oxidation occurs.

ATP production during degradation of fatty acids

Degradation of fatty acids produces a net gain of far more ATP to the organism than does combustion of carbohydrate. The main reason for this is the production of more AcCoA molecules and hydrogens than is produced during carbohydrate breakdown. Taking as an example the breakdown of stearic acid (18 carbons) we can calculate ATP production:

1. Four hydrogens are produced for each CoA formed in beta oxidation (see steps 2 and 4). Two of these are picked up by FAD, two by NAD.
2. To degrade an 18-carbon acid into 2-carbon units, 8 splits are required. Therefore, 8 × 4 or 32 carbons will be produced, 16 going to FAD, 16 to NAD.
3. Nine acetyl CoA molecules will be produced, which, going through the Krebs cycle, will produce 8 H and 1 ATP per molecule. Nine revolutions will take place, producing 72 H (9 × 8) and 9 ATP (9 × 1). Of the hydrogens, 18 will be picked up by FAD, and 54 by NAD.
4. A total of 34 H (16 + 18) will be carried on FAD to oxidative phosphorylation. The 34 H on FAD will give 2 ATP per pair, or 34 ATP.
5. The 70 H (16 + 54) on NAD will give 3 ATP per pair, or 105 ATP. A total of 148 ATP will be produced (9 from Krebs, 34 on FAD, 105 on NAD).

Formation of the ketone bodies

The combustion of excessive amounts of fat may cause "oversaturation" of the Krebs cycle. If Krebs cannot combust the AcCoA as fast as it is produced, a compound known as acetoacetic acid is formed, and, from this, acetone and beta-hydroxybutyric acid will be produced:

Formation of Acetoacetic Acid and Ketone Bodies

In uncontrolled diabetes mellitus (sugar diabetes), ketosis is common due to the reliance of the body on the combustion of fats. These acids (acetoacetic and beta-hydroxybutyric) are also threats to the body's buffering capacity. To buffer the organic acids may not leave enough buffer to handle the normal H_2CO_3 production. Hence, ketosis and acidosis are commonly seen to occur together.

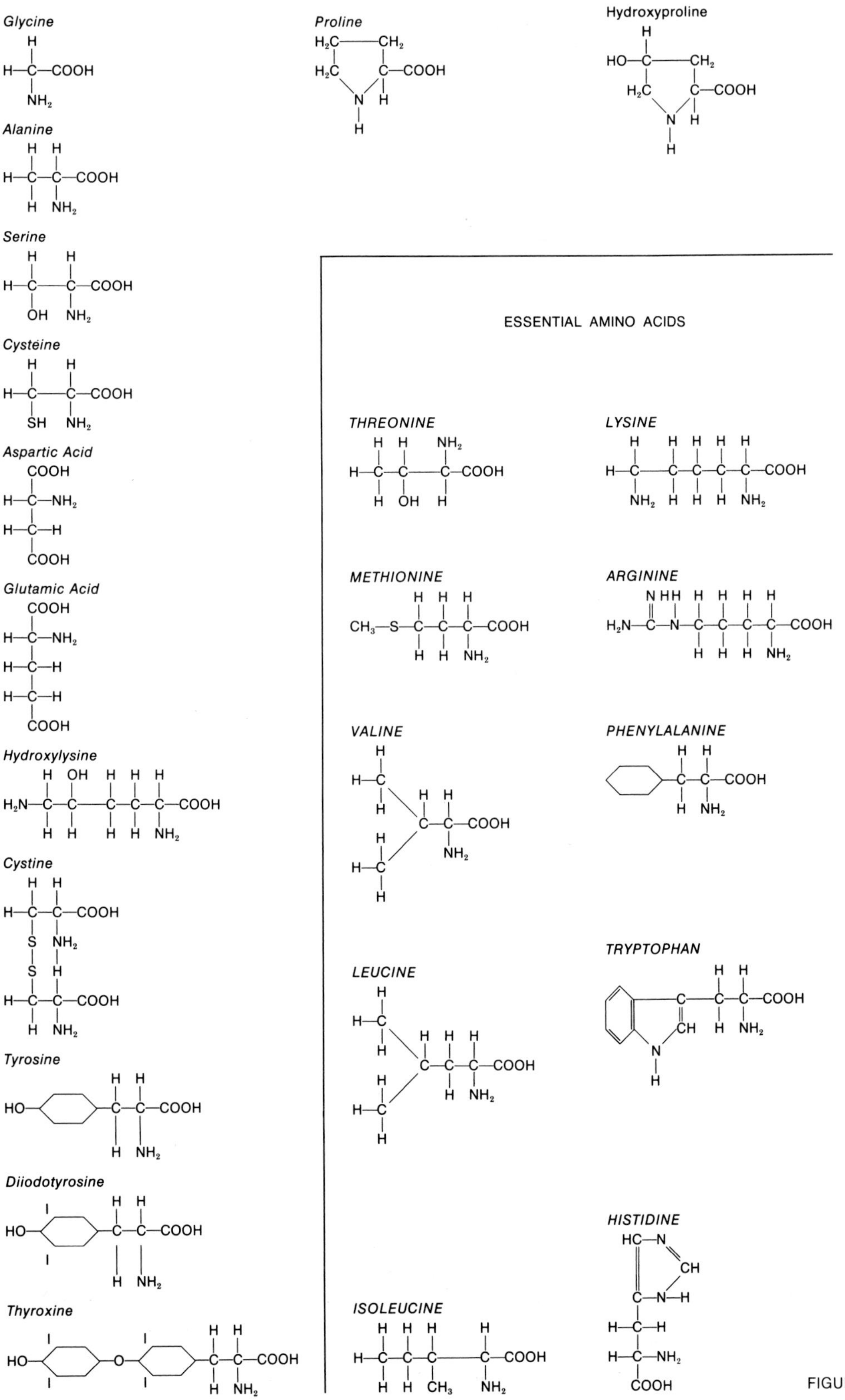

FIGURE 16–3. Amino acids.

PROTEIN AND AMINO ACID METABOLISM

Amino acids form the building units of proteins. These acids are linked by peptide bonds to form proteins which in turn are utilized as structural materials, enzymes, antibodies, some hormones, and as energy sources. A dynamic equilibrium exists between plasma amino acids and cellular proteins. There are about 20 different amino acids (Fig. 16–3) which fall into two categories:

1. Nonessential amino acids. These are 10 amino acids which can be synthesized in the body and need not be ingested as such in foods.
2. Essential amino acids. These are 10 acids which cannot be manufactured in the body or which are not produced in sufficient quantity to meet daily requirements. They must be ingested as such in the diet.

Synthesis of nonessential amino acids

Nonessential amino acids are synthesized from alpha-keto acids by a process of *transamination*. An alpha-keto acid is an acid which carried a carbonyl $\left(\begin{array}{c} O \\ \parallel \\ -C- \end{array}\right)$ group on the alpha carbon. Among such acids are:

α-keto Acids

pyruvic
$$CH_3-\overset{\overset{\displaystyle O}{\parallel}}{C}-COOH$$
α

oxaloacetic
$$COOH-CH_2-\overset{\overset{\displaystyle O}{\parallel}}{C}-COOH$$
α

α-ketoglutaric
$$COOH-CH_2-CH_2-\overset{\overset{\displaystyle O}{\parallel}}{C}-COOH$$
α

It may be recalled that such acids are produced in glycolysis and in the Krebs cycle. At the points in these cycles where such acids are produced, the metabolisms of carbohydrate and protein are interrelated. In the process of transamination, the amine group ($-NH_2$) from another compound (amine donor) is transferred to the alpha-keto acid, forming an amino acid. For example:

glutamic acid + pyruvic acid (an α-keto acid) →(transaminase)→ α-ketoglutaric acid + alanine (amino acid)

Amino acids so formed are rapidly synthesized into proteins. The synthesis involves the participation of the DNA of the nucleus, as indicated below:

Synthesis of proteins

The process of *transcription* involves the use of the genetic information in DNA to synthesize a particular sequence of nitrogenous bases into an RNA chain known as messenger RNA (m-RNA). The structure of DNA may be represented in several ways (Fig. 16–4). The molecule consists of a double helix held by complementary bonds between the nitrogenous bases adenine–thymine (A=T) and guanine–cytosine (G=C). Duplication of a DNA molecule occurs when one end of a double helix opens and the now freed bonds on the original chain attract complementary bases:

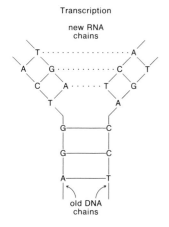

Since DNA does not leave the nucleus, the DNA serves as a template for the formation of m-RNA by the following process:

1. A single strand (not a double helix) of DNA attracts complementary bases to itself. Another nitrogenous base, uridine (U) may react with adenine to form an A=U combination. In RNA, U may replace C in the molecule and ribose replaces desoxyribose.

2. The strand of RNA so formed leaves the nucleus as m-RNA, moves to the ribosomes, and applies itself to the surface of one or several adjacent ribosomes.

   It may be noted that in the strand of m-RNA, the bases may be grouped by threes—for example A–A–G, C–U–U, U–G–A, and so on. A "genetic code" postulates that a given sequence of three bases specifies a given amino acid:

UUU = phenylalanine
UUA = isoleucine, leucine, tyrosine
UUC = leucine, serine
UUG = valine, cysteine, leucine
UAA = asparagine, lysine

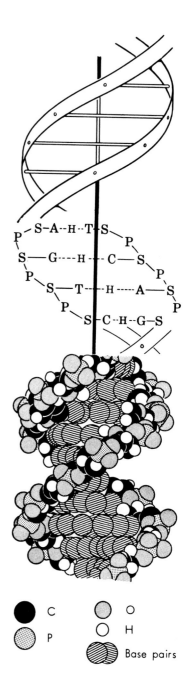

FIGURE 16–4. A three-way representation of DNA. (King and Showers, *Anatomy and Physiology,* courtesy of W. B. Saunders Co.)

We thus see that the m-RNA contains a series of directions for incorporation of amino acids in a specific order into a protein.

As these changes are taking place in the nucleus, another type of RNA is being formed in the cytoplasm. It is known as transfer RNA (t-RNA). It is prepared by the combination of cytoplasmic RNA with 3 molecules of the nucleotides cytidine triphosphate, guanine triphosphate, adenine triphosphate, or uridine triphosphate. It should be emphasized that the cytoplasmic RNA combines with only 3 of the possible nucleotides, establishing a "code" of three bases on the t-RNA. Next, a free amino acid is activated by combination with ATP. It will then attach to a specific t-RNA molecule. Remember that a series of three bases is thought to specify a given amino acid. For example:

activated phenylalanine will attach to a t-RNA molecule having the three bases UUC.

The t-RNA molecule with attached amino acid then moves to the ribosome where it attaches to the m-RNA at a point where the bases in the two molecules are complementary (i.e., A=T, G=U, G=C).

The complementary nature of the 2 sets of bases will determine the *position* of the amino acid in the protein chain. This is termed *translation.*

Last, the now adjacent amino acids latch together by the formation of peptide bonds, the new protein "peels off" from the t-RNA, and is either incorporated into the structure of the cell or leaves the cell. The t-RNA molecules detach from the m-RNA and are free to pick up other amino acids.

Use of amino acids for energy

If amino acids are not converted to protein, they usually undergo a process of *deamination,* whereby the amino group is removed, and ammonia (NH_3) and an alpha-keto acid are formed:

$$CH_3\text{—}CH\text{—}NH_2 \ (COOH) + \tfrac{1}{2}O_2 \xrightarrow[\text{oxidase}]{\text{amino acid}} CH_3\text{—}C{=}O \ (COOH) + NH_3$$

alanine   pyruvic acid   ammonia

The keto acids formed then enter the scheme of glycolysis or Krebs cycle and undergo degradation, or are converted to other materials, such as fatty acids or glucose.

Fate of Ammonia

Ammonia is extremely toxic to the body and is converted to urea before elimination by way of the ornithine cycle:

It may be noted that the cycle not only produces urea, but fumaric acid. The latter substance is a component of the Krebs cycle. In order that the student may more clearly see the interrelationships of the various metabolic schemes, a sheet is provided at the end of this Section. ("metabolic mill").

Ornithine Cycle – Production of Urea

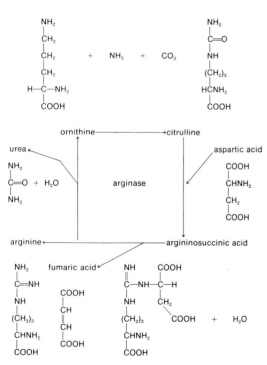

"The Metabolic Mill"

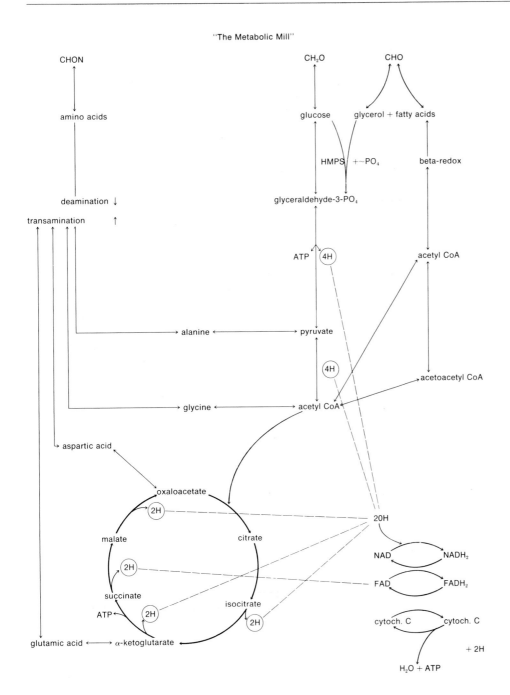

CALORIMETRY

The preceding Sections have emphasized the fact that foodstuffs are converted into energy for physiological activity. The measurement of energy requirements and release, and the determination of energy values of foodstuffs, is the area of study encompassed in calorimetry.

The calorie and heat values of foodstuffs

It may be useful to state again that energy is most easily measured by the large Calorie (kilocalorie), which is that amount of heat necessary to raise the temperature of 1 kilogram of water one degree centigrade. To determine the heat content of any given amount of a foodstuff, it may be ignited or combusted in a bomb calorimeter (Fig. 16–5), and the heat liberation measured directly. Such determinations made for carbohydrate, fat, and protein show heat values of 4.1, 9.4, and 5.6 kcal per gram of substance, respectively. Within the body, protein is not completely combusted, the urea being excreted. Since urea has a small energy content which is not available for use, the figure of 5.6 kcal/g must be reduced to 4.1 kcal/g of protein. For ease of calculation, the heat values are commonly rounded off to 4 kcal/g carbohydrate, 9 kcal/g fat, and 4 kcal/g protein.

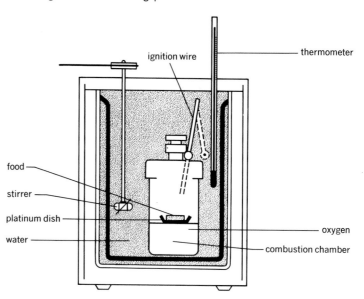

FIGURE 16–5. Bomb calorimeter.

Direct and indirect calorimetry

Since all of the energy produced by body activity is ultimately dissipated as heat, measurement of heat production by a living organism can give clues as to the amount of activity occurring and the caloric requirements which must be met to keep the animal in a state of caloric equilibrium. In short, caloric intake must balance output. Direct calorimetry involves placing an animal in a chamber similar to that shown in Fig. 16–6, and measuring its heat production, O_2 consumption, and CO_2 production, and determining the energy content of feces and urine. The apparatus is expensive and cumbersome. Indirect calorimetry is based upon the fact that the combustion of foodstuffs is attended by a more or less fixed requirement for oxygen to be used in the combustion, and by the production of a fixed amount of CO_2 by the combustion. Knowing the volume of O_2 required and the volume of CO_2 produced, one may calculate the respiratory quotient (RQ). $RQ = CO_2/O_2$.

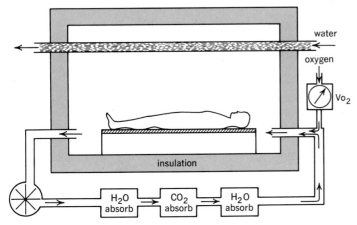

FIGURE 16–6. Direct calorimetry.

1. Carbohydrate (glucose) combustion may be represented as

$$C_6H_{12}O_6 \quad + \quad 6\ O_2 \longrightarrow 6\ CO_2 + 6\ H_2O$$
$$\text{glucose}$$

$$RQ = \frac{6}{6} = 1$$

2. Fats, having less oxygen in the molecule, require more from the outside in order to liberate CO_2.

$$2\ C_{57}H_{110}O_6 + 163\ O_2 \longrightarrow 114\ CO_2 + 110\ H_2O$$
$$\text{tristearin}$$

$$RQ = {}^{114}/_{163} = 0.70$$

3. Proteins have a calculated average RQ of 0.8. Because of unknown structure, most proteins cannot be specifically represented as to O_2 and CO_2 required and produced.

The particular RQ exhibited by the body as a whole is a reflection of the relative amounts of basic foodstuffs that are being combusted. If an RQ at rest was measured as 1.0, one could assume that the foodstuff being combusted was entirely carbohydrate. A similar conclusion could be reached for fats if the resting RQ was 0.7. A normal RQ for the body is about 0.82, indicating that a mixture of carbohydrate and fat is being combusted. Protein is not normally utilized for energy if carbohydrate and fat are present, so that one may assume that a given RQ is the result of CH_2O and fat combustion. Tables have been worked out relating the percentage of CH_2O and fat combusted at different RQs. Table 16–1 presents such data.

TABLE 16–1

Nonprotein RQ	1.0	0.95	0.9	0.85	0.8	0.75	0.71
Percent CH_2O combusted	100	82	65	47	29	11	0
Percent fat combusted	0	18	35	53	71	89	100

Knowing what the RQ is, we can next calculate a quantity known as the caloric equivalent of oxygen. It represents the number of calories produced by fuel combustion per liter of O_2 at a given RQ. This quantity, once determined, will enable calculation of caloric production, if O_2 consumption only is known, and thus we can express energy production and requirements in terms of calories. Again, tables have been produced relating oxygen equivalents to calories. Table 16–2 presents such data.

TABLE 16–2

Caloric Value per Liter of O_2

RQ	0.707	0.75	0.80	0.85	0.90	1.0
kcal	4.686	4.739	4.801	4.862	4.924	5.047

A determination of heat production by the indirect method would therefore require the following information:

1. The RQ under the particular conditions specified.
2. The caloric equivalent of O_2 at that RQ.
3. The liters of O_2 consumed during the time the experiment was proceeding.

As a sample calculation:

$$RQ = 0.8$$
caloric equivalent at $0.8 = 4.801$ kcal/l
$$O_2 \text{ consumption} = 20 \text{ l/hr}$$

Question: What is subject's caloric output per day?

O_2 consumption of 20 l/hr = 20×24 hr or 480 l/day
at RQ of 0.8, 1 l of $O_2 = 4.801$ kcal/l
480 l/day $\times$ 4.801 kcal/l = 2304.48 kcal/day

The basal metabolic rate (BMR)

If determinations, as have just been illustrated, were to be taken with the subject at complete rest, the resulting number of calories could be said to represent the subject's basal metabolic rate, or the rate of calories produced to maintain all basic body processes, extra activity eliminated. Actual BMR is determined by several factors:

1. *Age.* There is a progressive decrease in rate associated with aging.
2. *Sex.* Females have about 10 percent lower BMR than equivalent-sized males. This represents a smaller percent of muscle in the female body.
3. *Race.* Orientals have been shown to have a slightly lower BMR than Caucasians.
4. *Emotional state.* Anxiety or stress elevates BMR.
5. *Climate.* Individuals adapted to cold climates have higher BMRs.
6. *Hormone levels* in bloodstream. Thyroxin appears to be a substance which "sets the thermostat" of oxidative processes in the body. BMR rises and falls in direct proportion to circulating levels of this substance.

7. *Surface area.* Caloric expenditure appears best related to body surface, inasmuch as the skin represents a radiation surface for heat loss. Caloric release is usually expressed in kilocalories per square meter of surface per hour ($kcal/m^2/hr$).

Dietary considerations

To supply the necessary number of calories per day, one should select a diet which contains all three basic foodstuffs. Recommendations suggest that 10 to 15 percent of the daily caloric intake be in the form of protein, 55 to 70 percent as carbohydrate, and 20 to 30 percent as fat. These percentages are roughly equal to 33 to 42 grams of protein, 250 to 500 grams of CH_2O and 66 to 83 grams of fat. Such intake may be achieved by adequate intake of "the basic seven," which also provides adequate intake of necessary minerals and vitamins as well:

1. Milk. Two glasses per day for adults; 3 to 4 for children.
2. Vegetables. Two or more servings per day, one raw.
3. Fruits. Two or more servings per day, one a citrus fruit or tomato.
4. Eggs. Three to five per week.
5. Meats, cheese, fish. One serving per day.
6. Cereal or bread made from enriched or natural flours or grains.
7. Butter. Two tablespoons per day.

THE VITAMINS

The vitamins are a group of organic substances whose lack or deficiency in the diet can cause disruption of normal body processes. The term "vitamine" was originally coined to express the lack of "vital amines" as a cause of certain nutritional diseases. It is now known that such substances are not necessarily amines, and the term was shortened to vitamin. Vitamins are required in only small amounts to insure proper body function, and while some are synthesized in the body, a majority must be included as part of the diet. Table 16–3 presents the basic facts about the vitamins. Their formulas are presented in Fig. 16–7.

FIGURE 16–7. Formulas of vitamins (con't. on next page.)

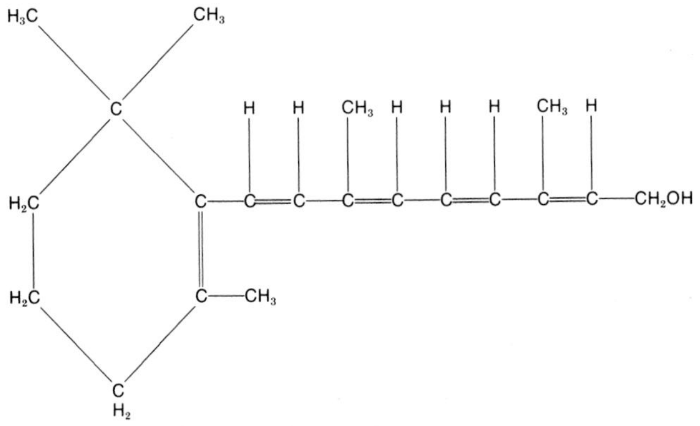

Vitamin A

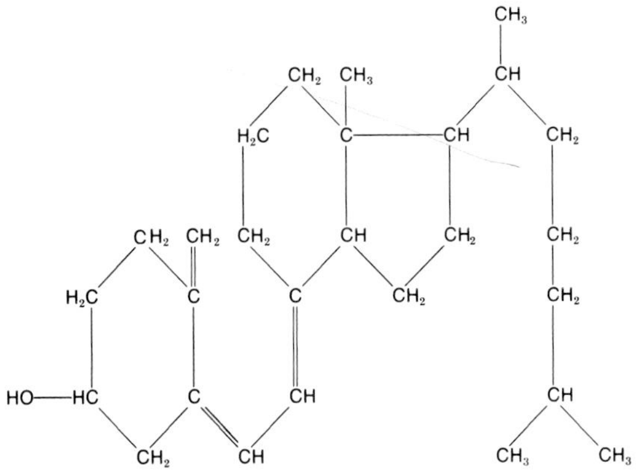

Vitamin D

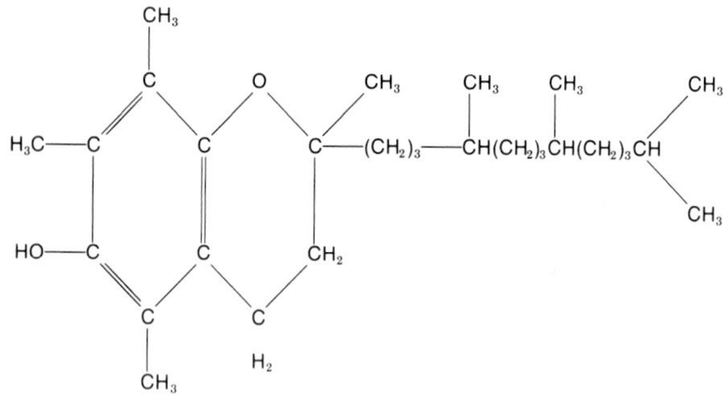

Vitamin E

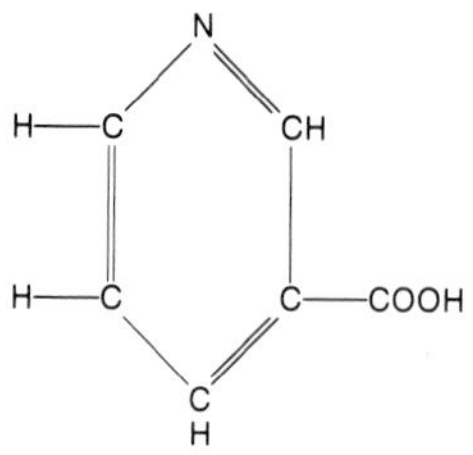

niacin

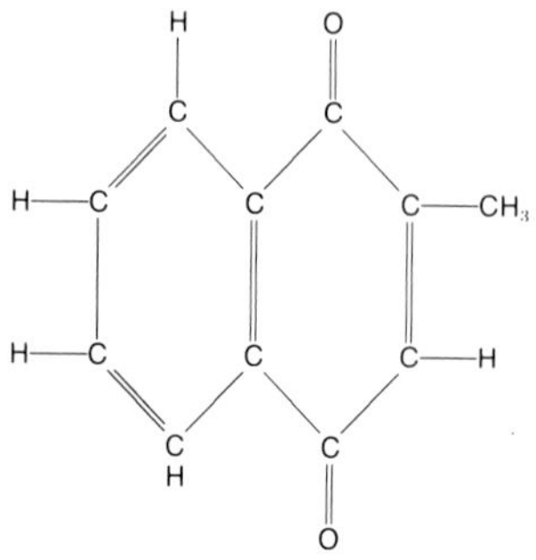

Vitamin K

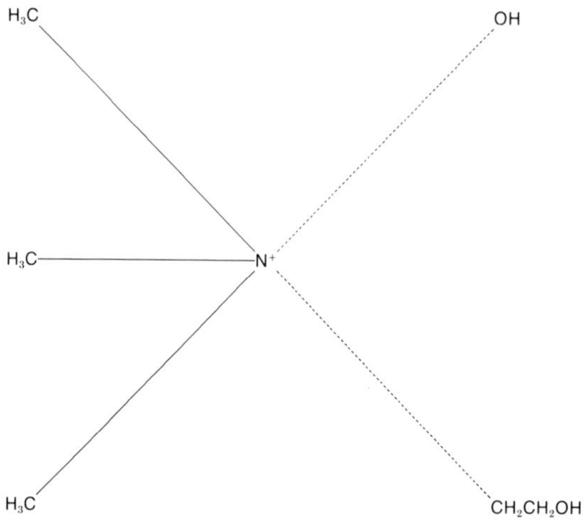

choline

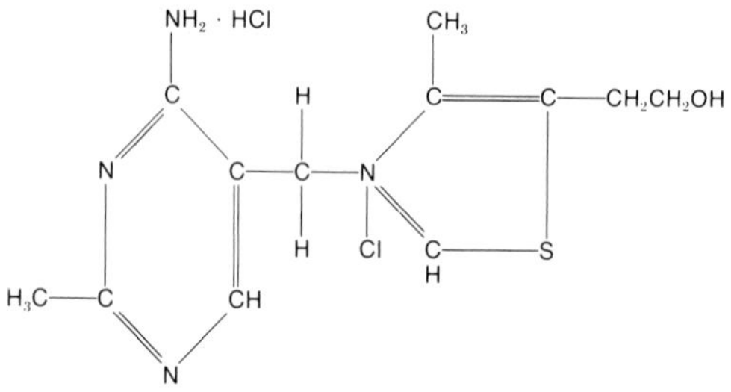

thiamine (Vitamin B_1)

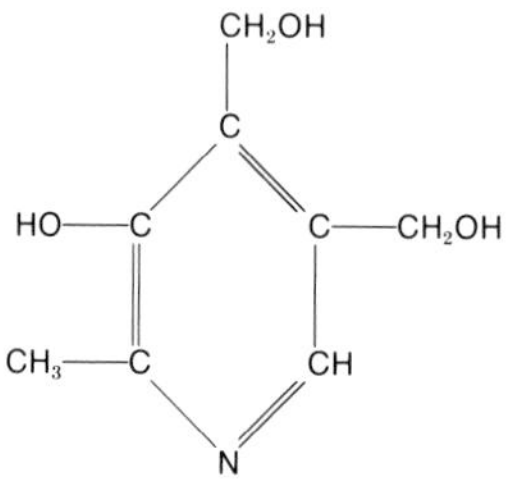

pyridoxine (Vitamin B_6)

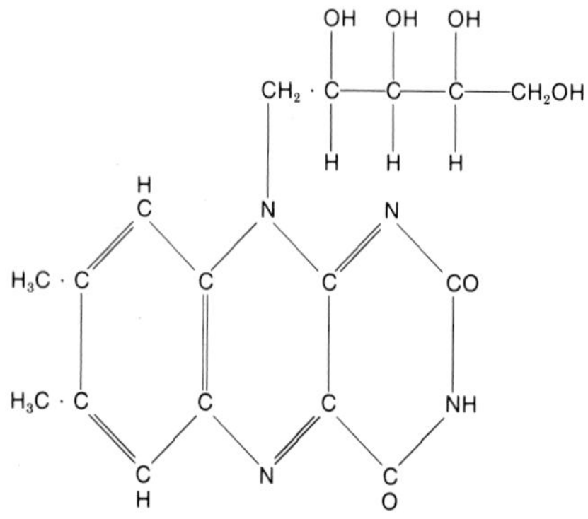

riboflavin (Vitamin B_2)

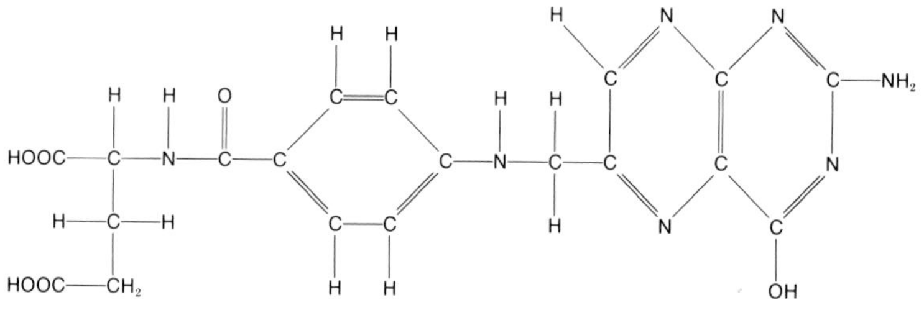

folic acid

FIGURE 16–7. Formulas of vitamins. *(con't.)*

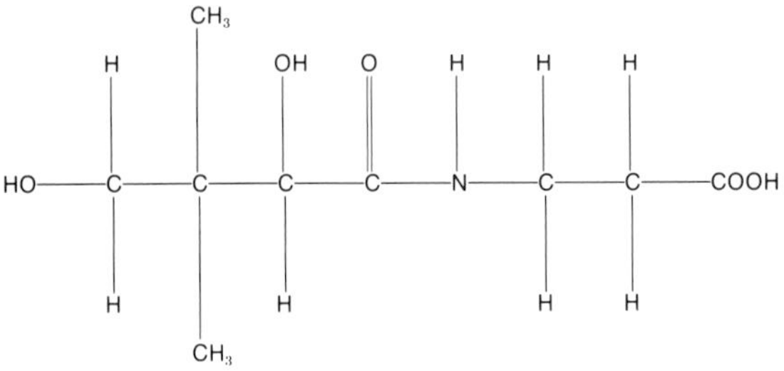

pantothenic acid

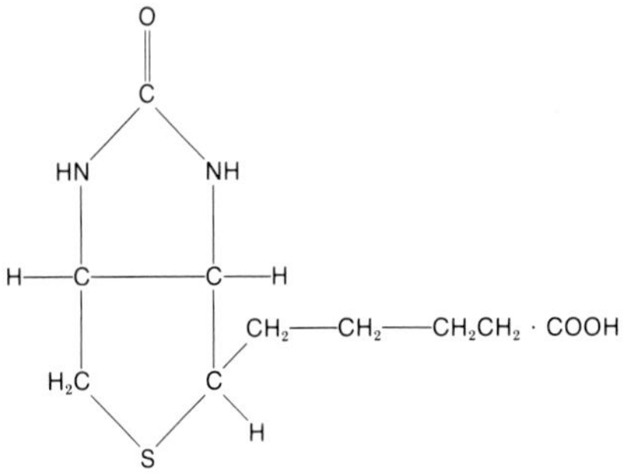

biotin

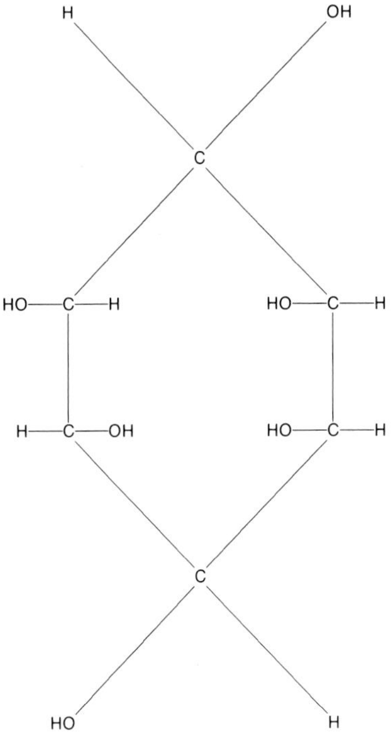

inositol

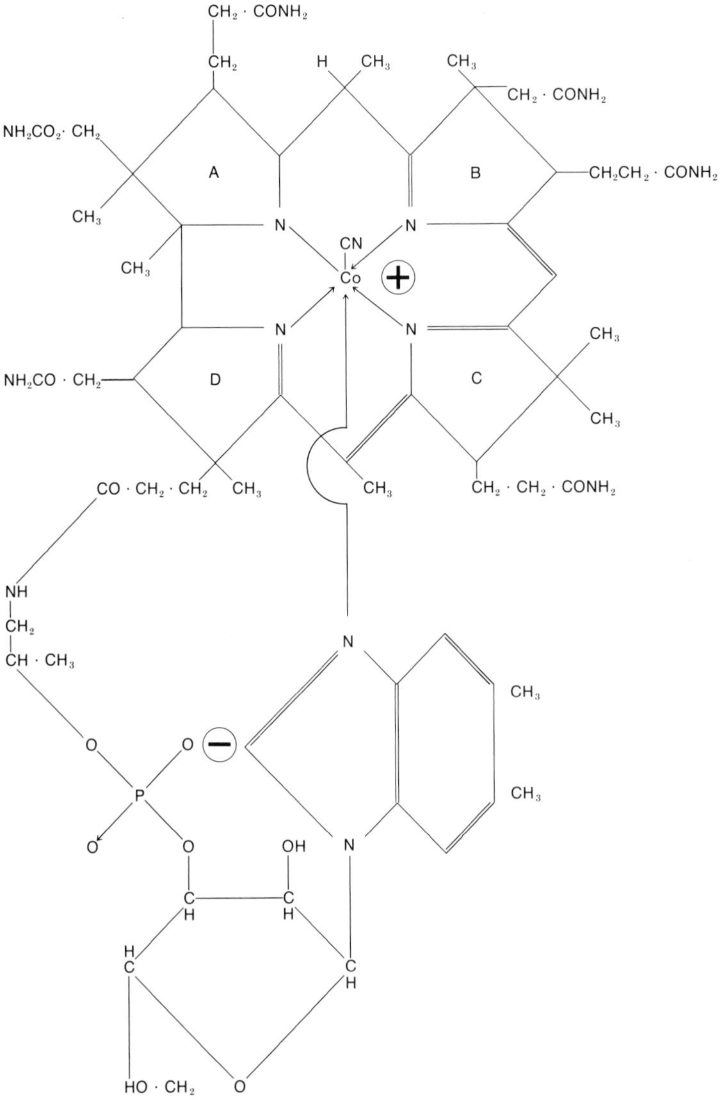

cyanocobalamin (Vitamin B_{12})

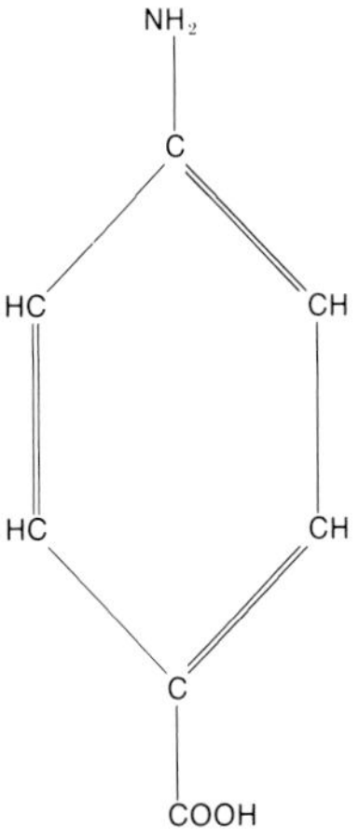

para-amino benzoic acid

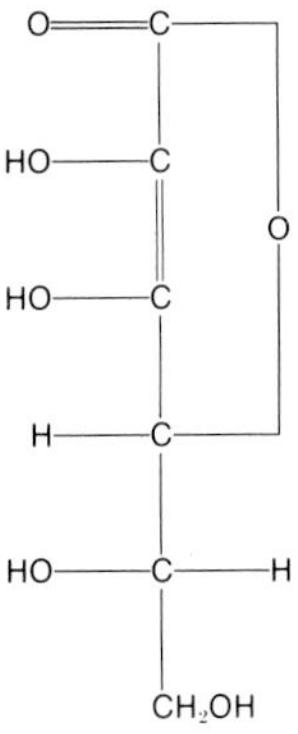

ascorbic acid (Vitamin C_1)

FIGURE 16–7. Formulas of vitamins.

TABLE 16–3

The Vitamins

Designation letter and name	Major properties	Requirement per day	Major sources	Metabolism	Function	Deficiency symptoms
A — Carotene	Fat — soluble yellow crystals, easily oxidized	5000 I.U.	Egg yolk, green or yellow vegetables and fruits	Absorbed from gut; bile aids; stored in liver	Formation of visual pigments; maintenance of normal epithelial structure	Night blindness, skin lesions
D₃ — Calciferol	Fat — soluble needlelike crystals, very stable	400 I.U.; much made through irradiation of precursors in skin	Fish oils, liver	Absorbed from gut; little storage	Increase Ca absorption from gut; important in bone and tooth formation	Rickets (defective bone formation)
E — Tocopherol	Fat — soluble yellow oil, easily oxidized	Not known for humans	Green leafy vegetables	Absorbed from gut; stored in adipose and muscle tissue	Humans — maintain resistance of red cells to hemolysis	Increased RBC fragility
					Animals — maintain normal course of pregnancy	Abortion, muscular wastage
K — Naphthoquinone	Fat — soluble yellow oil, stable	Unknown	Synthesis by intestinal flora; liver	Absorbed from gut; little storage; excreted in feces	Enables pro-thrombin synthesis by liver	Failure of coagulation
B₁ — Thiamine	Water — soluble white powder, not oxidized	1.5 mg	Brain, liver, kidney, heart; whole grains	Absorbed from gut; stored in liver, brain, kidney, heart; excreted in urine	Formation of cocarboxylase enzyme involved in decarboxylation (Krebs cycle)	Stoppage of CH_2O metabolism at pyruvate, beri-beri, neuritis, heart failure, mental disturbance
B₂ — Riboflavin	Water soluble; orange-yellow powder; stable except to light and alkalies	1.5-2.0 mg	Milk, eggs, liver, whole cereals	Absorbed from gut; stored in kidney, liver, heart; excreted in urine	Flavoproteins in oxidative phosphorylation (hydrogen transport)	Photophobia; fissuring of skin
Niacin	Water soluble; colorless needles; very stable	17-20 mg	Whole grains	Absorbed from gut; distributed to all tissues; 40% excreted in urine	Coenzyme in H transport, (NAD, NADP)	Pellagra; skin lesions; digestive disturbances, dementia
B₁₂ — Cyanocobalamin	Water soluble; red crystals; stable except in acids and alkalies	2-5 mg	Liver, kidney, brain. Bacterial synthesis in gut	Absorbed from gut; stored in liver, kidney, brain; excreted in feces and urine	Nucleoprotein synthesis (RNA); prevents pernicious anemia	Pernicious anemia; malformed erythrocytes

TABLE 16–3, con't.

The Vitamins

Designation letter and name	Major properties	Requirement per day	Major sources	Metabolism	Function	Deficiency symptoms
Folic acid (Vitamin B_c, M, pteroyl glutamate)	Slightly soluble in water; yellow crystals; deteriorates easily	500 micro-grams or less	Meats	Absorbed from gut; utilized as taken in	Nucleoprotein synthesis; for-mation of erythrocytes	Failure of erythrocytes to mature; anemia
Pyridoxine (B_6)	Soluble in water; white crystals; stable except to light	1-2 mg	Whole grains	Absorbed from gut; one-half appears in urine	Coenzyme for amino acid metabolism and fatty acid metabolism	Dermatitis; nervous disorders
Pantothenic acid	Water—soluble; yellow oil; stable in neu-tral solutions	8.5-10 mg	?	Absorbed from gut; stored in all tissues; urine	Forms part of coenzyme A (CoA)	Neuromotor disorders; cardiovascular disorders; GI distress
Biotin	Water soluble; colorless needles; stable except to oxidation	150-300 mg	Egg white; synthesis by flora of GI tract	Absorbed from gut; excreted in urine and feces	Concerned with protein synthe-sis, CO_2 fixation and transami-nation	Scaly dermatitis; muscle pains, weakness
Choline (maybe not a vitamin)	Soluble in water; color-less liquid; unstable to alkalies	500 mg	?	Absorbed from gut; not stored	Concerned with fat transport; aids in fat oxidation	Fatty liver; inadequate fat absorption
Inositol	Water soluble; white crystals	No recom-mended allowance	?	Absorbed from gut; metabo-lized	Aids in fat metabolism; prevents fatty liver	Fatty liver
Para-amino benzoic acid (PABA)	Slightly water soluble; white crystals	No evidence for requirement		Absorbed from gut; little storage; excreted in urine	Essential nutrient for bacteria; aids in folic acid synthesis	No symptoms established for humans
Ascorbic acid (Vitamin C)	Water soluble; white crystals; oxidizable	75 mg/day	Citrus	Absorbed from gut; stored; excreted in urine	Vital to collagen and ground substance	Scurvy—failure to form c.t. fibers

QUESTIONS

1. What compound acts as the storage form of energy readily available for insertion into physiological mechanisms? How is it different from other chemical compounds?

2. What is the first step necessary for the combustion of glucose? What compound is responsible for the process?

3. Describe, in general terms, how it is possible to convert galactose to glucose. What is the advantage of such a process?
4. Define glycogenesis, and describe the process.
5. Describe glycolysis.
6. What aerobic cycle completes the combustion of glucose? Describe.
7. What is oxidative phosphorylation? What does it produce?
8. List the products which result from the hexose monophosphate shunt.
9. How are fatty acids combusted? Describe the cycle.
10. What is deamination? What happens to the ammonia which results? Describe the cycle.
11. What is transamination? What compounds are required and what results?
12. List the five steps required to synthesize a protein from amino acids.
13. Someone claims 38 ATP will result from the complete combustion of a molecule of glucose. Right or wrong? Prove the answer, showing all calculations.
14. How can you make a fatty acid from glucose? Describe process.
15. What is acetoacetic acid? Where does it come from and what can it form?
16. How many ATP can be produced from the complete combustion of a 30-carbon fatty acid? Show all calculations.
17. What are the three hydrogen acceptors used in intermediary metabolism? From what cycles does each collect its hydrogens, and to what cycles does each carry the hydrogen?
18. What compounds act as "crossover points" between two or more cycles? List them, list the cycles each connects, and the compounds involved on either side.
19. What is the value in the ability to form lactic acid? Under what conditions is it formed and what happens to it afterwards?
20. What cycles are disturbed by the disease diabetes mellitus? List four compounds that would either not be formed or formed in excess in the disease.
21. What vitamins are essential to intermediary metabolism? Designate specifically the compound each vitamin becomes a part of.

CHAPTER 17

The Urinary System

Falstaff: "Sirrah, you giant, what says the doctor to my water?"
WM. SHAKESPEARE, IN KING HENRY IV

THE URINARY SYSTEM

Introduction

Pathways of excretion. Metabolism of nutrients by the body cells results in the production of wastes. Carbon dioxide, nitrogenous by-products (such as ammonia, urea, and uric acid), heat, and excess water must be eliminated lest they disrupt the homeostasis of the body. The lungs account for elimination of the greater part of the carbon dioxide, along with small quantities of water and heat. The alimentary tract accounts for the loss of some carbon dioxide, water, and heat, some salts, and the secretions of the digestive glands. The skin plays the major role in the elimination of excess body heat, but a minor role in elimination of solid wastes. Also treated as wastes, in that they will ulti-mately be eliminated from the body, are the substances present in greater amounts than are required for normal cellular function. Water, salts, hydrogen ion, sulfates, and phosphates are examples of physi-ologically important substances often present in excess amounts. Of

455

primary importance for the elimination of nitrogenous wastes of metabolism and excess materials are the organs of the urinary system.

Organs of the system.   The urinary system includes two kidneys, two ureters, a single urinary bladder, and a single urethra to the exterior (Fig. 17–1).

General functions of the system.   The kidneys are responsible for the removal of metabolic wastes and excess materials from the blood by the processes of excretion and urine formation. The kidney also regulates the composition and physical properties of the blood, and hence the whole internal environment. It monitors acid-base balance, osmotic relationships, and the content of organic and inorganic solutes in the blood. The other organs of this system are involved only in the relatively simple tasks of transport, storage and elimination of urine.

The kidneys

Size, location, and attachments.   The living human kidneys are bean-shaped, reddish organs, located to either side of the twelfth thoracic to third lumbar vertebral bodies. Lacking mesenteries, the organs are *retroperitoneal,* that is, they lie external to the lining of the abdominal cavity. The left kidney is usually placed slightly higher than the right. The kidneys measure about 11 cm long, 5 to 7.5 cm wide, and 2.5 cm thick (about $4 \times 2 \times 1$ inches).

Gross anatomy.   A medially directed concavity is termed the *hilus,* and leads to the *renal sinus.* The blood vessels, nerves, and ureters enter and exit the organ at this region. Affixing the kidneys in position behind the parietal peritoneum are the *adipose capsule (perirenal fat)* and the double layers of the *subserous (renal) fascia.* A *fibrous capsule* forms the external covering of the kidney itself. A frontal section of the organ (Fig. 17–2) reveals the upper expanded end of the ureter, the *renal pelvis,* lying within the renal sinus. The *major* and *minor calyces* are the primary and secondary subdivisions of the pelvis. An inner *medulla* is composed of 8 to 18 *renal (medullary) pyramids,* with bases directed towards the periphery of the kidney, and apices or *papillae* projecting into the minor calyces.

The *cortex* arches around the bases of the pyramids and projects between the pyramids, where it is designated as the *renal columns.* Cortical and medullary portions together constitute the *parenchyma* or cellular portion of the kidney. The parenchyma is formed, in each kidney, of at least one million microscopic *nephrons (renal tubules)* with their associated blood supply. It is the nephron which carries out the functions of the kidney, that is, formation of urine and regulation of the blood composition.

Blood vessels of the kidney (Fig. 17–3).   Originating from the aorta and passing to each kidney is a large *renal artery.* These arteries carry to the kidneys approximately one-fourth of the total cardiac output, a quantity (renal blood flow or RBF) averaging 1200 ml of blood per minute. Of this volume, 650 ml is plasma (renal plasma flow or RPF). Upon entering the kidney at the hilus, the renal artery branches to form several *interlobar arteries* which enter the parenchyma. The vessels pass between the pyramids and, near their bases, form a series of curved vessels, the *arcuate arteries.* The arcuate arteries in turn give rise to a series of radially directed vessels running through the cortical region of the kidney. These vessels are known as *interlobular arteries.*

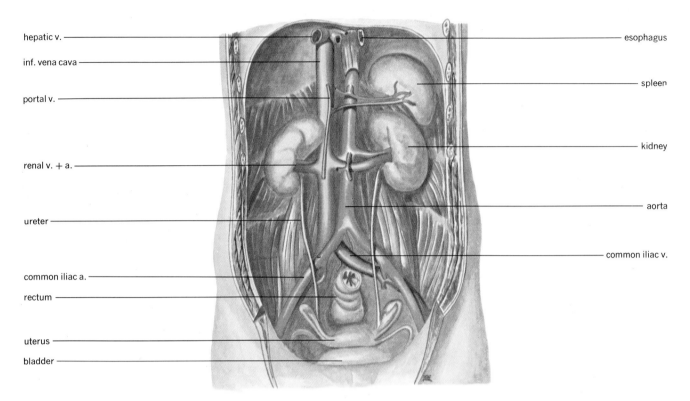

hepatic v.

inf. vena cava

portal v.

renal v. + a.

ureter

common iliac a.

rectum

uterus

bladder

esophagus

spleen

kidney

aorta

common iliac v.

FIGURE 17–1. Organs of the urinary system and associated structures.

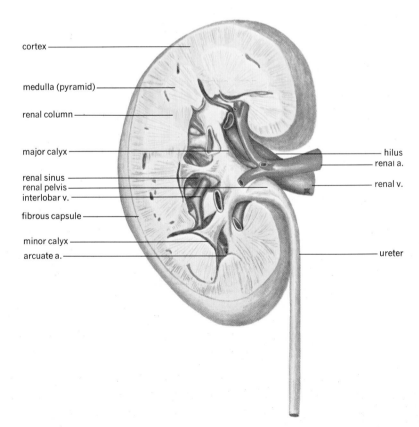

cortex

medulla (pyramid)

renal column

major calyx

renal sinus
renal pelvis
interlobar v.

fibrous capsule

minor calyx

arcuate a.

hilus
renal a.

renal v.

ureter

FIGURE 17–2. Frontal section of the kidney.

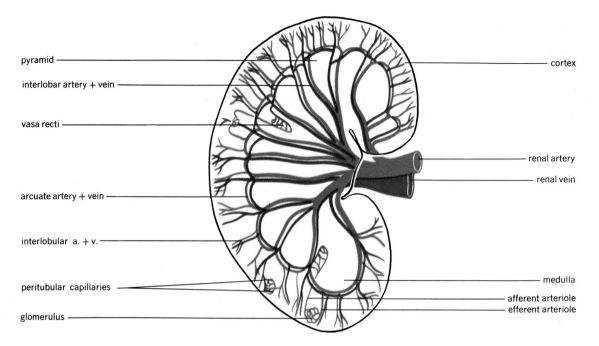

pyramid

interlobar artery + vein

vasa recti

arcuate artery + vein

interlobular a. + v.

peritubular capillaries

glomerulus

cortex

renal artery

renal vein

medulla

afferent arteriole

efferent arteriole

FIGURE 17–3. The blood vessels of the kidney.

The complex of arcuate arteries arising from any one given interlobar artery does not interconnect with similar vessels arising from other interlobar arteries. Because of this arrangement, the arcuate arteries are known as *end arteries.* This arrangement is one factor insuring the maintenance of a high level of pressure on the blood as it reaches the outer portions of the cortex. The interlobular arteries give off nutrient branches to the cortical substance and to the capsule of the kidney. Also, and most important, branches called *afferent arterioles* go to the glomerular capsules, where they form knots of capillaries designated *glomeruli.* From the glomeruli, *efferent arterioles* lead out to form either a network of capillaries around the renal tubules or a looped blood vessel which dips deeply into the medullary portion of the kidney and then returns toward the cortex. The efferent arteriole carries a smaller volume of blood than the afferent and is smaller in diameter. If a network of capillaries is formed, it is known as the *peritubular capillary network;* the vascular loop is known as the *vasa recti.* From here the blood vessels form a series of veins named in the same manner as the incoming arteries. *Interlobular veins* drain into *arcuate veins* which lie along the bases of the medullary pyramids. Unlike the corresponding arteries, the arcuate veins do connect with one another. These in turn form *interlobar veins* which form a single *renal vein* from each kidney, the latter emptying into the inferior vena cava.

The nephron.   Two types of nephron (Fig. 17–4) are generally recognized. Both have similar parts and differ from one another only in the length of certain portions and in what happens to the efferent arteriole as it leaves the glomerulus. The two types are known as cortical and juxtamedullary nephrons. The *cortical nephron* usually has its glomerulus in the outer portion of the cortex, and the rest of its parts rarely reach deeply into the medulla. The *juxtamedullary nephron,* on the other hand, has its glomerulus close to the corticomedullary junction, and has part of its structure located deep within the medulla.

A nephron begins with a double-walled cup termed the *glomerular (Bowman's) capsule.* The inner wall of this capsule is known as the *visceral layer* and it follows intimately the twists and turns of the glomerular capillary network. The outer or *parietal layer* of the capsule lies a short distance from the visceral layer so that an actual space between the two layers is created (Fig. 17–5). The capsule and the contained glomerulus form a unit designated as the *renal corpuscle.* The capillary endothelium and the two layers of the capsule are composed of simple squamous epithelium. Leading from the capsule is a *proximal convoluted tubule* in which the cells are cuboidal, have central nuclei, and have a brush border on the lumen side. The brush border consists of minute cytoplasmic extensions of the cell called microvilli. These serve to increase the surface area for reabsorption and/or secretion of ma-

FIGURE 17–4. The nephrons of the kidney and their associated blood vessels. A. Juxtamedullary nephron. B. Cortical nephron.

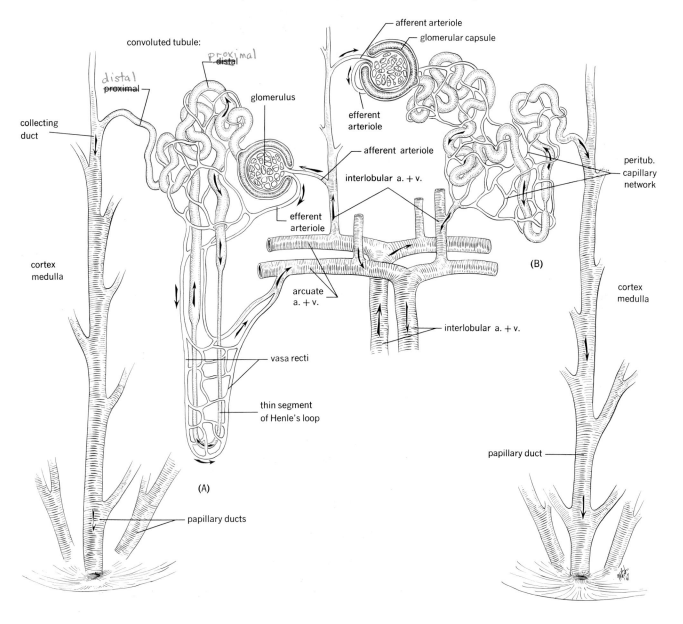

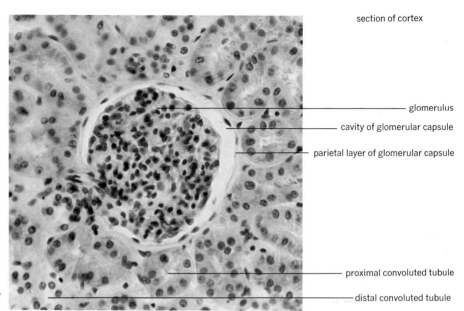

section of cortex

glomerulus

cavity of glomerular capsule

parietal layer of glomerular capsule

proximal convoluted tubule

distal convoluted tubule

FIGURE 17–5. Photomicrograph of kidney cortex.

terials by this part of the nephron. The proximal tubule changes abruptly as it nears the medullary region. The cells become flat, and the tube narrows and dips toward or into a pyramid as the *descending loop of Henle.* Then the tube bends back upon itself, enlarges, its cells become rectangular, and, as the *ascending loop of Henle,* it returns to or towards the cortical region (Fig. 17–6). In the cortex, the tube again becomes convoluted and is known as the *distal convoluted tubule.* Its cells are cuboidal, have central nuclei, have lighter staining cytoplasm than the proximal tubule, and carry no brush border. This portion joins a *collecting tubule.* The collecting tubules receive the distal termina-

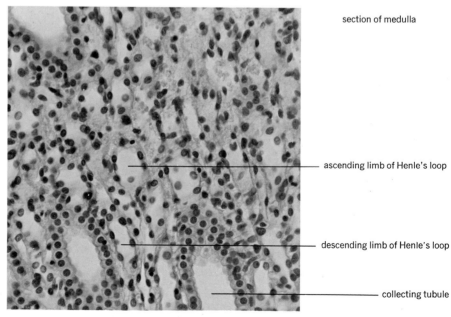

section of medulla

ascending limb of Henle's loop

descending limb of Henle's loop

collecting tubule

FIGURE 17–6. Photomicrograph of kidney medulla.

tions of many nephrons. They open into the calyces of the pelvis through the *papillary ducts*. In a cortical nephron, the loop of Henle is quite short and typically does not reach into the medulla. It is around the tubules of such a nephron that we find the peritubular capillary network. The loops of Henle of the juxtamedullary nephrons, on the other hand, dip very deeply into the medulla, and the loops of these nephrons are followed by the looped blood vessels known as the vasa recti. An anatomical difference such as described obviously suggests a functional difference between the two types of nephrons and, indeed, such a difference is found to exist. In general, it may be said that cortical nephrons are reabsorptive and secretory structures. Juxtamedullary nephrons also carry out these functions, but, in addition, are responsible for creating an osmotic gradient within the parenchyma which makes it possible to elaborate a hypertonic urine.

The formation of urine

The nephron forms urine which is hypertonic and usually more acid than the plasma. The nephron also governs, within narrow limits, the composition of the blood leaving the kidney. In accomplishing these tasks, the kidney is aided by hormones secreted by certain endocrine glands and by chemical substances originating within the organ itself.

The processes which the nephron employs in forming urine are filtration, tubular transport, the countercurrent multiplier and exchanger, and acidification.

Filtration.   The blood, with its contained load of wastes, excess materials, and substances the body wishes to conserve, is delivered to the glomerulus under a hydrostatic blood pressure (Pb) of approximately 75 mm Hg. The end artery structure of the arcuate arteries is instrumental in assuring maintenance of high pressure as the vessels branch and penetrate into the kidney. The glomerular capillaries and the closely applied visceral layer of the glomerular capsule act as a coarse sieve. Providing the blood pressure is higher than the sum of all opposing pressures (see below), all materials small enough to pass through these membranes will be forced through by the pressure differential in a process of filtration. The capillaries behave as though they possess pores of small diameter (100 Å) and the cells of the visceral layer do not form a continuous stratum (Fig. 17–7A, B, C). There is therefore little resistance to the passage of materials. It is in fact only the *basement membrane* lying between the two cellular layers which truly governs what will pass. In general, it may be stated that substances having a molecular weight up to 10,000 pass easily through these layers of tissue. Above this figure, materials pass with increasing difficulty, and a probable upper limit is reached at a molecular weight of about 200,000. Thus, all materials present in the plasma, except the formed elements, are filtered into the cavity of the glomerular capsule. The filtrate contains protein, chiefly albumin, to the extent of 10 to 20 mg/100 ml. A mathematical calculation would indicate that some 30 grams of protein is thus filtered per day. The material (filtrate) formed by this process resembles plasma (see Table 2), except that it lacks formed elements and has a lower protein concentration. Since the greater concentration of plasma proteins does not filter through, they exert an osmotic pull, or back pressure, which tends to cause water movement back into the glomerulus. This osmotic back pressure (P_o) normally amounts to

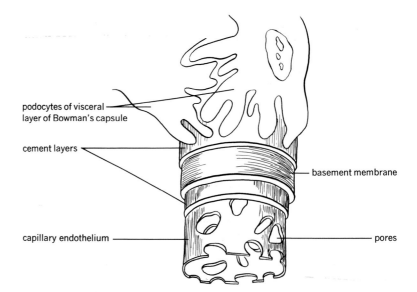

podocytes of visceral
layer of Bowman's capsule

cement layers

basement membrane

capillary endothelium

pores

FIGURE 17–7A. Diagram of the ultrastructure of the glomerular membranes.

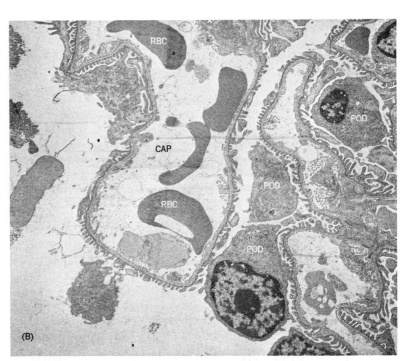

(B)

FIGURE 17–7B. Glomerular loop from a renal corpuscle of the mouse kidney, showing relationships of capillaries and visceral layer of glomerular capsule. CAP, capillary; POD, podocyte of visceral layer; RBC, red blood corpuscle. (Photograph supplied by Norton B. Gilula, Department of Physiology-Anatomy, University of California, Berkeley.) X 2,750.

about 30 mm Hg. Also opposing the filtration of material through the glomerulus is the renal interstitial pressure (P_{RIP}) or the pressure within the interstitial space around the kidney tubules, and the intra-tubular pressure (P_{it}). This latter quantity is the resistance the filtrate encounters as it attempts to push into the kidney tubules from the glomerulus. Because the tubules are already full of fluid, a resistance to the further movement of fluid is encountered. P_{RIP} and P_{it} amount to about 10 mm Hg each. By adding together the opposing forces and subtracting them from the original blood pressure, a quantity called the effective filtration pressure (P_{eff}) may be calculated. This pressure represents the net pressure, or the force which actually causes filtration through the glomerular capsule. Filtration will not proceed normally, nor will the wastes be removed effectively from the blood, unless a normal effective filtration pressure is maintained. The relationships described above may be expressed mathematically by the following formula:

$$P_{eff} = P_b - (P_o + P_{RIP} + P_{it})$$

$$P_{eff} = 75 - (30 + 10 + 10)$$

$$P_{eff} = 25 \text{ mm Hg}$$

■ The rate of glomerular filtration (GFR) can be measured experimentally and averages 120 ml per minute for both kidneys. The total plasma flow was 650 ml per minute. If only 120 ml per minute is filtered, a filtration fraction of 18.5 percent may be calculated ($^{120}/_{650} \times 100$). That is, only about one-fifth of the fluid arriving in the glomerulus is actually filtered. ■

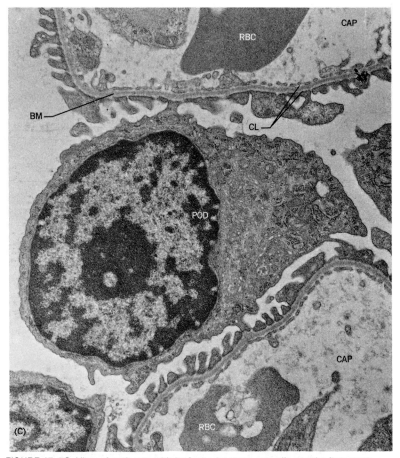

FIGURE 17–7C. View of podocyte with its foot processes extending to the basement lamina of capillaries in the renal corpuscle. CAP, capillary; POD, podocyte of visceral layer; RBC, red blood corpuscle; BM, basement membrane; CL, cement layers. (Photograph supplied by Norton B. Gilula, Department of Physiology-Anatomy, University of California, Berkeley.) X 10,650.

Tubular transport.   The filtrate contains not only the waste materials which must ultimately be eliminated from the body, but also quantities of substances which the body finds useful, such as water, hormones, vitamins, enzymes, glucose, inorganic salts, protein, and amino acids. In order that these materials not be lost from the body, a second process must occur at this time. It is called tubular transport, and involves the use of active processes to transport through the cells of the kidney tubule materials which would not otherwise be removed by passive processes. If the direction of tubular transport is from the lumen or cavity of the tubule through the cell and ultimately into the surrounding blood vessels, it is called *reabsorption*. The materials the body finds useful are reclaimed in this fashion. Some materials considered as wastes are not removed completely enough from the blood by filtration, and are actively transported from the blood through tubule cells and into the tubule lumen. Movement in this direction is termed *secretion*.

In the proximal convoluted tubule an average reabsorption of 80 to 90 percent of the physiologically important solutes is achieved. Among the materials which are actively transported through this region are glucose, amino acids and protein, phosphate, sulfate, uric acid, vitamin C, beta-hydroxybutyric acid, and calcium, sodium, and potassium ions. As these solutes are moved out by active transport, water follows osmotically, so that it too achieves an 80 to 90 percent reabsorption in this area. Chloride, being a negatively charged ion, follows positive ions (Na^+ and K^+) out of the tubule by electrostatic attraction. Because of the water movement which accompanies the solute transport, the net tonicity of the fluid at this point does not change. Its volume has been reduced, but its osmolarity and pH have not been affected.

The reabsorption of sodium ion by the proximal tubule is directly proportional to the concentration of aldosterone which is secreted by the adrenal cortex. Reabsorption of calcium ion is increased by parathyroid hormone. Glucose usually achieves a 100 percent removal in the proximal tubule because the carrier system transporting glucose has a transport maximum (TM_G) of 300 to 375 mg glucose per minute. The amount of glucose in the filtrate is normally 80 to 120 mg/100 ml. Therefore, the carrier system has no difficulty in removing all the filtered glucose. In certain diseases where the concentration of glucose in the blood is high, such as diabetes mellitus, the amount of filtered glucose may exceed the ability of the transport system to reabsorb it, and under these conditions glucose will spill over into the urine. Protein is also absorbed completely in the proximal tubule. As water passes out of the tubule, there occurs an increase in the concentration of urea in the tubule. Urea (a nitrogenous waste) does not possess a carrier system and would not normally be reabsorbed. However, its concentration rises to a level at which some will pass out of the tubule by simple diffusion. The reabsorption of urea is an example of *obligatory ("have to") reabsorption* over which the nephron has no control. The movement by carriers can be controlled and is referred to as *facultative (optional) reabsorption*.

The proximal tubule is an important area for secretion as well. A variety of organic acids (phenol red, hippuric acid, creatinine, para-aminohippuric acid, penicillin, Diodrast) and strong organic bases (choline, guanidine, histamine) are secreted into the filtrate by the proximal tubule cells.

The filtrate next enters the loop of Henle, and here, in the juxtamedullary nephron, is exposed to the countercurrent multiplier and the countercurrent exchanger. These mechanisms create the conditions necessary for achieving a concentrated or hypertonic urine.

■ *The countercurrent multiplier.* This physiological mechanism is so named because of the hairpin-like arrangement of Henle's loop and the fact that fluid flows in opposite directions in the two limbs (countercurrent). As the mechanism operates, the concentration of solute is increased (multiplied) in the descending limb and interstitium by the activity of the ascending limb. In order for the multiplying effect to be achieved, several conditions must be met. First, the ascending portion of the loop of Henle must be capable of actively transporting ions from the filtrate in its lumen into the interstitium and into the descending portion of the loop. A carrier system for sodium ion exists in these cells, and the predominant extracellular ion is sodium. Therefore one may say that it is the transport of sodium ion upon which the creation of the osmotic gradients depends. Second, the ascending portion of the loop must be impermeable to water, so that as sodium is transported out, water does not follow it osmotically. Third, the ascending portion of the loop must be capable of transporting sodium to the extent that it can achieve a 200-mOs difference in the concentration of sodium between it and the descending portion of the loop. Assuming an initial input of fluid to the top of the descending portion of the loop of 300-mOs concentration, and viewing what happens next much in the same fashion as a movie film run slowly enough that one can see the individual frames, let us describe what happens (Fig. 17–8).

Assuming the entire loop to first be filled with 300-mOs fluid (a), the ascending loop will transport sodium into the descending loop until a 200-mOs difference has been attained (b). This activity will place a 400-mOs solution in the descending loop opposite a 200-mOs solution in the ascending loop. If we now advance the whole mechanism one step (c), a new mass of fluid of 300 mOs concentration will enter the top of the descending loop and a 200-mOs solution will exit from the top of the ascending loop. In the lower part of the loop, moving a 400-mOs area around the tip and into the ascending loop will place it opposite another 400-mOs region, which descends into the space vacated by the one which moved around the tip of the loop. The 200-mOs difference has been lost at the top and tip of the loop. Sodium ion will again be transported (d) until a 200-mOs difference has been attained. In the lower part of the descending loop, a concentration of 500 mOs will be

FIGURE 17–8. The operation of the countercurrent multiplier.

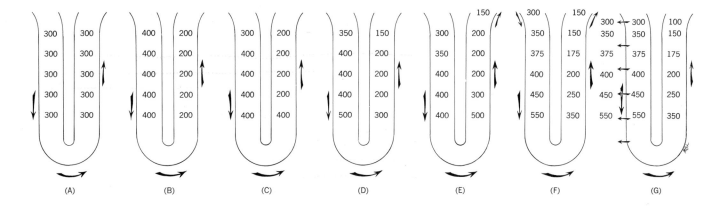

attained, with a 300-mOs area opposite it. Advancing the mechanism still another step (e) will place a 300-mOs solution opposite a 200-mOs solution in the ascending portion of the loop. Again sodium ion will be transported until the 200-mOs difference is attained (f). As this process progresses, it can be seen that there will be created a gradually increasing milliosmolar concentration as the descending loop is traversed, and in the ascending portion there will result a gradually decreasing milliosmolar concentration. Remembering that the descending portion of the loop permits free diffusion of materials, the changes created in the descending loop will result in similar changes in the interstitium. Therefore, an increasing concentration of solute (Na⁺) will be found the deeper one penetrates into the medulla. The maximum figure to which the concentration can rise, assuming that the previously mentioned conditions are met, is 1200 to 1400 mOs. The material thus delivered to the distal convoluted tubule is one which is hypotonic, containing about one-third the solute originally fed into the loop.

The countercurrent exchanger. If the sodium ion which diffused or was actively transported into the interstitium is to exert any osmotic activity, it must remain within this area of the kidney. The countercurrent exchanger operates between the loop of Henle and the vasa recta. Blood in the vasa recta is forced to pass through the medullary region which has the increasing concentration of solutes created by the multiplier. Into this vascular loop will diffuse sodium ion, and water will diffuse out, so that the changes which occur in the osmolarity of this blood follow the same pattern as the fluid in the loop of Henle. As the vascular loop climbs back out of the medulla, sodium ion diffuses out and water in. This tends to keep the sodium ion within the medullary region and maintains the concentration gradient. ■

Acidification. The distal convoluted tubule receives a hypotonic fluid from which it may reabsorb any remaining physiologically important solutes. Recall that the proximal tubule achieved only an 80 to 90 percent reabsorption of materials. The distal tubule is also an important area of secretion, primarily of hydrogen ion, ammonia, and potassium ion. The secretion of these three materials is important in the ability of the kidney to participate in the regulation of *acid-base balance*.

Acid-producing foods predominate over alkali-producing foods under normal dietary conditions. Much free hydrogen ion also results from the production of HCl by the stomach cells and subsequent reabsorption lower in the tract. The body therefore faces the problem of maintaining normal pH (7.4) in the face of forces tending primarily to lower the pH. Acids entering the extracellular fluids first react with the chief buffer in the extracellular fluid, sodium bicarbonate. Carbonic acid is formed, which is subsequently carried to the lungs and disposed of as carbon dioxide.

$$HA + NaHCO_3 \longrightarrow NaA + H_2CO_3 \xrightarrow[\text{in lungs}]{\text{carbonic anhydrase}} H_2O + CO_2 \uparrow$$

$$A = \text{unspecified anion } (SO_4^=, PO_4^\equiv, Cl^-)$$

The anion combined with sodium will eventually be excreted in the urine. It would be disadvantageous to the body to lose the anion as the sodium salt since sodium, in the form of sodium bicarbonate, is the primary buffering agent of the extracellular fluid. The kidney rids the body of the anion and reabsorbs the sodium, exchanging it for a hydrogen ion.

Ammonia, derived from the deamination of amino acids and glutamine, diffuses into the tubule lumen and there reacts with a hydrogen ion to form an ammonium ion. The ammonium ion then may react with a variety of anions and thus carry out not only an excess hydrogen ion, but an anion as well. Ammonia production normally occurs at a relatively low level unless the body is presented with large amounts of hydrogen ion. Ammonia production therefore acts as an additional means of disposing of hydrogen ion, in the form of ammonium ion. Fig. 17–9 illustrates the operation of some of these processes.

The urine leaving the distal tubule has thus been acidified, but is still hypotonic, since any exchange of materials has been made primarily on a one-for-one basis.

Production of a hypertonic urine. The collecting tubules receive the hypotonic solution from the distal tubule. These tubules run through the medulla toward the tips of the pyramids. Remember that in the medulla the countercurrent multiplier has operated to create an increasing degree of solute concentration towards the pyramid tip. The tubule therefore has around it a solution of much greater solute concentration than the fluid in the tubule. There is thus a tendency for water to leave the collecting tubule by osmosis, but this is not permitted unless the tubule becomes permeable to water.

The tubule becomes permeable to water in the presence of ADH or antidiuretic hormone. This substance is produced in the hypothalamus and is stored in and released from the posterior lobe of the pituitary gland (Fig. 17–10). It exerts a permissive effect upon the cells, causing them to allow passage of water into the interstitium. The amount of ADH secreted is determined by the osmotic pressure of the blood reaching the hypothalamus. As the fluid passes through the collecting tubules and into the papillary ducts, it loses more and more

FIGURE 17–9. Mechanisms of acidification.

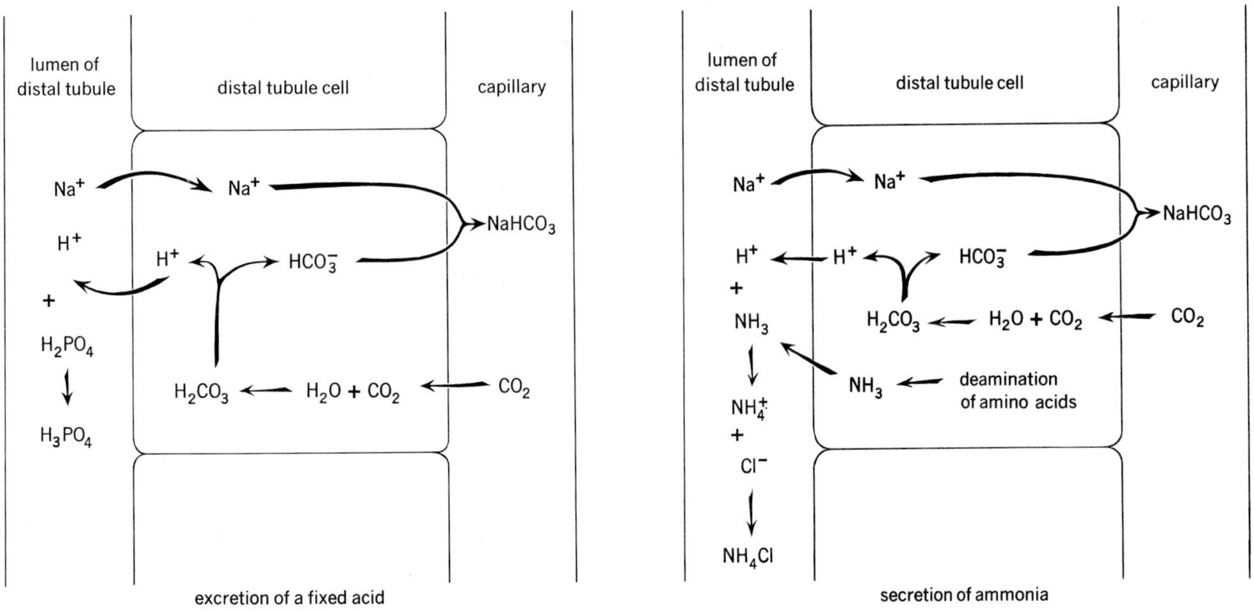

excretion of a fixed acid

secretion of ammonia

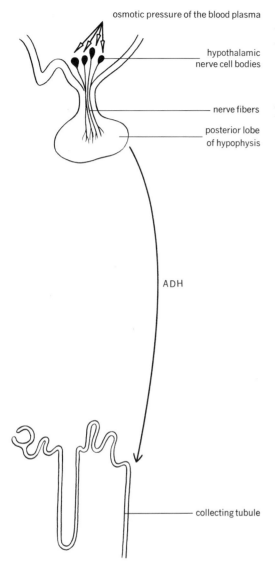

osmotic pressure of the blood plasma

hypothalamic nerve cell bodies

nerve fibers

posterior lobe of hypophysis

ADH

collecting tubule

FIGURE 17–10. The antidiuretic hormone mechanism.

water and will become hypertonic. The highest concentration to which the fluid can rise is 1200 to 1400 mOs, which is the maximum concentration that the multiplier can create. Normally, a urine of about 800 mOs is produced, indicating that the cells of the collecting tubules never become freely permeable to water. Table 17–1 summarizes the processes involved in urine formation.

Factors controlling urine volume. The volume of urine produced by the kidneys is dependent upon the balance between several of the basic physiological processes discussed in the previous Sections.

1. The amount of filtrate formed varies directly with the hydrostatic pressure of the blood. Assuming that osmotic back pressure remains constant, then the higher the blood pressure, the greater the rate of filtration. In shock or cardiac failure the blood pressure may fall below that required for adequate filtration. Uremia may result and require artificial removal of wastes. The principle of the artificial kidney was described in Chapter 2.

2. The volume of the filtrate may be increased by an increase in concentration of solutes in the filtrate, or a decrease in plasma protein content. In diabetes mellitus, for example, the large amount of glucose present in the filtrate draws water osmotically into the tubule, or prevents its loss as sodium is transported out.

3. The volume of urine formed varies inversely with the amount of antidiuretic hormone secreted, and in turn, ADH secretion varies directly with the solute concentration of the blood reaching the hypothalamus. Excess fluid intake poses a threat of dilution of the internal environment and is relieved by decrease in water reabsorption. Diabetes insipidus results from loss or greatly diminished production of ADH.

4. The volume of fluid excreted by the urinary system varies inversely with water loss by other systems, especially the digestive system, and water loss through perspiration. The body must maintain a reasonable balance between fluid intake and loss, and if vomiting, diarrhea, or excessive sweating increase fluid loss, urine volume will decrease. This again is effected mainly by variation in ADH secretion.

5. Alteration in the efficiency of operation of the countercurrent multiplier, particularly in the transport of sodium, will change the urine volume. Certain drugs, diuretics, increase urine volume by decreasing the active removal of sodium ion from the filtrate. This allows these ions to remain as osmotically active particles in the filtrate. Diuretics also cause the countercurrent multiplier to equilibrate at a lower maximum concentration, so that the amount of water passing through the walls of the collecting tubules will be less.

The urine

Normal urine is an amber or yellow transparent fluid having a pH between 5 and 7, a characteristic odor, and a specific gravity varying between 1.015 and 1.025.

The color is due to the presence of urobilinogen, a pigment derived from the destruction of hemoglobin by cells of the reticuloendothelial system.

TABLE 17-1

Summary of the Processes Occurring in Urine Formation

Process	Where occurring	Force responsible	Result
Filtration	Renal corpuscle	Blood pressure, opposed by osmotic, interstitial, and intratubular pressures	Formation of fluid having no formed elements and low protein concentration
Tubular transport reabsorption	Proximal tubule Distal tubule Loop of Henle	Active transport	Return to blood stream of physiologically important solutes
Secretion	Proximal tubule Distal tubule	Active transport	Excretion of materials Acidification of urine
Acidification (acid and base regulation)	Distal tubule	Active transport and exchange of alkali for acid	Excretion of excess H^+ Conservation of base (Na^+ and HCO_3^-)
Countercurrent multiplier and exchanger	Loop of Henle and vasa recta	(Multiplier) active transport (Exchanger) diffusion	Creates conditions for hypertonic urine formation
ADH mechanism	Collecting tubule	Osmosis of water under permissive action of ADH	Formation of hypertonic urine

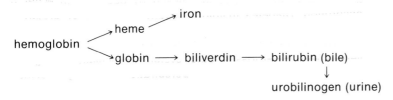

The acidic reaction is due primarily to secretion of H^+ into the fluid of the distal tubule. The odor of freshly voided urine is due chiefly to the presence of organic acids, while the odor of urine which has stood is ammoniacal due to the decomposition of urea with release of ammonia. The specific gravity is dependent upon the total solute concentration per unit volume of fluid, and accounts for the variability of this characteristic.

The composition of normal urine varies little in terms of materials present. Tables 17-2 and 17-3 indicate some comparisons between plasma, filtrate, and urine, and some normal constituents of urine, their origin, and amounts excreted per day.

Glucose normally does not appear in the urine except after a meal rich in carbohydrates because it is usually completely reabsorbed.

Protein is filtered only to a limited extent, and is also normally completely absorbed.

The ketone bodies, acetone and beta-hydroxybutyric acid, are present in the urine only when large amounts of fatty acids are being catabolized, as in diabetes mellitus. Only when these materials are present in the urine above trace levels can their presence be considered abnormal.

TABLE 17-2

Comparisons of the Properties of Plasma, Filtrate, and Urine*

Property or substance	Plasma	Filtrate	Urine	Degree of change between filtrate and urine
Osmolarity (mOs)	300	300	600–1200	2–4 fold
pH	7.4 ± 0.2	7.4 ± 0.2	5–7	300–30,000 fold
Specific gravity	1.008–1.010	1.008–1.010	1.015–1.025	—
Sodium	0.3	0.3	0.35	—
Potassium	0.02	0.02	0.15	7 fold
Chloride	0.37	0.37	0.60	2 fold
Ammonia	0.001	0.001	0.04	40 fold
Urea	0.03	0.03	2.0	60 fold
Sulfate	0.002	0.002	0.18	90 fold
Creatinine	0.001	0.001	0.075	75 fold
Glucose	0.01	0.01	0	—
Protein	7–9 gm. %	10–20	0	—

*Values, unless otherwise indicated are mg. %

TABLE 17-3

Some Normal Constituents, Origins, and Amounts Excreted per Day in Normal Urine

Constituent	Origin	Amount per day
Water	Diet and metabolism	1200–1500 ml
Urea	Ornithine cycle	30 g
Uric acid (purine)	Catabolism of nucleic acids	0.7 g
Hippuric acid	Liver detoxification of benzoic acid	Trace
Creatinine	Destruction of intracellular creatine phosphate of muscle	1–2 g
Ammonia	Deamination of amino acids	0.45 g
Chloride (as NaCl)	Diet	12.5 g
Phosphate	Diet and metabolism of phosphate containing compounds	3 g
Sulfate	Diet, metabolism of sulfate containing compound, formation of H_2SO_4 in kidney tubules	2.5 g
Calcium	Diet	200 mg

Casts may be found in the urine and are evidence of renal lesions. They are named according to composition, and include epithelial, fatty, pus, blood, or hyaline (clear) casts. The casts maintain the shape of the kidney tubules where they are formed.

Calculi (stones) are precipitated masses of inorganic material. They may grow to large size and bizarre shapes, and may require surgery for their removal.

Kidney disorders

Glomerulonephritis, nephrotic syndrome, and pyelonephritis are probably the most frequently encountered kidney disorders.

Glomerulonephritis is almost always associated with a previous infection by a beta-hemolytic streptococcus bacterium elsewhere in the body (usually the pharynx) and is generally regarded as an "autoimmune disease." One of the components of the basement membrane of the glomerular capsule, sialic acid, is thought to be hydrolyzed by the bacterial toxin. The body then recognizes the hydrolyzed acid as a "new" or "foreign" substance and produces antibodies to neutralize it. In the process, the basement membrane is injured and allows red cells and more protein to leak into the filtrate. Edema is also a common occurrence because the osmotically active protein molecules are diminished in the blood, resulting in a failure to return water to the circulation.

The *nephrotic syndrome* also involves an increased glomerular permeability, but appears to involve the tubules to a greater degree than the glomeruli. The chief difference from glomerulonephritis lies in the fact that there is a massive loss of protein from the blood, with cast formation in the tubules and a near shutdown of tubular function.

Pyelonephritis is an infection of the kidney which sometimes begins in the lower tract and spreads upward to involve the kidney. Females are more susceptible to this disease as a consequence of the short urethra and the greater possibility of bladder infection occurring. It is the most common disease affecting the urinary system.

Other functions of the kidney

At the point shortly before the afferent arteriole enters the glomerulus, there is an interesting group of modified smooth muscle cells called the *juxtaglomerular apparatus.* This particular region of the arteriole has been shown to be sensitive to changes in blood flow. It has been hypothesized that if the kidney becomes ischemic, this area secretes an enzyme-like material called *renin.* This particular substance works on a substrate in the plasma called *hypertensinogen (angiotensinogen),* and converts it into a weakly active material, *hypertensin I.* Hypertensin I is in turn converted to *hypertensin II* by a plasma enzyme. Hypertensin II is an active vasoconstrictor, bringing about a narrowing of arterioles over the entire body. This particular effect raises the blood pressure and insures a continuance of high blood pressure to the kidney, which is a necessary requisite for filtration to occur. Renal hypertension, a chronic elevation of the blood pressure, may arise if the kidney suffers continual ischemia and renin production is not diminished. Hypertensin II also affects the outer zone of the adrenal cortex which secretes aldosterone. The hormone is instrumental in governing reabsorption of sodium by the kidney tubules. Therefore, we may see that the kidney governs its own activity to a small degree.

A great degree of autoregulation is demonstrated by the kidney in controlling blood flow to the glomerulus. The afferent arteriole has been shown to undergo constriction or dilation in response to a wide range (90 to 220 mm Hg) of renal artery pressure so as to maintain renal blood flow fairly constant. The response occurs in the absence of all nerves to the kidney, and in a kidney removed from the body and artificially perfused. It is therefore an inherent mechanism within the kidney itself.

The ureters

Gross anatomy and relationships. The ureters extend from the hilus of the kidney to the urinary bladder, a distance of 28 to 35 cm (about 11 to 14 inches). They are retroperitoneal in placement and have an increasing diameter as they course toward the bladder.

Microscopic anatomy. Three coats of tissue form the wall of the ureter. A *mucosa,* composed of transitional epithelium and lamina propria, forms the inner layer. The central *muscularis* is composed of inner longitudinal and outer circular layers of smooth muscle throughout most of the length of the ureter. On the lower one-third of the organ, a third layer of muscle (outer longitudinal) is added. The outer *fibrous coat (adventitia)* blends without demarcation into the surrounding subserous fascia.

Innervation. The ureters in their upper part receive sympathetic fibers from the renal plexus; in the middle part, they receive fibers from the ovarian or spermatic plexus; and near the bladder, they receive fibers from the hypogastric nerves (Fig. 17–11). These fibers exert primarily a motor effect. They cause the ureters to exhibit rhythmical peristaltic contractions traveling at a speed of 20 to 25 mm per second and at a frequency of 1 to 5 per minute. The urine, therefore, enters the bladder, not in a continuous stream, but in separate squirts synchronous with the arrival of the peristaltic wave.

The urinary bladder

Gross anatomy and relationships. The urinary bladder serves as a reservoir for the urine until it is voided. The organ is located posterior to the symphysis pubis and is separated from the symphysis by a prevesicular space. The space, filled with loose connective tissue, allows for expansion of the filling bladder. Internally, three openings may be found in the bladder wall: the two ureters, and the urethra. An imaginary line drawn to connect these three openings outlines the trigone.

Microscopic anatomy. While similar to the ureter in structure, the bladder has more cell layers in the transitional epithelial lining, a submucous layer of loose tissue between mucosa and muscularis, and three heavy layers of smooth muscle in the muscularis, disposed longitudinally, circularly, and longitudinally. Around the urethral opening, a dense mass of circularly oriented smooth muscle forms the internal sphincter of the bladder. A serous layer is formed by the peritoneum over the superior surface of the organ.

Innervation. The efferent nerves to the bladder and urethra (Fig. 17–12) are from both sympathetic and parasympathetic divisions. The sympathetic fibers furnish inhibitory fibers to the muscle of the bladder; they furnish motor fibers to the trigone and internal sphincter, and the muscle of the upper part of the urethra. These fibers arise in the lumbar spinal segments and pass to the bladder, via the inferior hypogastric plexus. The parasympathetic nerves supply motor fibers to the detrusor muscle (the muscle of the bladder) and inhibitory

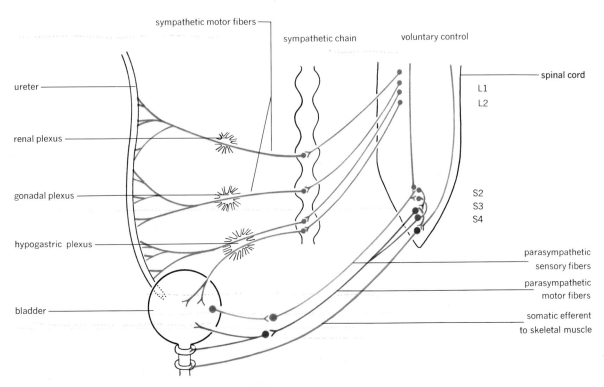

FIGURE 17–11. Innervation of the ureters, bladder, and urethra.

FIGURE 17–12. Nerves of the bladder illustrating genesis of bladder dysfunction.

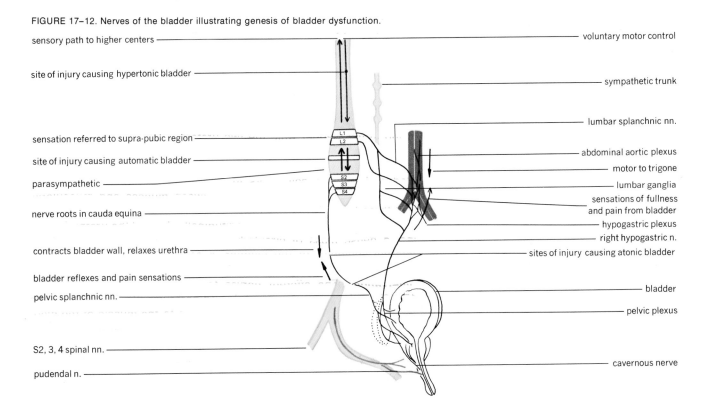

fibers to the internal sphincter. The desire to urinate occurs when a volume of 200 to 300 ml of urine has accumulated in the bladder. Stretch on the muscle of the bladder, brought about by filling, evokes an afferent impulse in the pelvic nerves. This impulse ascends through the spinal cord to a center in the hindbrain. An efferent discharge of motor impulses to the muscle of the bladder is accomplished through descending pathways in the cord and through the pelvic motor nerves. The same efferent or motor fibers bring about a simultaneous relaxation of the internal sphincter so that urine may be emptied from the bladder into the urethra (micturition). Urination may occur by reflex action not involving the hindbrain center. Filling evokes a reflex contraction of the detrusor and relaxation of the sphincter, which is served by lower cord segments only. The reflex is seen in infants, where the voluntary control over sphincters has not yet been achieved. Nerve injuries produce three types of bladder dysfunction.

1. *Atonic bladder.* Interruption of sensory supply results in loss of bladder tone, and the organ may become extremely distended with no development of an urge to urinate.
2. *Hypertonic bladder.* Interruption of the voluntary pathways results in excessive tone, and very small distensions create an uncontrolled desire to urinate.
3. *Automatic bladder.* Complete section of the cord above the first sacral nerve exit (S1) produces automatic emptying in response to filling, by the above-described cord reflex.

The urethra

The female urethra is about 4 cm (1¾ inches) in length, is completely separate from the reproductive system, and bears no regional differences. A mucosa is present, consisting of transitional epithelium near the bladder and stratified squamous elsewhere, and underlain by a lamina propria. A muscularis is present, consisting of circularly arranged fibers of smooth muscle. The female urethra opens just anterior to the vaginal orifice.

The male urethra is about 20 cm (8 inches) in length, serves as a common tube for the terminal portions of both reproductive and urinary systems, and does show regional variations. The *prostatic urethra* is the first 3 cm of the organ; it is surrounded by the prostate gland and lined with transitional epithelium. The *membranous urethra* is 1 to 2 cm long; it penetrates the pelvic floor, is very thin, and has a pseudostratified epithelial lining. The *cavernous urethra* is about 15 cm in length and lies within the penis.

Disorders associated with the lower urinary tract include cystitis, urethritis, and obstruction of the urethra.

Cystitis, or inflammation of the bladder, is usually secondary to infection of the prostate, kidney, or urethra.

Infection of the male urethra is most commonly associated with gonorrheal infections; in the female, almost any infection of the perineum may invade the urethra.

Blockage of the urethra is most commonly associated with calculi (stones) in the bladder. The stone may be voided in the urine, if small enough, or may be crushed by use of a cystoscope inserted through the urethra.

QUESTIONS

1. Describe the location of the kidneys and their supporting structures.
2. What structures enter or leave the hilus of the kidney?
3. Describe the distribution of blood vessels within the kidney.
4. What is a nephron and what are its functions?
5. What are the two types of nephrons and how are they different?
6. What are the processes the nephron uses in forming urine? Describe the main result of the operation of each process.
7. What measurements of kidney function can be made? Give average values for each.
8. List five factors controlling urine volume, and explain how each works.
9. Compare plasma, filtrate, and urine for the following properties: Specific gravity, water content, glucose content, protein content, pH. What accounts for any differences?
10. Compare the structure of ureters and bladder.
11. Describe the mechanism involved in emptying of the bladder.
12. Describe differences between the male and female urethra.
13. Describe two disorders of the kidney, their cause, and their effect on normal nephron functioning.

CHAPTER 18

The Reproductive Systems

The omnipresent process of sex, as it is woven into the whole texture of our man's or woman's body, is the pattern of all the process of our life. HAVELOCK ELLIS

THE REPRODUCTIVE SYSTEMS

Introduction

The term reproduction implies to most individuals the act of mating between two members of one species and of opposite sex, with consequent production of offspring. Yet, in its broadest sense, the term should be understood to include replacement of damaged or dead cells in many body areas. In either case, there ought not to be uncontrolled production of new units. The results of uncontrolled production of individual cells is only too evident in a variety of cancerous conditions. Similarly, uncontrolled reproduction of members of a species can lead to crowding and stress, shortage of food, unhealthy competition for living space, and in human terms, war and pestilence.

THE MALE REPRODUCTIVE SYSTEM

Organs

The organs of the male reproductive system (Fig. 18–1) may be conveniently grouped under three headings:

1. *The essential organs.* The *testes,* because of their production of male sex cells (spermatozoa or sperm), constitute the essential

479

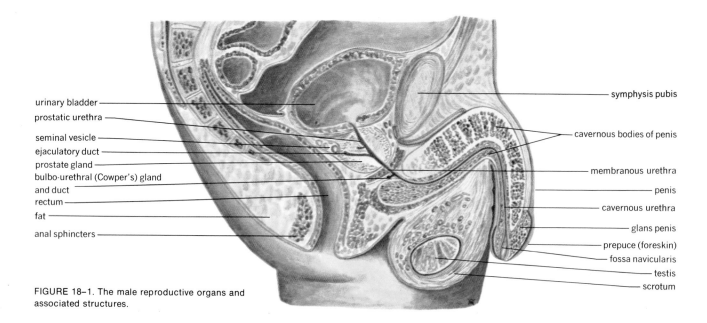

urinary bladder
prostatic urethra
seminal vesicle
ejaculatory duct
prostate gland
bulbo-urethral (Cowper's) gland
and duct
rectum
fat
anal sphincters

symphysis pubis
cavernous bodies of penis
membranous urethra
penis
cavernous urethra
glans penis
prepuce (foreskin)
fossa navicularis
testis
scrotum

FIGURE 18–1. The male reproductive organs and associated structures.

organs of the system. The testes also produce the male sex hormone (testosterone), which is responsible for development of certain of the remaining organs of the system and of the male secondary sex characteristics.

2. *The excretory ducts.* The straight tubules, rete tubules, efferent ductules, epididymis, ductus (vas) deferens, ejaculatory ducts, and urethra store sperm or convey sperm to the exterior.

3. *The accessory glands.* The *seminal vesicles, prostate,* and *bulbourethral (Cowper's) glands* contribute secretions, which together with the sperm, comprise the semen.

The scrotum

After about 8 months of intrauterine life, the testes are located within the scrotum (Fig. 18–1). The scrotum is an evagination of skin and superficial fascia from the lower anterior abdominal wall. It is divided into lateral portions by the superficial raphe. Within the fascia is found the dartos tunic, several layers of smooth muscle. Exposure to cold causes contraction of the tunic and the scrotum becomes wrinkled. Warmth reverses the process. Internally, the scrotum is divided into right and left compartments by the dartos and fascia. Each compartment houses a testis, epididymis, and other associated structures, including ducts, nerves, and blood vessels. Each compartment communicates with the abdominal cavity through the inguinal canal.

The testis

Basic structure. Each testis is an ovoid body, measuring 4 to 5 cm (1¾ to 2 inches) in length by about 2.5 cm (1 inch) in diameter. Each weighs 5 to 7 grams. The outer covering is formed by the *tunica vaginalis,* a portion of the abdominal lining *(mesothelium)* which was pushed ahead of the testis during its descent into the scrotum (Fig. 18–2). Forming a capsule around the testis is a layer of collagenous

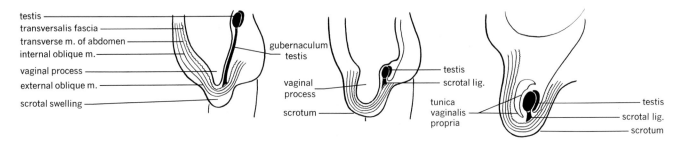

FIGURE 18–2. The descent of the testis.

tissue, the *tunica albuginea.* The albuginea is thickened on the medial side of the testis to form the *mediastinum testis.* A third and vascular coat, the *tunica erythroides,* lies deep to the albuginea. From the albuginea arise connective tissue partitions or *septae,* which penetrate radially into the testis and divide it into about 250 separate *lobules.* Each lobule contains a much-coiled single tubule, the *seminiferous tubule.* The seminiferous tubule connects with a *straight tubule* which runs to the mediastinum and joins the *rete tubules* located therein. These features are depicted in Fig. 18–3.

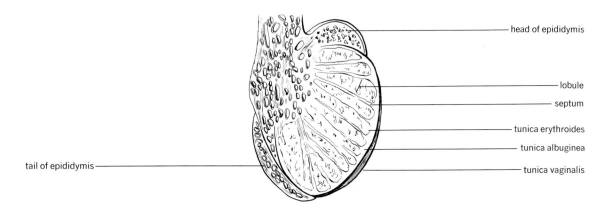

FIGURE 18–3. The testis, internal structure.

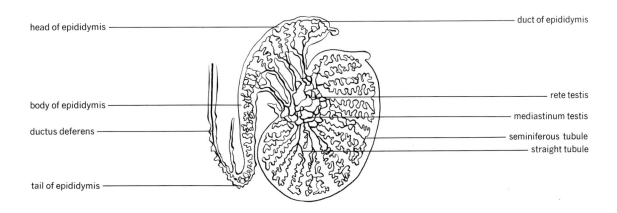

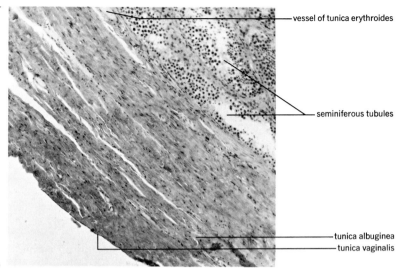

vessel of tunica erythroides

seminiferous tubules

tunica albuginea
tunica vaginalis

FIGURE 18–4A. Testis, general view.

spermatogonium

spermatid

primary
spermatocyte

secondary
spermatocyte

sperm head

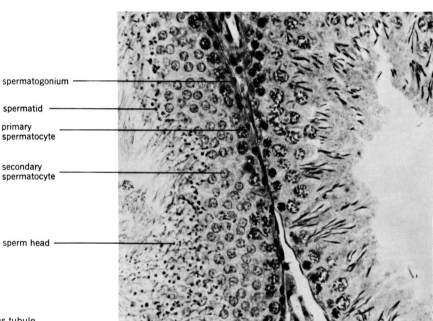

FIGURE 18–4B. Seminiferous tubule.

Microscopic anatomy. A section of the testis shows it to be composed of many cut seminiferous tubules with intertubular connective tissue, blood vessels, and cells (Fig. 18–4A and B). The seminiferous tubule is lined with the *germinal epithelium.* The epithelium contains two types of cells:

1. *Sertoli cells.* Recognizable by their oval nuclei with prominent nucleoli, the Sertoli cells are actually columnar cells which produce secretions designed to nourish the spermatozoa during their final stages of development.
2. *Spermatogenic cells,* arranged in several rows from the periphery of the tubule inwards, represent spermatozoa in various stages of

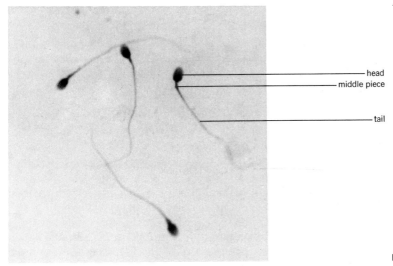

- head
- middle piece
- tail

FIGURE 18–5A. Spermatozoa.

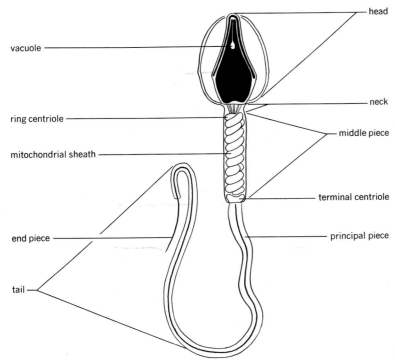

vacuole

ring centriole

mitochondrial sheath

end piece

tail

head

neck

middle piece

terminal centriole

principal piece

FIGURE 18–5B. A spermatozoon.

development. It may be noted that spermatogenesis occurs by meiosis (Fig. 2–16), so that the sperm cells are haploid. The most immature cells are located against the basement membrane and are called *spermatogonia*. Next is found a layer or two of large cells with large round nuclei, the *primary spermatocytes*. Smaller cells, having the same general appearance as the primary spermatocytes, are found still closer towards the tubule lumen and are *secondary spermatocytes*. The *spermatids* are more internal yet and show small dark nuclei. Spermatids metamorphose into *spermatozoa* (Fig. 18–5A and B). Sections of testes taken before sexual maturity show only cords of epithelial cells. The stages described above appear at puberty.

Between the coils of the seminiferous tubules are groups of *interstitial cells.* These cells produce the male sex hormone *testosterone,* which controls the appearance and development of the male secondary sex characteristics and the growth of some of the accessory organs of the male reproductive system.

Ducts

The seminiferous tubules connect with the straight tubules. Straight tubules are lined with a low, simple cuboidal epithelium. The junction between a seminiferous and straight tubule is marked by the presence of only Sertoli cells, which project into the straight tubule (Fig. 18–6). Straight tubules pass to the mediastinum testis where they join the rete tubules (Fig. 18–7). Rete tubules are irregular in diameter and are lined with simple squamous or simple cuboidal epithelium. All remaining portions of the duct system lie outside the limits of the testis.

Efferent ductules number 12 to 16, and convey spermatozoa to the epididymis for storage. They are lined with alternating groups of tall, ciliated cells and short, nonciliated cells. The lumen of the tubule thus *appears* folded (Fig. 18–8). A thin coat of smooth muscle appears around the tubule. The *epididymis* is a single, coiled tube about 7 meters (20 to 21 feet) in length, which is held to the testis by connective tissue. Its superior end, or head, receives the efferent ductules; the centrally placed body continues into the elongated, inferiorly placed tail. The lining of the epididymis is a pseudostratified epithelium, consisting of tall columnar elements bearing microvilli, the stereocilia, and basal cells which do not reach the lumen (Fig. 18–8). It is thought that the stereocilia act to convey nutrient materials to the sperm which are stored in the epididymis. Again, a thin layer of smooth muscle surrounds the epithelium. The *ductus (vas) deferens* (Fig. 18–9) is continuous with the tail of the epididymis. It passes from the scrotum through the inguinal canal, arches behind the urinary bladder, and joins the *ejaculatory duct* to empty into the male *urethra.* The deferens is lined by a folded mucous membrane consisting of pseudostratified epithelium and an underlying lamina propria. A heavy coat of smooth muscle is next, consisting of inner longitudinal, middle circular, and outer longitudinal layers. An adventitia of connective tissue surrounds the muscle. The heavy muscular coat propels the sperm through the deferens with considerable force.

The adventitia of the deferens not only serves as the outer coat of that duct, but also serves to bind the deferens and its accompanying arteries, veins, and nerves into the structure known as the *spermatic cord.* A small band of skeletal muscle, the *cremaster* muscle, is also found in the cord. It elevates the testis during sexual stimulation and upon exposure to cold.

Accessory glands (Figs. 18–10 and 18–11)

The *seminal vesicles* are paired, tortuous sacs lying on the posterior aspect of the urinary bladder. The vesicles do not, as was once thought, store sperm. Occasionally, a few sperm may be found within the lumina of the vesicles, but they have simply been "eddied" into the vesicles. The duct of the vesicle and deferens join to form the ejaculatory duct, emptying into the urethra. The *prostate gland* surrounds the upper

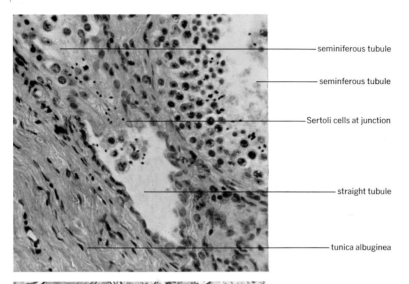

seminiferous tubule

seminiferous tubule

Sertoli cells at junction

straight tubule

tunica albuginea

FIGURE 18–6. Junction of seminiferous and straight tubules.

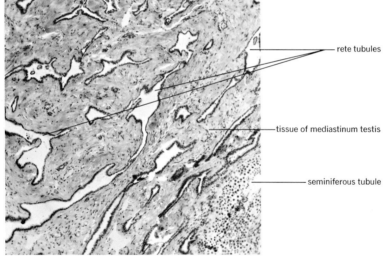

rete tubules

tissue of mediastinum testis

seminiferous tubule

FIGURE 18–7. Rete tubules.

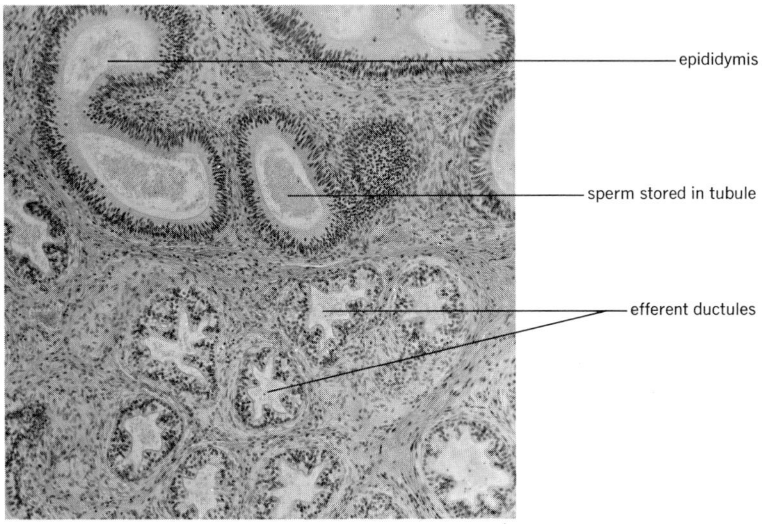

epididymis

sperm stored in tubule

efferent ductules

FIGURE 18–8. Efferent ductules and epididymis.

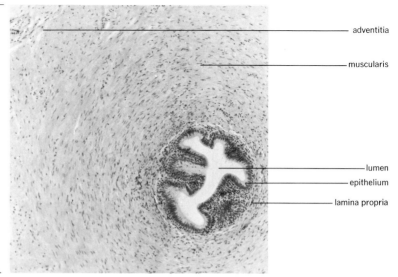

adventitia

muscularis

lumen

epithelium

lamina propria

FIGURE 18–9. Vas deferens.

urethra. It consists of 30 to 50 separate secretory units, emptying by 16 to 32 separate ducts into the urethra. The gland has much smooth muscle between the secretory portions, which is an aid to the emptying of the organ. The secretion of the prostate is alkaline, slimy, and has a characteristic odor. It is thought to activate the heretofore immobile sperm.

The *bulbourethral glands* empty into the urethra below the prostate. Their secretion is mucoid in nature, and its alkaline reaction aids in neutralization of any acid urine in the urethra.

Semen

The spermatozoa, normally numbering 300 to 400 million per ejaculate, are combined with the secretions of the seminal vesicles and prostate to form the semen. The average volume of the semen is about 3 ml, the

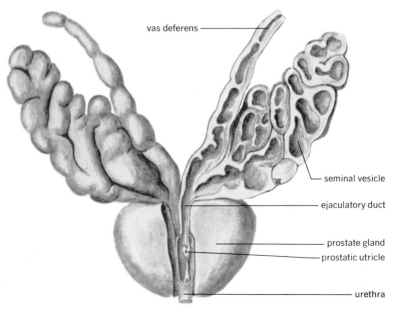

vas deferens

seminal vesicle

ejaculatory duct

prostate gland

prostatic utricle

urethra

FIGURE 18–10. The prostate, seminal vesicles, and associated structures.

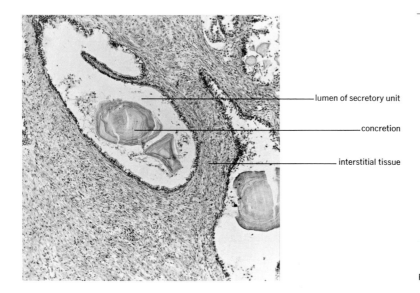

lumen of secretory unit

concretion

interstitial tissue

FIGURE 18–11. Prostate gland.

pH 7.35 to 7.50. Fructose and ascorbic acid are the major components of the seminal vesicle contribution, which comprises 60 percent of the total volume. The prostate contributes 20 percent of the total semen volume. Prostatic secretion is rich in cholesterol, phospholipids, buffering salts, and prostaglandins. The latter are fatty acids having vasopressor activity and possible central nervous system regulator roles. Semen as a whole gives bulk to suspend the sperm, aids in neutralization of the acidity in the female tract, and activates the sperm. Sperm placed in the female reproductive tract have a maximum life expectancy of 3 days.

Male urethra and penis

The male urethra is about 20 cm (8 inches) in length and is subdivided into three portions:

1. *Prostatic urethra (pars prostatica).* This portion is approximately the first 3 cm below the bladder, and is surrounded by the prostate gland. It is lined with transitional epithelium and receives the ejaculatory ducts and the prostatic ducts. A bulge on the posterior wall marks the position of the prostatic utricle, the male homolog of the uterus.
2. *Membranous urethra (pars membranacea).* This portion is 1 to 2 cm in length and passes through the pelvic floor. It is lined with a pseudostratified epithelium and receives the ducts of the bulbourethral glands.
3. *Cavernous urethra (pars cavernosa).* The remaining portion of the urethra lies within the penis, and is about 15 cm in length. It is lined with pseudostratified epithelium. Fissure-like evaginations from the lumen form the lacunae of Morgagni, into which empty ducts from mucus-secreting glands of Littré.

The penis (Fig. 18–12A, B) serves as a copulatory organ to introduce sperm into the vagina of the female. The basic structure of the organ is formed by three cylindrical masses of tissue known as the *cavernous bodies.* The bodies are supported by a *tunica albuginea,* and are bound together by connective tissue and covered by thin skin. The skin at the

glans penis

cremasteric muscle and fascia

ext. spermatic fascia

int. spermatic fascia

ductus deferens and a.

cremasteric m. and fascia

scrotal septum

pampiniform plexus

epididymis

testis

tunica vaginalis

ext. spermatic fascia

dartos muscle

scrotal skin

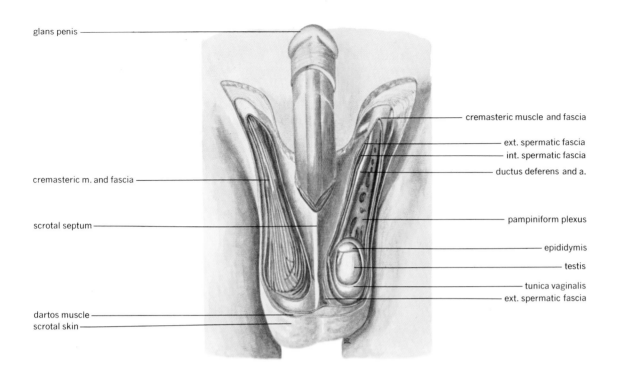

thin skin

connective tissue

corpora cavernosa penis

tunica albuginea

urethra

corpus spongiosum penis

FIGURE 18–12A. The penis and associated structures.

glans penis

corpus cavernosum penis

corpus spongiosum penis

bulb of penis

levator ani m.

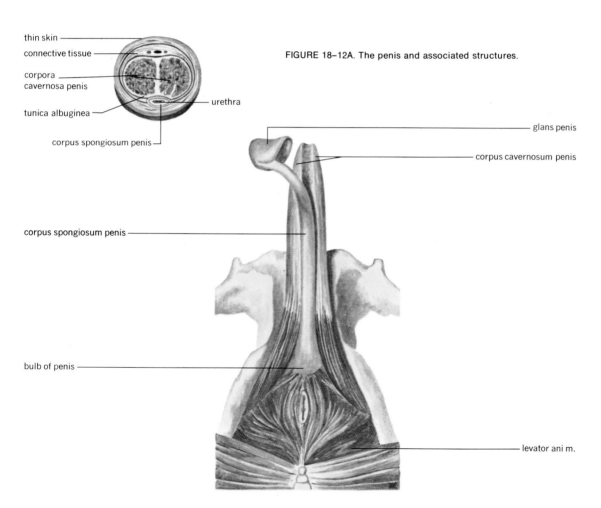

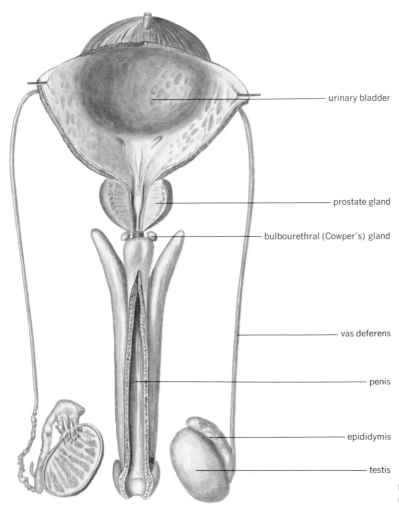

urinary bladder

prostate gland

bulbourethral (Cowper's) gland

vas deferens

penis

epididymis

testis

FIGURE 18–12B. The major organs of the male reproductive system.

neck of the penis folds back on itself to cover the head or *glans* of the penis and forms the *prepuce* or foreskin. Circumcision removes the prepuce.

Two of the cavernous bodies are dorsally placed. These are the *corpora cavernosa penis*. The single ventrally placed member is the *corpus spongiosum penis (corpus cavernosum urethrae).* It transmits the cavernous urethra. The distal end of the spongiosum is expanded into the glans. The blood vessels to the organ are muscular arteries which empty into the sinuses of the corpora. Dilation of these vessels allows a greater inflow of blood into the bodies. They swell, the veins draining the corpora are compressed, and the penis becomes turgid or erect, and an effective copulatory organ.

Endocrine relationships

Full discussion of endocrine control of the testis and the effects of testosterone will be given in Chapter 22. However, it may be stated at this time, that the follicle-stimulating hormone (FSH) of the anterior pituitary is the controlling factor of spermatogenesis. The interstitial cell stimulating hormone (ICSH), also a product of the anterior pituitary, controls production of testosterone. Testosterone in turn assures full development of the external aspects of maleness, such as hair growth, voice changes, and muscular development, and assures the complete growth of the ducts and accessory glands of the male reproductive system.

Anomalies

Failure of testicular descent is known as *cryptorchidism.* This condition results in sterility because the spermatogenic cells are destroyed by the higher body temperatures prevailing in the abdominal cavity. Interstitial cell function is not disturbed. *Hypospadias* and *epispadias* refer to urethral openings on the ventral and dorsal surfaces of the penis. Other anomalies will be considered later.

FEMALE REPRODUCTIVE SYSTEM

Organs

The organs of the female reproductive system (Fig. 18–13A, B) may be grouped under two headings:

1. *Internal organs,* including:
   (a) *Ovaries.* Like the testes, the paired ovaries are responsible for production of female sex cells or ova, and for the production of two female hormones.

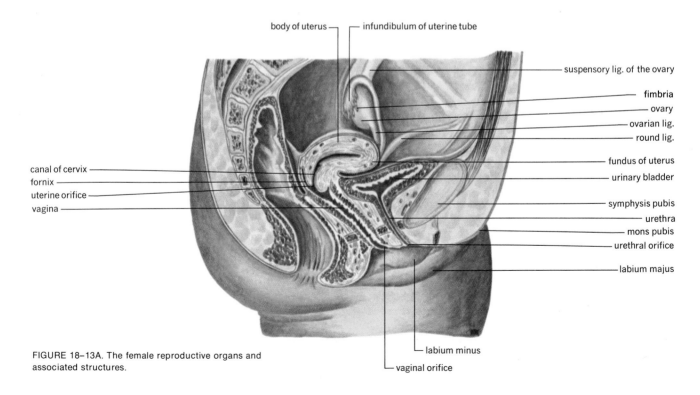

FIGURE 18–13A. The female reproductive organs and associated structures.

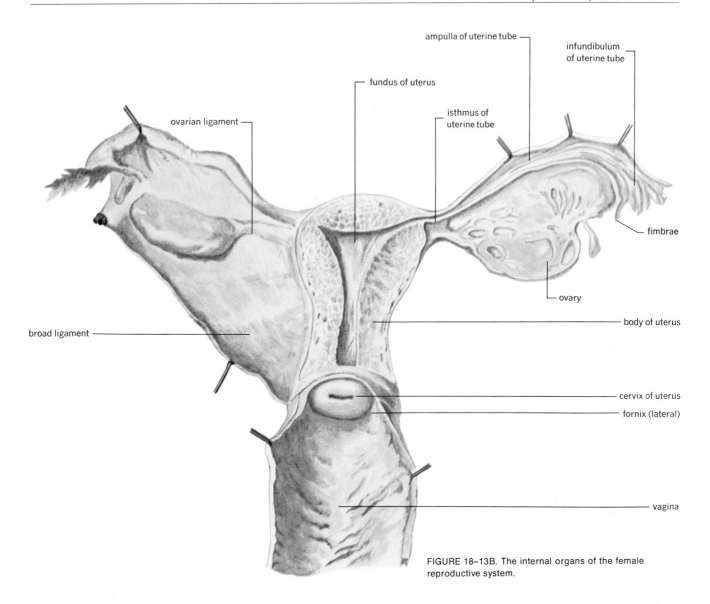

ampulla of uterine tube

infundibulum
of uterine tube

fundus of uterus

isthmus of
uterine tube

ovarian ligament

fimbrae

ovary

broad ligament

body of uterus

cervix of uterus

fornix (lateral)

vagina

FIGURE 18–13B. The internal organs of the female
reproductive system.

    (b) *Uterine tubes.* These paired structures convey ova to the
        uterus.
    (c) *Uterus.* The single organ wherein the development of new
        individuals begins.
    (d) *Vagina.* The externally opening tubular organ which serves
        as the receptacle for sperm and as the birth canal.
  2. *External organs,* including:
    (a) *Clitoris.* A single organ of sensory nature.
    (b) *Labia majora* and *minora.* Folds of skin and mucous mem-
        brane surrounding the vaginal and urethral openings.
    (c) *Mammary glands.* Actually modified skin glands, these organs
        are so closely related functionally to the other organs, they
        are usually discussed as parts of the reproductive system.

Ovaries

The ovaries are paired ovoid bodies measuring about 3 cm long, 2 cm wide, and 1 cm deep. They are pelvic in location, lying lateral to the uterus. A mesentery, the mesovarium, suspends the ovaries on the posterior aspect of the uterine broad ligament. Each ovary has a *hilus* marking the entrance of the blood vessels and nerves supplying the organ. The *ovarian ligament* attaches the ovary to the uterus, while the *suspensory ligament* attaches it to the pelvic wall.

Microscopically, the ovary may be seen to be composed of (1) a surface epithelium, the *germinal epithelium,* (2) a capsule of collagenous connective tissue, the *tunica albuginea,* (3) a *stroma* of connective tissue internal to the capsule, and (4) *ovarian follicles,* representing ova and surrounding tissues in various stages of development. The germinal epithelium is a layer of simple cuboidal or high squamous cells on the ovarian surface. During fetal life, it may be the source of some ova, but is functionless in this respect after birth.

The tunica albuginea confers some strength and support to the entire organ. The stroma may be subdivided into an outer, more dense layer, the cortex, and an inner, looser medulla, or zona vasculosa. The latter term reflects the vascular nature of the inner region.

The follicles (Fig. 18–14A and B) are estimated to number about 400,000 in the ovaries prior to birth. They remain in a quiescent condition until puberty, when hormonal changes initiate follicle development and menstruation. A total of about 400 ova will develop and be discharged over the reproductive life of the female. The remainder will undergo a degenerative process known as atresia.

Primary follicles are the most numerous of the follicles and are also the smallest. These follicles are 40 to 50 microns in diameter and may be found in the external portions of the cortex. They consist of an ovum surrounded by a single layer of stromal cells. Further development of the follicle results in the formation of a *secondary follicle.* Secondary follicles are larger, about 80 microns in diameter, and have several layers of what may now be designated as follicular cells around the ovum. Next, several small cavities develop with the layers of follicular cells. These cavities enlarge, fuse and eventually form a single follicular cavity. The ovum is forced to the periphery of the follicular cavity. The entire follicle increases to a maximum size of 10 to 13 mm, and is known as a *vesicular* or *Graafian follicle.* It possesses many parts, labeled in Fig. 18–15.

The developing follicle is the source of *estrogen,* a female sex hormone responsible for the development of female secondary sex characteristics and for the development of several of the organs of the female system. Ovulation, release of the ovum, occurs from the vesicular follicle. The follicle collapses and a small amount of bleeding may occur into the follicle, forming a *corpus hemorrhagicum.* The tissues remaining in the ovary are next caused to form a *corpus luteum* (Fig. 18–16), the source of the second hormone, *progestin.* The latter hormone maintains and advances the changes initiated by estrogen. The fate of the corpus luteum will be the same regardless of whether or not the ovum is fertilized. With no fertilization, the luteum undergoes degeneration in about 3 weeks; with fertilization and successful uterine implantation, the luteum remains functional for about 6 months of the

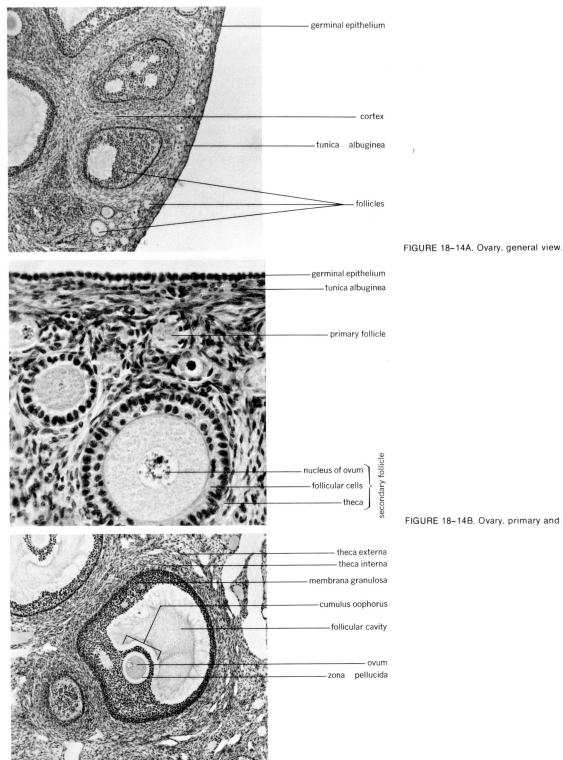

germinal epithelium

cortex

tunica albuginea

follicles

FIGURE 18–14A. Ovary, general view.

germinal epithelium
tunica albuginea

primary follicle

nucleus of ovum
follicular cells   secondary follicle
theca

FIGURE 18–14B. Ovary, primary and secondary follicles.

theca externa
theca interna
membrana granulosa
cumulus oophorus
follicular cavity

ovum
zona pellucida

FIGURE 18–15. Ovary, vesicular follicle.

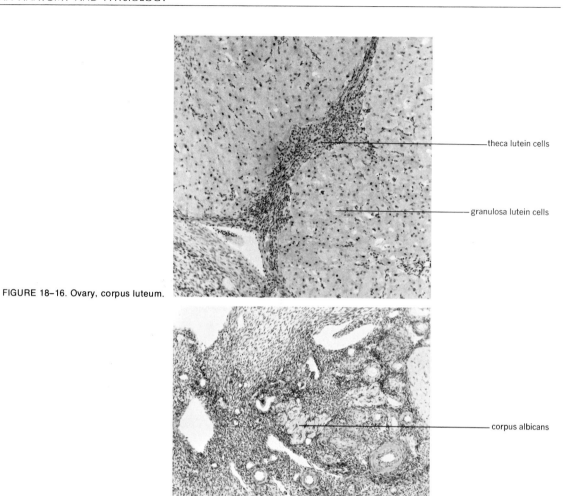

theca lutein cells

granulosa lutein cells

FIGURE 18–16. Ovary, corpus luteum.

corpus albicans

FIGURE 18–17. Ovary, corpus albicans.

pregnancy before degenerating. In either case, the luteum is replaced by collagenous connective tissue (scar tissue) and a *corpus albicans* (Fig. 18–17) if formed.

Uterine tubes

The paired uterine *(Fallopian)* tubes are about 10 cm (4 inches) in length and attach to the superiolateral aspect of the uterus. They are supported by the mesosalpinx. They consist of an expanded funnel-shaped outer end, the *infundibulum,* bearing many finger-like *fimbrae.* The *ampulla* lies between the infundibulum and the uterine wall; the *isthmus* of the tube penetrates the uterine wall to open into the uterine cavity. The uterine tube receives the discharged ovum and conveys it to the uterus. Microscopically (Fig. 18–18) the uterine tube shows three layers in its walls:

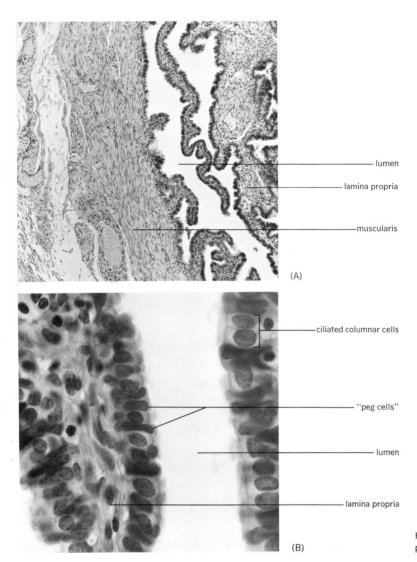

(A)

lumen

lamina propria

muscularis

ciliated columnar cells

"peg cells"

lumen

lamina propria

(B)

FIGURE 18–18. Uterine tube. A. Low power. B. High power of epithelium.

1. An inner *mucosa* consisting of a simple columnar epithelium having two types of cells: ciliated cells, believed responsible for creating a "current" in the fluids of the pelvic cavity which draws the egg into the tube; and nonciliated secretory cells (*peg cells*) which are inserted between the ciliated cells and which are thought to contribute nutrients to the ovum. A lamina propria underlies the epithelium.
2. A *muscularis,* consisting of a heavy, inner, circular, and a thin, outer, longitudinal coat of smooth muscle. The muscle is responsible for movement of the ovum through the tube.
3. An outer *serosa,* of typical structure.

■ Eggs ovulated from the ovary are capable of being fertilized and undergoing development for a maximum of 24 hours after release. The journey through the uterine tube requires about 3 days. If fertilization is to occur, it must therefore take place in the outer portion of the uterine tube. The implications of these facts for the sperm will be discussed later. Occasionally, a fertilized ovum may not be drawn into the tube, but will implant in the pelvic cavity. Or, it may implant in the uterine tube itself. Such implantations are designated ectopic. Pelvic implantations generally fail due to inadequate vascular connections; tubal implantations are usually surgically terminated (if discovered) before the growing embryo ruptures the tube. ■

Uterus

The single uterus is a pear-shaped organ measuring 7.5 cm in length, 5 cm wide and 2.5 cm thick (2½ × 2 × 1 inches) in the female who has never borne children. It is supported behind the bladder and anterior to the rectum by 8 ligaments. The paired *broad ligaments* run laterally from the uterus to the side walls of the true pelvis; the paired *round ligaments* pass anteriorly from the superior lateral margins of the uterus; four *uterosacral ligaments* pass from the uterus to the posterior pelvic wall.

The uterus itself has a broad upper *fundus,* a tapering *body,* and a *cervix,* or neck, which projects into the upper portion of the vagina.

Microscopically, the uterus shows three major layers of tissue (Fig. 18–19):

1. The *perimetrium* or outer serous coat covers all parts of the uterus except the cervix, and consists of a mesothelial lining supported by underlying connective tissue.
2. The *myometrium* or muscular coat, forms about 75 percent of the thickness of the uterine wall. It consists of four, ill-defined layers of smooth muscle.
3. The *endometrium,* or mucous membrane, shows cyclical changes in appearance, after puberty, with the assumption of the menstrual cycles. This layer is composed of two sublayers: a deep lying *basalis,* which is not shed during menstruation, and a superficial *functionalis* which is shed during menstruation.

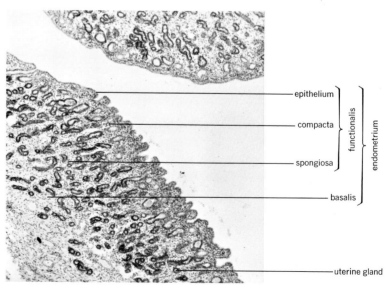

FIGURE 18–19. Uterus, early secretory phase.

The menstrual (estrus) cycle is hormonally controlled, depending upon ovarian production of estrogen and progesterone. The hormonal inter-relationships involved will be discussed in greater detail in Chapter 22.

The cycle is usually subdivided into four stages:

1. *Menstrual stage,* the first 3 to 5 days of the cycle. This stage com-mences with the first external show of blood and is characterized by involution of the blood vessels of the thickened endometrium, tissue degeneration, and sloughing of the lining.
2. *Proliferative stage,* the next 7 to 15 days. Under the influence of estrogen from the growing follicle, epithelial repair is begun, and there is multiplication of connective tissue elements and pro-liferation of glands and blood vessels. A thickness of about 2 mm is attained. Ovulation occurs during the latter portion of this stage.
3. *Secretory stage,* the next 14 to 15 days. After ovulation, corpus luteum development and progestin production cause increased glandular and vascular proliferation, and some glandular secre-tion. A thickness of 4 to 5 mm is attained preparatory to receiving the ovum if it is fertilized. Implantation of a fertilized ovum "sig-nals" the luteum to maintain itself. If no implantation occurs, the luteum degenerates and precipitates the fourth stage.
4. *The premenstrual stage.* Arteries begin to involute, tissue break-down is initiated. The stage terminates with external show of blood.

The length of a cycle is variable according to the individual. It may be as short as 24 days, or as long as 35 days. If there is one time span which remains more or less constant, it is the time from ovulation to menstruation, 14 to 15 days. It should not be assumed that all cycles are 28 days in length, or that ovulation always occurs 14 days after men-struation. Some interrelationships of hormones and the ovarian and uterine changes are shown in Fig. 18–20.

Vagina

The vagina is a tubular organ about 8 cm in length. Its lumen is nor-mally not maintained in an open state, but is collapsed. At the upper end, the cervix of the uterus projects about 1 cm into the vagina. A se-ries of moat-like fornices (posterior, anterior, lateral) surround the cervix.

Microscopically, the vagina may be seen to possess three coats of tissue:

1. An outer *adventitia* of connective tissue fixes the vagina to sur-rounding tissue.
2. A middle *muscularis* of circularly and longitudinally arranged smooth muscle is present.
3. An inner *mucous membrane* consists of a lamina propria and a stratified squamous epithelium. The lamina is rich in glycogen, which decomposes into organic acids. These create an acidic en-vironment in the vagina useful in retarding microorganism growth.

The vagina receives the penis, and sperm are deposited in its upper end. The acid nature of the vagina is spermicidal. Recall that semen is

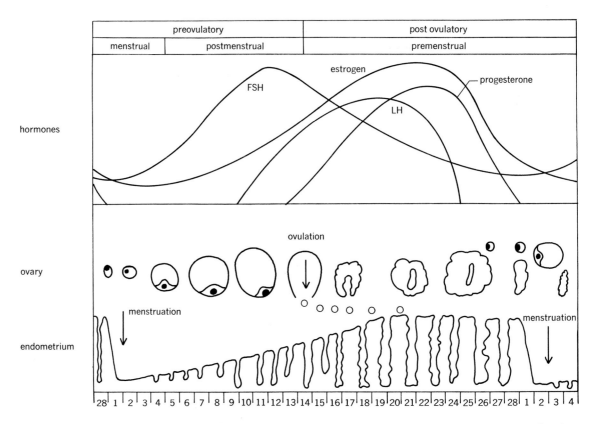

FIGURE 18–20. Interrelationships of hormones and ovarian and uterine changes.

alkaline. If sperm are to survive in the vagina, temporary neutralization of acidity is necessary.

Acidity is only the first hazard a sperm encounters if it is to fertilize an ovum in the uterine tube. It must next make a trip of some 6 inches through the uterus and "upstream" against the ciliary current in the uterine tube. Finally, only one sperm may fertilize the egg, the remainder dying. These hazards may account, in part, for the necessity of having many millions of sperm in the ejaculate.

External organs (Fig. 18–21)

Collectively, the external genital organs are designated as the *vulva* or *pudendum.* The *labia majora* are two, fat-filled, hair-covered (after puberty) folds of skin extending from a point about 1 cm anterior to the anus forwards to the pubis. They appear to be protective in function.

The *labia minora* are smaller folds of skin, lacking hair and fat, which intimately surround the vaginal and urethra orifices.

The *clitoris* is a single midline organ lying anterior to the urethra and surrounded by the anterior portion of the labia minora. It is composed of cavernous tissue similar to that of the penis. The clitoris becomes engorged with blood during sexual excitement and may contribute to the achievement of female orgasm.

The *mammary glands* (Fig. 18–22) are modified apocrine sweat glands. In infants, children, and men, they are present in rudimentary or undeveloped form. Under hormonal influences, chiefly that of estrogen, they begin their development in the adolescent female. The glands reach their maximum development in the pregnant female. Each gland lies superficial to the pectoralis major muscle, and is composed of some twenty lobes of secretory tissue. These lobes empty through 3 to 5 ducts at the apex of the nipple. The pigmented area around the nipple, the areola, lacks mammary ducts, but may have several small papillae representing the openings of several large

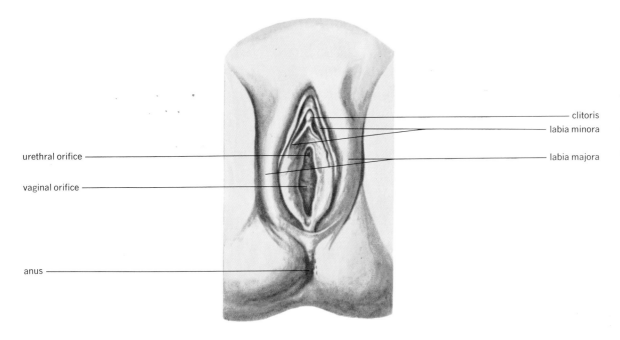

FIGURE 18–21. The external organs of the female reproductive system and associated structures.

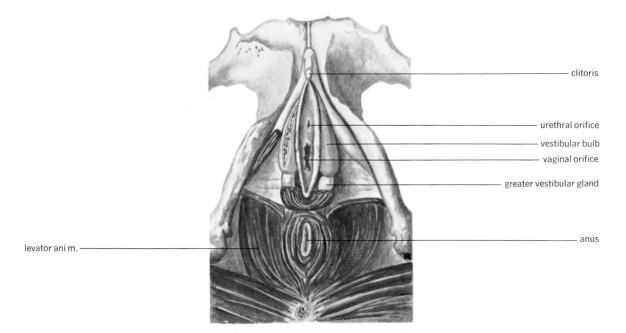

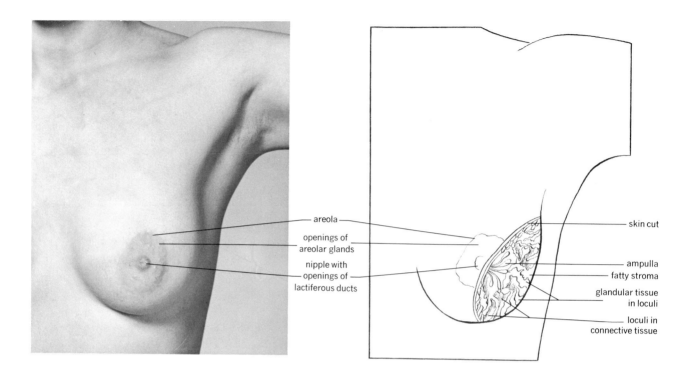

areola

openings of
areolar glands

nipple with
openings of
lactiferous ducts

skin cut

ampulla

fatty stroma

glandular tissue
in loculi

loculi in
connective tissue

FIGURE 18–22. The mammary gland.

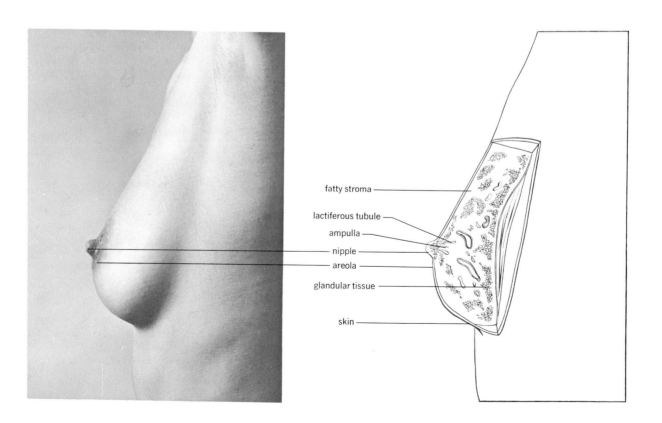

fatty stroma

lactiferous tubule

ampulla

nipple

areola

glandular tissue

skin

sebaceous glands. The glands are covered with skin, which is very thin over the areolae and nipples. Hair is absent in these two areas. The mature glands are very vascular organs, and have a lymphatic drainage towards the axillary, substernal and diaphragmatic lymph nodes. The extensive lymph drainage favors widespread metastases in malignancy of the breast.

Table 18–1 compares the male and female reproductive systems.

TABLE 18–1

A Comparison of Male and Female Reproductive Systems

	Male	Female
Essential organs	Testes—paired	Ovaries—paired
Position of essential organs	Scrotal—outside the abdominal cavity; descend to scrotum before birth	Pelvic—inside abdominal cavity

Structure of essential organ	Tunics:
	  Vaginalis—mesothelial outer covering
	  Albuginea—collagenous capsule around organ; thickened medially to form mediastinum
	
	  Erythroides—vascular coat deep to albuginea
	
	Subdivision:
	  Internal division into lobules by septae from albuginea; each lobule contains a seminiferous tubule lined with germinal epithelium
	
	Sex cell development:
	  After puberty, spermatogenesis proceeds in 5 stages:
	  1. Spermatogonia
	  2. Primary spermatocyte
	  3. Secondary spermatocyte
	  4. Spermatid
	  5. Spermatozoan
		Tunics:
		  Tunica albuginea forms capsule; covered by germinal epithelium
		
		Internal Structure:
		  Connective tissue stroma forms main mass of organ; may be subdivided into outer cortex, and inner vascular medulla; stroma contains follicles
		
		Sex cell development:
		  At birth, 400,000 immature eggs already present in ovary; at puberty, ovum development proceeds in 3 stages:
		  1. Primary follicle
		  2. Secondary follicle
		  3. Vesicular (Graafian) follicle
		  Ovulation occurs next (release of ovum from follicle); tissues remaining in ovary form corpus luteum, then corpus albicans

Ducts	Intratesticular:
	  Straight tubules connect seminiferous tubules to rete tubules
	  Rete tubules—in mediastinum; connect to efferent ductules
	Extratesticular:
	  Efferent ductules—12 to 16; connect to epididymis
	  Epididymis—20-ft tube for storage
	  Ductus deferens—ejaculatory duct
	  Urethra
		Uterine tubes:
		  Portions: infundibulum with fimbrae; ampulla; isthmus
		  Convey egg to uterus
		Uterus:
		  Single organ; parts include fundus, body, cervix; after puberty, undergoes cyclical changes
		Vagina: single organ opening to exterior

TABLE 18–1 *con't.*

A Comparison of Male and Female Reproductive Systems

	Male	Female

Accessory organ of the system	Seminal vesicles:
	    Paired structures contributing
	    alkaline secretion to semen
	
	Prostate:
	    Surrounds upper urethra; secretion activates
	    sperm
	Bulbourethral glands:
	    Secretes mucin
	
	Penis:
	    Erectile copulatory organ; three cavernous
	    bodies compose it
		Bulbourethral glands:
		    Empty into area of vaginal orifice
		    Secrete mucin
		
		External organs:
		    Labia majora—large folds of skin around
		    urethral and vaginal openings
		    Labia minora—small folds of skin closer
		    to vagina
		    Clitoris—single sensory and erectile
		    structure to urethra
		
		Mammary glands—modified skin glands
		    for nursing young

Cyclical changes after puberty	Not all seminiferous tubules are active
	    at the same time; waves of spermatogenesis
	    sweep over tubules
		Ovary ripens (on the average) one egg
		    per month
		Uterus shows menstrual cycle—phases:
		    1. Menstrual—shedding of tissue
		    2. Proliferative—repair and regrowth—
		        ovulation during this stage
		    3. Secretory—increase in glands and
		        vessels
		    4. Premenstrual—ischemia and tissue
		        degeneration
		Vagina shows changes associated with
		    uterine phases

Endocrine functions	Interstitial cells:
	    Produce testosterone
		Follicle:
		    Produces estrogen
		Corpus luteum:
		    Produces progestin

Endocrine control by pituitary	FSH:
	    Control spermatogenesis
	ICSH:
	    Controls interstitial cells
		FSH:
		    Control follicle maturation
		LH:
		    Controls corpus luteum formation
		    and secretion

Comments	Semen: Total secretions (including
	    sperm) of male tract. Vol: 3 ml—
	    300,000,000 sperms. Alkaline to
	    neutralize acidity in female tract.
	    Urethra common to reproductive and
	    urinary systems; 3 divisions
		Female reproductive system separate
		    from urinary

TABLE 18–2A

Methods of Contraception Currently Available

Method	Mode of action	Effective-ness if used correctly	Action needed at time of coitus	Requires resupply of materials used	Requires instruction in use	Requires services of physician	Suitable for menstrually irregular women	Side effects
Oral pill 21 day adminis-tration	Prevents follicle maturation and ovulation	Highest	None	Yes	Yes—timing	Yes—prescrip-tion	Yes	Early—some water retention, breast tenderness Late—possible embolism, hypertension
Intrauterine device (coil, loop)	Prevents implantation	High	None	No	No	Yes—to insert	Yes	Some do not retain device. Some have menstrual discomfort
Diaphragm with jelly	Prevents sperm from entering uterus, plus jelly spermicidal	High	Previous to coitus	Yes	Yes—must be inserted correctly each time	Yes—for sizing and instruction on use	Yes	None
Condom (worn by male)	Prevents sperm entry into vagina	High	Yes	Yes	Not usually	No	– –	Some deadening of sensation in male
Temperature rhythm	Determines ovulation time by noting body temp. at ovulation. T↑	Medium	No	No	Definitely. Must learn to interpret chart correctly	No. Physician should advise	Yes, if are skilled in reading graph	None. Requires abstinence during part of cycle
Calendar rhythm	Abstinence during part of cycle	Medium to low	No—no coitus	No	Definitely. Must know when to abstain	No. Physician should advise	No!	None (pregnancy?)
Vaginal foams	Spermicidal	Medium to low	Yes. Requires application before coitus	Yes	No	No	Yes	None usually. May irritate
Withdrawal	Remove penis before ejaculation	Low	Yes. With-drawal	No	No	No	Yes	Frustration in some
Douche	Wash out sperm	Lowest	Yes. Immedi-ately after	No	No	No	Yes	None

Birth control

Tables 18–2A and B are presented in the interests of acquainting the student with methods of contraception currently available.

TABLE 18–2B
Newer Methods of Contraception under Investigation

Method	Mode of action	Effectiveness if used correctly	Action needed at time of coitus	Requires resupply of materials used	Requires instruction in use	Requires services of physician	Side effects
"Mini-pill"— very low content of progesterone (½ mg.)	Inhibits follicle development	High	No	Yes	Yes	Yes— prescription	Irregular cycles and bleeding (25%)
"Morning-after pill"	Arrests pregnancy probably by preventing implantation. 50X normal dose of estrogen	By currently available data, high	No. For 1–5 days after coitus	Yes	No	Yes	Breast swelling, nausea, water retention
Vaginal ring— inserted in vagina; contains progesteroid in it	"Leaks" progesteroid into blood stream through vagina at constant rate. Thereby inhibits follicle maturation	Studies are "promising"	No	Yes. Perhaps at yearly intervals	Yes	Yes	Spotting, some discomfort
Once a month pill	Injected in oil base into muscle. Slow passage of birth control drug into circulation inhibits follicle maturation	Said to be 100 percent	No	Yes. On monthly basis	No	Yes	Similar to oral pill

QUESTIONS

1. In what ways do you believe that better understanding of the reproductive process could contribute to an increase in man's welfare and happiness?
2. Compare ovaries and testes in terms of structure, function, and hormone production.
3. Compare the ducts of the male and female systems in terms of structure and function.
4. What structures provide for nourishment of sperm and ova during their transit through the respective systems?
5. Describe the changes occurring in the uterus during the menstrual cycle.

CHAPTER 19

Basic Embryology of the Human

The student of Nature wonders the more and is astonished the less, the more conversant he becomes with her operations, but of all the perennial miracles she offers to his inspection, perhaps the most worthy of admiration is the development of a plant or animal from its embryo. HUXLEY

DEVELOPMENT OF THE INDIVIDUAL

Introduction

Spermatozoa and ova develop by the process of meiosis which results in the halving of the chromosome number characteristic for the species. In man, this number is 46, or 23 pairs. For a new individual to develop normally, a restoration of chromosome number must occur. Restoration is accomplished by the process of fertilization, whereby a single spermatozoan (23 chromosomes) penetrates the ovum (23 chromosomes), and the nuclei of the two cells are pooled. Fertilization normally also supplies the stimulus which initiates cell division in the combined cell *(zygote)*. Certain invertebrate eggs regularly develop without fertilization in the process known as parthenogenesis. Turkey and rabbit ova have also been induced to develop parthenogenically, but no scientifically documented cases of parthenogenesis have been recorded among higher mammals, including man. Experiments on lower animals have shown that a dead sperm can induce division, as can a live sperm, in an ovum from which the nucleus has been removed. Fertilization thus appears to provide a largely mechanical stimulus for division.

The events of fertilization (Fig. 19–1)

Sperm, moving independently, eventually contact an egg. A sperm penetrates the zona pellucida of the ovum, the tail is lost, and the head

507

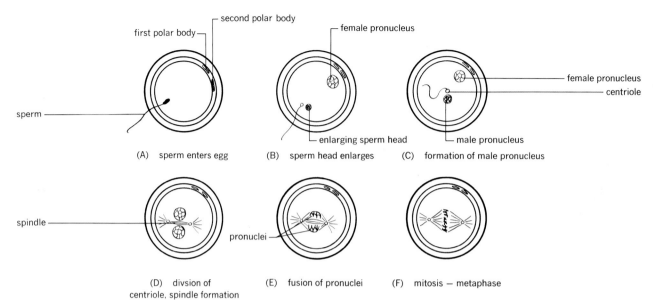

(A)   sperm enters egg   (B)   sperm head enlarges   (C)   formation of male pronucleus

(D)   divsion of
centriole, spindle formation   (E)   fusion of pronuclei   (F)   mitosis — metaphase

FIGURE 19–1. The events of fertilization.

enlarges to form the *male pronucleus.* After penetration of a single sperm, the membrane of the egg becomes impervious to the further entry of sperm. The single centriole which was carried in with the sperm divides, and the two centrioles migrate to opposite poles of the zygote. The ovum nucleus enlarges to form a *female pronucleus.* The chromatin of the two nuclei resolves itself into chromosomes which are pooled in the equatorial plane of the cell. The cell is now ready to undergo the first mitotic division of cleavage.

Cleavage (Fig. 19–2)

Cleavage refers to the repeated series of mitotic divisions which transforms the zygote into a solid mass of smaller cells, or *blastomeres.* The solid mass is referred to as a *morula,* because of its resemblance to a mulberry. The number of cells is usually doubled for each of the first 8 to 10 cleavages, and thereafter becomes irregular. Little time

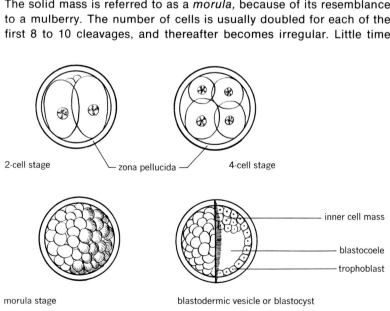

2-cell stage   zona pellucida   4-cell stage

inner cell mass

blastocoele

trophoblast

FIGURE 19–2. Cleavage.   morula stage   blastodermic vesicle or blastocyst

elapses between the first cleavages, so that the morula is little larger than the zygote; the cells do not grow between divisions. In man, the initial divisions are carried out as the egg traverses the uterine tube, a journey requiring about 3 days. A reorganization and differentiation of cells occurs next, to form a *blastocyst.* The solid morula develops a cavity within the mass of cells, which enlarges through fluid secretion into it, and the morula is transformed into a sphere. In the blastocyst, two regions may be recognized: an outer covering of cells, the *trophoblast,* and an *inner cell mass.*

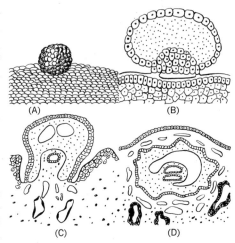

FIGURE 19–3. Implantation.

Implantation (Fig. 19–3)

The divisions and reorganization just described have been carried on utilizing the energy inherent in the zygote alone. Contact with an outside supply of nutrients must occur quickly or further development will fail due to energy depletion. The blastocyst adheres to the uterine endometrium, which is in the secretory stage of development. Enzymatically, it digests its way into the endometrium until it is completely buried. This process constitutes implantation, and enables absorption of nutrients from the vessels and glands of the endometrium.

The blastocyst is now ready to differentiate the three primary germ layers, ectoderm, endoderm, and mesoderm, from which all tissues and organs of the body will develop.

Germ layer formation (Fig. 19–4)

Within the inner cell mass of the blastocyst, two cavities form, separated by a two-layered plate of cells. The cavity in closest approximation to the trophoblast is the *amniotic cavity,* which is lined with presumptive *ectoderm.* The other cavity is the cavity of the *yolk sac,* which is lined with presumptive *endoderm.* The two-layered plate of cells is the *embryonic disc.* Simultaneously with these changes, cells arise from the trophoblast, and spread through the space between inner cell mass and trophoblast. These cells differentiate to form the *extraembryonic mesoderm.* Cavities appearing in this mesoderm fuse to form the *extraembryonic coelom.* The mesoderm immediately adjacent to the trophoblast, and the trophoblast itself, associate to form

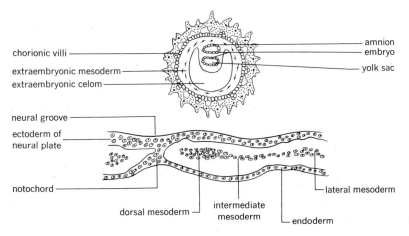

chorionic villi

extraembryonic mesoderm

extraembryonic celom

amnion

embryo

yolk sac

neural groove

ectoderm of neural plate

notochord

dorsal mesoderm

intermediate mesoderm

endoderm

lateral mesoderm

FIGURE 19–4. The formation of the germ layers.

the *chorion*. From the chorion, villi arise and grow into the endometrium to increase the surface for nutrient absorption. *Embryonic mesoderm* develops by cell proliferation from the center of the embryonic disc between the ectoderm and endoderm. The cells migrate peripherally and surround the yolk sac (gut). A cavity appears within this tissue, the *coelom,* or future thoracic and abdominal cavities. From these three germ layers will form the organs and the systems shown in Table 19–1. The processes occurring next establish the bases for the body systems, and will be considered in the Sections to follow. Further refinement produces, at 2.5 to 3 months, a *fetus* of clearly human form. At this time, all body systems are present, although not all are functional. From 3 months to birth (10 lunar months), growth and maturation of systems occurs, as well as addition of fine details such as fingernails, eyelashes and hair.

TABLE 19–1

Tissues Derived from Germ Layers in Man

Ectoderm	Mesoderm (including mesenchyme)	Endoderm

1. Epidermis of skin, including:
   Skin glands
   Hair and nails
   Lens of eye
2. Epithelium of:
   Nasal cavities
   Sinuses
   Mouth, including:
      Oral glands
      Enamel
   Sense organs
   Anal canal
3. Nervous tissues
   Hypophysis
      (neurohypophysis)
   Adrenal medulla
	1. Muscle:
	   Skeletal, cardiac,
	      smooth
	2. Connective tissue
	   including:
	   Cartilage
	   Bone
	   Blood
	   Bone marrow
	   Lymphoid tissue
	3. Epithelium of:
	   Blood vessels
	   Lymphatics
	   Coelomic cavities
	   Kidney and ureters
	   Gonads and ducts
	   Adrenal cortex
	   Joint cavities
		1. Epithelium of:
		   Pharynx
		   Auditory tube
		   Tonsils
		   Thyroid
		   Parathyroid
		   Thymus
		   Larynx
		   Trachea
		   Lungs
		   Digestive tube and
		      its glands
		   Bladder
		   Vagina and vestibule
		   Urethra and glands
		2. Hypophysis
		      (adenohypophysis)

DEVELOPMENT OF THE NERVOUS SYSTEM (Fig. 19–5)

Organization of the nervous system shows first as thickenings along the midline of the embryonic disc. Two folds *(neural folds),* separated by a groove *(neural groove),* appear. The folds increase in height and meet in the midline to enclose the *neural tube.* Within the walls of the tube, three layers are formed. The inner or *ependymal layer* will remain, and will form the lining of the cavities of the central nervous system. The middle or *mantle layer* represents the future gray matter of the central nervous system. The outer or *marginal layer* represents the future white matter of the system. *Neural crest* material, splitting off from the dorsolateral portion of the neural tube, will form the peripheral ganglia of the system, and medulla of the adrenal gland. Only slight further refinement of this structure is required to produce the characteristic form of the spinal cord.

The anterior end of the neural tube undergoes development to form three enlargements, the *forebrain (prosencephalon), midbrain (mesencephalon),* and *hindbrain (rhombencephalon)* as shown in Fig. 19–6. The anterior portion of the forebrain develops two lateral swellings, or *telencephalon,* which are forerunners of the cerebral hemispheres. The remaining portion of the forebrain, or *diencephalon,* forms the upper portion of the brain stem, including the thalamus, hypothalamus, and pineal body. The midbrain includes the corpora quadrigemina and cerebral peduncles. The hindbrain develops into the pons, medulla, and cerebellum. The peripheral portions of the nervous system, including ganglia and peripheral nerves, are derived by migration of neural crest material and by extensions of long processes from this material to the neural tube. From the mantle layer of the neural tube, processes also extend to the periphery.

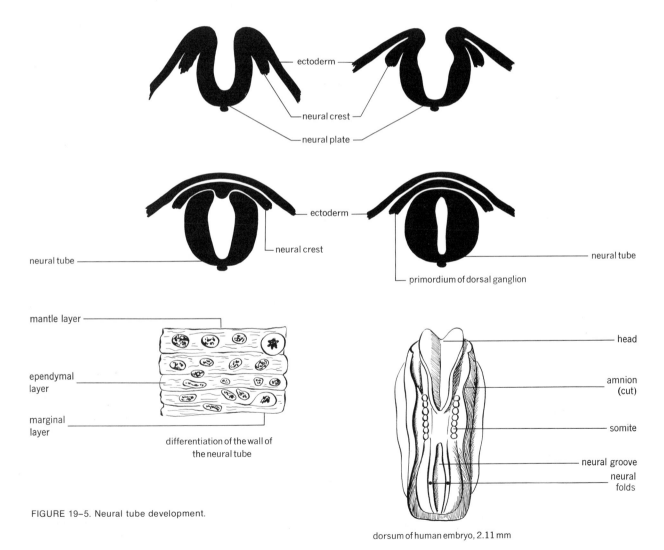

FIGURE 19–5. Neural tube development.

dorsum of human embryo, 2.11 mm

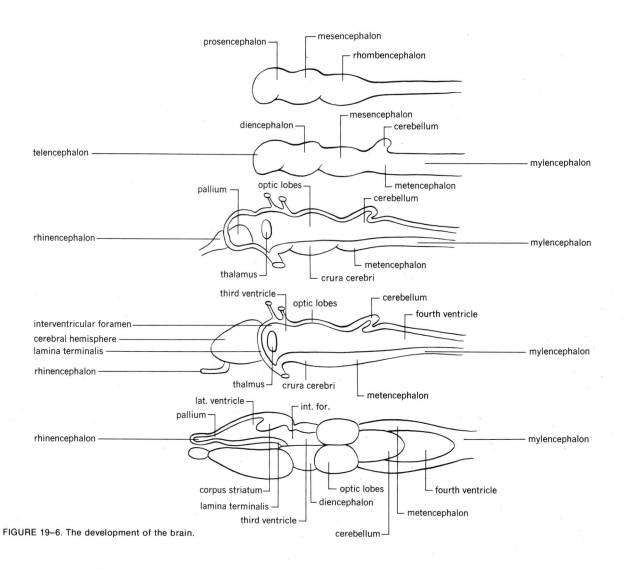

FIGURE 19-6. The development of the brain.

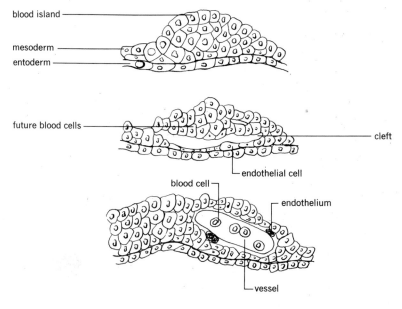

FIGURE 19-7. The development of blood vessels and blood cells.

DEVELOPMENT OF THE CIRCULATORY SYSTEM

Vessels (Fig. 19–7)

Within the extra embryonic mesoderm appear dense masses and cords of mesenchyme cells called *blood islands*. Cavities or lumina appear within these islands, and the peripheral cells become flattened into an endothelial lining. The cells within the lumina differentiate into primitive blood cells leading to the formation of erythrocytes. Growth and union of these islands provides a network of vessels in the entire embryo. Around the endothelium, mesenchyme cells will form the accessory coats of the intima, media, and adventitia of the larger arteries and veins. The first paired vessels to be formed in this process are the two large heart tubes and a dorsal and ventral aorta.

Heart (Fig. 19–8)

The paired heart tubes show slight constrictions marking, from anterior to posterior, the future bulbus, ventricle, and atrium. The bulbar and ventricular portions fuse, creating a single cavity, but leaving the atrial portion as paired sacs. The atria fuse during the next series of maneuvers. The entire tube now grows rapidly, and is caused to bend into an S, with the bulbus and ventricles being looped over the atria. Further growth forces the bulbus and ventricle caudally, reversing the original position of the three chambers. The atria enlarge laterally. The single ventricle incorporates the bulbus into itself and both atria and ventricles are subdivided by septae into paired chambers. The bulbus splits to form a pulmonary trunk from the right ventricle, and an aorta from the left ventricle. In the fetus, the foramen ovale provides an interatrial communication; no such opening is normally maintained between ventricles.

Blood

Some cells of the blood islands differentiate into erythrocytes for oxygen transport. Other cells secrete the plasma into the vessel lumen. The most immediate need of the embryo is for a transport fluid and for oxygen-carrying capacity. The development of the blood coincides with the formation of vessels and heart to pump it, so that a circulation is established very early. Development of leucocytes must wait until lymphoid tissue and bone marrow are formed.

DEVELOPMENT OF THE DIGESTIVE SYSTEM (Fig. 19–9 A, B)

Early development of the embryo results in the formation of a gut lined with endoderm. Elongation of the tube in an anterior-posterior direction produces an anteriorly directed, blind-ended *foregut,* a centrally placed, open *midgut,* and a posteriorly directed, blind-ended *hindgut.* Anteriorly and posteriorly, the gut contacts the ventrally placed ectoderm, and *oral* and *cloacal membranes* are formed. The oral membrane lies at the bottom of a depression known as the *stomodaeum,* or future oral cavity. At about the fourth week of development, the oral membrane is ruptured, and stomodaeum and foregut become contin-

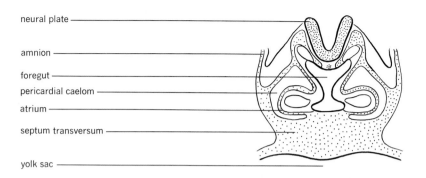

neural plate

amnion

foregut

pericardial caelom

atrium

septum transversum

yolk sac

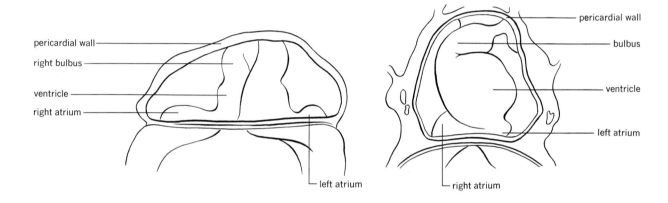

pericardial wall

right bulbus

ventricle

right atrium

left atrium

pericardial wall

bulbus

ventricle

left atrium

right atrium

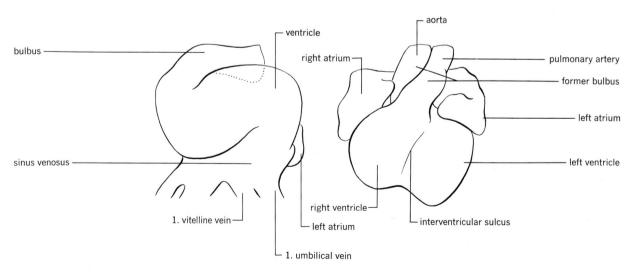

bulbus

sinus venosus

1. vitelline vein

ventricle

right atrium

right ventricle

left atrium

1. umbilical vein

aorta

pulmonary artery

former bulbus

left atrium

left ventricle

interventricular sulcus

FIGURE 19–8. The development of the heart.

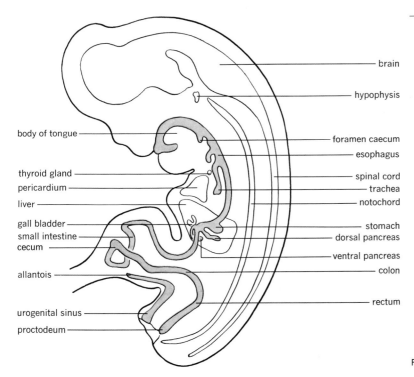

FIGURE 19–9A. Embryonic gut at 17 mm stage.

uous. The cloacal membrane also lies at the bottom of a depression, the *proctodaeum.* The most posterior portion of the hindgut is the *cloaca,* forming a common chamber for the terminal portions of digestive, urinary, and reproductive systems. The cloaca is a transitory structure in man, and is soon subdivided into a dorsal *rectum* and ventral *bladder* and *urogenital sinus.* Rupture of the cloacal membrane establishes two openings to the exterior: the anal and urogenital orifices.

The main portion of the gut differentiates into mouth, pharynx, and digestive tube. Only the linings of the tube are endodermal in origin, as are the linings of any evaginations which may occur from the tube. All muscle and connective tissue components are from the mesoderm which surrounds the tube.

The liver develops as an outgrowth from the ventral portion of the anterior midgut. Two solid masses of endoderm at the end of the outgrowth form the right and left lobes of the liver. Bile and cystic ducts develop from the original evagination.

The pancreas develops as dorsal and ventral outgrowths of the gut, which fuse to form the entire organ. The secretory acini develop as sidegrowths from the original diverticulum, while the endocrine tissue (islets) develop from the terminal portions of the duct system. The latter portions subsequently detach from the ducts and assume an intimate relationship with the circulatory system.

DEVELOPMENT OF THE RESPIRATORY SYSTEM (Fig. 19–10)

The nasal cavities and pharynx are derivatives of the *olfactory pits* and the foregut, respectively. The remainder of the system first appears as a midventral groove in the foregut. Deepening of the groove and fusion

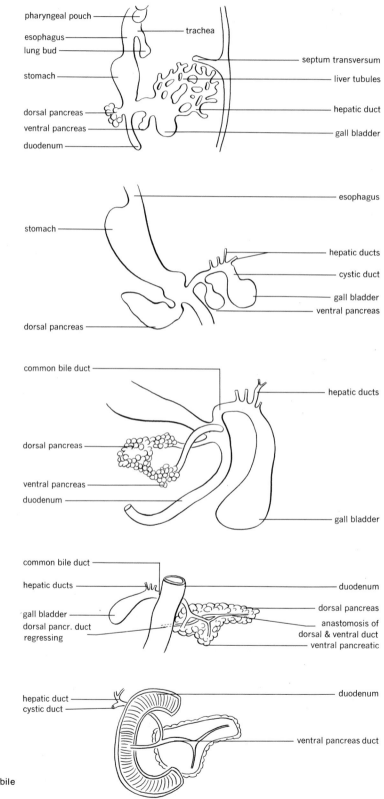

FIGURE 19–9B. Development of the pancreas and bile ducts.

of its walls creates the *laryngotracheal tube,* with a slit-like anterior connection, the *glottis,* being retained into the pharynx. The tube is therefore of endodermal origin.

The lower end of the tube, or *lung bud,* bifurcates to form the bronchi, and subdivides repeatedly to form the many bronchioles, alveolar ducts and sacs.

The cartilages, muscle, and connective tissue components are supplied by the mesoderm of the *branchial arches,* a pharyngeal derivative.

The pleurae are also formed from cells contributed by the splanchnic and somatic mesoderm.

FIGURE 19–10. Development of the respiratory system.

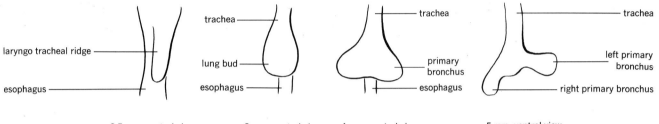

2.5 mm, ventral view   3 mm, ventral view   4 mm, ventral view   5 mm, ventral view

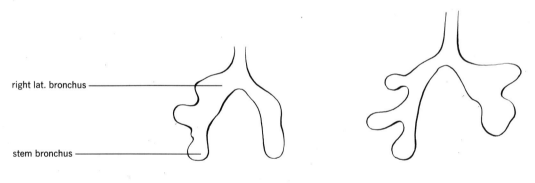

7 mm, ventral view   8.5 mm, ventral view

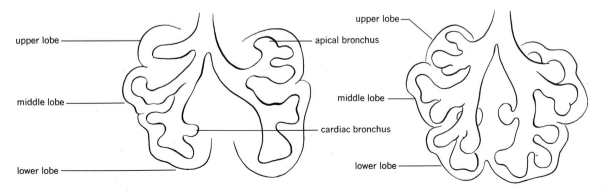

10 mm, ventral view   14 mm, ventral view

DEVELOPMENT OF THE URINARY SYSTEM (Fig. 19–11)

A mesodermal *urogenital ridge,* developing along the posterior half of the dorsal side of the embryo, gives rise to the urinary and reproductive systems. As the urinary system develops, it initially empties into the cloaca, and later into the urogenital sinus.

The human produces three kidneys during development of the uri-

FIGURE 19–11. Development of the urinary organs.

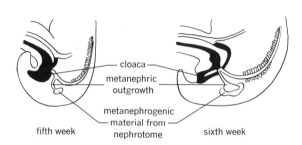

cloaca
metanephric outgrowth
metanephrogenic material from nephrotome

fifth week   sixth week

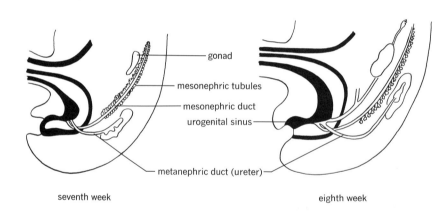

gonad
mesonephric tubules
mesonephric duct
urogenital sinus

metanephric duct (ureter)

seventh week   eighth week

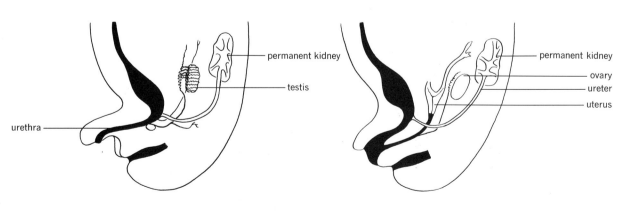

permanent kidney
testis

urethra

male   three months

permanent kidney
ovary
ureter
uterus

female   three months

nary system. A *pronephros* or "head kidney" is the first to be formed. It degenerates to be replaced by a middle *mesonephros,* which in turn is replaced by a *metanephros* as the most posterior derivative of the ridge.

Associated with the development of the pronephros is a duct which reaches the cloaca. It is known as the *pronephric duct.* A portion of this duct is retained to serve the mesonephros, and is here termed the *mesonephric duct.* As the mesonephros degenerates, the terminal portions of the duct are retained and give rise to an outgrowth which will become the renal pelvis, medulla, and collecting tubules of the kidney, and the ureter.

The bladder is a derivative of the urogenital sinus. Separation of the urethra from the reproductive system, in the female, occurs by a partition which divides the urogenital sinus into two parts.

DEVELOPMENT OF THE REPRODUCTIVE SYSTEMS (Fig. 19–12 A, B)

The gonads also develop within the urogenital ridge. The testes and ovaries form bulges into the coelom at about 6 weeks, and assume the

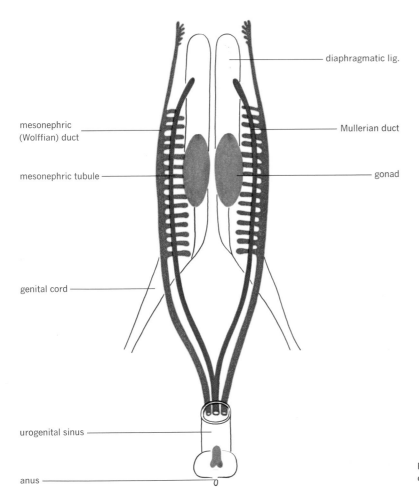

FIGURE 19–12A. Development of internal reproductive organs.

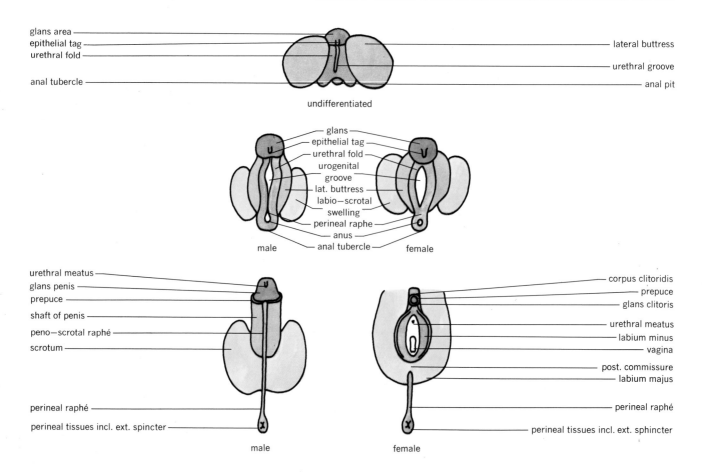

FIGURE 19-12B. Development of external genitalia.

internal structure characteristic of each organ at about 8 weeks. Portions of the cranial *mesonephric* tubules and of the mesonephric duct persist as the efferent ductules of the testis, and the head of the epididymis. The mesonephric duct alone contributes the remainder of the epididymis and the ductus deferens.

In the female, tubular *Mullerian ducts* develop on either side of the urogenital ridge; posteriorly, they connect to the cloaca. The terminal portions of the ducts fuse to form the primordia of the vagina and uterus; the unfused anterior portions will become the uterine tubes.

The external genitalia are not characteristic of the sexes until about the 8th week. A medially placed *phallus* has a lengthwise *urogenital groove,* and is flanked by *genital tubercles.* The phallus becomes the *penis* or *clitoris,* while the genital tubercles form *scrotum* or *labia majora.* In the female, the lips of the groove form the *labia minora.* Table 19–2 presents some of the homologies between the two sexes in the development of the reproductive systems.

DEVELOPMENT OF THE ENDOCRINES (Fig. 19–13)

During the 4th to 5th week of development, the embryo shows in the neck region a series of cartilagenous arches, the *branchial arches.* Between arches will develop outpocketings of the lining of the foregut,

TABLE 19-2

Some Homologies of the Reproductive System

Indifferent stage	Male	Female
1. Gonad	Testis	Ovary
2. Mesonephric tubules (cranial)	Efferent ductules and head of epididymis	Epoöphoron (vestige)
3. Mesonephric duct	Ductus epididymis	Duct of epoöphoron (vestige)
4. Mesonephric duct	Ductus deferens	Gartner's duct (vestige)
5. Mullerian duct, upper part	Appendix testis (vestige)	Uterine tube
6. Mullerian duct, middle part	—	Uterus
7. Mullerian duct, lower part	—	Vagina (upper)
8. Urogenital sinus	Prostatic urethra	Vestibule (middle)
9. Phallus	Penis	Clitoris
10. Lips of urogenital groove	Urethral penile surface	Labia minora
11. Genital swellings (tubercles)	Scrotum	Labia majora

the *pharyngeal pouches.* In the roof of the foregut, a single median outpocketing, *Rathke's pouch,* appears. The hypophysis (pituitary) is a gland whose anterior portion (anterior lobe) is derived from Rathke's pouch. This portion migrates up and backward to fuse with a downgrowth of the brain, the latter representing the posterior lobe of the gland.

The *thyroid gland* is derived from a single medial outgrowth of the floor of the foregut at the level of the second pharyngeal pouch. It migrates inferiorly to assume a final position around the larynx and upper trachea.

The *parathyroids* develop from the anterior portions of the 3rd and 4th pharyngeal pouches. These migrate inferiorly to assume a final position on the dorsal side of the thyroid.

The *thymus,* not a true endocrine, originates from the posterior portions of the 3rd and 4th pharyngeal pouches, and migrates to lie behind the sternum.

The *adrenal medulla* is a derivative of neural crest material. To this is applied a covering of urogenital ridge mesoderm to form the *cortex* of the gland.

The endocrine tissue of the pancreas *(islets)* arises from the ends of the endodermal outpocketings of the gut. The outpocketings form the ducts, and the islets lose their connection with the ducts.

The endocrine cells of *testis* and *ovary* differentiate from the mesodermal cells as the gonads develop.

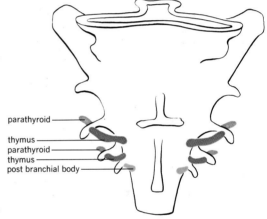

parathyroid
thymus
parathyroid
thymus
post branchial body

FIGURE 19-13. Origins of some of the endocrine organs from branchial pouch material.

FETAL MEMBRANES (Fig. 19-14 A, B)

Vertebrate embryos acquire during their development, a series of protective, nutritive, and excretory structures comprising the fetal membranes. A complete list of these would include the yolk sac, amnion, chorion, allantois, umbilical cord, and placenta.

Yolk sac. No true yolk mass is present in human embryos, but a yolk sac appears as the endoderm-lined cavity below the embryonic disc. It is a transitory structure, and is incorporated into the gut.

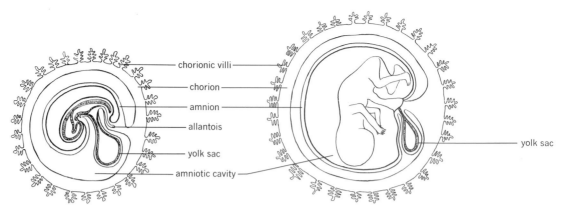

FIGURE 19–14A. The origin of some of the fetal membranes.

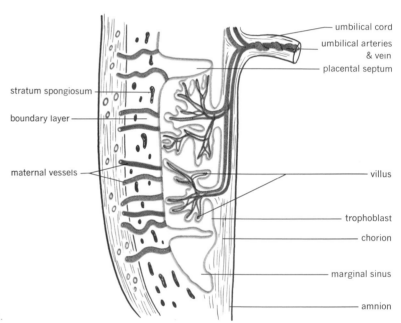

FIGURE 19–14B. The placenta.

Amnion. The amnion overlies and surrounds the embryonic disc. With growth of the embryo, it comes to surround the embryo on all sides. It is filled with amniotic fluid and, in effect, the embryo and fetus float in the fluid. The amnion and its contained fluid cushion the fetus, and are the so-called "bag of waters."

Chorion. The chorion is the trophoblast of the blastocyst and its associated mesoderm. At first surrounding the entire embryo, the chorion later comes to occupy only that portion of the trophoblast attached to the uterine wall. The chorion becomes the membrane for exchange between child and mother, and is the chief component of the human placenta.

Allantois. The allantois develops as an outpocketing of the cloaca. In the goat, the sac may reach two feet in length and serves as a storage area for excreted wastes. In man, the allantois remains small and essentially functionless. It becomes vascularized, and then degenerates. The vessels remain as an important part of the umbilical vessels to and from the placenta.

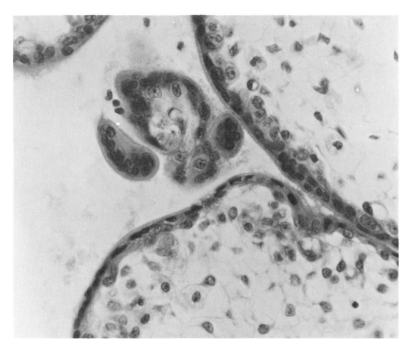

FIGURE 19–15. Chorionic villus.

Placenta. The placenta is comprised of a fetal component, chiefly the *chorion,* and a maternal component, the *endometrial wall.* The organ is a site of exchange of materials between mother and child. It also produces hormones required for continuation of pregnancy. Its villi (Fig. 19–15) are of two types: *anchoring villi* are embedded in the endometrium and hold the placenta to the uterus. *Free villi* lie in eroded blood spaces in the endometrium, and are the absorptive and excretory structures. Thus, the only layers of tissue separating maternal and fetal circulations are those of the villi.

Fate of the amnion and chorion. At birth, the amnion ruptures and the fetus emerges, trailing the umbilical cord. The cord is tied and cut. The afterbirth consists of detachment of the placenta from the uterine wall and its expulsion.

QUESTIONS

1. Compare and contrast oögenesis and spermatogenesis in terms of the genetic makeup of the product, and numbers of units formed.
2. What events are thought to occur within the uterine tube? At what stage is the zygote when it exits from the tube into the uterus?
3. What steps have been taken to prepare the uterus to receive the fertilized egg?
4. What occurs to insure adequate levels of nutrition for the embryo and fetus?
5. What tissues and organs develop from each germ layer?
6. Review the development of the urogenital system, and state what embryonic parts are retained in the fully developed system and their functions.
7. Describe the fetal membranes and the functions served by each in the fetus.

A Reference Table of Correlated Human Development.

Age in Weeks	Size (C R) in Mm.	Body Form	Mouth	Pharynx and Derivatives	Digestive Tube and Glands	Respiratory System	Coelom and Mesenteries
2.5	1.5	Embryonic disc flat. Primitive streak prominent. Neural groove indicated.	....	....	Gut not distinct from yolk sac.	....	Extra-embryonic coelom present. Embryonic coelom about to appear.
3.5	2.5	Neural groove deepens and closes (except ends). Somites 1 – 16± present. Cylindrical body constricting from yolk sac. Branchial arches 1 and 2 indicated.	Mandibular arch prominent. Stomodeum a definite pit. Oral membrane ruptures.	Pharynx broad and flat. Pharyngeal pouches forming. Thyroid indicated.	Fore- and hind-gut present. Yolk sac broadly attached at mid-gut. Liver bud present. Cloaca and cloacal membrane present.	Respiratory primordium appearing as a groove on floor of pharynx.	Embryonic coelom a U-shaped canal, with a large pericardial cavity. Septum transversum indicated. Mesenteries forming. Mesocardium atrophying.
4	5.0	Branchial arches completed. Flexed heart prominent. Yolk stalk slender. All somites present (40). Limb buds indicated. Eye and otocyst present. Body flexed; C-shape.	Maxillary and mandibular processes prominent. Tongue primordia present. Rathke's pouch indicated.	Five pharyngeal pouches present. Pouches 1-4 have closing plates. Primary tympanic cavity indicated. Thyroid a stalked sac.	Esophagus short. Stomach spindle-shaped. Intestine a simple tube. Liver cords, ducts and gall bladder forming. Both pancreatic buds appear. Cloaca at height.	Trachea and paired lung buds become prominent. Laryngeal opening a simple slit.	Coelom still a continuous system of cavities. Dorsal mesentery a complete median curtain. Omental bursa indicated.
5	8.0	Nasal pits present. Tail prominent. Heart, liver and mesonephros protuberant. Umbilical cord organizes.	Jaws outlined. Rathke's pouch a stalked sac.	Phar. pouches gain dors. and vent. diverticula. Thyroid bilobed. Thyro-glossal duct atrophies.	Tail-gut atrophies. Yolk stalk detaches. Intestine elongates into a loop. Caecum indicated.	Bronchial buds presage future lung lobes. Arytenoid swellings and epiglottis indicated.	Pleuro-pericardial and pleuro-peritoneal membranes forming. Ventral mesogastrium draws away from septum.
6	12.0	Upper jaw components prominent but separate. Lower jaw-halves fused. Head becomes dominant in size. Cervical flexure marked. External ear appearing. Limbs recognizable as such.	Lingual primordia fusing. Foramen caecum established. Labio-dental laminae appearing. Parotid and submaxillary buds indicated.	Thymic sacs, ultimobranchial sacs and solid parathyroids are conspicuous and ready to detach. Thyroid becomes solid and converts into plates.	Stomach rotating. Intestinal loop undergoes torsion. Hepatic lobes identifiable. Cloaca subdividing.	Definitive pulmonary lobes indicated. Bronchi sub-branching. Laryngeal cavity temporarily obliterated.	Pleuro-pericardial communications close. Mesentery expands as intestine forms loop.
7	17.0	Branchial arches lost. Cervical sinus obliterates. Face and neck forming. Digits indicated. Back straightens. Heart and liver determine shape of body ventrally. Tail regressing.	Lingual primordia merge into single tongue. Separate labial and dental laminae distinguishable. Jaws formed and begin to ossify. Palate folds present and separated by tongue.	Thymi elongating and losing lumina. Parathyroids become trabeculate and associate with thyroid. Ultimobranchial bodies fuse with thyroid. Thyroid becoming crescentic.	Stomach attaining final shape and position. Duodenum temporarily occluded. Intestinal loops herniate into cord. Rectum separates from bladder-urethra. Anal membrane ruptures. Dorsal and ventral pancreatic primordia fuse.	Larynx and epiglottis well outlined; orifice T-shaped. Laryngeal and tracheal cartilages foreshadowed. Conchae appearing. Primary choanae rupturing.	Pericardium extended by splitting from body wall. Mesentery expanding rapidly as intestine coils. Ligaments of liver prominent.
8	23.0	Nose flat; eyes far apart. Digits well formed. Growth of gut makes body evenly rotund. Head elevating. Fetal state attained.	Tongue muscles well differentiated. Earliest taste buds indicated. Rathke's pouch detaches from mouth. Sublingual gland appearing.	Auditory tube and tympanic cavity distinguishable. Sites of tonsil and its fossae indicated. Thymic halves unite and become solid. Thyroid follicles forming.	Small intestine coiling within cord. Intestinal villi developing. Liver very large in relative size.	Lung becoming gland-like by branching of bronchioles. Nostrils closed by epithelial plugs.	Pleuro-peritoneal communications close. Pericardium a voluminous sac. Diaphragm completed, including musculature. Diaphragm finishes its 'descent.'
10	40.0	Head erect. Limbs nicely modeled. Nail folds indicated. Umbilical hernia reduced.	Fungiform and vallate papillae differentiating. Lips separate from jaws. Enamel organs and dental papillae forming. Palate folds fusing.	Thymic epithelium transforming into reticulum and thymic corpuscles. Ultimobranchial bodies disappear as such.	Intestines withdraw from cord and assume characteristic positions. Anal canal formed. Pancreatic alveoli present.	Nasal passages partitioned by fusion of septum and palate. Nose cartilaginous. Laryngeal cavity reopened; vocal folds appear.	Processus (saccus) vaginales forming. Intestine and its mesentery withdrawn from cord.
12	56.0	Head still dominant. Nose gains bridge. Sex readily determined by external inspection.	Filiform and foliate papillae elevating. Tooth primordia form prominent cups. Cheeks represented. Palate fusion complete.	Tonsillar crypts begin to invaginate. Thymus forming medulla and becoming increasingly lymphoid. Thyroid attains typical structure.	Muscle layers of gut represented. Pancreatic islands appearing. Bile secreted.	Conchae prominent. Nasal glands forming. Lungs acquire definitive shape.	Omentum an expansive apron partly fused with dorsal body wall. Mesenteries free but exhibit typical relations. Coelomic extension into umbilical cord obliterated.
16	112.0	Face looks 'human.' Hair of head appearing. Muscles become spontaneously active. Body outgrowing head.	Hard and soft palates differentiating. Hypophysis acquiring definitive structure.	Lymphocytes accumulate in tonsils. Pharyngeal tonsil begins development.	Gastric and intestinal glands developing. Duodenum and colon affixing to body wall. Meconium collecting.	Accessory nasal sinuses developing. Tracheal glands appear. Mesoderm still abundant between pulmonary alveoli. Elastic fibers appearing in lungs.	Greater omentum fusing with transverse mesocolon and colon. Mesoduodenum and ascending and descending mesocolon attaching to body wall.
20-40 (5-10 mo.)	160.0- 350.0	Lanugo hair appears (5). Vernix caseosa collects (5). Body lean but better proportioned (6). Fetus lean, wrinkled and red; eyelids reopen (7). Testes invading scrotum (8). Fat collecting, wrinkles smoothing, body rounding (8-10).	Enamel and dentine depositing (5). Lingual tonsil forming (5). Permanent tooth primordia indicated (6-8). Milk teeth unerupted at birth.	Tonsil structurally typical (5).	Lymph nodules and muscularis mucosae of gut present (5). Ascending colon becomes recognizable (6). Appendix lags behind caecum in growth (6). Deep esophageal glands indicated (7). Plicae circulares represented (8).	Nose begins ossifying (5). Nostrils reopen (6). Cuboidal pulmonary epithelium disappearing from alveoli (6). Pulmonary branching only two-thirds completed (10). Frontal and sphenoidal sinuses still very incomplete (10).	Mesenteric attachments completed (5). Vaginal sacs passing into scrotum (7-9).

Urogenital System	Vascular System	Skeletal System	Muscular System	Integumentary System	Nervous System	Sense Organs	Age in Weeks
Allantois present.	Blood islands appear on chorion and yolk sac. Cardiogenic plate reversing.	Head process (or noto-chordal plate) present.	. . . .	Ectoderm a single layer.	Neural groove indicated.	. . . .	2.5
All pronephric tubules formed. Pronephric duct growing caudad as a blind tube. Cloaca and cloacal membrane present.	Primitive blood cells and vessels present. Embryonic blood vessels a paired symmetrical system. Heart tubes fuse, bend S-shape and beat begins.	Mesodermal segments appearing ($1-16\pm$). Older somites begin to show sclerotomes. Notochord a cellular rod.	Mesodermal segments appearing ($1-16\pm$). Older somites show myotome plates.		Neural groove prominent; rapidly closing. Neural crest a continuous band.	Optic vesicle and auditory placode present. Acoustic ganglia appearing.	3.5
Pronephros degenerated. Pronephric (mesonephric) duct reaches cloaca. Mesonephric tubules differentiating rapidly. Metanephric bud pushes into secretory primordium.	Hemopoiesis on yolk sac. Paired aortae fuse. Aortic arches and cardinal veins completed. Dilated heart shows sinus, atrium, ventricle, and bulbus.	All somites present (40). Sclerotomes massed as primitive vertebrae about notochord.	All somites present (40).	. . . .	Neural tube closed. Three primary vesicles of brain represented. Nerves and ganglia forming. Ependymal, mantle and marginal layers present.	Optic cup and lens pit forming. Auditory pit becomes closed, detached otocyst. Olfactory placodes arise and differentiate nerve cells.	4
Mesonephros reaches its caudal limit. Ureteric and pelvic primordia distinct. Genital ridge bulges.	Primitive vessels extend into head and limbs. Vitelline and umbilical veins transforming. Myocardium condensing. Cardiac septa appearing. Spleen indicated.	Condensations of mesenchyme presage many future bones.	Premuscle masses in head, trunk and limbs.	Epidermis gaining a second layer (periderm).	Five brain vesicles. Cerebral hemispheres bulging. Nerves and ganglia better represented. [Suprarenal cortex accumulating.]	Chorioid fissure prominent. Lens vesicle free. Vitreous anlage appearing. Octocyst elongates and buds endolymph duct. Olfactory pits deepen.	5
Cloaca subdividing. Pelvic anlage sprouts pole tubules. Sexless gonad and genital tubercle prominent. Müllerian duct appearing.	Hemopoiesis in liver. Aortic arches transforming. L. umbil. vein and d. venosus become important. Bulbus absorbed into right ventricle. Heart acquires its general definitive form.	First appearance of chondrification centers. Desmocranium.	Myotomes, fused into a continuous column, spread ventrad. Muscle segmentation largely lost.	Milk line present.	Three primary flexures of brain represented. Diencephalon large. Nerve plexuses present. Epiphysis recognizable. Sympathetic ganglia forming segmental masses. Meninges indicated.	Optic cup shows nervous and pigment layers. Lens vesicle thickens. Eyes set at 160°. Naso-lacrimal duct. Modeling of ext., mid. and int. ear under way. Vomero-nasal organ.	6
Mesonephros at height of its differentiation. Metanephric collecting tubules begin branching. Earliest metanephric secretory tubules differentiating. Bladder-urethra separates from rectum. Urethral membrane rupturing.	Cardinal veins transforming. Inf. vena cava outlined. Atrium, ventricle and bulbus partitioned. Cardiac valves present. Stem of pulm. vein absorbed into l. atrium. Spleen anlage prominent.	Chondrification more general. Chondrocranium.	Muscles differentiating rapidly throughout body and assuming final shapes and relations.	Mammary thickening lens-shaped.	Cerebral hemispheres becoming large. Corpus striatum and thalamus prominent. Infundibulum and Rathke's pouch in contact. Chorioid plexuses appearing. Suprarenal medulla begins invading cortex.	Chorioid fissure closes, enclosing central artery. Nerve fibers invade optic stalk. Lens loses cavity by elongating lens fibers. Eyelids forming. Fibrous and vascular coats of eye indicated. Olfactory sacs open into mouth cavity.	7
Testis and ovary distinguishable as such. Müllerian ducts, nearing urogenital sinus, are ready to unite as utero-vaginal primordium. Genital ligaments indicated.	Main blood vessels assume final plan. Primitive lymph sacs present. Sinus venosus absorbed into right atrium. Atrio-ventricular bundle represented.	First indications of ossification.	Definitive muscles of trunk, limbs and head well represented and fetus capable of some movement.	Mammary primordium a globular thickening.	Cerebral cortex begins to acquire typical cells. Olfactory lobes visible. Dura and pia-arachnoid distinct. Chromaffin bodies appearing.	Eyes converging rapidly. Ext., mid. and int. ear assuming final form. Taste buds indicated. External nares plugged.	8
Kidney able to secrete. Bladder expands as sac. Genital duct of opposite sex degenerating. Bulbo-urethral and vestibular glands appearing. Vaginal sacs forming.	Thoracic duct and peripheral lymphatics developed. Early lymph glands appearing. Enucleated red cells predominate in blood.	Ossification centers more common. Chondrocranium at its height.	Perineal muscles developing tardily.	Epidermis adds intermediate cells. Periderm cells prominent. Nail field indicated. Earliest hair follicles begin developing on face.	Spinal cord attains definitive internal structure.	Iris and ciliary body organizing. Eyelids fused. Lacrimal glands budding. Spiral organ begins differentiating.	10
Uterine horns absorbed. External genitalia attain distinctive features. Meson. and rete tubules complete male duct. Prostate and seminal vesicle appearing. Hollow viscera gaining muscular walls.	Blood formation beginning in bone marrow. Blood vessels acquire accessory coats.	Notochord degenerating rapidly. Ossification spreading. Some bones well outlined.	Smooth muscle layers indicated in hollow viscera.	Epidermis three-layered. Corium and subcutaneous now distinct.	Brain attains its general structural features. Cord shows cervical and lumbar enlargements. Cauda equina and filum terminale appearing. Neuroglial types begin to differentiate.	Characteristic organization of eye attained. Retina becoming layered. Nasal septum and palate fusions completed.	12
Kidney attains typical shape and plan. Testis in position for later descent into scrotum. Uterus and vagina recognizable as such. Mesonephros involuted.	Blood formation active in spleen. Heart musculature much condensed.	Most bones distinctly indicated throughout body. Joint cavities appear.	Cardiac muscle appearing in earlier weeks, now much condensed. Muscular movements in utero can be detected.	Epidermis begins adding other layers. Body hair starts developing. Sweat glands appear. First sebaceous glands differentiating.	Hemispheres conceal much of brain. Cerebral lobes delimited. Corpora quadrigemina appear. Cerebellum assumes some prominence.	Eye, ear and nose grossly approach typical appearance. General sense organs differentiating.	16
Female urogenital sinus becoming a shallow vestibule (5). Vagina regains lumen (5). Uterine glands appear (7). Scrotum solid until sacs and testes descend (7-9). Kidney tubules cease forming at birth.	Blood formation increasing in bone marrow and decreasing in liver (5-10). Spleen acquires typical structure (7). Some fetal blood passages discontinue (10).	Carpal, tarsal and sternal bones ossify late; some after birth. Most epiphyseal centers appear after birth; many during adolescence.	Perinal muscles finish development (6).	Vernix caseosa seen (5). Epidermis cornifies (5). Nail plate begins (5). Hairs emerge (6). Mammary primordia budding (5); buds branch and hollow (8). Nail reaches finger tip (9). Lanugo hair prominent (7); sheds (10).	Commissures completed (5). Myelinization of cord begins (5). Cerebral cortex layered typically (7). Cerebral fissures and convolutions appearing rapidly (7). Myelinization of brain begins (10).	Nose and ear ossify (5). Vascular tunic of lens at height (7). Retinal layers completed and light perceptive (7). Taste sense present (8). Eyelids reopen (7-8). Mastoid cells unformed (10). Ear deaf at birth.	20-40 (5-10 mo.)

CHAPTER 20

Nervous System

The use of intelligence is the highest privilege and the deadliest menace of humanity.   JOSEPH JASTROW

529

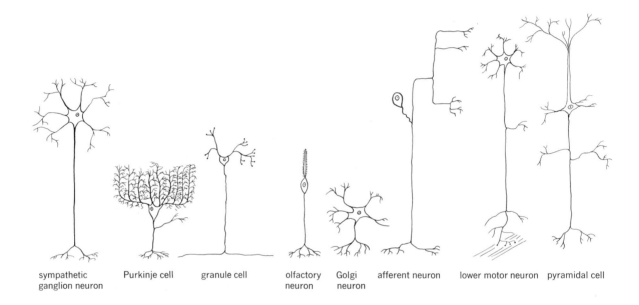

sympathetic
ganglion neuron

Purkinje cell

granule cell

olfactory
neuron

Golgi
neuron

afferent neuron

lower motor neuron

pyramidal cell

FIGURE 20–1. Some different forms of neurones from the human nervous system.

NERVOUS SYSTEM

Introduction

The activities of an organism and its parts must be integrated and controlled for the ultimate survival of both. Two systems of the body provide for most of this integration and control, the nervous system and the endocrine system. The nervous system exerts a rapid and more specific control over body function and is the interpretive system which enables us to appreciate the myriad stimuli which the body receives. The nervous system of man also possesses the ability to store information for later use. In many ways, our nervous system has many points in common with the computer.

The basic properties which all portions of the nervous system have developed are *irritability* and *conductivity*. Irritability implies the capacity to respond, by formation of a nerve impulse, to changes in the internal or external environment. Conductivity implies the ability to propagate or transmit the impulse along nerve fibers. The cells of the nervous system carrying out these functions are the neurones.

530

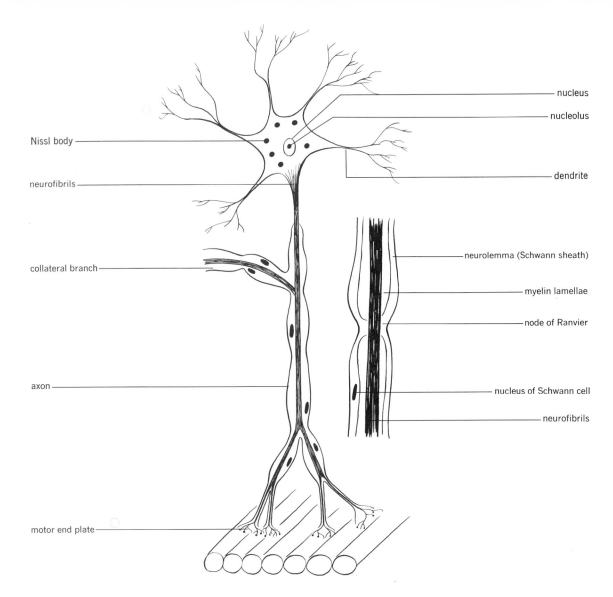

Nissl body

neurofibrils

collateral branch

axon

motor end plate

nucleus

nucleolus

dendrite

neurolemma (Schwann sheath)

myelin lamellae

node of Ranvier

nucleus of Schwann cell

neurofibrils

FIGURE 20–2. The structure of a multipolar neurone.

THE TISSUES OF THE NERVOUS SYSTEM

Neurones

There are numerous forms of neurones (Fig. 20–1). The most used classification describes their form according to the number of extensions (processes) which arise from the cell body. For example:

1. *Unipolar*—one process extending from the cell body
2. *Bipolar*—two processes extending from the cell body
3. *Multipolar*—more than two processes extending from the cell body

The multipolar neurone is the type most used to illustrate the parts of a neurone, for it has all of the features found in any neurone.

Structure of a multipolar neurone (Fig. 20–2)

A neurone is composed of three anatomically distinct parts:

1. The *cell body (perikaryon)* houses a large spherical nucleus con-

531

taining a prominent nucleolus. The cytoplasm contains the usual cell organelles; mitochondria, Golgi body, lysosomes, and reticulum. Conspicuously lacking, after about 16 years of age, is the cell center, instrumental in the process of cell division. The implication that mature neurones cannot divide or replace themselves is true. The cytoplasm contains two organelles unique to neurones, the *Nissl substance* and the *neurofibrils*. The Nissl substance is composed of darkly staining granules of RNA, and is regarded as the site of protein synthesis by the neurone. The neurofibrils are tiny tubular structures running through the cell body and into some of the processes. Their function is not known.

2. *Afferent processes.* Multiple, highly branched processes which extend only short distances from the cell body, and which conduct impulses toward *(afferent)* the cell body are termed *dendrites*. Their branches are studded with short spine-like extensions or *gemmules*.

3. *Efferent process.* Each neurone has a single, very long, sparsely branched process termed the *axon*. The axon conducts impulses away from *(efferent)* the cell body to some effector (organ which responds) or to another neurone. The axon is usually surrounded by a segmented fatty sheath *(myelin sheath)* and a thin Schwann sheath or *neurilemma* (Fig. 20–3A, B). The segments in the myelin sheath are termed the

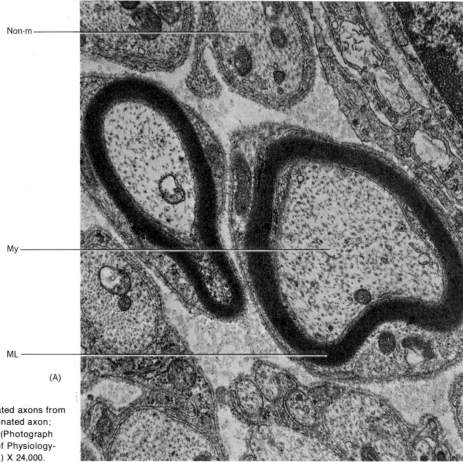

Non-m

My

ML

(A)

FIGURE 20–3A. Myelinated and nonmyelinated axons from the trachea of lab mouse. Non-m, nonmyelinated axon; My, myelinated axon; ML, myelin lamellae. (Photograph supplied by Norton B. Gilula, Department of Physiology-Anatomy, University of California, Berkeley.) X 24,000.

nodes (of Ranvier). An axon possessing such a sheath is said to be myelinated and can conduct an impulse at a faster rate than an axon lacking the sheath. The neurilemma is important in formation of the myelin and during regeneration of injured axons. At the junction with an effector or another neurone, the myelin sheath and neurilemma are lost, and the axon terminates in a number of fine branches known as the *telodendria*.

Neuroglia (Fig. 20–4)

Cells which serve the primary function of support within the nervous system are termed the neuroglia. The category includes the following types of cells:

1. *Ependyma* are short ciliated cells lining the cavities of the brain and spinal cord.
2. *Astrocytes* are star-shaped cells forming supporting networks within the brain and cord.
3. *Oligodendrocytes* are arranged in rows between the fibers of the brain and cord. They produce the myelin sheaths on fibers in the central nervous system.

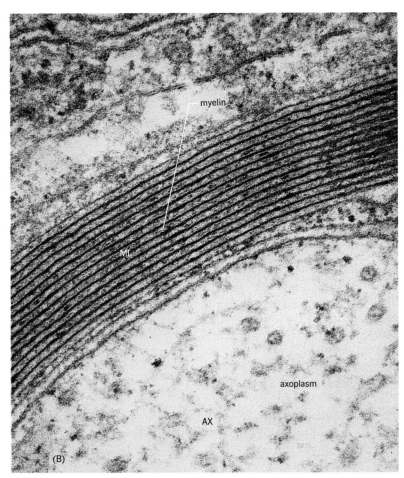

myelin

ML

axoplasm

AX

(B)

FIGURE 20–3B. Myelin surrounding an axon in the trachea of the lab mouse. AX, axon; ML, myelin lamellae. (Photograph supplied by Norton B. Gilula, Department of Physiology-Anatomy, University of California, Berkeley.) X 142,000.

4. *Microglia* are small, ameboid, and phagocytic cells which remove disintegrating elements within the brain and cord.
5. *Schwann cells* form the neurilemma.
6. *Satellite cells* surround the neurone cell bodies in outlying areas of the nervous system.

Glia are also regarded as contributing nutritive materials to the neurones and, in some cases, may become important in the formation of memory traces.

NEURONE FUNCTION

The basis of excitability

In order to create an excitable state, a cell must first achieve a difference in ionic concentrations on the inside and outside of its cell membrane. This ionic differential will create an electrical difference *(potential)* and render the cell excitable. The methods by which ionic differences are created are:

1. *Limiting free diffusion* of ions by the presence of a semipermeable membrane. The cell membrane acts selectively to place a barrier to free ionic movement across it. Potassium ion has a smaller hydrated diameter than does sodium ion, therefore it diffuses more rapidly across the membrane. The membrane thus tends to keep sodium out of the cell.

2. *Electrical charge* of the diffusing material. The proteins and fats of the membrane carry ionized groups which give the membrane an electrical charge. If a substance approaching the membrane has a like charge, it will be repelled.

3. *The Donnan-Gibbs equilibrium.* A special type of diffusion, this method results when there is present on one side of the membrane a large, nondiffusible substance. Ions will diffuse and at equilibrium an imbalance in ionic species will be achieved.

	Start		Equilibrium	
	$5 Na^+$	$10 Na^+$	$5 Na^+$	
Nondiffusible anion $5 A^-$		$10 Cl^-$	$5 A^-$	
			$4 Na^+$	$6 Na^+$
			$4 Cl^-$	$6 Cl^-$
	membrane		membrane	

It may be noted that at the start, there is a diffusion gradient for both Na^+ and Cl^- from right to left. Ions therefore diffuse from right to left in equal numbers, so as to maintain electrical equality. At equilibrium, two criteria must be met:

(a) The *product* of diffusible ions must be equal on both sides of the membrane

on left $5 Na + 4 Na = 9 Na^+$
                      $4 Cl^- = 4 Cl^-$ $9 Na^+ \times 4 Cl^- = 36$
on right $10 Na^+ - 4 Na^+ = 6 Na^+$ $6 Na^+ \times 6 Cl^- = 36$
                $10 Cl^- - 4 Cl^- = 6 Cl^-$

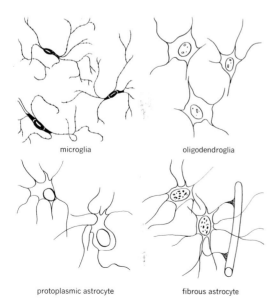

microglia   oligodendroglia

protoplasmic astrocyte   fibrous astrocyte

FIGURE 20–4. Neuroglia from the central nervous system.

(b) The side containing the nondiffusible substance will have a greater concentration of diffusible cations (9 Na$^+$) than the other side (6 Na$^+$). These processes operate in any cell to create an electrical potential between the inside and outside of the cell on the order of 20 to 30 mV. The electrical potential as measured on a neurone is 70 to 90 mV. Therefore, something more must be operating in the neurone to account for the difference. ∎

4. *Active processes.* Neurone membranes possess a carrier capable of actively transporting sodium ion to the exterior of the cell and potassium ion to the interior.

The mechanism may operate as follows:

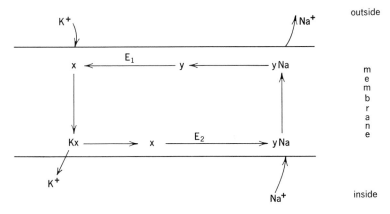

∎   Sodium combines with a specific carrier (y) and is transported to the interstitial fluid. Carrier y is enzymatically transformed (E$_1$) into carrier x which brings potassium in. Again, at the inside of the membrane, enzymatic transformation (E$_2$) changes x to y and the process is repeated.

The operation of this sodium-potassium pump creates a large imbalance of K$^+$ and Na$^+$ between the cell and its surrounding fluid (values in meq/l):

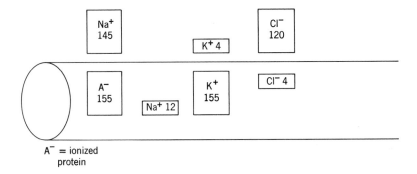

A$^-$ = ionized protein

Mathematical computation shows that the outside of the cell has a net positive charge [(145 + 4) − 120] or 29, while the inside has a net positive charge of [(155 − 155) + (12 − 4)] or 8.

The outside of the fiber is thus *more positive* than the inside, or to put it another way, the inside of the fiber is electrically negative to the outside.

This situation is represented as follows:

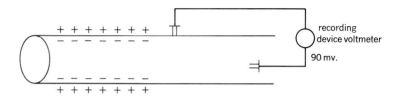

Measurement of the electrical difference shows a value of about 90 mV. The fiber is said to be *polarized,* and is excitable. ■

Formation of the nerve impulse

From the foregoing section, it should become apparent that the fiber is being maintained in the polarized state primarily by the sodium-potassium transport system, and that, as a result of the transport, there has been created a very large diffusion gradient for Na⁺ into the fiber. An interruption of the transport system, as by the application of a stimulus to the fiber, will allow the immediate inflow of Na⁺ by diffusion. Enough inflow of Na⁺ occurs to first equal, and then exceed, the net concentration of negative charges inside the fiber. The electrical potential thus first falls to zero, then reverses. The fiber is *depolarized*. A record of the electrical events occurring as Na⁺ inflow occurs is shown below:

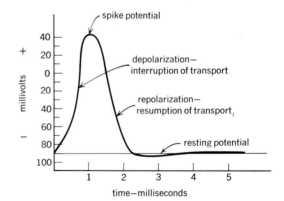

The electrical picture changes as follows:

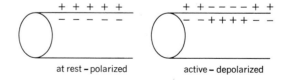

Until the original state has been achieved (*repolarization*), the fiber cannot conduct or form another impulse, and is in a *refractory state*. About 3 msec are required to repolarize.

Once the depolarized state has occurred, it is assumed that a battery effect is created between the polarized and depolarized sections. Current flows between the two areas and an electrical field results:

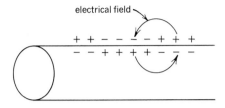

This field is actually the impulse.

Conductivity

The strength of the field decreases rapidly from its center. However, for a finite distance from its center, it is strong enough to cause depolarization of the next section of the fiber. The disturbance is advanced by that distance down the fiber, sodium-potassium transport resumes in the previously depolarized area, and repolarization occurs. The impulse thus advances itself down the fiber, that is, it is conducted. Propagation of the impulse in this manner is termed conduction by a core conductor.

Saltatory conduction

In myelinated fibers, the only regions where current flow can occur is at the nodes. These are spaced along the fiber at 1- to 3-mm intervals. In such a fiber, the impulse actually jumps from node to node, and thus progresses at a faster rate down the fiber. A myelinated fiber may conduct up to 20 times faster than an unmyelinated one.

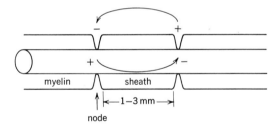

Another factor is also involved in determining the speed of impulse transmission: the diameter of the fiber. Three main types of fibers are recognized:

1. *A fibers.* These range in size from 1 to 20 microns in diameter, and have conduction velocities from 5 m/sec for the small fibers to 10 m/sec for the large ones. All A fibers are myelinated, and are found in the great motor and sensory nerves of the body.
2. *B fibers.* These fibers range in size from 1 to 3 microns in diameter, and conduct at speeds of from 3 to 14 m/sec. B fibers belong to the involuntary portion *(autonomic)* of the nervous system.
3. *C fibers.* These fibers are the smallest, being less than 1 micron in diameter. They are unmyelinated, and are found in skin and visceral nerves.

In general, an increase of 1 m/sec per micron of overall fiber diameter can be expected. The fastest fibers are therefore both large and myelinated.

Other physiological properties of neurones

In addition to being excitable and conductile, neurones:

1. *Follow the all-or-none law.* For a given strength of stimulus, the neurone either responds with a full sized impulse, or none at all. Apparently, the transport mechanism has some threshold for stoppage, and a stimulus either stops it or doesn't.
2. *Have a very short refractory period.* During depolarization, no additional impulses may be conducted. Repolarization occurs very rapidly (3 msec maximum), so that the neurone is capable of conducting a series of impulses.
3. *Exhibit a rheobase.* Rheobase refers to the *strength* of stimulus required to depolarize. Not all fibers have the same rheobase; some of them depolarize more easily than others.
4. *Exhibit a chronaxie.* Chronaxie is the *time* required for a 2x rheobase current to cause depolarization. It implies that any stimulus must last a certain length of time in order to cause depolarization.
5. *Show accommodation.* If a stimulus does not reach its peak value rapidly, no depolarization may occur even though the peak is above rheobase. The neurone accommodates, or increases its rheobase. This enables a neurone to be selective of stimuli. If we reacted to all stimuli which impinge upon us, we would literally be jumping all the time.

The properties of neurones may be seen to be similar to the properties of some other tissues which have been studied (notably muscle). As is true in many types of tissue, nerve tissue is better at the job than others.

Metabolism of nervous tissue

The chief fuel for continuance of neurone activity appears to be glucose, derived from the blood stream. The brain has an RQ (respiratory quotient) of 1.0, indicating a carbohydrate metabolism. If glucose supply diminishes, the brain is capable of utilizing lactic and pyruvic acids, and, occasionally, butyric acid. Peripheral nervous tissues, or those lying outside of the brain and spinal cord, have an RQ of about 0.8. This indicates some noncarbohydrate utilization. Fats may provide the noncarbohydrate energy.

The role of lipids in the metabolism of nervous tissue appears to be concerned with their synthesis and use in myelin sheaths, in phospholipids in cell membranes, and not as energy sources.

Amino acids are rapidly synthesized into synaptic transmitters (see below) and peptide hormones.

Nerve cells have the highest RNA content of any somatic body cell. The high levels of this nucleic acid may be associated with the neurones ability to store information as memory traces.

Oxygen utilization by the brain removes about 20 percent of the inspired gas. Measurements of total oxygen consumption by the nervous system puts it on a par with resting muscle. On a weight basis, the oxygen consumption of the brain is about 30 times as great as muscle.

Electrical discharge by neurones

In addition to the resting and spike potentials, spontaneous rhythmical electrical discharges may be recorded from the brain by placing electrodes on the scalp. The record of these waves is known as an electroencephalogram or EEG. Figure 20–5 shows an EEG recorded from a normal person at rest. Several basic types of rhythm may be detected upon analysis of the record:

1. *Alpha waves* may be recorded best in the inattentive person, such as during drowsiness or while at rest with the eyes closed. They are waves occurring about 10 to 12 times per second, and have magnitudes of about 50 μV. (1 μV = 1 millionth of a volt). Alpha waves diminish in magnitude when the eyes are opened, or when one's attention is directed toward a new or novel stimulus.
2. *Beta waves* occur 15 to 60 times per second and have a magnitude of 5 to 10 μV.
3. *Delta waves* occur during sleep at frequencies of 1 to 5 per second with voltages of 20 to 200 μV.
4. *Theta waves,* 5-8/sec., 10 μV.

Disruption of normal EEG patterns often supplies clues to deranged brain function. Figure 20–6 shows some abnormal EEGs in a variety of clinical disorders.

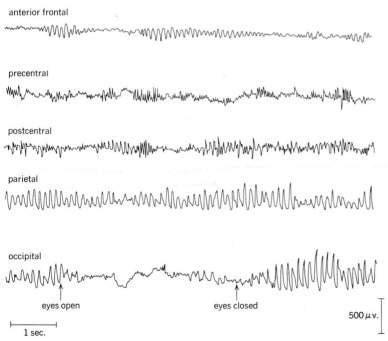

FIGURE 20–5. Normal EEG patterns from different regions of the cortex. Alpha waves predominate in parietal and occipital areas; beta waves in precentral area. Alpha waves are blocked when eyes are opened.

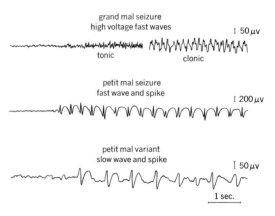

FIGURE 20–6. Some abnormal EEG's. Note large spikes and fast rhythm.

THE SYNAPSE

A junction between two neurones is a synapse. It is an area where two neurones come in close approximation to one another, but they are not anatomically continuous. The distal ends, telodendria, of an axon usually carry small (1 micron), round or oval expansions known as the *bouton terminaux* or *end feet.* These feet may contact the dendrites (usually), cell body, or axon of the next neurone in the chain. The electron microscope has shown that the feet contain numerous small granular structures known as *synaptic vesicles.* Analysis of the vesicles shows them to contain a chemical substance. The actual material may be acetylcholine, serotonin, norepinephrine, levarterenol, histamine, certain amino acids, or GABA (see below), depending upon the location of the synapse in the nervous system.

The transmission of an impulse across the synapse appears to occur chemically, not electrically as occurs on the axon itself. When an impulse arrives over an axon, it causes the rupture of the vesicles, and the contained chemical diffuses from the foot into the synapse and depolarizes the next neurone. Continued excitation of the second neurone is prevented by an enzyme which destroys the transmitter. For example, cholinesterase destroys acetylcholine; monamine oxidase destroys norepinephrine. Resynthesis of new vesicles occurs within the foot between impulses.

Due to the different mode of transmission, the synapse has a number of properties different from those of the neurones composing it.

1. *One-way conduction.* Since only axons possess the vesicles, and can liberate a chemical, transmission can only occur in an axon-synapse-dendrite direction. The value of such a control gate is obvious in preventing wrong-way conduction.
2. *Synaptic delay.* It takes 2 to 3 times longer for an impulse to cross a synapse than to pass an equivalent distance along an axon. The extra time is required for rupture of vesicles, diffusion of chemical, destruction and resynthesis of chemical.
3. *Inhibition.* A synapse does not automatically pass an impulse. It may block, facilitate, or actually alter the impulse. Inhibition involves blockage of impulse passage through the production of a nontransmitter agent, gamma-aminobutyric acid (GABA). Its value lies in enabling selection of important impulses from among the trivia which flood our nervous systems.
4. *Potentiation.* Passage of an impulse with greater ease than normal may occur at the synapse; this is potentiation. In the development of this property, the end feet may be seen to enlarge. The creation of preferred pathways for impulse passage could be carried out in this fashion, an important fact in learning.
5. *Repetitive discharge.* A single stimulation on one neurone in a series may result in a volley of impulses being sent down the second member of the chain. The impulse appears to have been multiplied by the synapse. This effect may be understood by assuming that neurone number 2 has a repolarization time faster than the ability of the enzyme to destroy the transmitter. The neurone is depolarized by the chemical, repolarizes, and is faced with still enough chemical to cause it to depolarize again. This situation

may continue until transmitter concentration is brought to less than that required to depolarize.

6. *Fatigue.* If stimulation, causing vesicle breakdown, is so rapid as to prevent resynthesis of vesicles between stimuli, the synapse may cease to transmit, that is, it shows fatigue. Apparently, synapses fatigue before neurones do, and thus act as a safety device.

7. *Effects of drugs.* The synapse is a region which is easily affected by drugs, particularly depressants. Many tranquilizing agents act on the synapse.

While the above list of synaptic properties is not exhaustive, it will serve to indicate that the synapse is not *only* an area of neuronal contact; it is an extremely important area of impulse control.

THE ORGANIZATION OF NEURONES INTO FUNCTIONAL UNITS

The reflex arc

Neurones do not operate alone. Two or more are placed in series to carry impulses to and from body organs. A complete unit which is capable of detecting an environmental change, and which causes an automatic, unconscious, and appropriate reponse to that change, is termed a *reflex arc* (Fig. 20–7). An arc always has five basic parts to it.

1. A *receptor.* This structure is usually some specialized organ capable of responding to an alteration of the environment by the production of a nerve impulse.
2. *An afferent neurone.* This neurone carries the impulse *toward* the central nervous system (brain and cord).
3. *A center or synapse.* This area is usually found within the central nervous system, and exerts control over the incoming impulse. The impulse may be blocked, transmitted or rerouted. Most reflex arcs have, inserted between the afferent neurone and the next neurone in the chain, an association or internuncial neurone. These provide additional pathways for the impulse to take.
4. *An efferent neurone.* This neurone leaves the central nervous system and carries the impulse towards the organ which will respond.
5. *An effector.* The effector is the structure which responds to the impulse. It is usually some type of muscle or secretory structure.

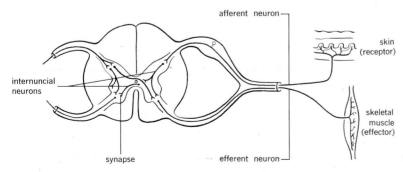

FIGURE 20–7. A reflex arc.

The reflex arc forms the most important unit for automatic control of the body's involuntary processes. Such activities as posture maintenance, control of heart rate, breathing, gastrointestinal activity, blood pressure, and many others are reflexly controlled.

Functionally, the reflexes themselves may be classed as *exteroceptive* if the receptor is located at or near the external body surface; *interoceptive* if the receptor is within a visceral organ or blood vessel; and *proprioceptive* if the receptor is located within a muscle or tendon.

SPECIFIC FUNCTIONS OF THE NERVOUS SYSTEM

Introduction

In an attempt to trace the functional aspects of the nervous system, we have departed from the usual consideration of areas of the nervous system such as brain, cord, and peripheral system. We shall present the next two Sections from the viewpoint of what pathways an impulse which results in movement of a skeletal muscle may follow, and the pathways an incoming sensory impulse may follow. As we trace these pathways, we shall, in time, describe the functions of all parts of the central nervous system. Some basic anatomy of the central nervous system is needed to understand the following Sections.

Parts of the nervous system

A simple classification of the system is presented below and diagrammed in Fig. 20–8.

  I. Central nervous system.
    A. Brain—The portion contained within the skull
      1. Brain stem—portion continuous with spinal cord; centrally located in brain.
        (a) Medulla oblongata—lower portion of stem
        (b) Pons—bulging portion above medulla
        (c) Midbrain—wedge shaped portion above pons
        (d) Thalamus—rounded area at apex of stem
        (e) Hypothalamus—area below thalamus
      2. Cerebellum—located dorsally on medulla
      3. Cerebrum—large, mushroom-shaped portion atop the stem, filling most of the skull.
    B. Spinal cord—the portion contained within the spinal column.

  II. Peripheral nervous system—all remaining nervous tissue. Chiefly the fibers to and from the central nervous system.
    A. Somatic nervous system—the fibers to skeletal muscle, and fibers from the receptors of the skin.
    B. Autonomic nervous system—fibers to smooth and cardiac muscle, and glands, and fibers from the same areas. Almost every body organ receives a double innervation of autonomic fibers, one set from each of the following subdivisions:
      1. Sympathetic division—generally raises level of activity and is most active during stress.
      2. Parasympathetic division—generally lowers activity and operates during normality.

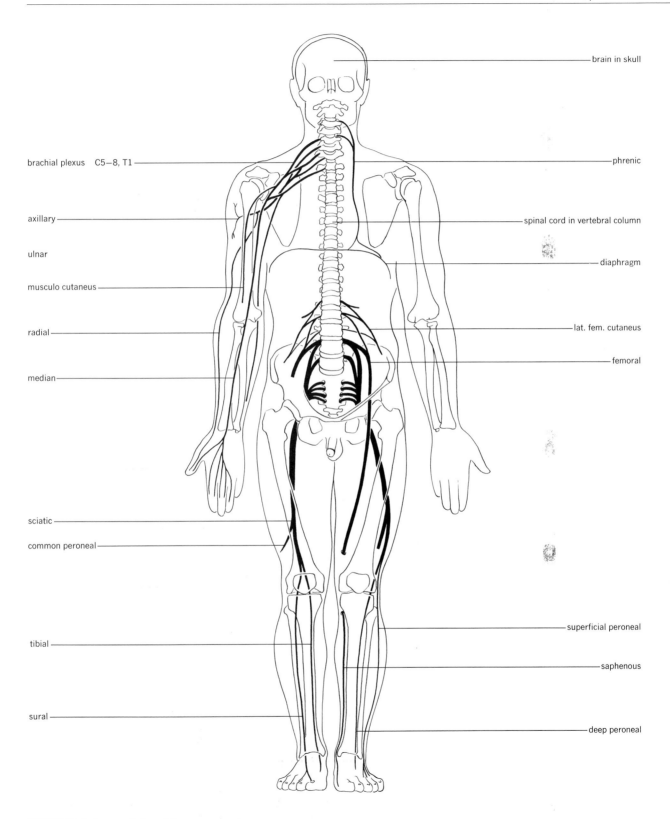

brain in skull

brachial plexus   C5—8, T1

phrenic

axillary

spinal cord in vertebral column

ulnar

diaphragm

musculo cutaneus

radial

lat. fem. cutaneus

median

femoral

sciatic

common peroneal

superficial peroneal

tibial

saphenous

sural

deep peroneal

FIGURE 20–8. A general view of the nervous system.

MOTOR FUNCTIONS OF THE NERVOUS SYSTEM

The ability to consciously move the skeletal muscles is implied in the words motor functions. We must realize that these same muscles may be reflexly controlled as well. In determining the pathway such an impulse may take, six parts of the nervous system appear to be involved:

1. The cerebrum; specifically, the outer covering of neurones known as the cerebral cortex.
2. The basal ganglia, masses of neurones buried deep within the cerebrum.
3. The reticular formation, a mass of neurones located within the brain stem.
4. The cerebellum.
5. The spinal cord, the highway over which the outflowing impulses pass.
6. The efferent neurones to the muscles themselves.

The contribution of each of these areas to the production of purposeful movement will be considered.

The cerebrum (Fig. 20–9)

The cerebrum is divided into right and left hemispheres by the *longitudinal fissure*. Each hemisphere is further subdivided into lobes by additional fissures: the *central fissure* separates a *frontal* from a *parietal* lobe; the *parieto-occipital fissure* separates the *parietal* from an *occipital* lobe; the *lateral fissure* separates a *temporal* lobe from the remaining lobes.

The surface of the cerebrum is convoluted and consists of upfolds *(gyri)* separated by downfolds *(sulci)*. Two gyri worth mentioning are the ones immediately in front of and behind the central fissure. These are known as the *precentral* and *postcentral gyri* respectively. The outermost layer of the cerebrum, or *cerebral cortex*, is 2 to 4 mm in thickness and contains the cell bodies (estimated to number 7 to 12 billion) for motor function and sensory function and interpretation. All areas are represented in each cerebral hemisphere, and coordination of activity is assured by the *corpus callosum* connecting the hemispheres.

Localization of motor function. It is proper to speak of cortical motor *areas,* for motor function is not restricted to a single region. Brodman, many years ago, attempted to correlate the structure and function of the cortex and arrived at a numerical designation of regions showing differential morphology (Fig. 20–10). Several of these many regions appear to be of primary importance to motor function. Area 4 occupies most of the precentral gyrus and is known as the *primary motor area*. It is the point of origin for those fibers which carry voluntary impulses to skeletal muscles. The typical cell of the motor area is the *pyramidal cell* (Fig. 20–11). It has been estimated that the fibers arising from the pyramidal cells of area 4 contribute, at maximum, only 3 percent of the fibers composing the motor pathways of the cord. Involuntary pathways therefore form the bulk of the motor fibers. In area 4, the body is represented upside down, with larger regions given over to those body areas requiring finer control (Fig. 20–12). Mild stimulation within area 4

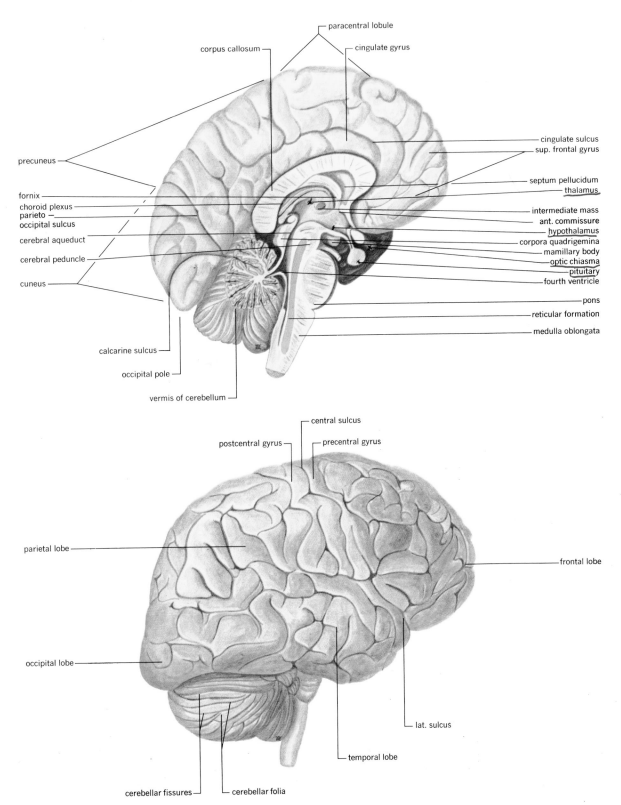

paracentral lobule
corpus callosum
cingulate gyrus
cingulate sulcus
sup. frontal gyrus
precuneus
septum pellucidum
thalamus
fornix
choroid plexus
parieto —
occipital sulcus
intermediate mass
ant. commissure
hypothalamus
cerebral aqueduct
corpora quadrigemina
mamillary body
cerebral peduncle
optic chiasma
pituitary
cuneus
fourth ventricle
pons
reticular formation
medulla oblongata
calcarine sulcus
occipital pole
vermis of cerebellum

central sulcus
postcentral gyrus
precentral gyrus
parietal lobe
frontal lobe
occipital lobe
lat. sulcus
temporal lobe
cerebellar fissures
cerebellar folia

FIGURE 20-9. The brain.

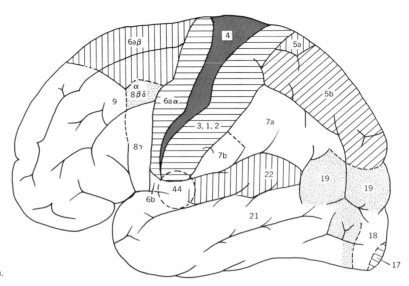

FIGURE 20–10. Some functional areas of the cerebrum.

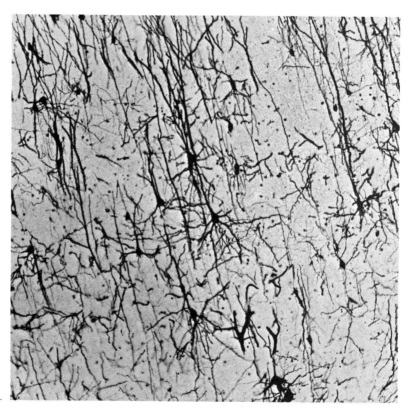

FIGURE 20–11. Pyramidal cells.

gives individual muscle contraction; stronger stimulation causes several related muscles to contract in a more purposeful movement.

Areas 6 and 8 are the derived motor areas. Area 6 fixes a body part to create a postural background against which skilled movements may be performed. Area 8 deals with fixation of the eyes upon objects being viewed.

The parietal lobe (areas 3, 1, 2, and 5) contains motor regions, which, when stimulated, cause a generalized movement associated with a sensation which may be localized to a given body area.

Area 44, Broca's speech area, includes the head portions of areas 4, 3, 1, 2, and a part of the temporal lobe. Damage in this region results in the inability to form words. Areas 20 and 21 have evolved simultaneously with area 44 to enable a person to choose the word he wishes to sound.

We may have implied in the foregoing paragraphs that the cortex acts on its own to initiate voluntary movement. This is not strictly true. The motor cortex receives sensory signals continuously from eyes, ears, and other peripheral receptors. It is probable that the cortex must be activated by some outside agent before a movement can be initiated.

Results of damage to the primary motor area include a flaccid (relaxed) paralysis of the associated body muscles on the side of the body opposite to the site of injury.

Basal ganglia

Included as members of the basal ganglia are groups of neurones forming what are termed *nuclei*. Specific nuclei are (Fig. 20–13):

1. Caudate nucleus
2. Lenticular nucleus, composed of:
    (a) Putamen
    (b) Globus pallidus
3. Substantia nigra
4. Subthalamic nucleus
5. Red nucleus

Connections exist between individual nuclei, and between the nuclei and cerebral cortex. The motor functions of the basal ganglia have been investigated primarily in lower animals. In such animals, the basal ganglia appear to perform nearly all the motor functions the animal possesses. The cortex is a minor contributor. In humans, the question of "how much function is left?" arises, because the cerebrum has taken over much of the functions of the ganglia. Results of animal investigations suggest the following conclusions.

Caudate and putamen.   These may initiate and regulate gross intentional movements performed unconsciously, for example, swinging the legs during walking. The cortex will then add to this movement the fine discrete motions which do require conscious thought.

Globus pallidus.   The globus appears to be a faciliatory nucleus, one whose stimulation increases muscle tone generally. It can act as an arousal area; for example, to bring an animal out of anaesthesia.

Subthalamic nucleus.   This area is essential for the establishment of rhythmical motions such as walking or chewing.

Substantia nigra.   This area appears to be concerned with control of tension in the muscle spindles (Fig. 20–14). If damaged, the nigra causes oscillatory movements of the limbs to occur.

Red nucleus.   This nucleus sends fibers to the muscles of the neck and trunk for postural maintenance.

Damage to the basal ganglia generally results in the genesis of abnormal movements, including:

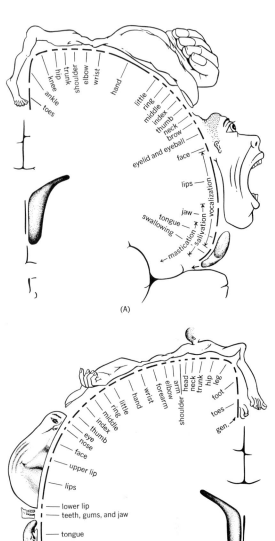

FIGURE 20–12. Cortical representation of motor and sensory functions. A. Motor homunculus. The figure represents, on a coronal section of the precentral gyrus, the location of the cortical representation of the various parts. The size of the various parts is proportionate to the amount of cortical area devoted to them. B. Sensory homunculus, drawn overlying a coronal section through the postcentral gyrus.

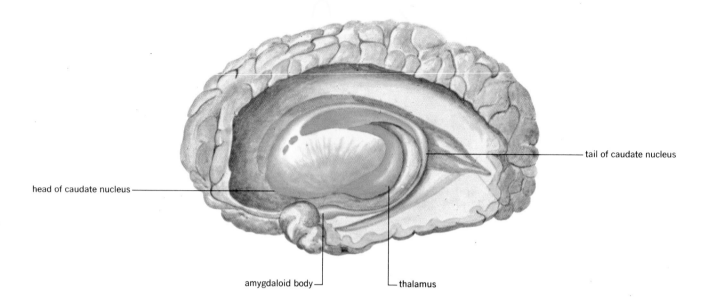

head of caudate nucleus

tail of caudate nucleus

amygdaloid body

thalamus

FIGURE 20–13. The basal ganglia.

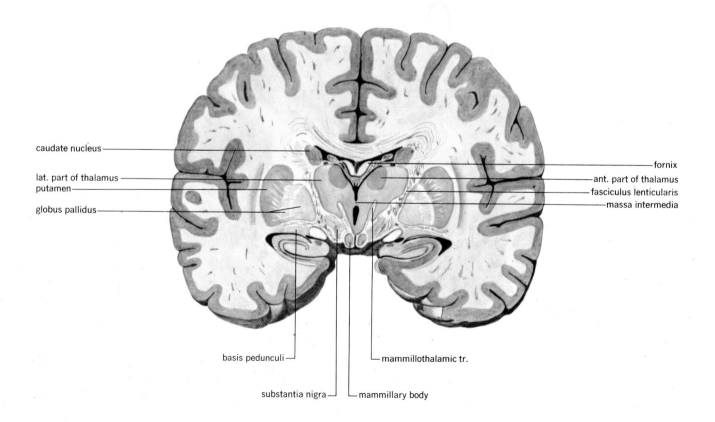

caudate nucleus

lat. part of thalamus

putamen

globus pallidus

fornix

ant. part of thalamus

fasciculus lenticularis

massa intermedia

basis pedunculi

mammillothalamic tr.

substantia nigra

mammillary body

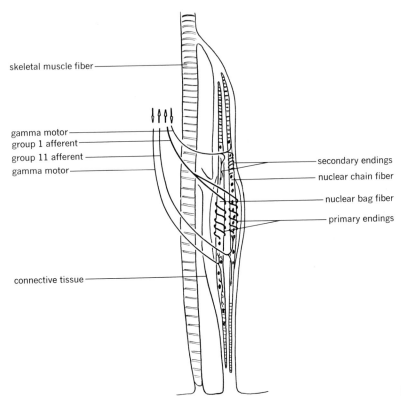

skeletal muscle fiber

gamma motor
group 1 afferent
group 11 afferent
gamma motor

secondary endings
nuclear chain fiber
nuclear bag fiber
primary endings

connective tissue

FIGURE 20–14. The muscle spindle.

Chorea (St. Vitus' dance)—a series of uncoordinated motions interrupting one another.

Athetosis—similar to chorea but at a slower rate.

Parkinson's disease—probably caused by basal ganglion injury. It is seen in older people and is characterized by tremor.

The reticular formation

Included in the reticular formation are all the neuronal bodies of the brain stem excluding those of the cranial nerves (Fig. 20–9). The formation receives input from the cortex and basal ganglia and sends its fibers down the cord. Stimulation of most parts of the reticular formation results in some type of excitation of the muscles: for example, increased tone, contraction of large or discrete groups of muscles, and stimulation of agonists while inhibiting the antagonists. The arousing nature of the formation has led to its inclusion as part of the *reticular activating system* which maintains the waking state of the organism.

Cerebellum (Fig. 20–15)

Every motor impulse originating anywhere else in the brain will pass through the cerebellum before reaching the muscles. Any muscle which changes its length, actively or passively, will send a sensory impulse to the cerebellum. The cerebellum thus integrates information which comes from the cortex, basal ganglia, and periphery. When this system operates, it can:

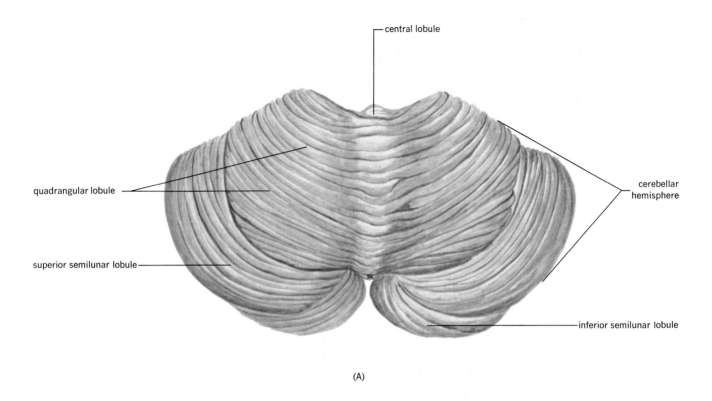

(A)

FIGURE 20–15. The cerebellum. A. Dorsal surface. B. Ventral surface.

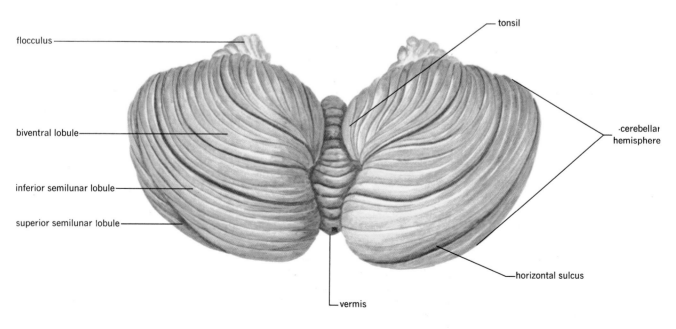

(B)

1. Exert *error control* on body movements. By comparing the intention of the cortex with actual body movement, the cerebellum insures that the limb will go where it is supposed to.
2. Exert a *damping action*, to stop the movement at the proper time.
3. Exert a *predictive function.* Detection of the rapidity of limb movement and prediction of when to stop the motion.

The cerebellum also receives fibers from the semicircular canals and maculae of the inner ear. These organs have to do with notification of the effects of motion and head position on the posture of the body. The cerebellar connections enable muscular adjustments to maintain balance and equilibrium.

Cerebellar damage results in improperly controlled and coordinated movements, and difficulty in maintaining equilibrium. The severity of the disturbances depends upon the amount of tissue destroyed, rather than the region.

Spinal cord (Figs. 20—16 and 20—17)

The spinal cord extends from the medulla some 45 cm (18 inches) to the level of the second lumbar vertebra. Internally, in the form of the letter H, is the gray matter of the cord, containing chiefly neuronal cell bodies. Surrounding the gray matter is the white matter, consisting chiefly of myelinated nerve fibers. The white matter is composed of functional regions known as *tracts* (Fig. 20—18A, B). The tracts may be descending, carrying motor impulses from the brain to the periphery, or ascending, carrying impulses from the periphery to the brain. The motor tracts of the cord are descending, and are derived from the various cortical areas and nuclei in the brain. Their names reflect their origin. The voluntary pathways are the *corticospinal tracts.* The lateral corticospinal tract is a tract which crosses in the medulla, so that the opposite side of the brain controls a given side of the body musculature.

All descending tracts form the so-called *upper motor neurones.* The multiple tracts converge upon single, lower motor neurones with cell bodies in the anterior gray column of the cord, and their axons pass to the muscles through the ventral root. This type of arrangement is called the *final common pathway,* and enables a wide variety of influences to be brought to bear upon the muscle itself.

The neuromuscular junction

Connection of the lower motor neurone to a skeletal muscle is made by a specialized structure known as a motor end plate (Fig. 20—19A, B). The plate is located beneath the sarcolemma of a skeletal muscle fiber, and, like the synapse, contains vesicles of acetylcholine. The nerve impulse arriving at the plate increases its permeability to calcium ion, which enters and causes liberation of acetylcholine. The chemical diffuses out and initiates the sequence of events leading to depolarization and muscle contraction. Cholinesterase destroys the chemical so that its effect is not prolonged. The junction fatigues before the muscle fibers do, and protects the fibers from passing the point of no return.

It should not be assumed that the cord serves only as a transmitter for

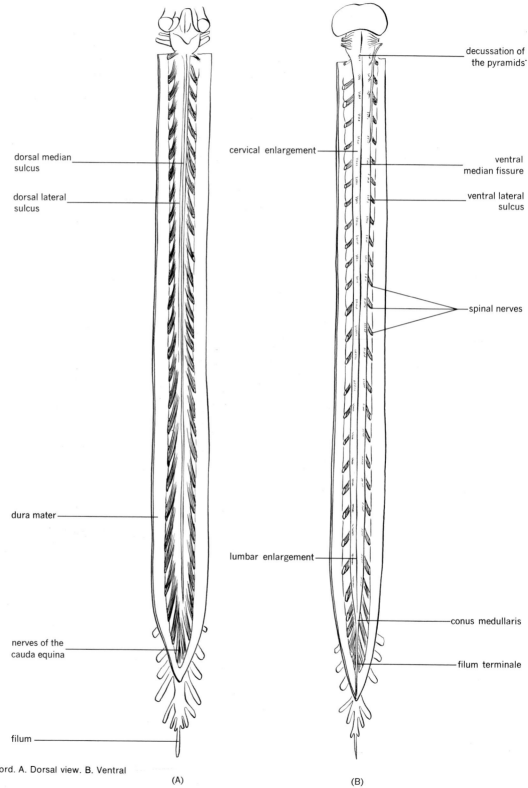

dorsal median
sulcus

dorsal lateral
sulcus

dura mater

nerves of the
cauda equina

filum

decussation of
the pyramids

cervical enlargement

ventral
median fissure

ventral lateral
sulcus

spinal nerves

lumbar enlargement

conus medullaris

filum terminale

FIGURE 20–16. The spinal cord. A. Dorsal view. B. Ventral
view.   (A)   (B)

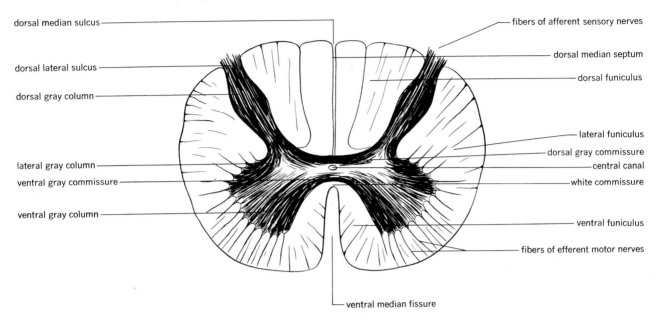

dorsal median sulcus

dorsal lateral sulcus

dorsal gray column

lateral gray column

ventral gray commissure

ventral gray column

fibers of afferent sensory nerves

dorsal median septum

dorsal funiculus

lateral funiculus

dorsal gray commissure

central canal

white commissure

ventral funiculus

fibers of efferent motor nerves

ventral median fissure

FIGURE 20–17. Cross section of the spinal cord.

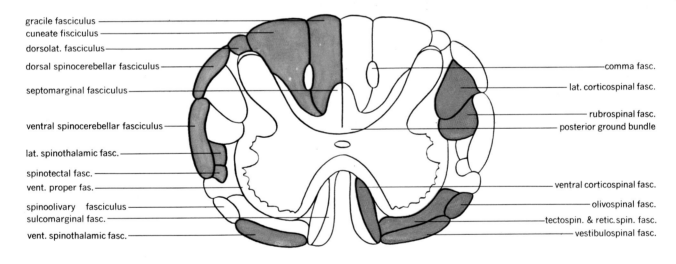

gracile fasciculus

cuneate fisciculus

dorsolat. fasciculus

dorsal spinocerebellar fasciculus

septomarginal fasciculus

ventral spinocerebellar fasciculus

lat. spinothalamic fasc.

spinotectal fasc.

vent. proper fas.

spinoolivary   fasciculus

sulcomarginal fasc.

vent. spinothalamic fasc.

comma fasc.

lat. corticospinal fasc.

rubrospinal fasc.

posterior ground bundle

ventral corticospinal fasc.

olivospinal fasc.

tectospin. & retic. spin. fasc.

vestibulospinal fasc.

ascending fibers   descending fibers

FIGURE 20–18A. The major spinal tracts or fasciculi.

impulses. It is an important center for monitoring reflex activity. Among the types of reflexes controlled by the cord alone are:

1. Babinski reflex—stroking the sole of the foot causes a fanning of the toes.
2. Knee jerk—tapping patellar tendon causes knee extension.
3. Ankle jerk—tapping Achilles tendon causes ankle extension.
4. Abdominal reflex—stroking abdomen causes abdominal contraction.
5. Biceps jerk—tapping biceps tendon causes elbow flexion.
6. Flexor reflex—strong stimulus to a limb involves withdrawal by flexor contraction. Extensors are simultaneously inhibited.

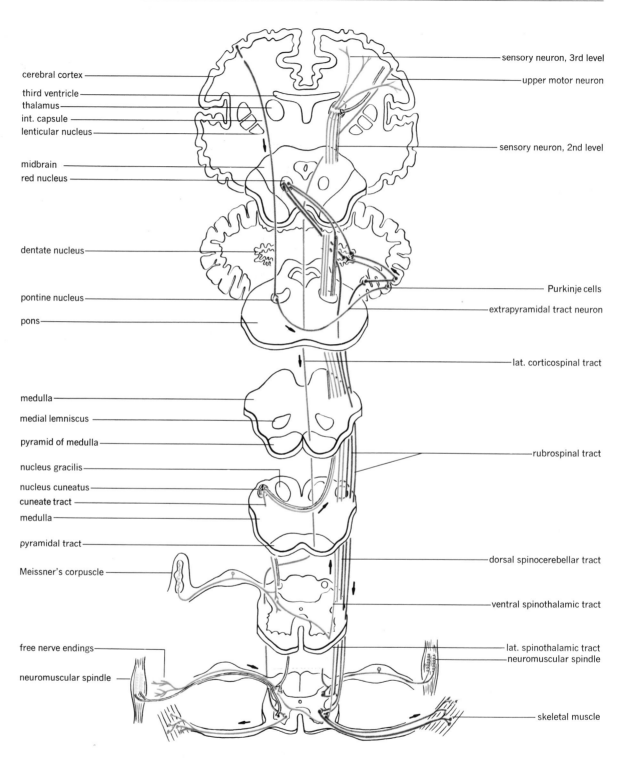

FIGURE 20–18B. course of some of the major spinal tracts.

The previous list is not exhaustive, but serves as an illustration of the types of activity involved. The neurologist uses many of these reflexes to test cord function, because most of them are served by different levels of the cord.

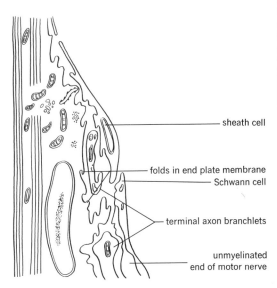

sheath cell

folds in end plate membrane
Schwann cell

terminal axon branchlets

unmyelinated
end of motor nerve

FIGURE 20–19A. Diagram of the neuromuscular junction.

FIGURE 20–19B. Myoneural junction. ME, motor end plate; SN, synaptic vesicles; SK, skeletal muscle fibrils. (Electron micrograph by Allen Clark and Alexander Mauro.) X 20,000.

SENSORY FUNCTIONS OF THE NERVOUS SYSTEM

Introduction

Sensory functions are those concerned with reception of stimuli, and their transmission to the central nervous system for interpretation and response. As indicated on page 541, receptors form the essential starting point for reflex arcs. Receptors of all types follow certain rules of operation. Among these are:

1. They follow the *law of adequate stimulus*. Each receptor responds best (not exclusively) to some one form of energy. Thus, the eye responds best to light, the touch receptors to mechanical pressure, baroreceptors to stretch.
2. They follow the *law of specific nerve energies*. The impulse generated by the stimulation of *any* receptor is the same. Subjective interpretation of specific sensations depends upon the central connection the fiber makes. The depolarization of a receptor apparently occurs in the same fashion (i.e., Na^+ inflow) as occurs in a nerve.
3. Receptors can, to a degree, *communicate intensity of stimulation*. The frequency of impulse discharge from a receptor increases with stronger stimulation. This is interpreted by the brain as increased intensity. Discrimination is apparently limited to increments of about 10 percent in stimulation intensity (*Weber-Fechner law*). For example, 1 pound added to 10 pounds held in the hand would be determined to be heavier; $1/2$ pound would not.
4. Receptors show *adaptation*. Frequency of discharge is initially rapid, then decreases.

The following discussion of receptors and their pathways within the nervous system will not include the "special senses" (eye, ear, taste, smell). These will be described in a separate chapter.

Classification of receptors

There are many different ways of grouping receptors. Two are shown below.

1. By Stimulus		2. By Location	
Stimulus	Examples of receptor	Location	Examples of receptor
Chemical (chemoreceptor)	Taste bud, olfactory epithelium	At or near body surface (exteroceptive)	Eye, ear, taste bud, olfactory epithelium, Pacinian and Meissner's corpuscles, heat and cold receptors
Movement (kinoreceptor)	Cristae of semicircular canals		
Gravity (statoreceptor)	Maculae of inner ear		
Stretch (baroreceptor)	Receptors of aorta, carotid sinus, lungs, muscle spindle		
Mechanical pressure	Meissner's corpuscle (touch)	In viscera (enteroceptive)	All receptors in heart, lungs, GI tract
	Pacinian corpuscle (pressure)		
Light (photoreceptor)	Rods and cones of retina	In muscle and tendon (proprioceptive)	Muscle spindles, Golgi organs
Sound (phonoreceptor)	Organ of Corti of cochlea		
Change in temperature (thermoreceptor)	Ruffini corpuscle (warmth), Krause corpuscle (cold)		
Strong stimulation of any kind (nociceptor)	Pain fibers		

Connection with the cord

Fibers from all but the special sense organs pass to the cord and enter through the dorsal root. Here they undergo a synapse and may then ascend to the brain on the same or opposite side of the cord as their fibers entered. Ascending to the brain, the fibers pass through specific tracts (Fig. 20–18). The main tracts and the sensations carried are shown below:

Sensory Tracts of the Cord

Tract	Crossed	Sensation
Spinocerebellar (dorsal and ventral)	No	Muscle sense
Gracile, cuneate	No	Muscle sense, touch, pressure
Spinothalamic (lateral, ventral)	Yes	Pain, thermal sensation

Interpretation of sensations

The fibers listed above ascend to the thalamus. Pain fibers stop here, and pain is apparently interpreted by the thalamus. Other fibers undergo a synapse with the cortical relay fibers which pass to areas 3, 1, 2, 5, and 7 in the parietal lobe. Interpretation of all sensations except pain is thus cortical. In the sense that pain may be an important modifying influence on behavior and learning, the thalamus becomes an important behavior center.

FUNCTIONS OF OTHER PARTS OF THE BRAIN

The medulla oblongata

In addition to housing the crossing of the lateral corticospinal tract, and the nuclei of origin of cranial nerves IX to XII, the medulla contains many centers for the automatic control of breathing, heart rate, blood vessel caliber, vomiting, coughing, salivation, and swallowing.

The pons

The pons contains the origin of cranial nerves V to VIII, and serves as distributing point for fibers to the cerebellum.

The midbrain

The most conspicuous feature of the midbrain are four small elevations on its dorsal surface. These are known as the *corpora quadrigemina*. They act as relay points to connect visual and auditory fibers to the motor system, thus enabling muscular response to sight and sound. The *cerebral peduncles* pass from the lower portion of the midbrain to the cerebrum, for transmission of descending impulses.

The hypothalamus

The hypothalamus lies below the thalamus and forms the floor of the third ventricle. Each half of the hypothalamus is divided into a supra-optic, tuberal, and mammillary portion, and within these larger areas are smaller groups of hypothalamic nuclei. The anterior, lateral, and ventromedial nuclei are three of the more prominent nuclei. The hypothalamus receives fibers from the thalamus, midbrain, and amygdala and sends fibers to the pituitary gland, blood vessels, eccrine sweat glands, and skeletal muscles. These connections suggest that the area is involved in some basic functions of the body.

1. An area in the anterior hypothalamus causes sweating, panting, and cutaneous vasodilation when stimulated. A drop in body temperature occurs; for this reason, the area is sometimes designated as the *heat loss center*. Stimulation of the posterior hypothalamus results in shivering, cutaneous vasoconstriction, and a rise in body temperature; that is, we have a *heat gain center*. The two areas function to maintain a nearly constant body temperature.

2. In the supraoptic region are cells sensitive to the osmolarity of the blood. These cells control the production of antidiuretic hormone and are thus involved in water balance control.

3. The lateral hypothalamus contains a feeding center governing food intake, and a satiety center regulating appeasement of appetite.

4. The hypothalamus plays a role in controlling anterior pituitary secretion and the metabolic implications of those hormone secretions.

5. Last, it may be said that the hypothalamus is involved in such psychological functions as motivation and emotional behavior.

If there were only one brain area most concerned with maintenance of body homeostasis, the hypothalamus would be that area.

SUBJECTIVE BRAIN FUNCTIONS

Emotion and motivation

Notable body alterations occurring as we respond to a stimulus is one definition of emotion. Motivation involves a drive to attain a goal, with reduction of drive when the goal is reached. These subjective functions of the brain share almost the same structures, the organs of the *limbic system* (Fig. 20–20). Of these structures, the hypothalamus and amygdala appear to play important roles.

Electrical stimulation of the lower central portion of the hypothalamus induces "sham rage" in an experimental animal, with development of fear and anger responses. Damage to the amygdala results in a placid animal. One may conclude that these areas exert a definite control over the expression of emotional behavior.

Motivating areas, which can cause an animal to perform some task for a pleasurable or gratifying reward, are found in the posterior

FIGURE 20–20. The limbic system.

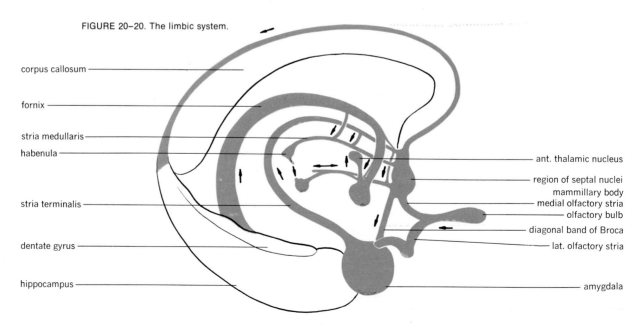

corpus callosum

fornix

stria medullaris

habenula

stria terminalis

dentate gyrus

hippocampus

ant. thalamic nucleus

region of septal nuclei
mammillary body
medial olfactory stria
olfactory bulb
diagonal band of Broca
lat. olfactory stria

amygdala

hypothalamus, fornix, hippocampus, and anterior hypothalamus. One experimenter suggests that the brain is designed for pleasure as well as for negative behavior.

Memory

This ability of the human may be defined as being able to recall or perform a specific task in the absence of the stimulus which originally caused the response. Such performance implies the storage within the nervous system of a "trigger" which can initiate the response, in other words, a memory.

Two types of memory appear to exist. *Short-term memory* is that which may be measured immediately following a learning experience. It persists for no longer than 30 minutes and is postulated to occur by impulses following circular neurone chains within the brain. *Long-term* or permanent memory traces appear to require chemical transformations within the nervous system. Ribonucleic acid (RNA) has been postulated to act as a template for the synthesis of a specific protein which acts as the memory trace. Recovery of the memory is theorized to involve chemical changes in the protein. Approximately three days are required for "solidification" of a memory trace.

Sleep

Sleep is an interruption of wakefulness and may be one of the most essential needs of higher mammals. We normally spend about one-third of our life sleeping. A person's sleep cycle appears to be entangled with a variety of metabolic changes, all of which follow some internal biological clock (circadian rhythm). Hormone production, body temperature, mental outlook and the environment are all interrelated in production of sleep. The brain area which induces sleep, and the mechanisms involved, are not definitely known, but if an individual is deprived of sleep, profound changes may occur in his behavior and efficiency, including temporary psychosis.

Sleep progresses in stages:

1. *Drowsiness.* A state of relaxed wakefulness in subdued surroundings seem to set the stage for sleep.
2. *Stage 1.* Respiration slows, heart rate slows, and a drifting sensation occurs. This stage lasts only a few minutes, and the person may easily be awakened.
3. *Stage 2.* A deeper stage, this one is associated with a fundamental change in brain function. The subject may show eye movements, and is usually not disturbed by ordinary stimuli.
4. *Stage 3.* Muscles are very relaxed, and the sleeper is unresponsive to external events.
5. *Stage 4.* Oblivious sleep occurs here, where no awareness is seen. However, a sleeper may record external stimuli in his brain. The brain is "awake," or at least receptive. Sleepwalking may occur. A gradual lightening of sleep occurs from this stage, until wakefulness is again achieved.

Several theories have been advanced to account for onset of sleep, all involving chemicals. Gamma-hydroxybutyric acid (a relative of

GABA) induces sleep, as does serotonin. The chemicals appear to act on the base of the brain stem, stimulate it, and trigger a series of events culminating in sleep.

Much investigation is currently being conducted on sleep. Within a relatively few years, it may become possible to control our sleep to suit our wishes. Whether this proves to be beneficial or not remains to be seen.

THE SPINAL NERVES

The spinal cord gives rise to 31 pairs of spinal nerves, which form the major portion of the peripheral nervous system. There are 8 cervical, 12 thoracic, 5 lumbar, 5 sacral, and 1 coccygeal nerve pairs. Each nerve arises, in segmental fashion along the length of the cord, by a dorsal and ventral root which join as the nerves pass through the intervertebral foramina. Passing to the periphery, the nerves give rise to dorsal and ventral rami, and communicating rami to the autonomic system (Fig. 20–21). The dorsal and ventral rami supply somatic afferent (sensory) and efferent (motor) fibers to skin and skeletal muscles, distributed in a segmental fashion to the body (Fig. 20–22). Areas served by adjacent nerves overlap to preserve some function within an area in the case of damage to a specific nerve.

In the regions of the limbs, the ventral rami regroup into plexuses before being distributed peripherally. Three major plexuses are recognized: cervical, brachial, and lumbosacral. Their components are diagrammed in Fig. 20–23.

THE CRANIAL NERVES (Fig. 20–24)

The brain gives rise to 12 pairs of cranial nerves which supply motor and sensory fibers to the head and neck. These nerves constitute the remaining portion of the peripheral nervous system. The nerves are numbered, using Roman numerals, and are named as well.

I. *Olfactory nerve.* The nerve originates as a mass of fine rootlets in the apex of the nasal cavity (olfactory region). Passing through the ethmoid cribriform plate, the fibers form the olfactory nerve. These terminate in the olfactory bulb, and from here fibers pass to the hippocampus of the brain. The nerve contains only afferent (sensory) fibers.

II. *Optic nerve.* The nerve originates from the inner layers of the retina and pass to the optic chiasma where a partial crossing of fibers occurs. A synapse occurs in the lateral geniculate body and the optic radiation passes from there to the occipital lobe. Like the olfactory nerve, the optic nerve is purely sensory.

III. *Oculomotor nerve.* This nerve contains both sensory and motor fibers. The sensory component passes from four of the six extrinsic eye muscles (the superior rectus, medial rectus, inferior rectus, and inferior oblique), the iris, and ciliary body. This portion conveys muscle sense, and information as to amount of light entering the eye. The motor component passes to these same areas and governs movement, pupillary constriction and dilation, and focusing of the eye.

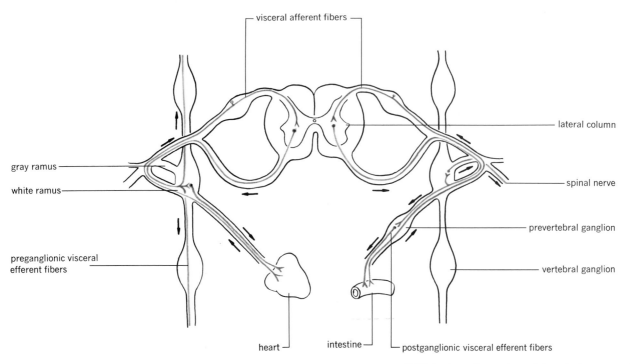

visceral afferent fibers

lateral column

gray ramus

spinal nerve

white ramus

prevertebral ganglion

preganglionic visceral
efferent fibers

vertebral ganglion

heart

intestine

postganglionic visceral efferent fibers

FIGURE 20–21. Connections of the spinal nerves to the
cord.

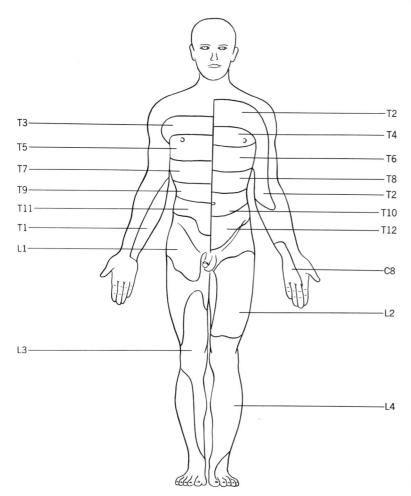

T3

T2

T5

T4

T7

T6

T9

T8

T11

T2

T1

T10

L1

T12

C8

L2

L3

L4

FIGURE 20–22. The overlapping distribution of the spinal
nerves to the body.

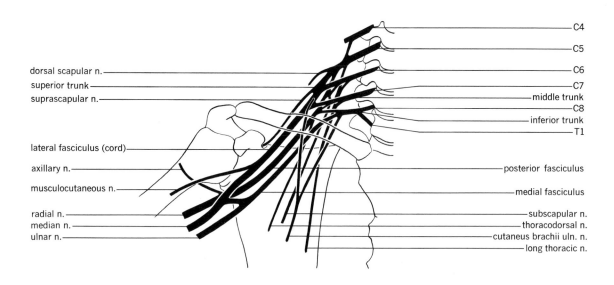

dorsal scapular n.
superior trunk
suprascapular n.

lateral fasciculus (cord)
axillary n.
musculocutaneous n.
radial n.
median n.
ulnar n.

C4
C5
C6
C7
middle trunk
C8
inferior trunk
T1

posterior fasciculus
medial fasciculus
subscapular n.
thoracodorsal n.
cutaneus brachii uln. n.
long thoracic n.

FIGURE 20–23. Components of the cervical, brachial, and lumbosacral plexuses.

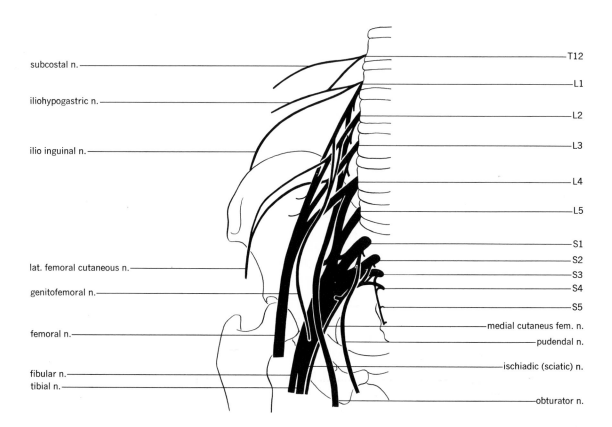

subcostal n.

iliohypogastric n.

ilio inguinal n.

lat. femoral cutaneous n.

genitofemoral n.

femoral n.

fibular n.
tibial n.

T12

L1

L2

L3

L4

L5

S1
S2
S3
S4
S5
medial cutaneus fem. n.
pudendal n.

ischiadic (sciatic) n.

obturator n.

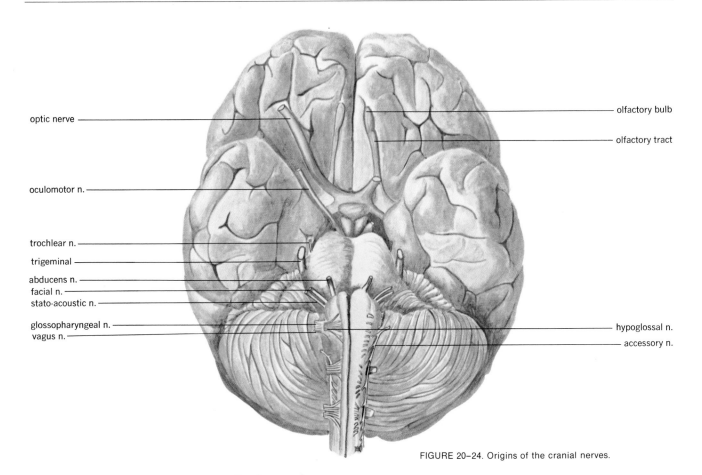

optic nerve

oculomotor n.

trochlear n.
trigeminal
abducens n.
facial n.
stato-acoustic n.

glossopharyngeal n.
vagus n.

olfactory bulb

olfactory tract

hypoglossal n.

accessory n.

FIGURE 20–24. Origins of the cranial nerves.

IV. *Trochlear nerve.* This nerve provides sensory and motor fibers for one of the six extrinsic eye muscles, the superior oblique.

V. *Trigeminal nerve* (Fig. 20–25). The largest of the cranial nerves, the trigeminal nerve consists of a large sensory and a small motor division arranged in three major branches.

The *ophthalmic branch* conveys sensory fibers from most of the head above the level of the eyes.

The *maxillary branch* conveys sensory fibers from the area between the upper teeth and eyes.

The *mandibular branch* conveys sensory impulses from the mandibular area and supplies motor fibers to the muscles of mastication.

VI. *Abducens nerve.* Containing both sensory and motor fibers, this nerve supplies the remaining extrinsic eye muscle, the lateral rectus.

VII. *Facial nerve* (Fig. 20–25). The facial nerve is composed of a small sensory and a large motor division. The sensory portion conveys impulses of taste from the anterior two-thirds of the tongue. The motor division supplies the muscles of facial expression.

VIII. *Statoacoustic (acoustic) nerve.* Another purely sensory nerve, nerve VIII has two branches:

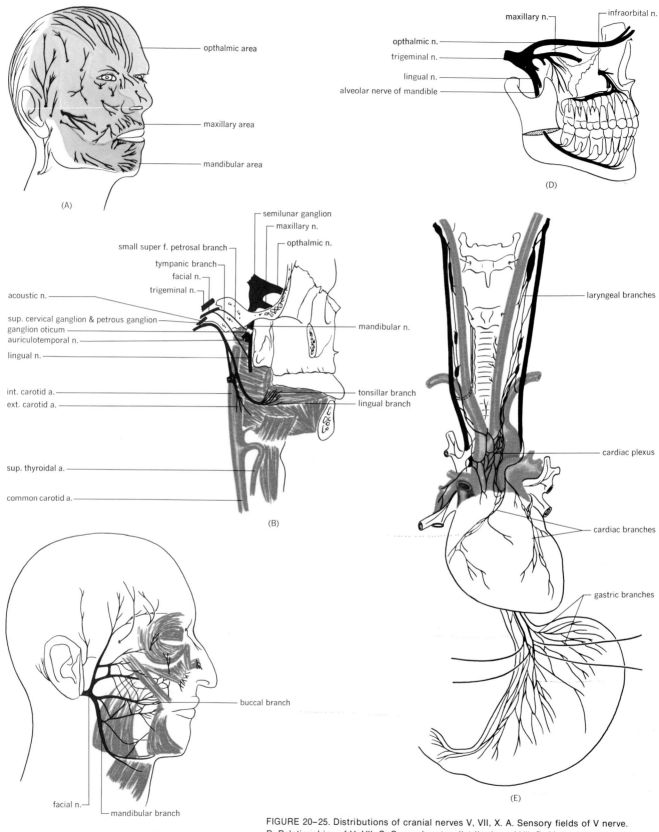

(A)

opthalmic area

maxillary area

mandibular area

maxillary n.

infraorbital n.

opthalmic n.

trigeminal n.

lingual n.

alveolar nerve of mandible

(D)

semilunar ganglion

maxillary n.

opthalmic n.

small super f. petrosal branch

tympanic branch

facial n.

trigeminal n.

acoustic n.

sup. cervical ganglion & petrous ganglion

ganglion oticum

auriculotemporal n.

lingual n.

int. carotid a.

ext. carotid a.

sup. thyroidal a.

common carotid a.

mandibular n.

tonsillar branch

lingual branch

(B)

laryngeal branches

cardiac plexus

cardiac branches

gastric branches

(E)

buccal branch

facial n.

mandibular branch

(C)

FIGURE 20–25. Distributions of cranial nerves V, VII, X. A. Sensory fields of V nerve.
B. Relationships of V, VII. C. General motor distribution of VII. D. Major branches of V.
E. Vagus (X) distribution in black. Note distribution to heart, lungs, stomach.

The *cochlear nerve* arises in the cochlea and conveys impulses for hearing to the cerebrum.

The *vestibular branch* is derived from the organs of balance and equilibrium in the inner ear, and conveys its impulses to the cerebellum.

IX. *Glossopharyngeal nerve.* Containing both motor and sensory fibers, nerve IX conveys sensory fibers from the posterior one-third of tongue for taste, and general sensation from the pharynx and tonsils. The motor fibers supply the swallowing muscles of the pharynx.

X. *Vagus nerve* (Fig. 20–25). The vagus nerve is a major component of the autonomic system, and detailed description of vagal distribution will be given in the following Section. The nerve supplies sensory and motor fibers to nearly all the organs of the thoracic and abdominal cavities.

XI. *Accessory nerve.* This nerve supplies motor fibers to the muscles of the throat, larynx and soft palate, and to the sternocleidomastoid and trapezius.

XII. *Hypoglossal nerve.* A motor nerve, this nerve supplies the extrinsic and intrinsic muscles of the tongue and most of the infrahyoid muscles.

Table 20–1 summarizes the cranial nerves.

AUTONOMIC SYSTEM (Fig. 20–26)

That portion of the peripheral nervous system which is primarily concerned with the automatic unconscious control of the operation of the body viscera, is the autonomic nervous system. The fibers of the system are derived from certain cranial and spinal nerves, and are designated as visceral afferent and visceral efferent fibers.

Visceral afferents have their cell bodies in the same ganglia as serve the somatic system, that is, in the dorsal root ganglia, and in the cranial nerve nuclei. The visceral efferent fibers have their cell bodies in the lateral gray column of the spinal cord and certain cranial nerve nuclei, and their axons pass through the communicating rami for peripheral distribution.

Receptors of the autonomic system are sensitive to pressure and tension, as well as chemical changes, and are located within the walls of blood vessels, and thoracic and abdominal viscera. The effectors of the system are chiefly smooth muscle and visceral glands. The makeup of the system suggests that it operates primarily at the reflex level, as indeed it does. Two neurones usually make the connection between the spinal cord or brain and the organ. A preganglionic neurone with cell body in the brain or cord extends to a synapse outside the brain or cord, and a postganglionic neurone extends from synapse to organ.

Divisions of the system

The *parasympathetic* or *craniosacral* portion of the system is composed of cranial nerves III, VII, IX, and X, and sacral spinal nerves 2 to 4. In this division, preganglionic fibers are long, the synapse is made in a

TABLE 20-1

Summary of the Cranial Nerves

Nerve	Composition M = motor; S = sensory	Origin	Connection with brain or peripheral distribution	Function
I. Olfactory	S	Nasal olfactory area	Olfactory bulb	Smell
II. Optic	S	Ganglionic layer of retina	Optic tract	Sight
III. Oculomotor	MS	M—Midbrain	4 of 6 extrinsic eye muscles (superior rectus, medial and inferior rectus, inferior oblique)	Eye movement
		S—Ciliary body of eye	Nucleus of nerve in midbrain	Focussing, pupil changes, musele sense
IV. Trochlear	MS	M—Midbrain	1 extrinsic eye muscle (superior oblique)	Eye movement
		S—Eye muscle	Nucleus of nerve in midbrain	Muscle sense
V. Trigeminal	MS (B)	M—Pons	Muscles of mastication	Chewing
		S—Scalp and face	Nucleus of nerve in pons	Sensation from head
VI. Abducens	MS	M—Nucleus of nerve in pons	1 extrinsic eye muscle (lateral rectus)	Eye movement
		S—1 extrinsic eye muscle	Nucleus of nerve in pons	Muscle sense
VII. Facial	MS B	M—Nucleus of nerve in lower pons	Muscles of facial expression	Facial expression
		S—Tongue	Nucleus of nerve in lower pons	Taste
VIII. Statoacoustic or auditory	S	Internal ear: balance organs, cochlea	Vestibular nucleus, cochlear nucleus	Posture, hearing
IX. Glossopharyngeal	MS B	M—Nucleus of nerve in lower pons	Muscles of pharynx	Swallowing
		S—Tongue, pharynx	Nucleus of nerve in lower pons	Taste, general sensation
X. Vagus	MS B	M—Nucleus of nerve in medulla	Viscera	Visceral muscle movement
		S—Viscera	Nucleus of nerve in medulla	Visceral sensation
Spinal XI. Accessory	M	Nucleus of nerve in medulla	Muscles of throat, larynx, soft palate, sternocleidomastoid, trapezius	Swallowing, head movement
XII. Hypoglossal	M	Nucleus of nerve in medulla	Muscles of tongue and infrahyoid area	Speech, swallowing

plexus close to the organ innervated, and the postganglionic fiber is short. The vagus nerve provides innervation for most body organs, and is the single most important nerve of this division.

The *sympathetic* or *thoracolumbar* division is composed of all thoracic and lumbar spinal nerves. In this division, preganglionic fibers are short, the synapse is usually made in the sympathetic ganglion, and postganglionic fibers are long. These nerves supply the same organs as are served by the parasympathetic division.

The two divisions provide an opposite effect on the organs innervated, as shown in Table 20–2. The differential effects may be accounted for, in part, by the fact that most parasympathetic nerves produce acetylcholine at their endings, while most sympathetic nerves produce norepinephrine. The nerves are known as cholinergic or adrenergic according to the chemical produced.

FIGURE 20–26. Autonomic nervous system.

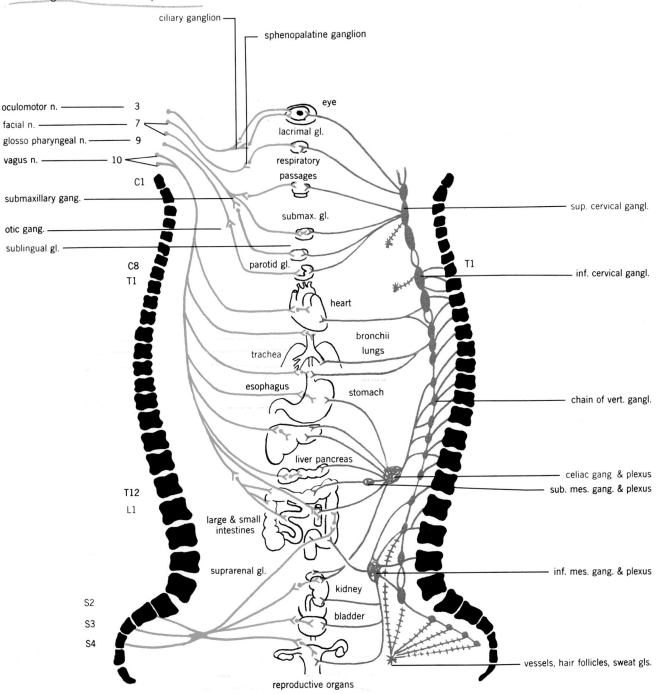

cranio-sacral division

thoraco-lumbar division

TABLE 20-2

Effects of Autonomic Stimulation

Organ affected	Parasympathetic effects	Sympathetic effects
Iris	Contraction of sphincter pupillae; pupil size decreases	Contraction of dilator pupillae; pupil size increases
Ciliary muscle	Contraction; accommodation for near vision	Relaxation; accommodation for distant vision
Lacrimal gland	Secretion	Excessive secretion
Salivary glands	Secretion of watery saliva in copious amounts	Scanty secretion of mucus—rich saliva
Respiratory system:		
Conducting division	Contraction of smooth muscle; decreased diameters and volumes	Relaxation of smooth muscle; increased diameter and volumes
Respiratory division	Effect same as on conducting division	Effect same as on conducting division
Blood vessels	Constriction	Dilation
Heart:		
Stroke volume	Decreased	Increased
Stroke rate	Decreased	Increased
Cardiac output and blood pressure	Decreased	Increased
Coronary vessels	Constriction	Dilation
Peripheral blood vessels:		
Skeletal muscle	Constriction	Dilation
Skin	Dilation	Constriction
Visceral organs (except heart and lungs)	Dilation	Constriction
Stomach:		
Wall	Increased motility	Decreased motility
Sphincters	Inhibited	Stimulated
Glands	Secretion stimulated	Secretion inhibited
Intestines:		
Wall	Increased motility	Decreased motility
Spincters		
Pyloric, iliocecal	Inhibited	Stimulated
Internal anal	Inhibited	Stimulated
Liver	Promotes glycogenesis, promotes bile secretion	Promotes glycogenolysis, decreases bile secretion
Pancreas (exocrine and endocrine)	Stimulates secretion	Inhibits secretion
Spleen	Little effect	Contraction and emptying of stored blood into circulation
Adrenal medulla	Little effect	Epinephrine secretion
Urinary bladder	Stimulates wall, inhibits sphincter	Inhibits wall, stimulates sphincter
Uterus	Little effect	Inhibits motility of nonpregnant organ; stimulates pregnant
Sweat glands	Normal function	Stimulates secretion (produces "cold sweat" when combined with cutaneous vasoconstriction)

Perusal of Table 20–2 indicates that the parasympathetic division functions primarily to assure normal operation of the viscera. The sympathetic division prepares the body for "fight-or-flight," and resistance to stressful situations.

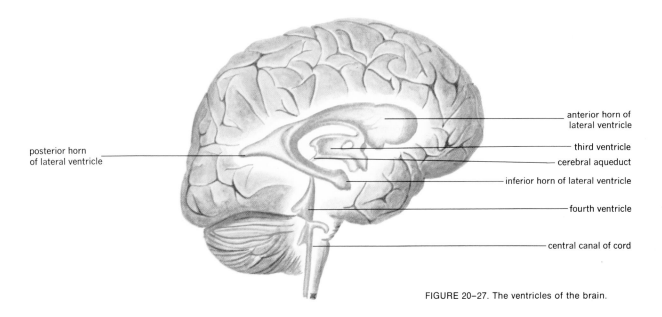

FIGURE 20–27. The ventricles of the brain.

posterior horn
of lateral ventricle

anterior horn of
lateral ventricle

third ventricle

cerebral aqueduct

inferior horn of lateral ventricle

fourth ventricle

central canal of cord

VENTRICLES AND MENINGES

The brain contains hollow cavities known as *ventricles* (Fig. 20–27). The *lateral ventricles* lie within the cerebral hemispheres. The *third ventricle* lies between the thalami, and connects, by way of the *cerebral aqueduct,* with the *fourth ventricle,* lying in the medulla. The ventricles contain *choroid plexuses,* vascular structures which secrete cerebrospinal fluid (CSF). The fluid traverses the ventricles and exits through the roof of the fourth ventricle. It surrounds the cord throughout its length.

Compared to plasma (Table 20–3), cerebrospinal fluid is deficient in protein, amino acids, and cholesterol, and is richer in sodium chloride and magnesium.

TABLE 20–3

Plasma and CSF Compared for Some Major Constituents

Constituent or property	Plasma (values in mg %)	CSF
Protein	6400–8400	15–40
Cholesterol	100–150	0.06–0.20
Urea	20–40	5–40
Glucose	70–120	40–80
NaCl	550–630	720–750
Bicarbonate (as vol % of CO_2)	40–60	40–60
pH	7.35–7.4	7.35–7.4
Magnesium	1–3	3–4

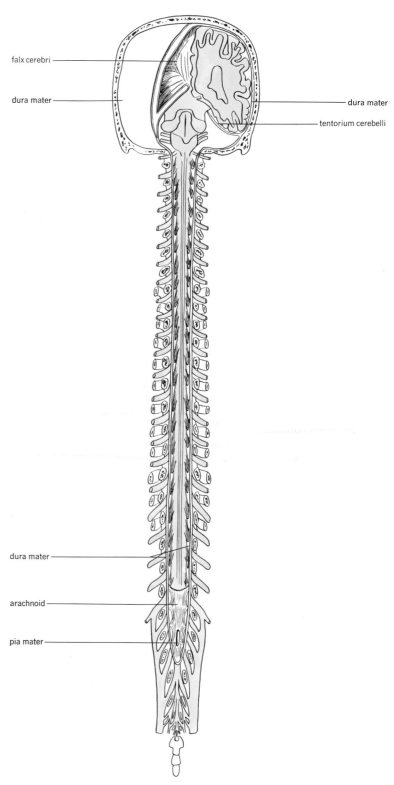

falx cerebri

dura mater

dura mater

tentorium cerebelli

dura mater

arachnoid

pia mater

FIGURE 20–28. The meninges of the central nervous system. Dural membranes of skull shown also.

Total volume of the fluid in the adult is about 200 ml. Rates of secretion have not been determined, but available evidence suggests replacement every 3 to 4 hours. The fluid is present under a pressure of 110 to 175 mm CSF. The fluid functions as a shock absorber for the nervous system, and compensates for changes in blood volume within the skull, thus keeping cranial volume constant.

Hydrocephalus (water on the brain) may result if the flow of CSF is blocked within the brain. Inflation of the ventricles results, and compression of the brain may occur.

The *meninges* (Fig. 20–28) are the several membranes which surround the central nervous system. The *pia mater* is a delicate vascular membrane applied to the surface of the CNS. A space *(subarachnoid space)* separates pia mater from the *arachnoid.* The subarachnoid space contains cerebrospinal fluid. A narrow space *(subdural space)* separates the arachnoid from the outer, tough, protective *dura mater.* Within the skull, the dura mater forms the *tentorium cerebelli,* lying between the cerebellum and cerebrum, and the *falx cerebri,* lying between the cerebral hemispheres. The blood sinuses of the skull also lie within the dura.

QUESTIONS

1. Compare the morphology and functions of neurones and glia.
2. How is a nerve impulse formed and transmitted?
3. What determines the rapidity of impulse conduction over a nerve fiber? Explain saltatory conduction.
4. Compare neuronal and synaptic properties, explaining the reason for differences encountered.
5. Describe the lobes of the cerebrum and the functions centered in each.
6. What does each of the following contribute to a motor impulse from the cerebral cortex:
    (a) Cerebellum
    (b) Reticular formation
    (c) Final common pathway
7. Describe transmission of impulses from nerve fiber to skeletal muscle. Compare to the synapse.
8. What rules do receptors follow? Are they capable of discriminating anything other than the fact that a stimulus has triggered them? Explain.
9. For each of the following sensations, describe the receptor, spinal pathway, final termination in the brain:
    (a) Touch
    (b) Heat
    (c) Pain
10. Describe the areas responsible for temperature regulation, heart rate control, and coupling of visual and auditory sensation to the motor system.
11. Compare origins and functions of spinal and cranial nerves.
12. Compare morphology and functions of the two subdivisions of the autonomic system. Is there any justification for separating the system into two parts? Defend your answer.
13. Describe the cerebrospinal fluid and trace its circulation.

CHAPTER 21

The Special Senses

What can give us more sure knowledge than our senses?
How else can we distinguish between true and false?

LUCRETIUS

SPECIAL SENSES

The organs serving the special senses are the eye (for sight), the ear (for hearing, balance, and equilibrium), the taste bud (for taste), and the olfactory epithelium (for smell).

SMELL AND TASTE

Smell and taste are both chemoreceptors, requiring substances which are dissolved as stimuli. The *olfactory epithelium* (Fig. 21–1) is located in the apex of the nasal cavities. Current research on the sense of smell suggests that the stimulating chemical fits like a key into locks provided by the nerve cells of the epithelium. A knowledge of molecular shape thus permits one to predict what an unknown substance will smell like if a known substance has the same shape. The classification of smells or odors is quite subjective, including goaty, burnt, ambrosial, and several others. The olfactory cells connect with cells located in the olfactory bulb of the brain. The olfactory tract arises from the bulb and passes ultimately to the temporal lobe where interpretation occurs.

    The receptors for taste are the *taste buds* (Fig. 21–2) located chiefly on the tongue. Some buds are found on the soft palate and pharyngeal wall. Though all taste buds have the same structure, four basic taste

575

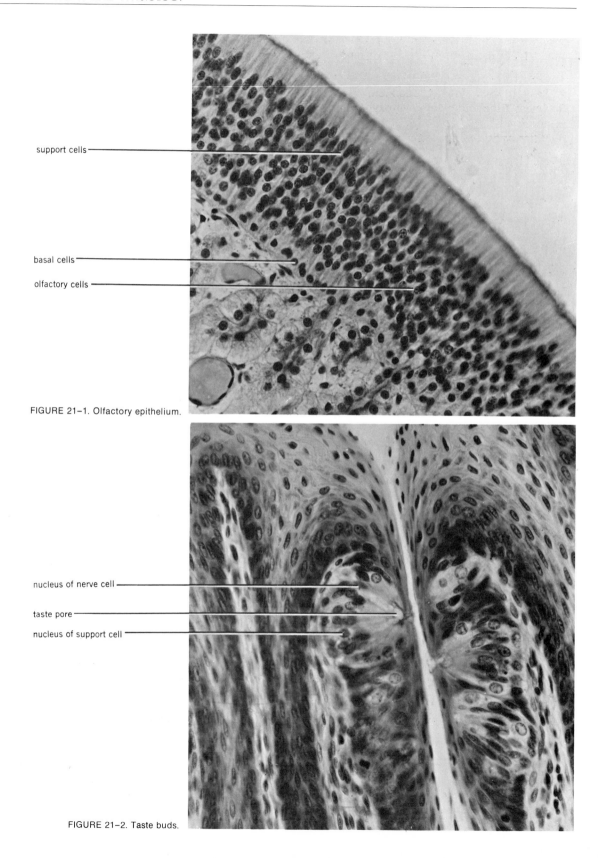

support cells

basal cells

olfactory cells

FIGURE 21–1. Olfactory epithelium.

nucleus of nerve cell

taste pore

nucleus of support cell

FIGURE 21–2. Taste buds.

sensations are described, and these sensations are not uniformly distributed over the tongue (Fig. 15–6). The sensations are:

1. *Salt*, triggered by metallic cations, such as Na, K.
2. *Sour*, triggered by hydrogen ion.
3. *Sweet*, triggered by the hydroxyl group, usually within a sugar or an alcohol.
4. *Bitter*, triggered by alkaloids such as quinine.

The various shades of taste we may experience is due to blending of impulses from the four types of buds and to reinforcement by the sense of smell. One has only to contract a cold to know how tasteless his food may become. This illustrates the interdependence of these two senses.

The impulses of taste are carried from the tongue over the VIIth and IXth cranial nerves, and the vagus for buds located elsewhere. These terminate in the brain stem. Second-order fibers convey the impulses to the postcentral gyrus for interpretation.

THE EAR

The ear is usually considered to be composed of three portions: outer, middle, and inner ears (Fig. 21–3).

The outer ear

This part is composed of the *pinna*, or fleshy flap attached to the skull, and the *external auditory tube* leading into the skull. These parts tend to gather and direct sound waves towards the structures of hearing. The external auditory tube is lined with thin skin and lubricated by the brownish waxy *cerumen*, produced by ceruminous glands in the wall.

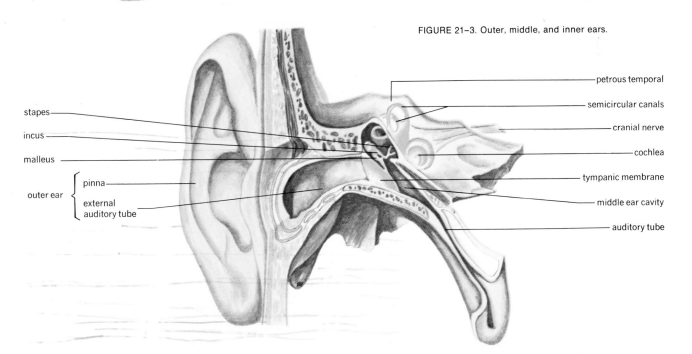

FIGURE 21–3. Outer, middle, and inner ears.

petrous temporal
semicircular canals
cranial nerve
cochlea
tympanic membrane
middle ear cavity
auditory tube

stapes
incus
malleus
outer ear { pinna
external auditory tube

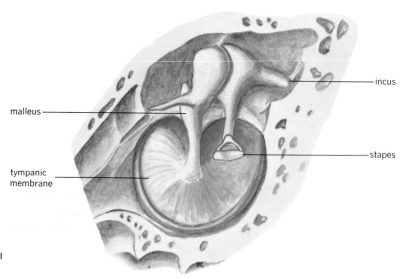

incus

malleus

stapes

tympanic
membrane

FIGURE 21–4. The ear ossicles viewed from the medial
aspect.

The middle ear

This part is an actual air-filled cavity which contains the *eardrum* or
tympanum, three small ear bones or *ossicles,* and the tendons of two
small muscles *(tensor tympani* and *stapedius).* It communicates with
the throat via the *auditory (Eustachian) tube.*

The eardrum forms the lateral wall of the middle ear cavity. It is composed of collagenous tissue, covered laterally by thin skin and medially
by the mucous membrane of the middle ear. It is set into vibration by
sound waves striking it. The ear ossicles (Fig. 21–4), *malleus (hammer),
incus (anvil),* and *stapes (stirrup)* act as a system of levers to transmit
eardrum vibrations to the fluid-filled cavities of the inner ear. In per-
forming this transmission, the bones decrease the amplitude of the
eardrum oscillation by a factor of about 10, and increase the pressure
at the inner ear by a factor of about 13. These tasks are essential in
order to prevent damage at the inner ear, and to transmit vibrations
in air to vibrations in fluid. The tensor tympani tendon attaches to the
malleus, the stapedius to the stapes. Contraction of these muscles
limits movement of the ossicles and prevents damage when sounds
are very loud. The auditory tube allows equalization of air pressure
on the two sides of the eardrum, preventing rupture.

The inner ear

The inner ear consists of a series of cavities and channels *(osseous
labyrinth)* hollowed out in the petrous temporal, and living tissues
(membranous labyrinth) lining the channels. All channels are fluid
filled; the osseous labyrinth is filled with *perilymph,* the membranous
with *endolymph.* The structures composing the inner ear include the
vestibule (utriculus and sacculus), the semicircular canals, and the
cochlea (Fig. 21–5).

The vestibular subdivisions *utriculus* and *sacculus* contain the
maculae (Fig. 21–6A), structures for sensing static posture. The essential mechanism of operation of the maculae is the force of gravity acting

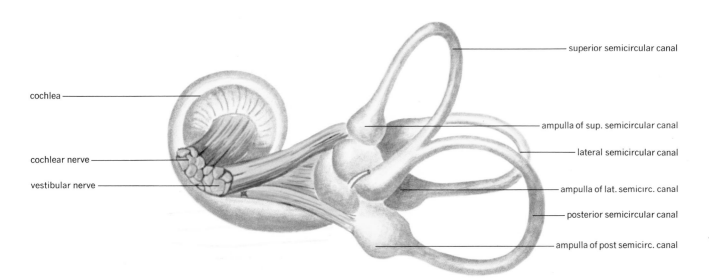

FIGURE 21–5. The membranous labyrinth of the inner ear.

on the *otoliths* and bending the hair cells. The nerve fibers from the maculae pass to the cerebellum over the VIIIth cranial nerve, where appropriate muscular responses are initiated to correct an abnormal head position.

The *semicircular canals* are three in number, and are placed in three mutually perpendicular planes. Movement of the head, or centrifugal force, causes the fluid in the canals to move towards or away from the *cristae* (Fig. 21–6B), which are located in the ampullae of the canals. The cristae are thus bent, and impulses pass over the VIIIth cranial nerve to the cerebellum. As before, muscular response corrects the body position. It may be noted that the maculae signal a held (static) position, while the canals are triggered by motion.

The *cochlea* (Fig. 21–7) is a coiled tube serving the sense of hearing. A central pillar of bone, the modiolus, has 2½ turns of the cochlea around it. The membranous structures of the cochlea divide the organ into three compartments or scalae. The *basilar membrane* is a tapered structure composed of some 25,000 strands of connective tissue, upon which rests the *organ of Corti* (Fig. 21–8). The sound waves transmitted through the ear ossicles to the cochlea create shock waves in the fluid of the scalae. The basilar membrane exhibits a more or less selective up and down motion which causes bending of the hair cells in certain regions of the organ of Corti. The bending apparently serves as an adequate stimulus for nerve impulse generation.

Any sound has three qualities: *pitch*, determined by frequency (vibrations per second) of the sound wave, *intensity* or loudness, and *timbre* or quality. Pitch, if frequency lies between about 16 cps to 20,000 cps, appears to be analyzed in the cochlea, due to the selective vibration of different areas of the organ of Corti for different pitches. Intensity is analyzed partially in the cochlea due to the amplitude of the bounce of the basilar membrane. Timbre appears to be a function of the brain.

The *auditory pathway* passes from the cochlea over the VIIIth cranial nerve to the temporal lobe. Failure of some part of the middle or inner ear structure, or brain damage, can lead to deafness.

Deafness is of three types: (1) transmission, a defect of eardrum or

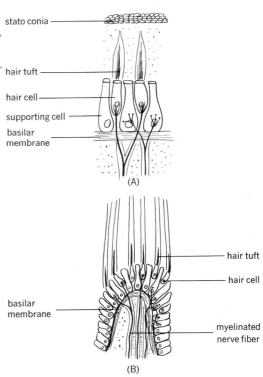

FIGURE 21–6. Balance and equilibrium receptors of the inner ear. A. Macula. B. Crista ampullaris.

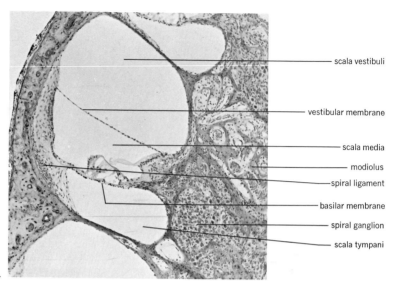

scala vestibuli

vestibular membrane

scala media

modiolus

spiral ligament

basilar membrane

spiral ganglion

scala tympani

FIGURE 21–7. Cochlea. Section of cochlear tube.

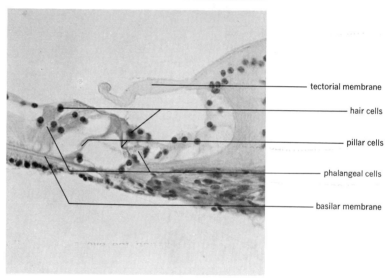

tectorial membrane

hair cells

pillar cells

phalangeal cells

basilar membrane

FIGURE 21–8. Organ of Corti.

ossicle activity (for example, otosclerosis, where the stapes becomes fixed in the oval window); (2) cochlear, a defect in organ of Corti; (3) central, injury to auditory cortex.

THE EYE

The eye (Fig. 21–9) is an organ capable of transducing light to visual impulses. Three layers of tissue compose the organ.

1. The outer *fibrous tunic,* composed of:
   (a) *Sclera* or white of the eye. This is a tough protective structure affording attachment for the six eye muscles which rotate the eyeball.
   (b) *Cornea.* The anterior, transparent portion, serving as a fixed lens to initially bend light rays to a near focus on the retina.

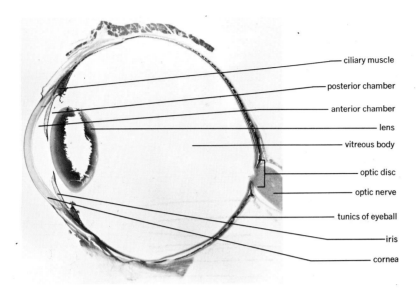

ciliary muscle
posterior chamber
anterior chamber
lens
vitreous body
optic disc
optic nerve
tunics of eyeball
iris
cornea

FIGURE 21–9. Eye.

2. The middle *uvea,* composed of:
   (a) *Choroid.* The posterior vascular coat of the eye.
   (b) *Ciliary body,* in turn containing:
      (1) *Ciliary muscle.* For fine focusing of the lens.
      (2) *Iris.* For light control.
      (3) *Lens* and supporting ligaments.
3. The inner *retina* (Fig. 21–10A-D). A ten-layered structure containing the photosensitive or visual receptors, the *rods* and *cones,* and the neurones leading to the optic nerve.

Additionally, the eye is protected by the *eyelids,* and lubricated by the *lacrimal apparatus* (Fig. 21–11).

Physiology of vision

Bending or refraction of light rays entering the eye occurs at the cornea and lens. If these areas are capable of forming sharp images on

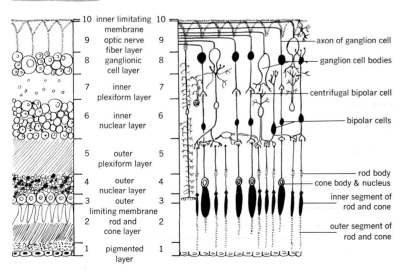

10 inner limiting membrane
9   optic nerve fiber layer
8   ganglionic cell layer
7   inner plexiform layer
6   inner nuclear layer
5   outer plexiform layer
4   outer nuclear layer
3   outer limiting membrane
2   rod and cone layer
1   pigmented layer

axon of ganglion cell
ganglion cell bodies
centrifugal bipolar cell
bipolar cells
rod body
cone body & nucleus
inner segment of rod and cone
outer segment of rod and cone

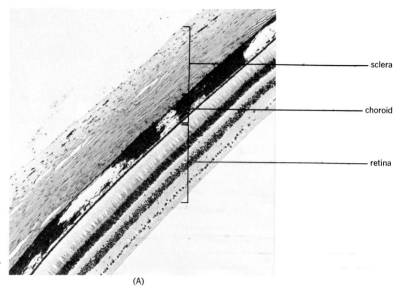

sclera

choroid

retina

(A)

FIGURE 21–10 A and B. Tunics of eye. A. Tunics. B. Retina.

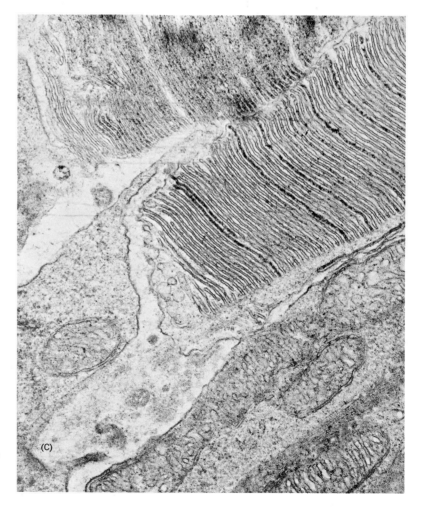

(C)

FIGURE 21–10C. Rod outer segment of the mouse retina. Note: mouse does not have any cones. (Photograph supplied by Norton B. Gilula, Department of Physiology-Anatomy, University of California, Berkeley.) X 35,750.

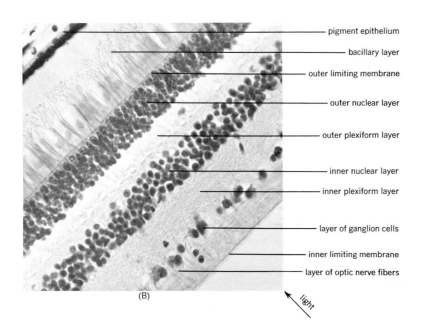

pigment epithelium

bacillary layer

outer limiting membrane

outer nuclear layer

outer plexiform layer

inner nuclear layer

inner plexiform layer

layer of ganglion cells

inner limiting membrane

layer of optic nerve fibers

(B)

light

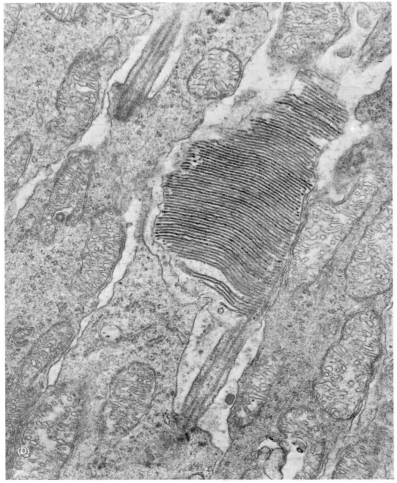

(D)

FIGURE 21–10D. Rod outer segment of the mouse retina. (Photograph supplied by Norton B. Gilula, Department of Physiology-Anatomy, University of California, Berkeley.) X 3,575.

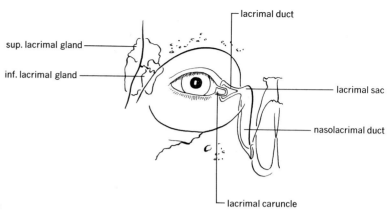

FIGURE 21–11. The lacrimal apparatus.

the retina, the eye is normal or *emmetropic*. In *nearsightedness (myopia)* the optical system brings the image to a focus ahead of the retina, while in *farsightedness (hypermetropia)* the focal point lies behind the retina. The latter two conditions may be alleviated by the placement of additional lenses (glasses or contacts) ahead of the cornea.

The term *accommodation* refers to the ability of the optical system to maintain sharp focus regardless of the distance of the object from the eye. Accommodation involves at least two things: the change in lens curvature to achieve focus, and change in pupil diameter to eliminate refractive errors. Fig. 21–12 shows the changes involved in accommodation.

Binocular vision, resulting from two eyes placed side by side, results in each eye receiving a slightly different view of an object. Integration of these images by the brain results in ability to perceive depth.

Visual receptors

Two types of visual receptors, rods and cones form the second layer of the retina. Rods are *scotopic receptors,* operating at low intensities of light and possessing no color function. A visual pigment, *rhodopsin,* formed from Vitamin A and a protein, is found in the rods. Light, falling upon the rhodopsin, decomposes it and creates a nerve impulse. The cones are *photopic* (or high intensity), and are color receptors. They are fewer in number than the rods, and are more numerous in the posterior region of the retina. The fovea contains only cones. There appear to be 3 types of cones with 3 different pigments in them. *Iodopsin,* a blue-sensitive pigment occurs in one type; *erythrocruorin,* a red-sensitive pigment is found in a second; *chlorocruorin,* a green-sensitive pigment occurs in the third. By mixing these three wavelengths of light (as in a color TV set) any color may be created. Color analysis thus appears to be a function of the retina.

Visual pathways

The fibers of the optic nerve leave the eyes and undergo a partial crossing at the optic chiasma. The optic tract passes to the lateral geniculate bodies of the thalamus, and the optic radiation passes to area 17 of the occipital lobe.

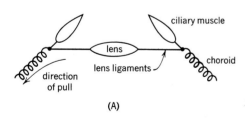

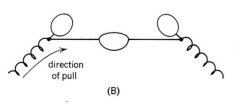

FIGURE 21–12. Changes occurring during accommodation. A. Ciliar muscle relaxed. Elastic pull of choroid places tension on lens ligaments, flattening lens. B. Ciliary muscle contracted. Tension removed from lens ligaments, lens rounds up of its own elasticity. Choroid is stretched.

Color blindness may be explained by assuming that a genetic defect has led to the absence of an enzyme system responsible for synthesis of one or more visual pigments. Fig. 21-13 (on following page) shows a common test for color blindness.

QUESTIONS

1. Compare the ear and the eye in terms of adequate stimulus to activate the receptor, the pathways to the brain and termination in the brain.
2. How do smell and taste reinforce one another?
3. After considering this chapter and the preceding one, why is the concept of ''the five senses'' outmoded?
4. Construct a comparative table for all receptors studied.

Legend key for FIGURE 21-13. (on following page)
Although the illustration is an accurate rendering of an actual pseudoisochromatic plate, its use should not be considered a valid color perception test, but rather an indication only. Study the drawing, then refer to the key below. If the test indicates any color perception difficulty, consult an optometrist.

If you are not color-blind you will see a bear in the upper right-hand side of the page, a deer on the upper left, nothing in the center left, a rabbit and squirrel in the lower portion of the page.

If you are totally color-blind you will see nothing in this picture. If you are red-green blind you will not see the bear, the deer would now be a cow, and in the center left you would see a fox.

If you are completely green blind you will see the rabbit but not the squirrel, and if you happen to be red blind, you will see the squirrel only.

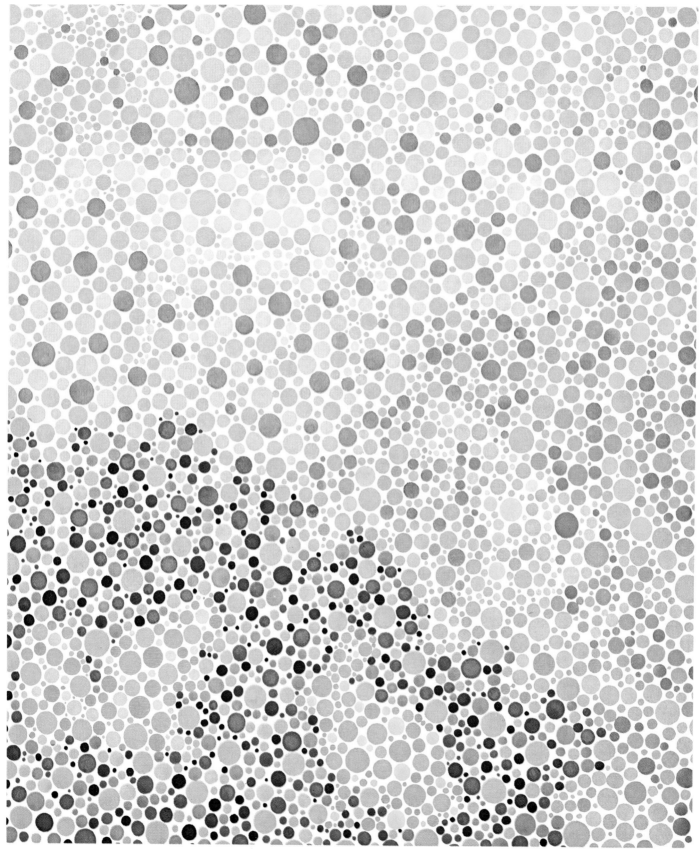

FIGURE 21–13. Color test. (From *Field & Stream*.)

CHAPTER 22

The Endocrine System

589

GENERAL PRINCIPLES

Introduction

We have examined the role of the nervous system in stimulus-response and homeostatic control. The endocrine system is primarily concerned with governing body processes such as growth, differentiation, and metabolism. Endocrine glands utilize the blood stream to distribute their chemical messengers, the hormones, and in this way achieve contact with all body cells. It is becoming increasingly evident that the two "control systems" operate together to control body activity. The nervous system may, for example, respond to environmental changes, and influence an endocrine to secrete a product that has the effect of restoring homeostasis.

Criteria for determining endocrine status

Endocrine glands are designated as being circumscribed or specific groups of cells, which produce specific chemical substances having well-defined effects on body function. Removal of a group of cells suspected of being an endocrine causes clearly defined alterations of body function, and administration of the active chemical of the gland likewise alters body function. Utilizing these criteria, we recognize the endocrine structures shown in Fig. 22–1.

590

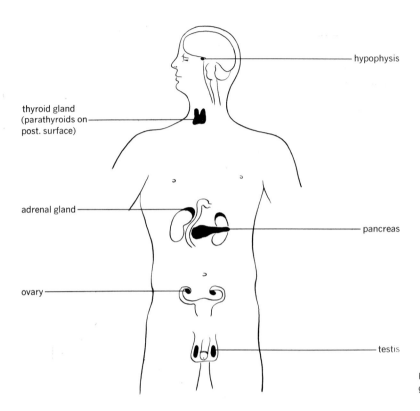

FIGURE 22–1. Locations and names of the endocrine glands.

Hormones and parahormones

The product of an endocrine gland (as defined above) may properly be called a hormone. There are many chemical substances that have specific effects on the body, but are not traceable to specific cells. Examples include carbon dioxide, gastrin, secretin, and enterocrinin. Such substances are commonly called "hormones"; the term para-hormone, however, serves to distinguish such substances from true hormones.

Chemical nature of hormones

Mammalian hormones fall into three classes: proteins or polypeptides, amines, and steroids. A general correlation exists between the germ layer of origin of the endocrine and the class of hormone produced. Most amine hormones are produced by ectodermally derived endo-crines, steroids are produced by mesodermal derivatives, and poly-peptide or protein hormones are produced by endodermal derivatives.

Secretion, storage, transport, and use of hormones

A given endocrine cell synthesizes its hormones from raw materials present in the blood reaching the gland. Some glands (for example, thyroid) produce more hormone than the body requires at a time, and may store excess hormone within the gland. Secretion into the blood stream occurs as the result of a chemical or nervous stimulus. In the blood stream, most steroid and amine hormones attach to a specific plasma protein "carrier," and are transported to the body cells in this combined fashion. At the cell, the hormone may attach to the cell surface and enter by pinocytosis, or pass directly through the membrane. As the cell utilizes the hormone, it generally alters the hormone's structure, partially or completely destroying the physiological activity of the chemical. This process is termed "inactivation." The inactivated product passes from the cell to the blood stream, and is usually excreted. Inactivation of hormones by the organ it affects (target organ) results in the requirement that hormones be continually secreted by the endocrine. In short, hormones are not *reused*, but *are used up* by the target organ.

Function, effect, and action of hormones

The function of a hormone is what we conclude the purpose of a hormone to be, in terms of whole body function. Effect refers to measurable alterations in function occasioned by deprivation or administration of a hormone. Action refers to the cellular mechanism influenced by the hormone. For example, thyroxin, the product of the thyroid gland, is known to increase glycolysis within the mitochondria (action), and to bring about an increase in BMR and heat production (effect). We conclude that the hormone controls BMR (function).

Hormone action appears to depend upon the presence within the cell of a chemical reaction whose rate can be influenced by the hormone. Some hormones are without action on certain cells, suggesting the absence of an influenceable reaction.

Control of secretion of hormones

In general terms, six different methods exist by which both rate and quantity of hormone secretion may be determined:

1. *Neurohumors.* A nerve cell or group of nerve cells produces a chemical (neurohumor) which is sent to the endocrine and causes secretion.
2. *Direct nervous connection.* The endocrine receives nerve fibers from the autonomic nervous system and responds to a nerve impulse.
3. *Reciprocal or feedback control.* One endocrine produces a hormone which affects a second endocrine. The second endocrine produces its hormone, which influences secretion of the first endocrine. If the effect of the second hormone is to inhibit secretion of the first endocrine, the effect is said to be a negative feedback. If the effect is to increase secretion, it is a positive feedback.
4. *Blood levels of organic substances* other than hormones.
5. *Blood levels of solutes* in general.
6. *Blood levels of specific inorganic substances.*

Understanding of the discussions of individual endocrines will be aided if the remarks on the previous pages are well understood. Table 22–1 summarizes the preceding sections.

TABLE 22-1

General Summary of Endocrine Glands

Item	Comments
Functions of endocrines	To integrate, correlate, and control body processes by chemical means.
Criteria for establishing function as endocrine	Cells are morphologically distinct.
	Cells produce specific chemicals, not produced elsewhere.
	Chemicals exert specific effects; effect is lost if gland removed, restored if chemical administered.
Hormones vs. parahormones	Hormones are produced by cells qualifying as endocrines.
	Parahormones are chemicals having specific effects but not produced by endocrines.
Chemical nature of hormones	Steroid—mesodermal origin (e.g., gonads).
	Polypeptides—endodermal origin (e.g., pancreas)
	Small MW amines—ectodermal origin (e.g., posterior pituitary, adrenal medulla).
"Life history" of hormones	Secreted into blood, go to cells and there (probably) influence chemical reactions. Are inactivated as used, inactivated product eliminated.
Methods of control available	
1. Neurohumor	Nerve cells produce a chemical, it goes to endocrine and controls secretion. Stimulus is chemical.
2. Nerves	Nerve fibers pass to endocrine. Stimulus is electrical (nerve impulse).
3. Feedback	Target organ hormone influences secretion of another endocrine which stimulated target organ.
4. Nonhormonal organic substances in blood	Glucose acting on pancreatic islets. Rise causes increased secretion (usually).
5. Total osmolarity of blood	Requires nervous system to detect it. Nervous system then signals endocrine.
6. Inorganic substances in blood	Ca^{++} on parathyroid. Effect usually direct, not inverse.

THE HYPOPHYSIS (PITUITARY)

The hypophysis is a gland of double origin lying in the sella turcica of the sphenoid bone. It measures about 10x13 mm., and weighs approximately 1/2 gram. The gland is composed of a portion derived from the roof of the oral cavity, the *adenohypophysis,* and a portion derived from

TABLE 22–2

Divisions of the Hypophysis

Major divisions	Subdivisions
Adenohypophysis	Pars distalis Pars tuberalis Pars intermedia
Neurohypophysis	Infundibulum (stalk) Neural lobe

the hypothalamus, the *neurohypophysis.* Each region has several subdivisions, as shown in Fig. 22–2, and Table 22–2.

Both major divisions of the gland maintain connections with the hypothalamus (Fig. 22–2). The adenohypophysis is connected to the hypothalamus by a system of blood vessels originating in the hypothalamus and terminating as sinusoids in the pars distalis. This system is designated as the *hypothalamico-hypophyseal portal system* (HHPS). The neurohypophysis is connected to the hypothalamus by nerve fibers originating in the hypothalamus and terminating in the neural lobe. These fibers form the *hypothalamico-hypophyseal tract* (HHT).

These blood and nervous connections emphasize at least two important facts relative to the activity of the hypophysis; first, these connections enable nervous and endocrine systems to work in an integrated fashion to control body activity; second, endocrine response to nerve input may be achieved rapidly.

Pars distalis

With ordinary stains, three cell types, designated *acidophils, basophils,* and *chromophobes,* may be recognized in the distalis. These cells are

FIGURE 22–2. The hypophysis.

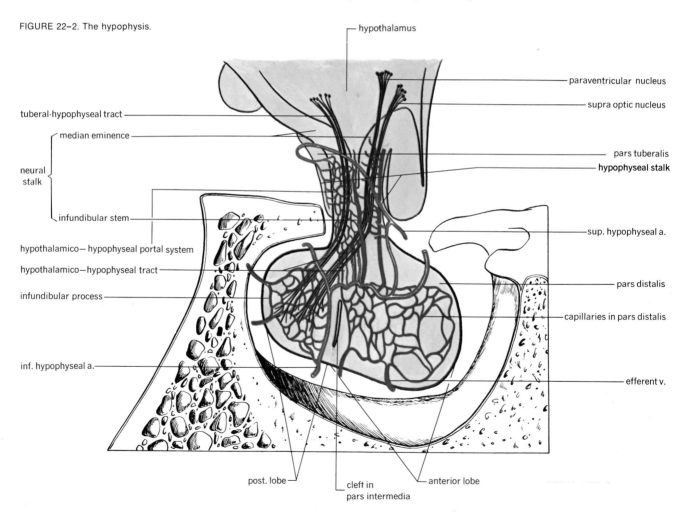

named according to their preference for red, blue, or no stain, respectively. With special stains, particularly PAS (periodic acid Schiff, a glycoprotein stain) and aldehyde thionine (mucopolysaccharide stain), seven cell types may be differentiated. The production of six hormones has been attributed to the distalis, and a tentative assignment of hormones to a particular cell type has been made. The morphology and function of these cells is shown in Fig. 22–3 A and B, and described in Table 22–3.

Hormones of the distalis

Growth hormone [*somatotropin*, *somatotrophic hormone* (STH), *human growth hormone* (HGH)]. Growth hormone from human hypophysis is a single chain of 188 amino acids linked in two places by disulfide (–s–s–) bonds (Fig. 22–4). The normal pituitary contains 4 to 10 mg of hormone and there is little difference in hormone content with age. Secretion of growth hormone is such as to maintain a blood level of about 3 μg (3/1,000,000 g) per ml.

The hormone exerts its effect primarily upon the hard tissues of the body, secondarily upon the soft tissues. In these two tissues, growth

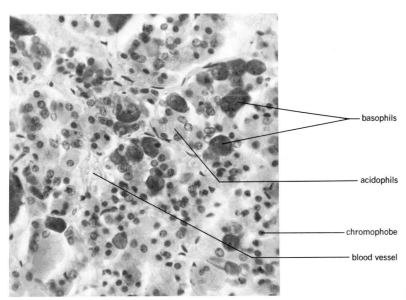

— basophils

— acidophils

— chromophobe

— blood vessel

FIGURE 22–3A. Anterior lobe (pars distalis) of hypophysis.

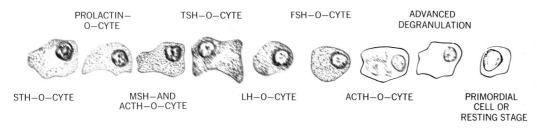

PROLACTIN–
O–CYTE

TSH–O–CYTE

FSH–O–CYTE

ADVANCED
DEGRANULATION

STH–O–CYTE

MSH–AND
ACTH–O–CYTE

LH–O–CYTE

ACTH–O–CYTE

PRIMORDIAL
CELL OR
RESTING STAGE

azocarmine–orange–G stain

FIGURE 22–3B. Morphology of distalis cells.

TABLE 22–3

Distalis Cell Types

Names of cells under:				
Ordinary stain	Special stain	Description	Staining preference	Hormone produced
Acidophils	Somatotropic cell (α_1-orangophil)	Round or ovoid shape with small round acidophilic cytoplasmic granules	Orange G	Human growth hormone (HGH)
	Lactotropic cell (α_2-carminophil)	Round or ovoid, contains large round or oval acidophilic cytoplasmic granules	Azocarmine	Prolactin
Basophils	Corticotropic cell ($\beta1$)	Irregular shape, eccentric nucleus	PAS	Adrenocorticotropic hormone (ACTH)
	Thyrotropic cell ($\beta2$)	Irregular shape, eccentric nucleus	PAS and aldehyde thionine	Thyroid stimulating hormone (TSH)
	Leuteotropic cell (Δ_1)	Round outline, nearly central nucleus	PAS and aldehyde thionine	Leuteinizing hormone (LH)
	Follicle stimulating hormone cell (Δ_2)	Round outline, nearly central nucleus	PAS	Follicle stimulating hormone (FSH)
Chromophobe	Chromophobe	Irregular outline, eccentric nucleus, very pale staining cytoplasm	None, stains little if at all	ACTH?

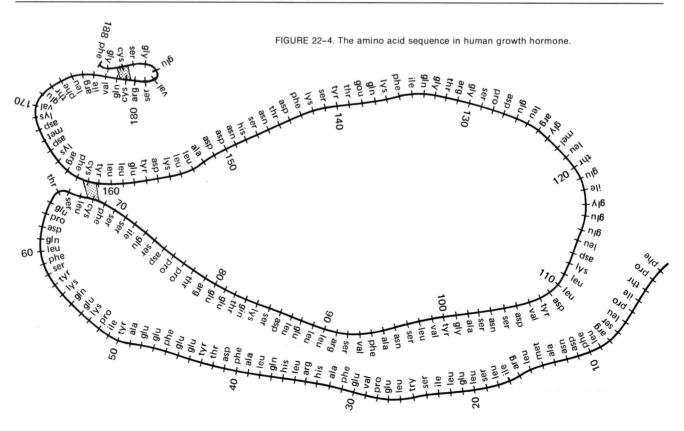

FIGURE 22–4. The amino acid sequence in human growth hormone.

hormone increases rate of growth, and maintains the size of body parts once maturity has been reached. Metabolic effects include: increasing the conversion of carbohydrates to amino acids (transamination); increasing the cellular uptake of amino acids; mobilization of fats from storage areas; increasing fat metabolism, as evidenced by a fall in respiratory quotient (RQ); increasing blood sugar level (diabetogenic effect).

Prolactin (lactogenic hormone). Prolactin is a polypeptide hormone of 205 amino acids. It is one of several hormones involved in milk production by the mammary glands. If the gland has been "primed" by sex hormones, thyroid hormone, and adrenal hormones, prolactin will cause milk secretion (Fig. 22–5). In birds, other effects of the hormone include aiding the expression of maternal behavior (nesting, egg hatching).

Adrenocorticotropic hormone (ACTH). ACTH is a single unbranched chain of 39 amino acids. It is synthesized as the body requires it, since the gland has less than 1/4 mg in it. ACTH controls the synthetic and secretory activity of the two inner zones of the adrenal cortex. Metabolic effects on cells in general include mobilization of fats, production of hypoglycemia, and increase in muscle glycogen. Acting through the adrenal cortex, ACTH is also involved in body resistance to stress.

Thyroid stimulating hormone (thyrotropin, TSH). TSH is a glycoprotein having a molecular weight of about 25,000. TSH influences all phases of thyroid gland activity: accumulation of material, synthesis, and secretion. Acting through the thyroid, TSH becomes involved in regulation of basal metabolic rate. Effects on other tissues include increasing the breakdown of fat, working with growth hormone, and increasing water content of loose connective tissue.

Leuteinizing hormone (LH, leuteotropin; in the male, it is called ICSH for interstitial cell stimulating hormone). LH is a glycoprotein with a molecular weight of about 30,000. In the female, LH occasions changes leading to the formation of the corpus luteum in the ovary, and is involved in readying the mammary gland for secretion. It is necessary for ovulation and implantation of the zygote. In the male, ICSH controls secretion of the interstitial cells of the testis.

Follicle stimulating hormone (FSH). FSH is also a glycoprotein with molecular weight about 30,000. In the female, FSH controls the maturation of primary follicles to vesicular follicles; in the male, it controls spermatogenesis. Table 22–4 presents a summary of the characteristics and effects of pars distalis hormones.

Control of secretion of pars distalis hormones

Release of distalis hormones appears to be under the control of neurohumors (designated *releasing factors*) which are produced in the hypothalamus and which reach the distalis via the blood vessels connecting these areas. Factors which stimulate secretion of all hormones except prolactin have been postulated. A factor which inhibits prolactin secretion has been described. All releasing factors appear to be produced in the median eminence of the hypothalamus. Table 22–5 summarizes the current state of knowledge concerning these factors.

With such a "chain of command" from nervous system to hypophysis, one might conclude that primary control of hypophyseal secretion depends on nerve input to the hypothalamus. The nervous system can

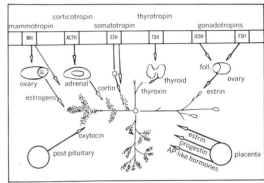

FIGURE 22–5. Hormones involved in lactation.

TABLE 22–4

Hormones of the Pars Distalis

Hormone	Characteristics	Effects on body
Growth Hormone (GH)	Molecular weight 21,500; unbranched peptide chain	Control body hard and soft tissue growth
Prolactin	Molecular weight about 23,500	Causes milk secretion in "primed" gland
Adrenocorticotropic hormone (ACTH)	Molecular weight 4567; contains 39 amino acids in single peptide chain	Controls activities of adrenal cortex
Thyroid stimulating hormone (TSH)	Molecular weight 26,000 to 30,000; composed of 200 to 300 amino acids	Controls all aspects of thyroid activity
Leuteinizing hormone (LH) or Interstitial cell stimulating hormone (ICSH)	Molecular weight about 30,000; a glycoprotein	Female: Corpus luteum formation, ovulation, implantation Male: Controls interstitial cell activity
Follicle stimulating hormone (FSH)	Molecular weight about 30,000; a glycoprotein	Female: Controls follicular maturation Male: Controls spermatogenesis

TABLE 22–5

Hypothalmic Factors

Name of factor	Characteristics	Hormones affected	Nature of effect	Comments
Corticotropin releasing factor (CRF)	Polypeptide, molecular weight about 1700, 13 amino acids	ACTH	Stimulation of secretion	Two forms exist α-CRF, β-CRF; α most active
Thyrotropin releasing factor (TRF)	Small, nonpolypeptide molecule	TSH	Stimulation of secretion	Effective in 1-μg doses
Growth hormone releasing factor (SRF)	Small molecular weight polypeptide	GH	Stimulation of secretion	More SRF in young individuals
Leuteinizing hormone releasing factor (LRF)	Basic polypeptide, molecular weight estimated at 1200 to 2000; 14 amino acids	LH	Stimulation of secretion	Cyclic changes occur in LRF with menstrual cycle
FSH releasing (FRF)	Amine?	FSH	Stimulation of secretion	
Prolactin inhibiting factor (PIF)	Not characterized	Prolactin	Inhibition of secretion	Present continually except after childbirth (when lactation starts)

detect change and signal the hypothalamus, which in turn causes selective secretion of distalis hormones appropriate to meeting the challenge.

Other factors than the hypothalamus appear to be involved in distalis secretion. For example, growth hormone secretion is increased by a fall in blood sugar level. This effect *may* be direct upon the gland (or through the hypothalamus). Negative feedback exists between thyroxin and thyroid stimulating hormone, cortical steroids and adrenocorticotropic hormone, progestin and leuteinizing hormone, and estrogen and follicle stimulating hormone. Again the problem is whether the effect is direct on the distalis or through the hypothalamus.

The pars intermedia

In the human hypophysis, the pars intermedia is virtually nonexistent, being reduced to a few cells and spaces *(cysts)* between the distalis and neural lobe. No known hormone is produced by human intermedia. In lower vertebrates (fish, amphibians), the intermedia is the source of *melanocyte stimulating hormone* (MSH), which causes expansion of pigment cells *(melanophores)* in the skin of such animals. A protective effect of "matching" the color of the animal to the environment is the result. In man, adrenocorticotropic hormone has a function similar to MSH.

The pars tuberalis. No known hormone is produced by the cells of the tuberalis.

Neurohypophysis

The neurohypophysis cannot be considered a separate endocrine gland, for it produces no hormones. The hypothalamus produces the hormones of the neural lobe, and passes them over the hypothalamico-hypophyseal nerve tract to merely be stored in and released from the neural lobe.

The functional cell of the neural lobe appears to be the *pituicyte* (Fig. 22–6).

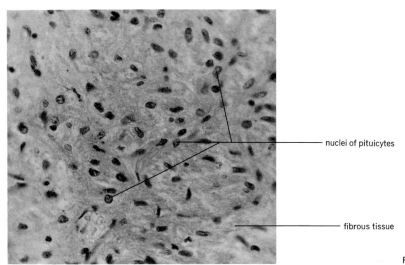

nuclei of pituicytes

fibrous tissue

FIGURE 22–6. Posterior lobe (pars nervosa) of hypophysis.

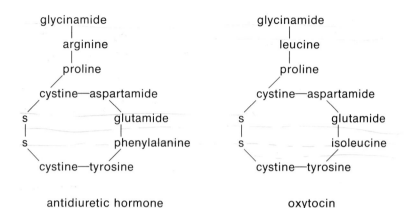

FIGURE 22-7. Amino acid sequences in antidiuretic hormone and oxytocin.

Hormones. Two hormones may be isolated from the neural lobe, vasopressin-ADH (VADH) and oxytocin. Both are small molecules containing 8 amino acids in a ring (Fig. 22-7).

Vasopressin—ADH exerts a stimulating effect on smooth muscle of arteries, causing vasoconstriction and an elevation of blood pressure. The hormone also exerts an antidiuretic effect (hence ADH for antidiuretic hormone) on the kidney, resulting in increased reabsorption of water from the kidney tubules. Control of VADH secretion is determined by osmosensitive cells within the hypothalamus which continually monitor blood osmotic pressure.

Oxytocin exerts its effect primarily upon the smooth muscle of the pregnant uterus and the contractile cells (myoepithelial cells) around the ducts of the mammary glands. It stimulates the contraction of both types of cells. Oxytocin is not essential to either function, but apparently aids childbirth and milk ejection. Control of secretion has not been definitely established. Suckling by the infant increases its secretion.

Disorders of the hypophysis

Excess secretion of hormones from the adenohypophysis may, in theory, result in overproduction of all hormones, or of only one. The usual cause of oversecretion of any hormone is a tumor involving the cells in which a particular hormone is produced. Chromophobe tumors account for 85 percent of all hypophyseal tumors; alpha cell tumors for 10 to 14 percent; beta and delta cell tumors are rare. Visual disturbances may be the first sign of a tumor, as the enlarging gland presses upon the neighboring optic tract.

Perhaps the most interesting tumors are those involving the growth hormone producing cells. *Giantism* (Fig. 22-8) is the result of excess growth hormone secretion occurring before skeletal maturity has been achieved. The body continues to grow and may reach heights in excess of 8 feet and weights in excess of 400 pounds. Excess growth hormone production after maturity results in *acromegaly* (Fig. 22-9). The bones can no longer grow in length, but can increase in width. The jaw, hands, and feet are most commonly affected. Treatment of hypophyseal tumors may be carried out by radiation delivered to the gland from the outside; the gland is nearly inaccessible to surgery.

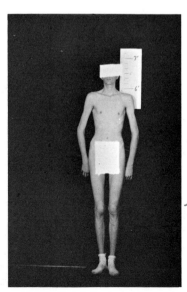

FIGURE 22-8. Giantism.

FIGURE 22-9. Acromegaly.

Deficient secretion of growth hormone results in *hypophyseal infantilism* (Fig. 22-10). The body is well proportioned but juvenile in appearance. Sexual characteristics are minimally developed and stature is small. There is apparently no effect on mental capacity. Treatment of this condition requires administration of human growth hormone. Hormones from other species are ineffective.

Excess secretion of vasopressin, secondary to hypothalamic tumor, results in excessive water retention, dilution of body fluids, and weight gain. Limitation of water intake and administration of diuretics aid treatment. Failure of vasopressin secretion as a result of hypothalamic damage results in *diabetes insipidus.* Excessive dilute urine production (to 56 l per day in severe cases) is characteristic. Apparently there are no disorders associated with oxytocin.

Table 22-6 presents a summary of hypophyseal disorders. The table includes disorders not described in the text.

THE THYROID GLAND

Introduction

The thyroid gland (Fig. 14-4) is a *two-lobed* organ lying on the lateral aspect of the lower larynx and upper trachea. A small *isthmus* of thyroid tissue connects the two lobes across the midline. The gland weighs about 20 grams in the adult and is one of the most vascular of all endocrines. The gland receives 80 to 120 mls of blood per minute, a fact important in maintaining adequate supplies of building blocks to the organ. The gland is composed of many small spherical units known as *thyroid follicles.* Each follicle is lined with a simple cuboidal epithelium carrying microvilli, and is filled with *thyroid colloid.* The colloid represents a storage form of hormone. Between follicles are reserve cells, the *interfollicular cells.* These cells can, if required, develop into functional follicles. These features are shown in Fig. 22-11.

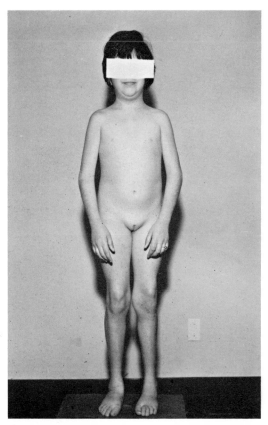

FIGURE 22-10. Hypophyseal infantilism.

TABLE 22-6
Some Disorders of the Hypophysis

Condition	Cause	Hormone(s) involved	Secretion is: Excess	Secretion is: Deficient	Characteristics	Comments
Giantism	Acidophil tumor before maturity	Growth hormone (GH)	X		Large body size	Proportioned but large
Acromegaly	Acidophil tumor after maturity	Growth hormone (GH)	X		Misshapen bones, hands, face, feet	Disproportional growth
Hypophyseal infantilism	Destruction of acidophils by disease, accident, etc.	Growth hormone (GH)		X	Juvenile appearance, sexually and in height	Body properly proportioned
Water retention + *demonstration of high VADH blood levels*	Hypothalamic tumor	Vasopressin-antidiuretic hormone	X		Dilution of body fluids, weight gain	*Must* differentiate from H_2O retention due to other causes
Diabetes insipidus	Hypothalamic damage	Vasopressin-antidiuretic hormone		X	Excessive urine excretion	No threat to life
Sheehan's syndrome	Atrophy of distalis	All		X	Sexual characteristics involute, no menses, anemic	Occurs mostly in the female. Give target organ hormones
Hypophyseal myxedema	Basophil degeneration	Thyroid stimulating hormone		X	Dry flaky skin, lethargy, anemia	May be controlled by thyroxin

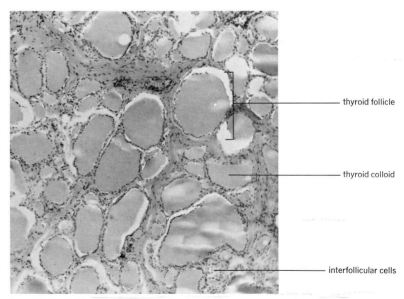

FIGURE 22-11. Thyroid gland.

thyroid follicle

thyroid colloid

interfollicular cells

Hormone synthesis

The substance generally regarded as the true hormone of the thyroid is thyroxin. Thyroxin (or tetraiodothyronine) is an iodinated amino acid with the formula:

HO⟨benzene ring⟩—O—⟨benzene ring⟩—CH_2—$CHNH_2$—COOH

formula for thyroxin

Thyroxin is synthesized only after the gland has accumulated iodide and the amino acid tyrosine. By iodinating the molecule of tyrosine, then coupling two acids together, thyroxin is formed. Depending upon body demands for hormone, the hormone may be released into the blood stream, or linked to the protein *thyroglobulin* in the colloid. If linked, the thyroxin must be hydrolyzed from the thyroglobulin before entering the blood. Once into the blood, thyroxin combines with a specific plasma protein designated *thyroid-binding globulin* (TBG) for transport.

Control of the thyroid

Accumulation of the necessary building blocks, the synthesis, and the release of thyroxin, are all controlled by pituitary thyroid stimulating hormone (TSH). Thyroxin in turn exerts a negative feedback on TSH production.

Effects of thyroxin

Thyroxin has no particular target organ; those cells possessing catabolic systems of enzymes respond to the hormone, and this means cells in general. The effects of the hormone may be divided into four general categories:

1. *Calorigenic effect.* Thyroxin accelerates the catabolic reactions of glycolysis, Krebs cycle, beta oxidation and oxidative phosphorylation. One mg of thyroxin can raise heat production by 1000 calories and CO_2 production by 400 g.
2. *Growth and differentiation.* Working with growth hormone, thyroxin insures proper development of the brain. Deficiency of thyroxin during development leads to smaller and fewer neurones, and defective myelination.
3. *Metabolic effects.* Thyroxin acts as a diuretic, increasing urine production. It increases protein breakdown, increases uptake of glucose by cells, enhances glycogenolysis, and depresses blood cholesterol levels.
4. *Muscular effects.* Thyroxin in excess interferes with ATP synthesis, and may thus speed exhaustion of energy in muscle. Both skeletal and cardiac muscle are affected in this manner.

Other factors influencing thyroid activity

A wide variety of influences may alter the basic activity of the gland. Among these are:

1. *Antithyroid agents.* Generally called goitrogens, because they cause goiter or enlargement of the thyroid, such materials inhibit one or more of the steps in synthesis of hormone. Included are the thioureas, phenols, and thiocyanates. The thiocyanates are present in cabbage, turnips, rutabaga, and mustard, hence these foods are said to be goitrogenic.
2. *Gonadal hormone levels.* Excessive gonadal hormone appears to decrease thyroxin transport.
3. *Pregnancy.* Pregnancy increases all aspects of thyroid activity, due apparently, to fetal competition for maternal iodide and amino acids.
4. *Age.* Activity of the thyroid decreases with age, though the decrease is small.
5. *Stress.* Stress, particularly that of a cold environment, increases thyroid activity.

Thyrocalcitonin (TCT, calcitonin)

In 1965, a hormone from the thyroid gland was shown to lower plasma concentrations of calcium and phosphate. It is secreted in response to a high blood calcium level. The name *calcitonin* was applied to the substance, and it was originally thought to be secreted by the parathyroids. Subsequently, the name *thyrocalcitonin* was applied, reflecting its thyroidal origin. Thyrocalcitonin apparently works with parathyroid hormone to control body calcium and phosphate balance. The effect is mediated primarily through the kidneys, (increased absorption) and possibly on bones (increase mineralization).

Disorders of the thyroid

Overproduction of thyroxin due to thyroid tumor, or overstimulation by TSH, brings about the condition known as *hyperthyroidism (Grave's disease)* (Fig. 22–12). There are different degrees of a basically similar set of symptoms in the disease, so the following list will not attempt to separate the various specific diseases.

1. Excessive nervousness and excitability.
2. Elevated heart rate and blood pressure.
3. Elevated BMR (40 to 100 percent).
4. Weakness.
5. Weight loss.
6. Bulging of the eyes; exophthalmus.

The first five symptoms may be accounted for by considering the catabolic stimulating effect of the hormone; in short, everything is accelerated. The exophthalmus is due to increases in mass of tissue behind the eye, pushing it forward. Deficient thyroid activity results in a wider variety of disorders, depending upon the cause of the deficiency.

Simple goiter is generally the result of insufficient dietary iodide. The gland enlarges "hoping to trap more of the available iodide," which, of course, creates more demand, a greater enlargement, and so on. Un-

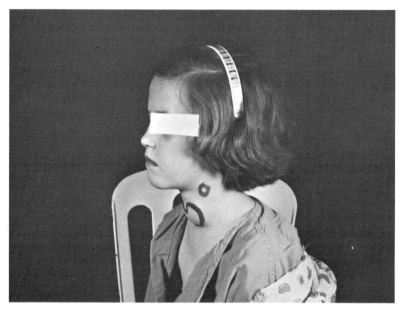

FIGURE 22–12. Hyperthyroidism.

treated goiters can reach extreme sizes. Dietary deficiency of iodide in a pregnant woman may result in insufficient thyroxin production by the embryo, and cretinism may occur.

Cretinism results in idiocy (remember the role of thyroxin in nervous system development) and retarded growth. The mental damage cannot be overcome. Myxedema is seen after birth, in both children and adults (Figs. 22–13 and 22–14). Myxedema is associated with dry skin, low BMR, and intolerance to cold.

Treatment of these conditions usually involves the administration of dried thyroid gland, by mouth, in the form of a tablet. Only cretinism will fail to respond to such treatment.

Table 22–7 summarizes facts about the thyroid gland.

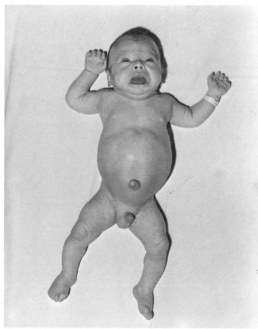

FIGURE 22–13. Childhood myxedema.

FIGURE 22–14. Adult myxedema.

TABLE 22-7

Summary of the Thyroid

Item	Comments
Location and parts; weight	Two-lobed, plus connecting isthmus; located on lower larynx and upper trachea; 20 grams in weight
Requirements for hormone synthesis	The amino acid tyrosine; iodide in diet; TSH to control all steps in synthesis
Hormones produced and effects: Thyroxin Thyrocalcitonin	 Main controller of catabolic metabolism Lowers blood Ca^{++} and $PO_4^=$ levels
Control of hormone secretion Thyroxin Thyrocalcitonin	 By thyroid stimulating hormone By blood Ca^{++} level
Disorders Hypersecretion Hyposecretion	 Creates hyperthyroidism: 1. Elevated BMR 2. Elevated heart action 3. Exophthalmus Creates: 1. Goiter—enlarged gland 2. Hypothyroidism (cretinism, myxedema) (a) Low BMR (b) Low heart rate, blood pressure, body temperature

THE PARATHYROID GLANDS

Introduction

Calcium, phosphate, magnesium, and citrate are ions which assume an extremely important role in the body economy. As constituents of the bones and teeth, these ions assume structural roles in the body. The phenomenon of muscular contraction apparently depends upon calcium ion, as does normal membrane permeability. Calcium is also involved in nervous irritability, and in the clotting of the blood. Regulation of the body content of these ions therefore becomes very important in terms of whole body function.

Calcium. The average adult body contains about 1100 g of calcium, 99 percent of which is found in bone. Intake occurs by absorption of the ion from the gut, and it is then placed in the extracellular fluids (plasma, interstitial fluid). Removal of calcium from extracellular fluid occurs by filtration in the kidney (99 percent reabsorbed), exchange with tissues (chiefly muscle and bone), and by secretion of digestive fluids into the gut and out via the feces. In the adult, about 1 g of calcium per day will keep all forces in balance.

Phosphate.   The body contains about 500 g of phosphate, 85 percent of which is in the skeleton. Other compounds containing phosphate include ATP, ADP, and creatine phosphate, all important energy sources. DNA and RNA also contain phosphate. Phosphate is absorbed through the gut, and is apparently absorbed in inverse proportion to calcium. The absorbed phosphate is also placed into the extracellular fluids, from which it is withdrawn by cells, bony tissue, and the kidney. The kidney route constitutes the major path for excretion of phosphate.

Magnesium.   Total body content of magnesium is about 200 meq, with 50 percent of this in the skeleton. Absorption occurs in the gut, apparently in inverse proportion to calcium absorption. Magnesium is essential as a cofactor in many enzymatic reactions, particularly those involving ATP synthesis and degradation, and ribosomal protein synthesis. Renal excretion is the major pathway for magnesium loss.

Citrate.   Citrate ion is an important intermediate in glycolysis and the Krebs cycle. It forms a major source of energy for ATP synthesis.

This brief discussion of these ions indicates their importance in body function. The regulation of cellular and fluid levels of these substances is under the primary control of the parathyroid glands.

Parathyroid glands

There are usually four parathyroid glands in the human, located upon the posterior aspect of the thyroid (Fig. 14–4). They measure about 5 mm in diameter, and have a combined weight of about 120 mg. Histologically, the glands consist of densely packed masses of *principal (chief)* cells, and *oxyphil* cells (Fig. 22–15). The principal cells are regarded as the source of *parathyroid hormone (PTH),* while the oxyphil cells are thought to be reserve cells, capable of assuming hormone production if need arises. PTH is a polypeptide chain containing 74 to

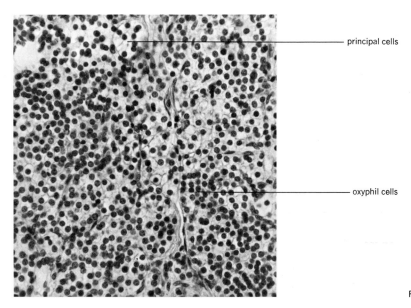

principal cells

oxyphil cells

FIGURE 22–15. Parathyroid gland.

80 amino acids. Regulation of PTH secretion is determined by the blood calcium level, a fall of blood calcium increasing PTH secretion, and vice versa.

Effects of the hormone

If the hormone controls the metabolism of the previously mentioned ions, it must be involved in at least three phases of this metabolism: (1) absorption, (2) exchange between cells and body fluids, (3) excretion. Accordingly, it is suggested that the gut, bones, and kidney are the three tissues or organs which are primarily affected.

1. *Effects upon the gut.* PTH increases absorption of calcium and magnesium from the intestine, provided adequate amounts of vitamin D are present. Phosphate absorption is also increased by the hormone.

2. *Effects upon bone.* PTH acts directly upon bone. The effect is thought to be mediated via *osteoclasts* (bone destroying cells) and the connective tissue fibers of bone. PTH causes increase in numbers of osteoclasts and occasions proliferation of fibers. The result will be removal of the inorganic phase of bone (with consequent rise of blood Ca^{++} and $PO_4^=$) and increase in organic content. Action of the hormone on bone may revolve around lysosome dissolution in bone cells and consequent enzymatic destruction of bony tissue.

3. *Effects upon the kidney.* PTH controls reabsorption of calcium and magnesium in direct proportion to PTH levels, while increasing phosphate excretion.

The combination of effects on intake, use, and excretion tends to keep plasma levels of these materials within normal limits.

Effects on other organs and cells

The mammary gland responds to PTH by lowering secretion of calcium in milk. On cells such as those of the kidney, liver, or intestine grown in tissue culture, PTH increases decarboxylation, release of H^+, causes swelling of mitochondria, and stimulates the Krebs cycle. A caution: PTH is not the only hormone exerting an effect on Ca^{++} metabolism. It might be predicted that any hormone concerned with growth would lead to, at the very least, increased absorption of Ca^{++} and $PO_4^=$. Thus growth hormone, thyroxin, and certain sex hormones work with PTH on the absorptive aspect. If one is concerned with maintenance, rather than growth, PTH and thyrocalcitonin are the two most concerned with *controlling* metabolism of these ions.

Disorders of the parathyroids

Hyperparathyroidism (von Recklinghausen's disease) usually results from tumor formation in the gland. Excess PTH results in destruction of bone tissue and subsequent formation of excess fibrous tissue and cysts in the bones (osteitis fibrosa cystica). The bones may become softened to the point that they collapse. Blood Ca^{++} levels are very high, resulting in muscular weakness, mental disorder, and cardiac irregularities. The kidneys may form calculi, and vomiting, constipation, and other intestinal disturbances are usually present.

Hypoparathyroidism (tetany) usually results from damage to or removal of parathyroids during thyroid surgery or destruction of parathyroids by disease, infection or hemorrhage. The term tetany reflects the primary symptom of the disease; muscular rigidity and paralysis due to low blood calcium levels. Characteristic positions of the hand *(carpospasm)* are seen. The primary consideration in treatment is to achieve an elevation of the blood calcium level. This may be achieved immediately by intravenous injection of a solution such as 5 percent Ca gluconate. This relieves the symptoms, and further treatment may then be instituted for control of the diseases.

THE PANCREAS

Introduction

The endocrine portion of the pancreas, the *islets (of Langerhans)*, develop from the terminal portions of the ducts of the exocrine (digestive portion) and subsequently separate from the ducts. Estimates of the number of islets vary from 500 thousand to 2 million. The tissue becomes functional at about 3 months *in utero*. Microscopically, the islets show two major cell types, designated alpha and beta. Alpha cells comprise about 25 percent of the islet population, and to them is attributed the production of the hormone *glucagon (hyperglycemic glycogenolytic factor* or HGF). Glucagon is a single, long chain polypeptide with a molecular weight of 3485. Its secretion from the alpha cells is caused by a fall in blood sugar. The hormone exerts its action on the enzymes stimulating conversion of glycogen to glucose *(glycogenolysis)*. Such conversion raises the concentration of blood glucose, and may, if secretion continues, create a hyperglycemia (excess blood sugar level).

The beta cells, constituting about 75 percent of the islet cells, produce the more familiar hormone *insulin*. Insulin is a complex protein hormone consisting of four protein chains of two types. It has a molecular weight of about 6000. Insulin is secreted in response to a rise in blood sugar level (as after a meal), and lowers the blood sugar by stimulating the conversion of glucose to glycogen *(glycogenesis)* and by stimulating cellular uptake of glucose.

Two conclusions may be drawn at this point: (1) the blood sugar level reflects a balance of the effect of the two hormones, and (2) the effects of both hormones are exerted primarily upon the metabolism of carbohydrate. Recalling that the metabolism of carbohydrates, fats, and proteins are interrelated, we may further conclude that a disturbance of carbohydrate metabolism must cause disruption in the metabolism of the other substances as well.

Insulin deficiency

Diabetes mellitus. Insulin deficiency results from the inability of the beta cells to produce enough insulin for body requirements. A series of reactions are set in motion by the deficiency:

1. Blood sugar level rises *(hyperglycemia)*. Glucose polymerization into glycogen is reduced, as is cellular uptake.

2. More sugar is filtered in the kidney than the tubule cells can reabsorb. Glucose appears in the urine *(glucosuria)*.

3. The cells of the body, unable to acquire glucose, shift their energy source to fats. The RQ decreases; excess acetyl-CoA (from beta oxidation) is produced; the Krebs cycle cannot handle it all and acetoacetic acid production is increased. The ketone bodies are increased in concentration, and ketosis may result. Also, the buffering systems of the body may be overtaxed and acidosis may occur.

4. The loss of glucose through the urine carries excess quantities of water with it. Urine volume is increased *(polyuria)*.

5. Excess urinary loss of water dehydrates the body and a great thirst develops, with rise of water intake *(polydipsia)*.

6. Glucose loss triggers the desire to eat and excess quantities of food may be consumed *(polyphagia)*.

7. High levels of fats in the blood may lead to narrowing of blood vessel caliber through *atherosclerosis*. Occlusions of vessels in the legs may give rise to diabetic gangrene as tissues are deprived of blood, die, and become infected. The series of events described, along with their causes, constitute the symptoms of diabetes mellitus.

Treatment of diabetes may depend upon the severity of the disease. Mild disease may be controlled by dietary measures, while severe disease may require insulin administration.

Insulin as the treatment for diabetes

Table 22–8 summarizes some types of insulin available for treatment of diabetes.

TABLE 22–8

Types of Insulin Available for Use in Diabetes

Type	pH	Time of maximum effectiveness (hours)	Duration of effect (hours)
Semilente	7.1–7.5	4–6	12–16
Lente	7.1–7.5	8–12	18–24
Ultralente	7.1–7.5	16–18	30–36
Crystalline Zn*	3.0	4–6	6–8
NPH†	7.1–7.4	8–12	18–24
Protamine Zn‡	7.1–7.4	14–20	24–36
Globin insulin	3.6	6–10	12–18

*Almost pure insulin—quick results, often used in combination with longer acting form.
†*"Neutral Protamine Hagedorn"*—Insulin and equivalent amount of protein.
‡Excess protein; stretches effect.

Insulin *must* be given by injection inasmuch as it is a protein and would be destroyed by the digestive enzymes if given by mouth. One must also recognize that it is virtually impossible to attain normal insulin levels by injection, and that only control, rather than cure, can be achieved in the disease.

Oral treatment of diabetes

If any beta cells survive in the islets, the administration by mouth of certain substances may control diabetes. Most belong to the group of chemicals known as sulfonylureas (Fig. 22–16). They act in several ways:

1. Tend to inhibit glycogen breakdown, an effect exerted directly upon the liver.
2. Increase secretion of insulin by any surviving beta cells
3. Increase cellular uptake of glucose

Hypoglycemia

Low blood sugar levels may result from starvation, failure of sugar absorption, high activity levels, or hyperinsulinism due to islet tumors (insulin shock), or overdosing with insulin. Regardless of cause,

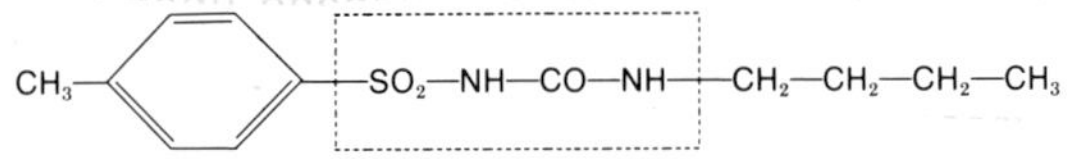

CH_3—⟨benzene ring⟩—SO_2—NH—CO—NH—CH_2—CH_2—CH_2—CH_3

tolbutamide
(1-butyl-3-p-tolylsulfonylurea)

CH_3—⟨benzene ring⟩—SO_2—NH—CO—NH—N⟨(CH_2—CH_2—CH_2)(CH_2—CH_2—CH_2)⟩

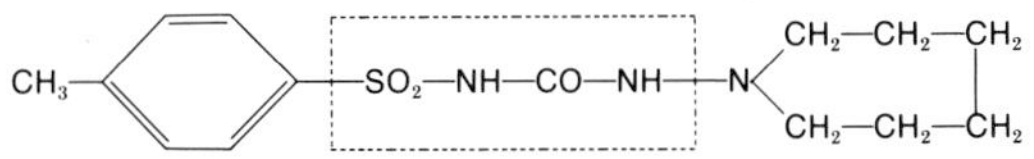

tolazamide
(1-(hexahydro-1-azepinyl)-3-p-tolylsulfonylurea)

CH_3—CO—⟨benzene ring⟩—SO_2—NH—CO—NH—⟨cyclohexyl ring⟩

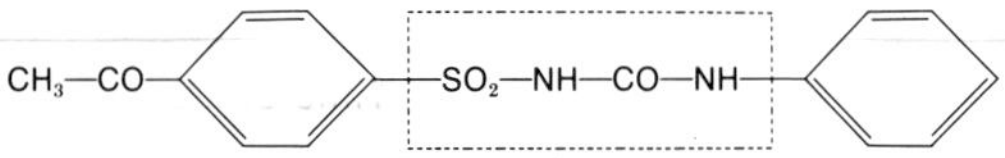

acetohexamide
(N-(p-acetylbenzenesulfonyl)-N¹-cyclohexylurea)

Cl—⟨benzene ring⟩—SO_2—NH—CO—NH—CH_2—CH_2—CH_3

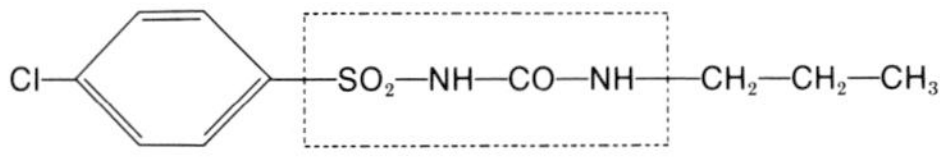

chlorpropamide
(1-propyl-3-p-chlorobenzenesulfonylurea)

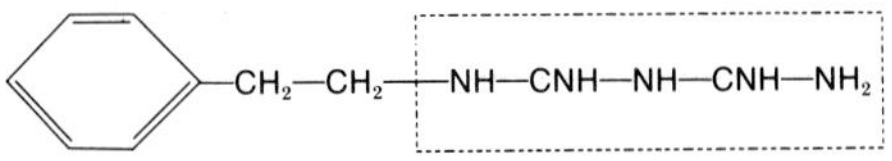

⟨benzene ring⟩—CH_2—CH_2—NH—CNH—NH—CNH—NH_2

phenformin
CN¹-B-phenethylformamidinyliminourea
or phenethylbiguanide)

FIGURE 22–16. Formulas of some orally active antidiabetic agents. Active portion enclosed in rectangle.

hypoglycemia affects one body area primarily, the brain. Brain cells normally utilize only glucose, and are therefore immediately affected by low blood sugar levels. The higher brain centers are affected first, causing:

1. Disturbances in locomotion
2. Mental confusion
3. "Dull wittedness"
4. Decreased muscle tone

Effects on lower centers include:
1. Disturbances in respiration
2. Loss of consciousness
3. Depression of reflexes
4. Depression of body temperature

If the blood sugar level continues to decline, death will result. Treatment involves raising the blood sugar level. If the individual can recognize the onset of symptoms, consumption of sugar or a candy bar may arrest their development. If unconscious, intravenous injection of glucose solution is necessary.

THE ADRENAL GLANDS

Introduction

The adrenal glands are paired, hat-shaped organs located superior to the kidneys. Each gland consists of an inner medulla of ectodermal origin, and an outer cortex of mesodermal origin (Fig. 22–18). The medulla is not essential for life; at least one-sixth the total cortical substance is necessary to maintain life.

The medulla

The hormone producing cells of the medulla are known as the chromaffin cells, and are arranged in the form of cords. The cords of cells are separated by large venous sinuses. Upon stimulation, the chromaffin cells secrete into the sinuses which empty the secretion very rapidly into the circulation. Control of medullary secretion is by way of preganglionic neurones of the sympathetic nervous system which terminate without synapse upon the medullary cells. Response is therefore very rapid.

Hormones of the medulla

Two active substances, norepinephrine and epinephrine, (Fig. 22–17) may be isolated from the medulla. *Norepinephrine* composes about 20 percent of the medullary secretion, *epinephrine* the remaining 80 percent. Both hormones are "sympathomimetic," that is, their effects are similar to that obtained by stimulation of the sympathetic nervous system. The effects of the two hormones are presented in Table 22–9.

The list of effects of epinephrine appears to create a set of conditions

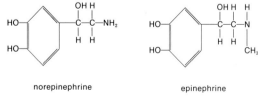

FIGURE 22–17. Chemical formulas of norepinephrine and epinephrine.

TABLE 22-9

Effects of Adrenal Medullary Hormones

Hormone	General effect	Mechanism of production of effect
Norepinephrine	Raises blood pressure	Causes constriction of muscular arteries
Epinephrine	Raises blood pressure	1. Increases rate and strength of heart beat
		2. Causes constriction of all muscular arteries except those to heart and skeletal muscles, the latter dilating
		3. Causes contraction of spleen
	Stimulates respiration	Increases rate and depth of breathing by direct effect upon respiratory centers
	Dilates bronchi and bronchioles	Causes relaxation of smooth muscle of respiratory system
	Slows digestive process	Inhibits muscular contraction of stomach, intestines
	Postpones fatigue in skeletal muscle, and increases efficiency of contraction	1. Increases muscle glycogenolysis 2. Creates more ATP
	Causes hyperglycemia and glucosuria	Increases liver glycogenolysis; kidney tuble Tm_g is exceeded
	Increases O_2 consumption, CO_2 production	Stimulates general metabolic activity of cells
	Miscellaneous effects:	
	Stimulates sweating	Stimulates eccrine secretion
	Increases lacrymation	Stimulates secretion lacrimal gland
	Dilates pupil	Causes contraction of dilator pupillae
	Increases salivary secretion	Stimulates secretion (nervous)
	Increases coagulability of blood	Accelerates clotting reaction

to enable the body to "run or fight." Fortunately, the effects, though great, are shortlived. The hormone is inactivated by the liver in about 3 minutes.

Disorders of the medulla

No disease of hypofunction exists. As stated before, the medulla is not essential for life. Its functions can be taken over by the sympathetic nervous system. Occasionally, a medullary tumor *(phaeochromocytoma)* may occur, resulting in overproduction of hormone. This is an extremely dangerous condition, because of the extremely high blood pressures which may be achieved (to 300 mm Hg systolic pressure). Surgical removal of the tumor is mandatory before circulatory damage occurs.

The cortex

The adrenal cortex is subdivided into three zones, based primarily upon cellular arrangement (Fig. 22–18). The outer *zona glomerulosa* contains cortical cells in ball- or knot-like masses *(glomeruli)*. The central *zona fasciculata* contains radially arranged cords *(fascicles)* of cells. The inner *zona reticularis* has branching cords of cells. Each zone produces a different type of steroid hormone having widely different effects on the body. The glomerulosa produces mineralocorticoids, affecting sodium metabolism; the fasciculata produces gluco-

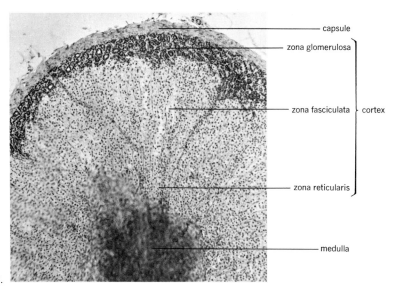

FIGURE 22-18. Adrenal gland.

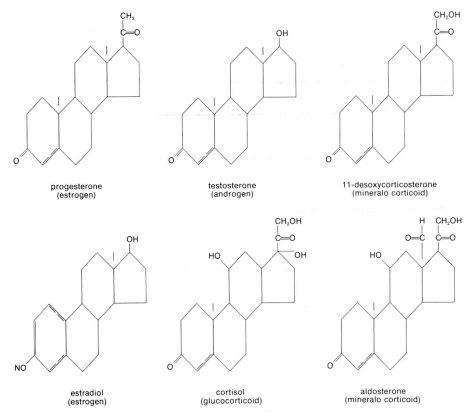

FIGURE 22-19. Formulas of some important cortical steroids (carbon atoms present at all lines and junctions but not shown).

corticoids, affecting foodstuff metabolism (chiefly carbohydrates); the reticularis produces sex hormones, chiefly androgens (male).

Table 22-10 presents some of the specific effects of these hormones. Fig. 22-19 shows the formulae of the more important hormones in each group.

TABLE 22-10

Adrenal Cortical Hormones

Zone of production	General category of hormones	Specific hormonal examples	Effects of hormone—how produced
Zona glomerulosa	Mineralocorticoids	Aldosterone	Increases tubular transport of sodium, decreases transport of potassium; sodium reabsorption causes chloride, bicarbonate, and H_2O reabsorption by physical attraction.
Zona fasciculata	Glucocorticoids	Cortisone Cortisol Corticosterone	1. Increases gluconeogenesis, stimulates deamination 2. Maintains muscular strength; keeps adequate glucose and ATP available 3. Maintains proper nerve excitability 4. Exerts anti-inflammatory effects, possibly by rendering lysosome more resistant to disruption by disease 5. Increases resistance to stress
Zona reticularis	Cortical sex	Androgens Estrogens	Androgens exert antifeminine effects; accelerate maleness; estrogen, just the reverse

Glucocorticoids and stress

Chronic stress of any sort causes increased production of glucocorticoids. This increase in hormone levels may serve as a call-to-arms to equip the body to resist the stress. As a result of the increased production, amino acids are redistributed in the body, anti-inflammatory effects are produced, and all of the repair mechanisms of the body are stimulated. Continued high levels of glucocorticoids are damaging to the body. For example, ulcer formation, high blood pressure, and atrophy of lymphatic tissue will occur. The individual may, in the long run, be rendered more susceptible to disease. In conclusion, we may state that increased hormone production is observed during stress, but the exact role this plays in the body's adaptive mechanisms is in doubt.

Control of cortical secretion

Activity of the glomerulosa appears to be controlled by the blood sodium level, and by angiotensin. Increase in either factor stimulates aldosterone secretion, leading to increased sodium reabsorption, water retention, and chloride and bicarbonate reabsorption. ACTH plays little or no role in glomerulosa activity. The fasciculata and reticularis are under ACTH control. The hormones produced from the latter two zones also exert a negative feedback upon ACTH production.

Disorders of the cortex

Hypofunction, particularly of the fasciculata, results in *Addison's disease*. Anemia, muscular weakness, fatigue, elevated blood potassium, and lower blood sodium evidence deficiency of glucocorticoids and mineralocorticoids. A peculiar bronzing of the skin is apparently due to the excess ACTH production, occasioned by the removal of the negative feedback system as cortical hormone production decreases. (Recall that ACTH has melanocyte expanding properties in the human.) Administration of cortisol controls the condition. Hyperfunction

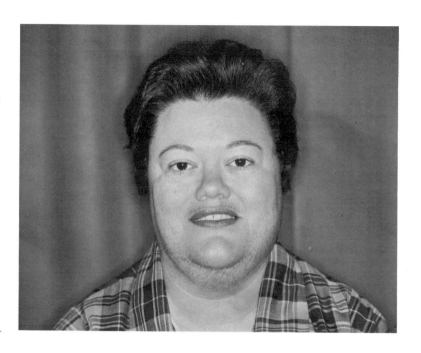

FIGURE 22–20. Cushing's disease.

creates several disorders, depending upon the zone involved. In most cases, hyperfunction is the result of a cortical tumor. Two of these conditions are of interest. *Cushing's disease* (Fig. 22–20) is due to overproduction of glucocorticoids. The individual shows regional adiposity sparing the extremities, stripes on the abdomen, and diabetic tendencies. Surgical removal of the tumor followed by cortisol administration controls the disease. The *adrenogenital syndrome* results from a fasciculata tumor with resultant overproduction of sex hormones. In general, overproduction of this category of hormones will masculinize a female, or accelerate the sexual development of a male. If the excess production occurs *in utero,* when the child's reproductive organs are developing, bizarre alterations in appearance of the organs may result. For example, if a genetic female is subjected to excess male hormone, the result will be conflicting orders to the developmental mechanism. Fig. 22–21 shows a sufferer from adrenogenital syndrome.

ENDOCRINE ORGANS OF SEX

Introduction

The anatomy of the ovary and testis has been described elsewhere (Chapter 18). Because of the similarity in effects which gonadal hormones have upon the human body, the ovarian and testicular hormones will be described together. An individual's maleness or femaleness depends upon his chromosomal constitution, reinforced by hormones. The hormonal influence is most apparent on the so-called accessory organs of sex (all organs except gonads) and upon the secondary sexual characteristics (fat distribution, hair growth, voice), which confer the external characteristics of sex. The sex hormones develop

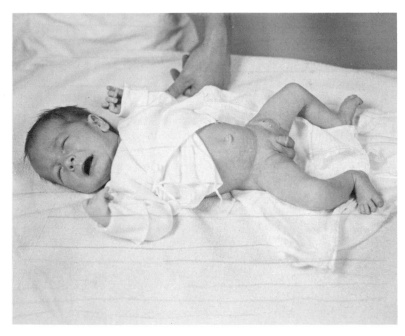

FIGURE 22–21. Adrenogenital syndrome.

and maintain the organs and characteristics, and any cyclical changes in these organs (menstrual cycle in the female).

Ovarian hormones

The ovary produces three hormones.

1. *Estrogen (estrin)* is a steroid hormone produced by the vesicular follicle. Estrogen causes the proliferative phase of the menstrual cycle (see chap. 18), increases mammary growth, stimulates contraction of the uterus, prevents atrophy of the accessory organs, and develops the female secondary characteristics. It also primes the body for;
2. *Progestin (progesterone)* produced by the corpus luteum. Progestin is responsible for the secretory phase of the menstrual cycle, further develops the mammary glands, is neccessary for placental formation, and is required for ovulation.
3. *Relaxin,* produced in very small quantities before birth, softens the pelvic ligaments. This softening theoretically allows enlargement of the birth canal. In the human, the effect is negligible.

Control of ovarian secretion

Follicle stimulating hormone (FSH) from the anterior pituitary initiates development of follicles and, as a consequence, estrogen levels rise. Estrogen inhibits FSH production.

Leuteinizing hormone (LH) from the pituitary causes corpus luteum formation with progestin production. Progestin also inhibits FSH production. The four hormones work together as shown in Fig. 18–20. We may note the reciprocal relationships of these substances.

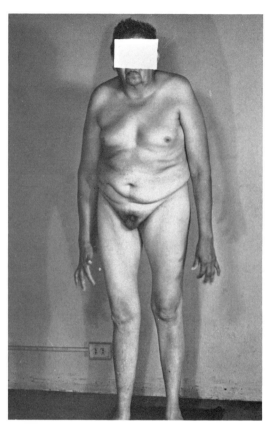

FIGURE 22–22. Eunuchoidism.

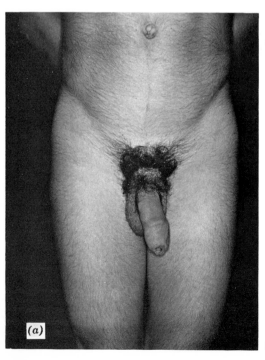

(a)

FIGURE 22–23. Precocious puberty.

Oral contraception

The use of "the pill" to control childbearing is based upon the above stated effects of estrogen and progestin in inhibiting FSH production. The first substances utilized were similar to progestin. Although they inhibited FSH, and therefore follicle development, they also created a number of uncomfortable side effects. These effects included mammary swelling, water retention, and a state of "pseudopregnancy." Newer compounds are estrogen-like in effect, and create far fewer side effects.

The testis

The interstitial cells of the testis produce *testosterone,* a steroid. This hormone serves the same function in the male that estrogen serves in the female, that is, it develops and maintains the accessory organs and the secondary sexual characteristics.

Control of testicular secretion

Interstitial cell stimulating hormone (ICSH) from the pituitary governs testosterone secretion. The latter hormone in turn exerts a negative feedback on ICSH production.

Disorders of the ovaries and testes

Both sexes may suffer from deficient production of hormones, in the condition known as *eunuchoidism* (Fig. 22–22). As expected, the most obvious symptoms will be atrophy of accessory organs, and disappearance of secondary characteristics. Complete loss of hormones produces *eunuchism,* similar to eunuchoidism, but presenting more severe symptoms. Treatment with the appropriate hormone will control symptoms.

Hyperfunction, due to gonadal tumors, creates *precocious puberty.* In this condition, accessory organ development and characteristics appear at earlier than normal ages. Precocious females may actually become fertile at very tender ages (the earliest on record became pregnant at 5 years old). Figure 22–23 shows male and female precocious puberty patients. Surgical removal of the tumor is required, and the development may be arrested and partially reversed.

Climacterics

Both males and females, as they age, undergo a cessation or diminution of gonadal hormone production. In the female, this perid is often called the menopause and it usually occurs during the fourth decade of life. In males, similar slowdown is termed the male climacteric, and it occurs gradually, at about 60 years of age.

In both sexes the following symptoms may develop:

1. Psychic symptoms (more common in female). Nervousness irritability, crying spells, some loss of mental acuity.
2. Vasomotor symptoms. Sweating, headache, hot and cold flashes.

3. Constitutional symptoms. Fatigue, muscular weakness, "lack of ambition."
4. Sexual symptoms. Loss of sex drive; in female, ultimate sterility.

It should be emphasized that the climacterics will occur as aging proceeds. Understanding of the condition and acting accordingly is often the most important factor in getting through the period.

QUESTIONS

1. The hypophysis has sometimes been designated as "the master gland" of the endocrine system. What, in your mind, justifies the exalted position?
2. What hormones are required for normal body growth and development? Give contribution of each to these processes.
3. What hormones are concerned with carbohydrate metabolism, and how is each involved?
4. What endocrines exhibit a negative feedback mechanism of control? Which are controlled by blood levels of substances other than hormones?
5. What hormones enable the body to meet situations which are stressful or meet sudden demands for accelerated activity?
6. What hormones are concerned with bone and tooth formation? Where are they produced and how are they involved?
7. Describe hypothalamic control of the hypophysis.
8. What justifies the statement that nervous and endocrine function are related?
9. What hormones insure proper sexual development? Discuss contribution of each.

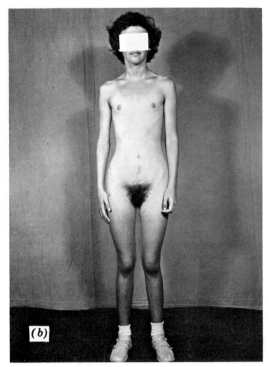

FIGURE 22–23. Precocious puberty.

Epilogue

As you have studied the human body, you have undoubtedly become aware of the complexity of its structure and functions. You must have also recognized that inherent within the limits of the body are controlling devices (cybernetics) which assure, in most cases, that a properly balanced environment (homeostasis) is maintained for the operation of the body's individual cells.

The most severe problems that man currently faces, however, appear to be those of his external environment, natural, social and political. He has exploited the natural environment for individual gain to the extent that he now faces problems of pollution, shortage of natural resources, and an inadequate food supply. He has bred himself into overpopulation which threatens his capacity to lead a good life. Man is adaptable and possessed of a technology sophisticated enough to place men on the moon, and ultimately on the planets; this leads one to the suspicion that no problem is too great to be solved, providing the concern to solve it is present.

Man will solve the problems of relating to his environment only if he acquires a better understanding of himself, recognizes his oneness with his environment, and assumes more individual and social responsibility for its maintenance. We must be as objective about the resolution of our environmental and social problems as we have been of our technological ones.

Creative individuals must be allowed the freedom to pursue their tasks in all areas of human endeavor. They must also assume some responsibility, along with others, to insure that their works are used to benefit rather than destroy mankind. The question remains, can man realize as effective a relationship with his external environment as nature has provided for regulating and stabilizing his internal environment?

621

Additional Readings

CHAPTER 1

Corner, George W. 1944. *Ourselves Unborn.* Yale University Press, New Haven. Page 131.

Harrison, R. J. 1958. Man the Peculiar Animal. Penquin Books, Inc., Baltimore. Chapters 1–4.

Nomina Anatomica. 1961. 2nd. Ed. Revised by the International Anatomical Nomenclature Committee and approved by the Seventh International Congress of Anatomists, New York. 99+ pages. Excerpta Medica Foundation, Amsterdam.

Simpson, George Gaylord. 1949. The Meaning of Evolution. A Mentor Book, New York. Pages 136–181.

Singer, Charles. 1957. *A Short History of Anatomy and Physiology from the Greeks to Harvey.* 2nd. Ed. Dover Publications, New York.

Stacey, Ralph W., and John A. Santolucito. 1966. *Modern College Physiology.* C. V. Mosby Co., St. Louis, Mo. Chapter 7, pages 131–142.

Weiner, Norbert. 1948. *Cybernetics.* John Wiley and Sons, New York.

Weisz, Paul B. 1963. *The Science of Biology.* 2nd. Ed. McGraw-Hill Book Co. Inc., New York. Pages 454–464.

CHAPTER 2

Wald, George. August, 1954. The Origin of Life. *Scientific American.*

Readings from Scientific American. 1965. *The Living Cell.* W. H. Freeman and Co., San Francisco and London.

Life Science Library. *The Cell.* Time Inc., New York.

Robertson, J. D. April, 1962. The Membrane of the Living Cell. *Scientific American.*

Solomon, A. K. December, 1960. Pores in the Cell Membrane. *Scientific American.*

Hokin, L. E., and M. R. Hokin. October, 1965. The Chemistry Of Cell Membranes. *Scientific American.*

de Duve, Christen. December, 1963. The Lysosome. *Scientific American.*

Rich, Alexander. December, 1963. Polyribosomes. *Scientific American.*

Hurwitz, J., and J. J. Furth. February, 1962. Messenger RNA. *Scientific American.*

Doty, Paul. September, 1957. Proteins. *Scientific American.*

Stumpf, P. K. April, 1953. ATP. *Scientific American.*

Crick, F. H. C. September, 1957. Nucleic Acids. *Scientific American.*

Bloom, W., and D. W. Fawcett. 1968. *Textbook of Histology.* 9th Ed. W. B. Saunders Co., Philadelphia, Pa.

Loewy, A. G., and Philip Siekevitz. 1963. *Cell Structure and Function.* Modern Biology Series. Holt, Rinehart, and Winston, New York.

Brian, F. C. C., and K. A. Marcker. January, 1968. How Proteins Start. *Scientific American.*

Mirsky, A. E. June, 1968. The Discovery of DNA. *Scientific American.*

Allison, Anthony. November, 1967. Lyosomes and Disease. *Scientific American.*

CHAPTER 3

Bloom, W. D., and D. W. Fawcett. 1968. *A Textbook of Histology.* 9th Ed. W. B. Saunders Co., Philadelphia, Pa. Chapters 3, 4, 6, 7, 10.

DiFiore, Mariano S. H. 1960. *An Atlas of Histology.* Lea and Febiger, Philadelphia, Pa. Plates 2–15.

Fawcett, D. W. 1966. *An Atlas of Fine Structure: The Cell.* W. B. Saunders Co., Philadelphia, Pa.

Reith, Edward J., and Michael H. Ross. 1965. *Atlas of Descriptive Histology.* Hoeber Medical Division. Harper and Row, Publishers, 49 E. 33rd St., New York. Pages 1–39.

CHAPTER 4

Abbott Laboratories, North Chicago, Ill. 1960. *Fluid and Electrolytes.*

Ganong, William F. 1967. *Review of Medical Physiology.* Lange Medical Publications, Los Altos, Calif. Chapters 1, 40.

Pitts, Robert F. 1966. *Physiology of the Kidney and Body Fluids.* Year Book Medical Publishers Inc., Chicago, Ill. Chapters 2, 10, 12.

Winters, Robert W., Knud Engle, and Ralph B. Dell. 1969. *Acid-base Physiology in Medicine, A Self Instruction Program.* 2nd Ed. The London Co., 811 Sharon Drive, Westlake, Ohio.

CHAPTER 5

Francis, Carl C. 1968. *Introduction to Human Anatomy.* C. V. Mosby Co., St. Louis, Mo. Chapter 2.

Gray, Henry. 1966. *Anatomy of the Human Body.* 28th Ed. Lea and Febiger, Philadelphia, Pa. Pages 61–118.

Lockhart, R. D., G. F. Hamilton, and F. W. Fyfe. 1965. *Anatomy of the Human Body.* J. B. Lippincott Co., Philadelphia, Pa.

Lockhart, R. D. 1966. *Living Anatomy.* 5th Ed. Faber and Faber Ltd., 24 Russell Square, London.

Royce, Joseph. 1965. *Surface Anatomy.* F. A. Davis Co., Philadelphia, Pa.

CHAPTER 6

Structure

Bloom, W. D., and D. W. Fawcett, 1968. *Textbook of Histology.* W. B. Saunders Co., Philadelphia, Pa. Chapter 22.

Reith, Edward J., and Michael H. Ross. 1965. *Atlas of Descriptive Histology.* Hoeber Medical Division, Harper and Row, Publishers, New York. Page 87 *et. seq.*

Function

Only four references are quoted as being sufficiently detailed to further the student's knowledge.

Rothman, Stephen. 1954. *Physiology and Biochemistry of the Skin.* University of Chicago Press, Chicago, Ill.

Shelley, W. B., and R. P. Arthur. 1958. *Physiology of the Skin.* Annual Reviews, Inc., Palo Alto, Calif.

Montagna, William. 1964. The Skin. *Scientific American.*

Harrison, Richard, J., and William Montagna. 1969. *Man.* Appleton Century Crofts, New York.

CHAPTER 7

Crouch, James E. 1967. *Functional Human Anatomy.* Lea and Febiger, Philadelphia, Pa. Chapter 7.

Francis, Carl C. 1968. *Introduction to Human Anatomy.* 5th Ed. C. V. Mosby Co., St. Louis, Mo. Chapter 5.

Duvall, E. N. 1959. *Kinesiology, The Anatomy of Motion.* Prentice Hall Inc., Englewood Cliffs, N.J.

Rasch, Philip J., and Roger K. Burke. 1963. *Kinesiology and Applied Anatomy.* 2nd. Ed. Lea and Febiger, Philadelphia, Pa.

Wells, Katherine F. 1960. *Kinesiology.* 3rd. Ed. W. B. Saunders Co., Philadelphia, Pa.

Sobotta, Johannes, and Frank Figge. 1963. *Atlas of Descriptive Human Anatomy.* 8th. Ed. Vol. 1. Hafner Publishing Co., Inc., New York.

CHAPTER 8

Cunningham, D. J. 1964. *Textbook of Anatomy.* G. J. Romanes (editor). 10th. Ed. Oxford University Press, London.

Gray, H. 1966. *Anatomy of the Human Body.* C. M. Goss (editor). 28th. Ed. Lea and Febiger, Philadelphia, Pa.

Krogman, Wilton M. 1962. *The Human Skeleton.* Charles C Thomas, Springfield, Ill.

Romer, Alfred S. 1959. *The Vertebrate Story.* 4th. Ed. University of Chicago Press, Chicago, Ill.

Romer, Alfred S., 1962. *The Vertebrate Body.* 3rd. Ed. W. B. Saunders Co., Philadelphia, Pa.

Romer, Alfred S. 1966. *Vertebrate Paleontology.* 3rd. Ed. University of Chicago Press, Chicago, Ill.

Steindler, Arthur. 1955. *Kinesiology.* Charles C Thomas, Springfield, Ill.

Woodburne, Russell T. 1965. *Essentials of Human Anatomy.* 3rd. Ed. Oxford University Press, New York.

Zukerman, Sir Solly. 1961. *A New System of Anatomy.* Oxford University Press, London.

CHAPTER 9

Sobotta, Johannes, and Frank Figge. 1963. *Atlas of Descriptive Human Anatomy.* 8th. Ed. Vol. 1. Hafner Publishing Co., Inc., New York.

Rasch, Philip J., and Roger K. Burke. 1963. *Kinesiology and Applied Anatomy.* 2nd. Ed. Lea and Febiger, Philadelphia, Pa.

Grant, J. C. B. 1962. *An Atlas of Anatomy.* 5th. Ed. Williams and Wilkins Co., Baltimore, Md.

Gray, Henry. 1966. *Anatomy of the Human Body.* 28th. Ed. Lea and Febiger, Philadelphia, Pa.

CHAPTER 10

Bloom, William, and D. W. Fawcett. 1968. *Textbook of Histology.* W. B. Saunders Co., Philadelphia, Pa. Chapter 5.

Best, Charles H., and Norman B. Taylor. 1966. *Physiological Basis of Medical Practice.* 8th. Ed. Williams and Wilkins Co., Baltimore, Md. Chapters 22, 23, 26, 27, 30, 31, 33.

Diggs, L. W., Dorothy Sturm, and Ann Bell. 1954. *The Morphology of Blood Cells.* Abbott Laboratories, North Chicago, Ill.

Mountcastle, Vernon B. (Editor). 1968. *Medical Physiology,* Vol. 1. C. V. Mosby Co., St. Louis, Mo. Chapters 1, 2.

Ganong, William F. 1967. *Review of Medical Physiology.* Lange Medical Publications, Los Altos, Calif. Chapter 27.

CHAPTER 11

Ganong, William F. 1967. *Review of Medical Physiology.* Lange Medical Publications, Los Altos, Calif. Chapters 28, 29, 31.

Mountcastle, Vernon B. (Editor). 1968. *Medical Physiology.* Vol. 1. C. V. Mosby Co., St. Louis, Mo. Chapters 4, 6.

Crouch, James E. 1964. *Functional Human Anatomy.* Lea and Feiber, Philadelphia, Pa. Chapter 12.

Sobotta, Johannes, and Frank Figge. 1963. *Atlas of Descriptive Human Anatomy.* 8th. Ed. Vol. 2. Hafner Publishing Co., Inc., New York.

CHAPTER 12

Ganong, William F. 1967. *Review of Medical Physiology.* Lange Medical Publications, Los Altos, Calif. Chapters 30, 31, 32.

Crouch, James E. 1964. *Functional Human Anatomy.* Lea and Febiger, Philadelphia, Pa. Chapter 12.

Mountcastle, Vernon B. (Editor). 1968. *Medical Physiology.* Vol. 1. C. V. Mosby Co., St. Louis, Mo. Chapters 7, 8, 9, 10, 11, 12, 13.

Sobotta, Johannes, and Frank Figge. 1963. *Atlas of Descriptive Human Anatomy.* 8th. Ed. Vol. 2. Hafner Publishing Co., Inc., New York.

Best, Charles H., and Norman B. Taylor. 1966. *Physiological Basis of Medical Practice.* 8th. Ed. Williams and Wilkins Co., Baltimore, Md. Chapters 42, 43, 44.

Bloom, William, and D. W. Fawcett. 1968. *Textbook of Histology.* 10th Ed. W. B. Saunders Co., Philadelphia, Pa. Chapter 13.

CHAPTER 13

Best, Charles H., and Norman B. Taylor. 1966. *Physiological Basis of Medical Practice.* 8th. Ed. Williams and Wilkins Co., Baltimore, Md. Chapter 25.

Crouch, James E. 1964. *Functional Human Anatomy.* Lea and Febiger, Philadelphia, Pa. Pages 349 *et. seq.*

Ganong, William F. 1967. *Review of Medical Physiology.* Lange Medical Publications, Los Altos, Calif. Pages 427 *et. seq.*; 467 *et. seq.*

Bloom, William, and D. W. Fawcett. 1968. *Textbook of Histology.* 9th. Ed. W. B. Saunders Co., Philadelphia, Pa. Chapters 14–16.

CHAPTER 14

Comroe, Julius H. 1965. *Physiology of Respiration.* Year Book Medical Publishers, 35 E. Wacker Drive, Chicago, Ill.

U.S. Department of Health, Education and Welfare. Public Health Service Publication No. 1715. *Emphysema, the Battle to Breathe.* U.S. Govt. Printing Office, Washington, D.C.

Best, Charles H., and Norman B. Taylor. 1966. *Physiological Basis of Medical Practice.* 8th. Ed. Williams and Wilkins Co., Baltimore, Md. Chapters 50–53.

Mountcastle, Vernon B. (editor). 1968. *Medical Physiology.* Vol. 2. C. V. Mosby Co., St. Louis, Mo. Part V.

Ganong, William F. 1967. *Review of Medical Physiology.* Lange Medical Publications, Los Altos, Calif. Chapters 34–36.

Crouch, James E. 1964. *Functional Human Anatomy.* Lea and Febiger, Philadelphia, Pa. Chapter 14.

Bloom, William, and D. W. Fawcett. 1968. *Textbook of Histology.* 9th. Ed. W. B. Saunders Co., Philadelphia, Pa. Chapter 29.

CHAPTER 15

Davenport, H. W. 1966. *Physiology of the Digestive Tract.* 2nd. Ed. Year Book Medical Publishers, 35 E. Wacker Drive, Chicago, Ill.

Crouch, James E. 1964. *Functional Human Anatomy.* Lea and Febiger, Philadelphia, Pa. Chapter 13.

Mountcastle, Vernon B. (editor). 1968. *Medical Physiology.* Vol. 1. C. V. Mosby Co., St. Louis, Mo. Part III.

Best, Charles H., and Norman B. Taylor. 1966. *Physiological Basis of Medical Practice.* Williams and Wilkins Co., Baltimore, Md. Section VII, Chapters 56–62.

Bloom, William, and D. W. Fawcett. 1968. *Textbook of Histology.* W. B. Saunders Co., Philadelphia, Pa. Chapters 23–28.

Netter, Frank H. *Ciba Collection of Medical Illustrations.* Vol. 3. Ciba Pharmaceutical Products, Summit, N.J. Part 1, 1959, Upper Digestive Tract. Part 2, 1962, Lower Digestive Tract. Part 3, 1957, Liver, Biliary System, Pancreas.

CHAPTER 16

Harper, Harold A. 1963. *Review of Physiological Chemistry.* Lange Medical Publications, Los Altos, Calif. Chapters 13–15.

White, Emil H. 1964. *Chemical Background for the Biological Sciences.* FMB Series. Prentice-Hall Inc., Englewood Cliffs, N.J.

Barry, J. M., and E. M. Barry. 1969. *Introduction to the Structure of Biological Molecules.* Prentice-Hall Inc., Englewood Cliffs, N.J.

Abbott Laboratories, North Chicago, Ill. 1962. *The Vitamins.*

Best, Charles H., and Norman B. Taylor. 1966. *Physiological Basis of Medical Practice.* Williams and Wilkins Co., Baltimore, Md. Section VIII, Chapters 63–73.

CHAPTER 17

Pitts, Robert F. 1966. *Physiology of the Kidney and Body Fluids.* Year Book Medical Publishers Inc., 35 E. Wacker Drive, Chicago, Ill.

Mountcastle, Vernon B. (editor). 1968. *Medical Physiology.* C. V. Mosby Co., St. Louis, Mo. Chapter 17.

Bloom, William, and D. W. Fawcett. 1968. *Textbook of Histology.* W. B. Saunders Co., Philadelphia, Pa. Chapter 30.

Best, Charles H., and Norman B. Taylor. 1966. *Physiological Basis of Medical Practice.* Williams and Wilkins Co., Baltimore, Md. Section IX, Chapters 79, 80.

Crouch, James E. 1964. *Functional Human Anatomy.* Lea and Febiger, Philadelphia, Pa. Chapter 15.

Ganong, William F. 1967. *Review of Medical Physiology.* Lange Medical Publications, Los Altos, Calif. Chapters 38, 39.

CHAPTER 18

Bloom, William, and D. W. Fawcett. 1968. *Textbook of Histology.* W. B. Saunders Co., Philadelphia, Pa. Chapters 31–33.

Turner, C. Donnell. 1966. *General Endocrinology.* 4th. Ed. W. B. Saunders Co., Philadelphia, Pa.

Crouch, James E. 1964. *Functional Human Anatomy.* Lea and Febiger, Philadelphia, Pa. Chapter 16.

Netter, Frank H. 1953. *Ciba Collection of Medical Illustrations.* Vol. 2, Reproductive Systems. Ciba Pharmaceutical Products, Summit, N.J.

CHAPTER 19

Arey, L. B. 1965. *Developmental Anatomy.* 7th. Ed. W. B. Saunders Co., Philadelphia, Pa.

Balinsky, B. I. 1962. *Introduction to Embryology.* W. B. Saunders Co., Philadelphia, Pa.

Patten, B. M. 1964. *Foundations of Embryology.* 2nd. Ed. McGraw-Hill Book Co., New York.

Patten, B. M. 1953. *Human Embryology.* 2nd. Ed. McGraw-Hill Book Co., New York.

CHAPTER 20

Gardner, Ernest. 1968. *Fundamentals of Neurology.* 5th. Ed. W. B. Saunders Co., Philadelphia, Pa.

Eyzaguirre, Carlos. 1969. *Physiology of the Nervous System.* Year Book Medical Publishers, 35 E. Wacker Drive, Chicago, Ill.

U.S. Department of Health, Education and Welfare. Public Health Service Publication No. 1389. *Current Research on Sleep and Dreams.* U.S. Govt. Printing Office, Washington, D.C.

Ranson, Stephen W., and Sam L. Clark. 1964. *The Anatomy of the Nervous System.* 10th. Ed. W. B. Saunders Co., Philadelphia, Pa.

Netter, Frank H. 1957. *Ciba Collection of Medical Illustrations.* Vol. 1 and supplement, The Nervous System and Functional Relationships of the Hypothalamus. Ciba Pharmaceutical Products, Summit, N.J.

CHAPTER 21

Bloom, William, and D. W. Fawcett. 1968. *Textbook of Histology.* 9th. Ed. W. B. Saunders Co., Philadelphia, Pa. Chapters 34, 35.

Best, Charles H., and Norman B. Taylor. 1966. *Physiological Basis of Medical Practice.* Williams and Wilkins Co., Baltimore, Md. Section II, Chapters 13–20.

Annals, New York Academy of Sciences. Vol. 116. July 30, 1964. Article 2, Recent Advances in Odor.

Crouch, James E. 1964. *Functional Human Anatomy.* Lea and Febiger, Philadelphia, Pa. Chapter 19.

CHAPTER 22

Tepperman, Jay. 1967. *Metabolic and Endocrine Physiology.* 2nd. Ed. Year Book Medical Publishers, 35 E. Wacker Drive, Chicago, Ill.

Lisser, H., and Roberto F. Escamilla. 1964. *Atlas of Clinical Endocrinology.* 2nd. Ed. C. V. Mosby Co., St. Louis, Mo.

Netter, Frank H. 1965. *Ciba Collection of Medical Illustrations.* Vol. 4, The Endocrine System and Selected Metabolic Diseases. Ciba Pharmaceutical Products, Summit, N. J.

Turner, C. Donnell. 1966. *General Endocrinology.* 4th. Ed. W. B. Saunders Co., Philadelphia, Pa.

Williams, Robert H. 1968. *Textbook of Endocrinology.* 4th. Ed. W. B. Saunders Co., Philadelphia, Pa.

Subject Index

Entries in Roman type indicate pages where an item is described in text material. Italicized entries indicate that the item appears on an illustration.

Absorption in digestion, 388, 389, 407, 410, 412
Acetabulum, 186, 196, *197, 198,* 206, *207*
Acetylcholine, 339, 373, 540, 551
Acetyl-coA, 432, 436–438
Acid, definition, 97
  strong, 97
  weak, 98
Acid-base balance, 18, 97–101
  disturbances of, 100, 101
  role, of kidney in, 466
    of lung in, 378
Acidosis, 438
Acromegaly, 601, 602
Acromion processes, 187, *187, 188*
Active transport, 48, *48*
''Adam's apple,'' 364
Adaptation, 17, *17*
Addison's disease, 615
Adenoids, 354, 364; *see also* Pharyngeal tonsil
Adenosine triphosphate, 222, 223, 426, 427, 437, 438
Adrenal glands, 612–616, *614*
  control of cortical secretion, 615

cortex of, 613, *614,* 615
  disorders, of cortex of, 615, 616
    of medulla of, 613
  glucocorticoids and stress, 615
  hormones of, 615
  medulla of, 612
    hormones of, 613
Adrenocorticotropic hormone (ACTH), 597, 598, 615
Adrenogenital syndrome, 616, *617*
Agglutinins, 296
Agglutinogens, 296
Air, composition of respired, 375
Airway resistance, 373
Albumin, 283, 285
Aldosterone, 95, 615
Alimentary tract, summary of motility in, 394; *see also* Digestive system
Allantois, 522, *522*
All-or-none law, 312
Alveolar ducts, 363, *368,* 369
Alveolar process, *168*
Alveolar margin, *168, 171*
Alveolar sacs, 363, *368*
Alveoli, 369, 373